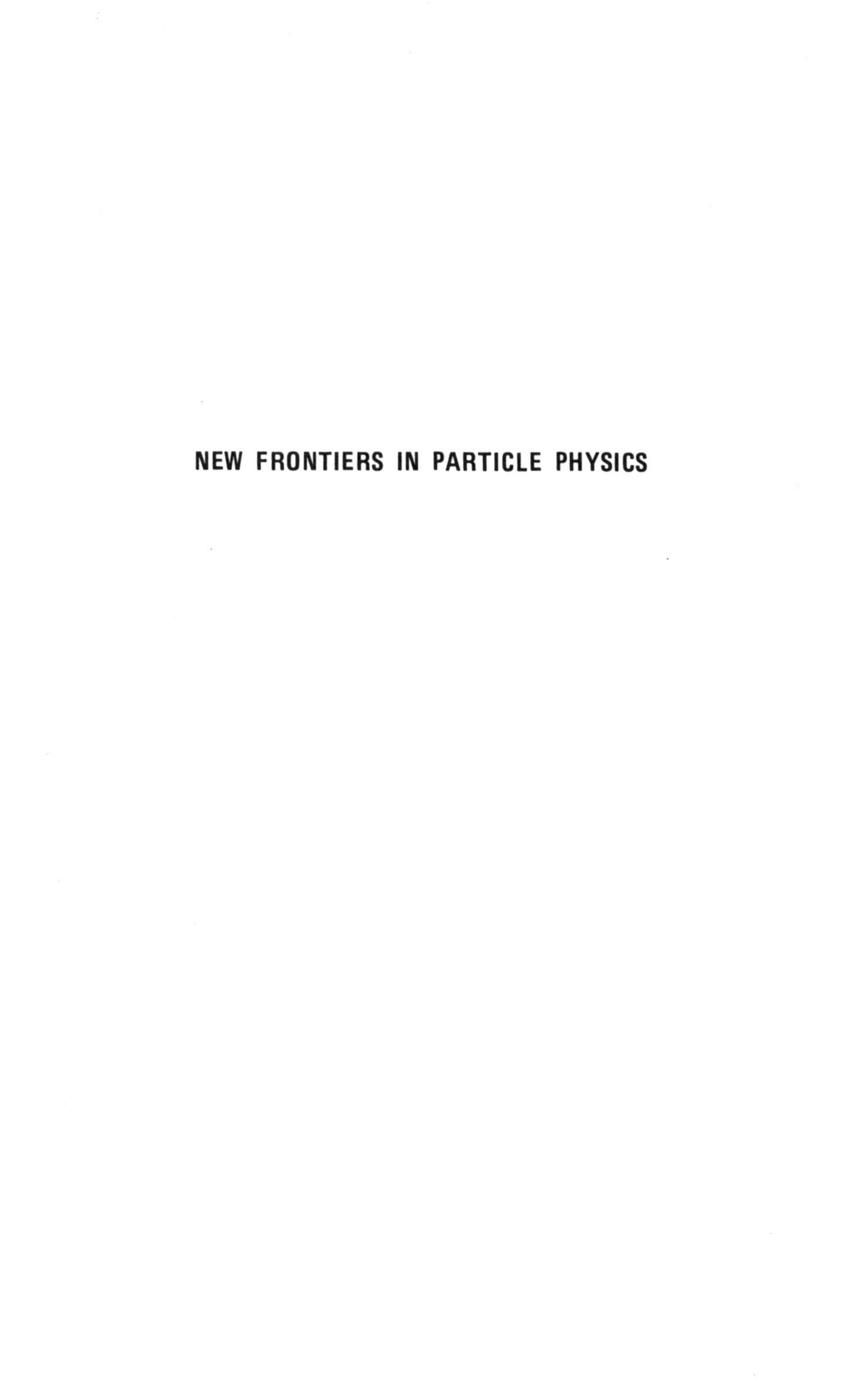
NEW FRONTIERS IN PARTICLE PHYSICS

NEW FRONTIERS IN PARTICLE PHYSICS

Lake Louise, Canada   16-22 February 1986

Editors

J M CAMERON

B A CAMPBELL

A N KAMAL

F C KHANNA

World Scientific

Published by

World Scientific Publishing Co Pte Ltd.
P.O. Box 128, Farrer Road, Singapore 9128
242 Cherry Street, Philadelphia PA 19106-1906, USA.

Library of Congress Cataloging-in-Publication Data

New Frontiers in particle physics.

1. Particles (Nuclear physics) — Congresses.
I. Campbell, B. A.   II. Kamal, Abdul N., 1935-
III. Khanna, F. C.
QC793.N43   1986   539.7'21   86-24642
ISBN 9971-50-135-X

Printed in Singapore by General Printing and Publishing Services Pte Ltd.

PREFACE

The 1986 Lake Louise Winter Institute inaugurates a new annual series of Winter Institutes in the Canadian Rockies. In 1986 the subject of discussion was the new developments in particle physics under the title: <u>New Frontiers in Particle Physics</u>. The Institute started on the 16th of February, 1986 with a series of invited lectures by: A. Astbury on 'Hadron collider physics', M. Dine on 'A superstring primer', J. Dorfan on 'Future frontiers for e^+e^- physics', J. Ellis on 'From the Higgs to superstring', and G. Wolf on 'HERA: physics, machine and experiments'. The Institute was followed by a three-day topical conference which started on the 20th of February, 1986. The format of the topical conference included one-hour review talks by: D. Bryman on 'Searches for lepton flavor violation', J. Lo Secco on 'Physics at the underground laboratories', K. Olive on 'Dark matter as a probe of the early universe', and short presentations (30 minutes) from many of the participants.

The local organizing committee was ably assisted by Ms. Greta Tratt in making detailed arrangements. Several graduate students are also to be thanked for their assistance with the reception and the transportation of the participants.

We acknowledge with thanks the generous support from the following organizations: The University of Alberta, Institute of Particle Physics, TRIUMF, and TRIUMF Users Group. Our special thanks go to the University of Alberta for their very generous support that made it possible to transform the idea of a winter institute into a reality.

Local Organizing Committee: J.M. Cameron (Chairman)
B.A. Campbell
A.N. Kamal
F.C. Khanna

Edmonton, July 1986

TABLE OF CONTENTS

NEW FRONTIERS IN PARTICLE PHYSICS

HADRON COLLIDER PHYSICS

Alan Astbury

Physics Department, University of Victoria
Victoria, B.C. V8W 2Y2

The outline of the three lectures is as follows:

Lecture 1

(a) **Hadron Collisions** – The main features of hadron collisions
are compared with those of e^+e^- collisions.

(b) **Hadron Colliders** – The example chosen is the CERN SPS
Antiproton Proton Collider, with only a few comments devoted to
the FNAL Tevatron Collider. This is purely personal knowledge or
ignorance, whichever way you look at it.

(c) **Detectors** – We have learned, through experience, some of the
demands placed on detectors at Hadron Colliders, and what the main
aims must be in order to select and study constituent (i.e. quark
and gluon) physics in the potentially complicated collisions.

(d) **The Immediate Future** – A few indications are given of what
integrated luminosities might be anticipated from hadron colliders
before LEP and SLC become the "Z° factories" of the next decade,
capable of checking the Standard Model with infinite precision – of
course one hopes for more from them.

Lecture 2

QCD at the Collider – The comparison of SPS Collider results with
Perturbative QCD is very impressive, overshadowed by the W and Z
discoveries, and possible New Physics, but nonetheless a major lasting
contribution, which has in some sense transformed the status of QCD in
the eyes of the experimentalist. This is perhaps surprising if one
recalls that the SPS Collider came into operation in the era of "Where
are the jets?". We will discuss the emergence of jets, the inclusive
jet cross section, the angular distribution of parton collisions,
structure functions, and the fragmentation of jets.

2

<u>Lecture 3</u>

(a) **Intermediate Vector boson physics** – The prime reason for the
conversion of the SPS machine to a Collider was the possibility of
the weak boson discovery. This physics is behind us, and already the
precision measurements of the properties of the W and Z are placing
significant constraints on the parameters of the Standard Model.

(b) **Heavy Flavour at the Colliders** – This will be mainly a few
comments on how to look for heavy flavours in hadron collisions; it
will not be a review of the current status of the "top" quark, which
remains an unresolved issue in the UA1 experiment.

(c) **The Longterm Future** – The distant future of Hadron Colliders
appears to rest with the proposed Superconducting Super Collider (SSC)
in the USA and a possible Large Hadron Collider (LHC) in the LEP
tunnel at CERN. Both appear challenging, if not downright hostile, to
most experimentalists, but offer perhaps the only known practical way
ahead, of achieving multi TeV collisions between constituents, at
least at this moment.

Throughout all of these lectures, I apologize for a very heavy bias
towards the UA1 experiment; this reflects personal laziness and is, I
hope, somewhat excusable since I believed the aim of the lectures to
be mostly pedagogical in nature and not a review of the Physics of
Hadron Colliders.

<u>Lecture 1</u>

HADRON COLLISIONS: A General Introduction

The hadron is a collection of quarks, antiquarks and gluons,
(partons), each carrying some fraction x_i of the hadron four momentum
such that $p_i = x_i p$ and $\sum x_i = 1$. For the moment it is assumed that the
transverse momentum of these partons is zero. A collision between a
pair of monoenergetic hadrons is therefore a collision between broad
band beams of partons.

Suppose we consider the inclusive reaction $a + b \rightarrow c + \text{anything}$,[1]) which can be represented by the pair of diagrams given in Figure 1.

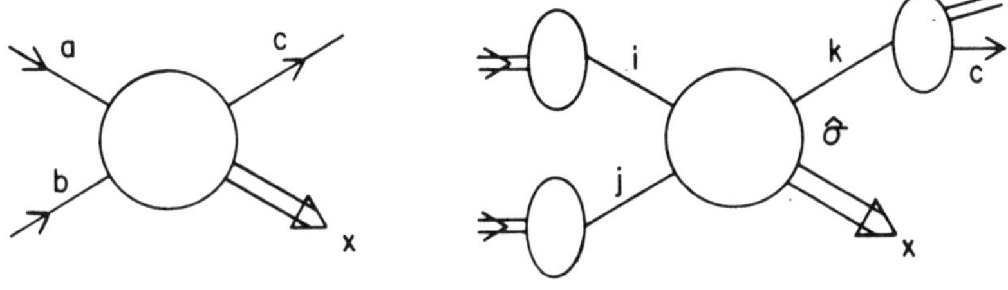

Figure 1

The cross section $d\sigma(a+b \rightarrow c+X)$ can be considered to have the form

$$\sum_{ij} f_i^{(a)}(x_i) f_j^{(b)}(x_j) d\sigma(i+j \rightarrow c+X)$$

where $f_i^{(a)}(x_i)$ represents the probability of finding constituent i with fraction x_i of the four momentum of hadron a, similarly for j in hadron b. The cross section $d\sigma$ represents the constituent cross section for $i+j \rightarrow k+\text{"X"}$ followed by k "dressing itself" as hadron c.

In the centre of mass of the a-b collision $s = (p_a + p_b)^2 = 4p^2$, while the centre of mass energy squared for the i-j collision is $\hat{s} = (x_i^a p + x_j^b p)^2$, which becomes $\hat{s} = x_i^a x_j^b s = \tau s$. A knowledge of the quark or constituent distributions in the hadrons f_i^a, f_i^b can be obtained from deep inelastic lepton scattering where, for example, $F_2(x) = \sum_i e_i^2 x f_i(x)$ – the structure function for deep inelastic electron scattering. Crudely speaking, at the current energies of the proton-antiproton hadron colliders in operation, SPS at $\sqrt{s} = 630$ GeV and FNAL at $\sqrt{s} = 800$ TeV, the most likely value for x_i and x_j is ~.16, and therefore for quark-antiquark collisions $\sqrt{\hat{s}} \simeq \sqrt{s}/6$. If the details of the appropriate structure functions and the constituent cross section $\hat{\sigma}$ are known, then the hadron cross section $a+b \rightarrow c+x$ can be evaluated; we should also, of course, have a knowledge of how constituent "k" becomes a hadron "c". It remains the task of the experimenter to isolate the i-j collision from the a-b collision in order to perform the "constituent physics".

4

A Comparison with e^+e^- Collisions

This comparison follows very closely that of Bjorken[2] in the proposal for the Tevatron Collider at FNAL. In direct contrast to the hadron collision, in the e^+e^- machine the centre of mass of the particles in the machine is the same as that of the constituents, the leptons e^+ and e^-. In this sense the lepton colliders represent a much more efficient way of probing high masses than the hadron colliders.

e^+e^- Collisions (SLC, LEP)	$\bar{p}$-p Collisions (SPS, FNAL)
Luminosity ~ $10^{32}\text{cm}^{-2}\text{sec}^{-1}$	~ $10^{30}\text{cm}^{-2}\text{sec}^{-1}$
e^+e^- couple through charge	$\bar{q}q$ and gg couple through "colour"
the strength of the coupling is ~$(1/137)^2$	strength α_s^2 ~ $(0.1 - 0.2)^2$
the annihilation is into charged fermions	the annihilation is into coloured fermions
the threshold for producing a new "object" Q is given very roughly by $m_Q \simeq E_{beam}$	the threshold $2m_Q \simeq \sqrt{\hat{s}} \simeq [2E_{beam}]/6$
the initial state is clean, well defined in momentum $\sum \vec{p} = 0$	the initial state is complicated with $\langle p_T \rangle$ only approximately zero, and $\sum_{ij} P_{ij_L} \neq 0$, mostly
direct channel resonances are sharp, e.g. J/ψ	direct channel resonances are smeared by the initial state
typical collision rates are low, e.g. the "point cross-section for the electromagnetic production of $\mu^+\mu^-$ pairs ~ 86 nb/s	total cross sections for strongly interacting particles (for the a-b collision) are large ~ 100 mb

Both types of colliders can, of course, participate in the weak interaction; e.g., Z° production at LEP/SLC, and W and Z° production at the $\bar{p}$-p colliders. Here the distinction in cross section has largely disappeared; e.g., σ_{Z° at $e^+e^-(\sqrt{s} \simeq 100$ GeV$) \simeq 41$ nb while $\cdot\sigma_{\bar{W}}$ at $\sqrt{s}$ $\bar{p}p$ $630 \simeq 7$ nb.

The lesson learned so far, concerning the ease of observing specific final states produced in hadron collisions, is that if one concentrates on "high mass" where decay products have a large transverse momentum with respect to the incident beams, then hadron collisions can be viewed almost as cleanly as the e^+e^- collision. There remains, of course, the complication of the "underlying event" since the constituents i-j are not the only ones around during the collision! At the very least the experimentalist at the hadron collider should select a mass range for which his detector is optimized; this task is performed by the machine at the lepton colliders.

Hadron Colliders:

The table below lists some of the properties of the hadron colliders
past and present, with a few future possibilities.

Table 1.

Machine	$\sqrt{s}$ GeV	E (TeV) equivalent fixed target energy	£ $cm^{-2}sec^{-1}$	Fate
ISR two rings $p p(\bar{p}p)$ D.C. beams	62	2	$10^{31}(10^{29}\bar{p}p)$ 10^{32} super- conducting · low β	closed 1983
CERN SPS $\bar{p}p$ one ring	540, 630 (ramped to 900)	~150	~6×10^{29}	first collision 1981, to be up- graded in 1986 to £ ~ 10^{31}
CBA pp two rings	800	340	10^{33}	cancelled 1983
FNAL TeV1 one ring $\bar{p}p$	2,000	2,000	10^{30}	first collisions 1985
LHC two rings pp, $\bar{p}p$?	≦20 TeV	~50,000	10^{31-32}	may happen in LEP tunnel 199?
SSC two rings pp	~40 TeV	~800,000	~10^{33}	proposed for U.S.A. 199?

The CERN SPS Antiproton-Proton Collider[3]

We will use this machine as our example in order to introduce some
of the quantities which are important to the experimentalist. The
machine has the special feature that one of the beams consists of
antiprotons - particles not readily available to us. This introduces
the need for accumulation and beam cooling in an antiproton source.

Traditionally the magnets of the SPS were ramped to ~2.0 T and protons
accelerated to 450 GeV before being ejected onto production targets.
The proposal of Rubbia et al.[4] pointed out the feasibility of using
protons and antiprotons counter rotating within the single ring of
magnets maintained in D.C. operation, after accelerating the beams
to ~270 GeV.

The first major problem is the accumulation of sufficient antiprotons
with a phase space density high enough to be useful for injection into
the machine. The following ratio sets the scale of the task since the
antiprotons must be produced as the secondaries of a high energy
interaction.

$$n_{\bar{p}}/n_p = \frac{\text{number of } \bar{p}\text{'s produced into the useful aperture of the accumulator}}{\text{number of protons incident on the production target}} \sim 10^{-6}.$$

$\therefore 10^{13}$p/s would yield $\sim 10^7 \bar{p}$/s. The machine used in CERN to produce
the antiprotons is the 26 GeV PS which delivers $\sim 10^{13}$p/2.4s. The
principle of the accumulation and the need for "cooling" is shown very
schematically in Figure 2. In the CERN project the method of cooling
which increases the density of antiprotons in phase space is
stochastic[5] and was invented and implemented by Van der Meer[6].

The luminosity of a colliding beam machine operating with bunched
beams and head-on collisions is given by

$$\mathcal{L} = \frac{N_p \, N_{\bar{p}} \, f}{n_b (\varepsilon_H \, \beta_H^*)^{1/2} \cdot (\varepsilon_v \, \beta_v^*)^{1/2}}$$

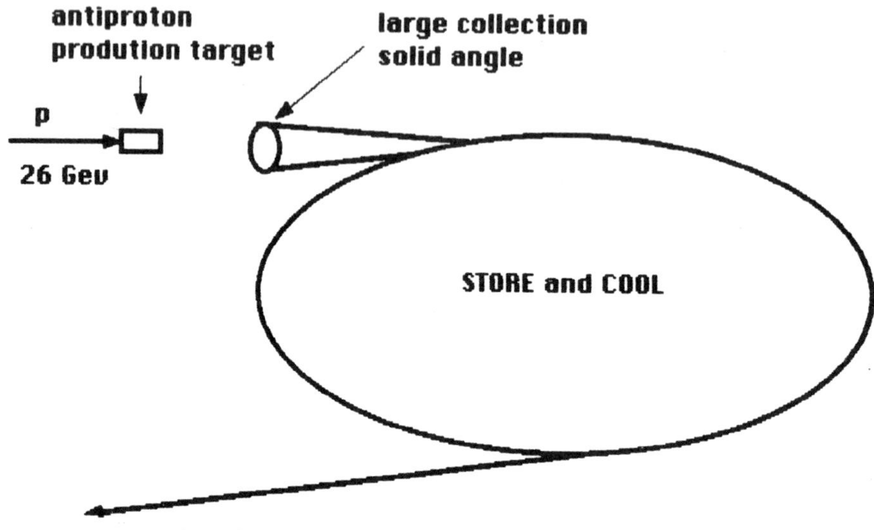

Figure 2

where N_p and $N_{\bar{p}}$ are the total numbers of protons and antiprotons in the machine, f is the rotation frequency, and n_b is the number of bunches of protons and antiprotons.

The quantity $n_b(\epsilon_H \, \beta_H^*)^{1/2} \cdot (\epsilon_V \, \beta_V^*)^{1/2}$ is essentially the size of the beam at the interaction point. At the SPS, the design luminosity was $\simeq 10^{30} \text{cm}^{-2}\text{sec}^{-1}$ with 6 bunch operation, and $N_p = N_{\bar{p}} = 6\times10^{11}$.

A Digression

Phase space is usually defined in terms of the two variables, momentum p and position q, and Liouville's theorem states that the area $\int p dq$ is constant. If we consider the plane (x, p_x) with $p_x = m\gamma c\beta_x$ perpendicular to a beam of momentum p and longitudinal coordinate s, then it is possible to replace p_x by the quantity $x' = dx/ds = \dot{x}/\dot{s} = \beta_x/\beta$ and the $\int p dq$ becomes $mc\beta\gamma\int x' dx$ which is constant. In fact, $\int x' dx = \epsilon\pi$ where ϵ is called the emittance[7]). Often there is reference to normalized emittances $\epsilon^* = \beta\gamma\epsilon$.

Suppose we assume that the motion of a particle in an accelerator undergoes oscillations (betatron oscillations) about some reference orbit such that $x = \sqrt{\varepsilon\beta}\,\sin(\omega t + \lambda)$. β and ε are "free" constants of the machine, and this β should not be confused with the ratio of v/c! - it is a constant of the machine lattice. If $\omega t = \omega(s)$, the phase of oscillation as it proceeds round the circumference then $x' = \psi'\sqrt{\beta\varepsilon}\,\cos\!\big(\psi(s)+\lambda\big)$. We can set $\psi' = 1/\beta$ with $\psi = \int ds/\beta$. Therefore $x = \sqrt{\varepsilon\beta}\,\sin\!\big(\psi(s)+\lambda\big)$ and $x' = \sqrt{\varepsilon/\beta}\,\cos\!\big(\psi(s)+t\big)$. The accelerator "phase space" diagram now appears as in Figure 3, and the quantity $\sqrt{\beta\varepsilon}$ represents the amplitude of the oscillation. In the machine β is a function of s, and near the interaction point $\beta(s) \simeq s^2/\beta^*$. The value of β^* at the intersection point should be as small as possible to minimize $\sqrt{\varepsilon\beta^*}$; however the amplitude of oscillation will blow up rapidly as one moves away from the interaction point. Typically at the CERN SPS the "low-β" values are $\sim 0.5 - 1.0$ m. Even if one can have large apertures to accommodate the blow-up of the beam as it enters and leaves the intersection region, one cannot increase the luminosity indefinitely by decreasing β; at some point the finite length of the bunch must be taken into account.

Figure 3 also shows the principle of how beam cooling can work. The area $\int x'dx$ can be shrunk if the position and divergence of individual particles can be measured and a correction applied. This cannot be achieved practically but the mean position, or momentum, of an assembly of particles can be measured, a correction derived and applied to the same assembly. The particles in the ring will subsequently mix, and the process can be repeated. This very crudely is the principle of stochastic cooling. In the Antiproton Accumulator at the CERN Collider the beam is cooled in momentum space by measuring the deviation from the mean revolution frequency, and in "betratron" space by sensing the deviation from the mean orbit.

x'
√(ε/β)
√(εβ)
COOL
x
x'
x
Area = √(ε/β) · √(εβ) · π = πε.
The Accelerator Physicists view
of phase space.
Ideal world of individual particles
"stacked" into a smaller ellipse.

Antiproton Accumulator (3.5 GeV/c)
has   Momentum Cooling   and ·   "Betatron oscillation" cooling
longitudinal pick-up
Preamp.
Power amp.
filter
longitudinal "kicker"
Transverse pickup
Amp.
Transverse Kicker
Both employ Stochastic Cooling Methods.

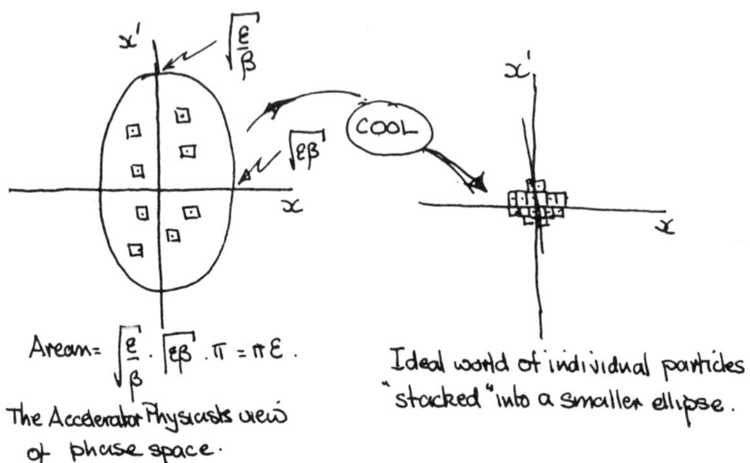

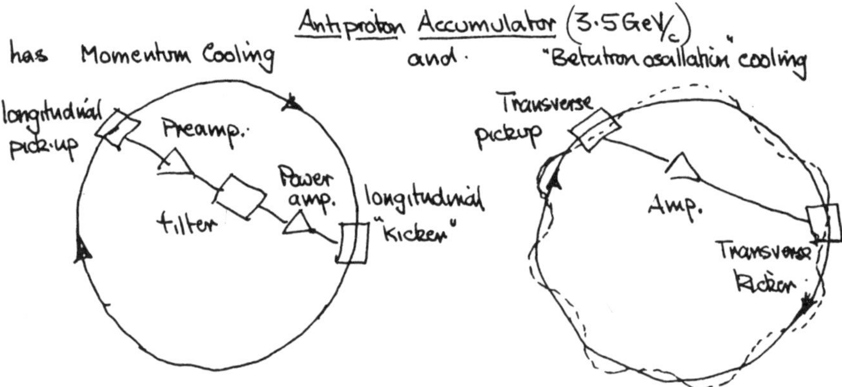

Figure 3

Figure 4 shows schematically the complex of accelerators used to produce the collisions in the SPS rings. The oversimplified sequence of events runs roughly as follows:

1. $\sim 10^{13}$ protons at 26 GeV pass from the PS to the production target

2. $\sim 5 \times 10^6$ antiprotons are produced at 3.5 GeV/c into the useful phase space of the Antiproton Accumulator (AA) $\Delta p/p \pm .75\%$ $\varepsilon_V = \varepsilon_H = 100\pi$ mm·mrad

3. the momentum bite is cooled to $\sim \pm .083\%$ in ~ 2 sec and the bunch added to the stack of antiprotons already accumulated and continually cooled. These steps are repeated leading to accumulation rates $\sim 5 \times 10^9$ $\bar{p}$/hr which provides $\sim 10^{11}$ $\bar{p}$/day – a sufficiently large number for injection into the SPS.

4. The antiprotons are extracted from the AA and reinjected into the PS where they are accelerated to 26 GeV and transferred to the SPS.

5. In the SPS the antiprotons and the protons, already injected prior to the antiproton transfer, are accelerated from 26 GeV to 310 GeV where the magnet ramp goes D.C. and collisions ensue. Clearly the lifetime of the luminosity should be ~ 24 hours in order to give enough time for a new refill of antiprotons to be accumulated while the current fill is used for physics.

SPS Ring · Collisions {310 GeV * 310 GeV.}
26 GeV protons to SPS.
26 GeV antiprotons to SPS.
ANTIPROTON ACCOMULATOR 3.5 GeV/c
3.5 GeV/c anti protons returned to PS.
P.S. BOOSTER
ANTI PROTON PRODUCTION TARGET
P.S. 26 GeV
PS antiprotons
PS protons

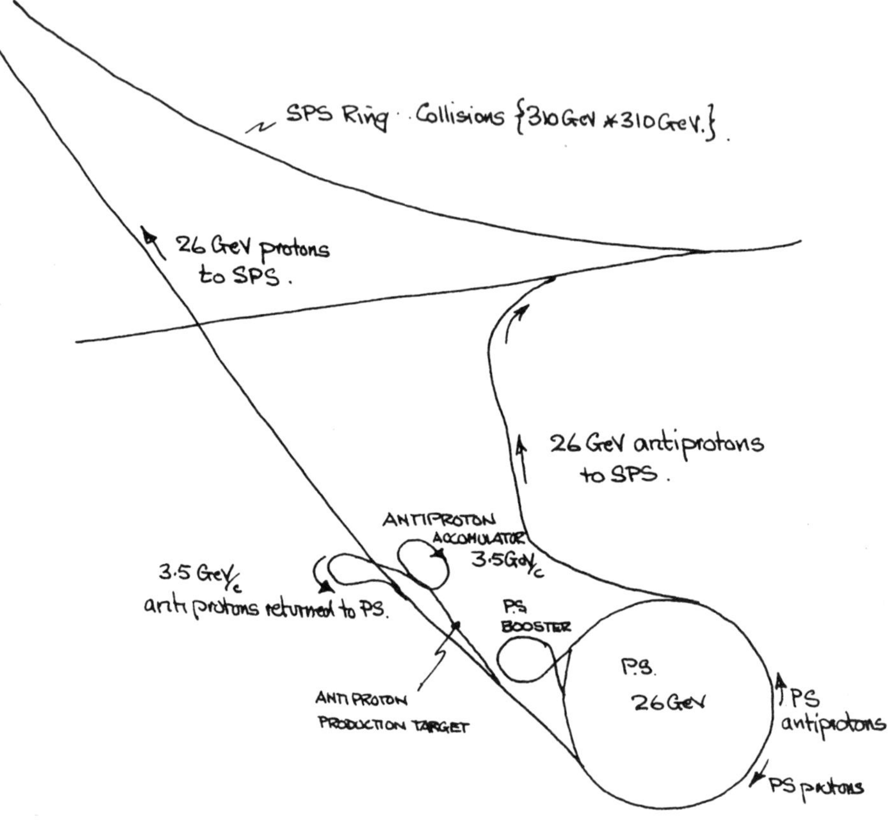

Figure 4

Table 2 gives some of the major parameters of the CERN SPS Collider
compared with the design goals.

Table 2

	Design	Operation
A.A. accumulation rate $\bar{p}$/hr:	2.5×10^{10}	6×10^9
Transfer efficiency AA→SPS	100%	~75%
Number of bunches	6 * 6	3 * 3
Protons/bunch	10^{11}	1.5×10^{11}
Antiprotons/bunch	10^{11}	0.15×10^{11}
Normalized emittances p	20	~18
($10^{-6} \pi$ mrad) $\quad \bar{p}$	15	~14
β_H^*(m) × β_V^*(m)	2×1	1×0.5
Luminosity cm^{-2}sec^{-1}	10^{30}	~5×10^{29}
Lifetimes: proton	>24 h	~150 h
antiproton	>24 h	~40 h
luminosity		~25 h
Tune shift/crossing	.003	.003

In the above table the tune shift[8] represents the electromagnetic
effect of one bunch on the other as they cross. The effect is like a
nonlinear component in the machine lattice and cannot be analytically
calculated for real accelerator bunches. Prior to operation a number
of accelerator physicists confidently predicted that the CERN Collider
would not work, or at best produce luminosities with lifetimes too
short to be useful as the beams blow up due to the effects of
tuneshift. This was not the case, so in addition to providing the
means for the discovery of the W and Z bosons, the SPS machine broke
new ground in the operation of bunched hadron colliders.

The only other bunched hadron collider which has worked is at FNAL. The cooling method for the antiprotons is again stochastic but the antiproton source has separate rings for the "fast cool" stage and the main stack. Antiprotons are produced by the Main Ring at 120 GeV and the projected accumulation rate is ~$1 \times 10^{11} \bar{p}$/hour, thus making refills possible in ~2 hours. On acceleration – towards collisions at 150 GeV – protons and antiprotons are passed from the "normal" main ring of the magnets to the superconduction ring of the Tevatron, which accelerates them to ~1 TeV before going to DC operation for collisions. The machine produced events in 1985 and should start serious physics production in 1986. The projected luminosity is $> 10^{30} \mathrm{cm}^{-2} \mathrm{sec}^{-1}$.

Detectors:

<u>The principles of parton and lepton detection:</u>

The detectors around the Hadron Colliders register, of course, the stable hadrons, e.g. pions, kaons, protons, lambdas, etc., photons which are mostly the debris from neutral meson decays, e.g. π^0, η, and the leptons e, μ, ν. The physics which we wish to discuss in the future lectures is largely that of quarks and gluons, along with the leptons; i.e., constituent physics.

The detection of quarks and gluons becomes the detection of jets, collimated groups of hadrons and photons limited in p_T with respect to some jet axis. Since some jets are entirely neutral in character, their detection puts emphasis on good calorimetry, hadronic and electromagnetic. The distinction between electromagnetic and hadronic calorimetry is to some extent "washed out" in the reconstruction of jet energies, since the quarks and gluons fragment into many soft hadrons. A large fraction of the jet energy is obtained from the electromagnetic section of the calorimeter (usually ~1 interaction length thick) which sits inside the hadron section; e.g., for a 50 GeV jet on average ~70% of the energy is measured in the electromagnetic section.

The principles of lepton detection are shown schematically in
Figure 5; we will discuss the three kinds e, μ and ν.

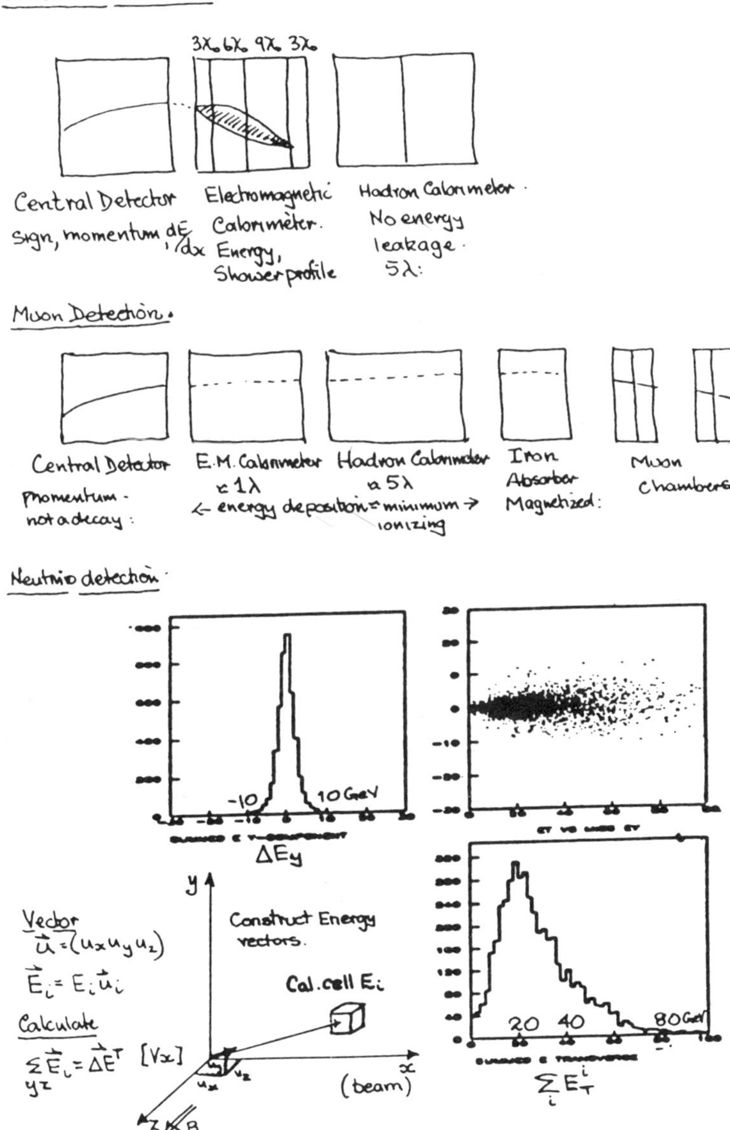

Figure 5

<u>electrons</u> - the coordinate device at the centre of the detector measures the electron track, and with the aid of a magnetic field determines the sign and momentum. The energy deposition in the electromagnetic calorimeter must then match the track prediction in position and magnitude. In addition the longitudinal segmentation of the calorimeter provides a profile of the energy deposition which must be characteristic of an electromagnetic shower, with no energy leakage into the hadronic section. These, broadly speaking, are the most common tools available to identify electrons and, for example at ~40 GeV, can yield a pion rejection of ~10^4:1. They can be augmented by other methods, e.g. dE/dx in the central detector of Transition Radiation Detectors.

<u>muons</u> - at modest energies (< a few tens of GeV) the distinguishing feature of the muon is its ability to penetrate matter losing energy only by dE/dx with no nuclear interaction. The central coordinate device should have the resolution to reveal decay muons (i.e., k-μ) and the external muon detector should have the ability to measure the direction of the penetrating muon, which should match - apart from the multiple scattering in some eight interaction lengths of iron - with the central track. The external measurement of a track reduces the chances of "shower particle" leakage from the calorimeter masquerading as a muon. An external momentum measurement is also extremely valuable in discriminating against μ's from π and k decay. At higher energies the muon energy loss is not as innocuous, and radiative effects play a role such that precise muon measurement necessitates the addition of significant amounts of energy seen in the calorimeter.

<u>neutrinos</u> - the neutrino, being uncharged and interacting only through the weak interaction, has the ability to "leak" large amounts of undetected energy from the event; it is this feature which is used for its detection. We remarked that for hadron-hadron collisions one is rarely in the true centre of mass of the colliding partons. This, plus the non-negligible leakage of undetected energy through the beam pipes at the ends of the detector, immediately suggests that any measured lack of conservation of "transverse energy" is the way to

look for neutrinos. Each calorimeter cell yields a vector in the plane transverse to the beams and a test of closure is applied; i.e., $\sum_i E_i^T = 0$. In the UA1 detector the precision on the ΔE^T measurement is approximately $\sigma_{\Delta E}^T \simeq 0.6\sqrt{\sum_i |E_i^T|}$. For typical events, with an E^T requirement in the trigger (e.g., $\sum_i E_i^T \simeq 80$ GeV), a "3σ requirement" for missing energy means one is detecting ≥ 17 GeV of missing transverse energy; i.e., relatively high energy neutrinos. In order to have "neutrino detection", the apparatus evidently must be as hermetic as possible, over as large a solid angle as possible.

Some typical Detectors at Hadron Colliders

We comment very briefly on three detectors currently in use at the Hadron Colliders. Their similarities and principal differences are collected in Table 3.[9]

<u>UA1</u>: A section through UA1 is shown in Figure 6. The central detector, a cylindrical drift chamber, occupies a volume of 5.8 m × 2.3 m diameter around the interaction point and yields a "bubble chamber like" view of the charged tracks. Its spatial precision is in the range 100 – 350 μm depending on the drift distance (max ~ 18 cm). The curvature in a transverse magnetic field of 0.7 T yields a momentum precision of 25% at 50 GeV/c for a track perpendicular to the field. The central detector is surrounded by the electromagnetic calorimeters, based on lead-scintillator "sandwiches". These are in turn surrounded by the hadron calorimeter built by placing scintillator in the iron return yoke of the magnet. Because of the transverse magnetic field acting on the beams, the experiment has compensating magnets upstream and downstream of the apparatus. These have been calorimetrized and, along with the very forward calorimeters, give coverage to within ~0.2° to the incident beams. The detector is clad with additional iron shielding; at 90° this shield is magnetized and contains 3 planes of proportional tubes. Outside of this, muon chambers provide good spatial (300 μ) and angular definition of any penetrating tracks.

Table 3.

Experiment	Magnetic Field	Tracking	Electromagnetic Calorimetry	Hadronic Calorimetry	Rapidity range of full Calorimetry	Muon Capability
CDF(FNAL)	Solenoid 1.5 T	Cylindrical DC10°<θ<170° Vertex TPC's	Pb/Sc: $\sigma/E \simeq 14\%/\sqrt{E}$ Tower geometry behind coil. Forward Pb/ proportional tubes $\sigma/E \simeq 24\%/\sqrt{E}$	Fe/Sc: $\sigma/E \simeq 70\%/\sqrt{E}$ Tower geometry. Forward Fe/ proportional tubes $\sigma/E \simeq 100\%/\sqrt{E}$	± 6 units	Yes – 90° Forward toroids.
UA1(CERN)	Dipole 0.7 T	Cylindrical DC 5°<θ<175° $\sigma \simeq 250\mu m$ $\sigma(dE/dx) \simeq 6\%$	Pb/Sc: 25° < θ < 155° $\sigma/E \simeq 17\%/\sqrt{E}$ $\Delta\theta \simeq 2°\Delta\phi \simeq 180°$ 5° < θ < 25° $\sigma/E_T \simeq 12\%/\sqrt{E_T}$ $\Delta\theta \simeq 20°\Delta\phi \simeq 11°$	Fe/Sc: $\sigma/E \simeq 75\%/\sqrt{E}$ $\Delta\theta \simeq 15\%\Delta\phi \simeq 18°$	± 7.4 units	Yes. "Full" solid angle
UA2(CERN)	NONE: Toroid Spec 20° < θ < 37° Forward/backward spectrometer with DC and shower counters	Cylindrical DC and MWCP 20°<θ<160°	Pb/Sc: 40° < θ < 140° $\sigma/E \simeq 14\%/\sqrt{E}$ $1.5X_0$ preshower detector Tower geometry $\Delta\theta \simeq 10°\Delta\theta \simeq 15°$	Fe/Sc: 40° < θ < 140° $\sigma/E \simeq 60\%/\sqrt{E}$ Tower geometry $\Delta\theta \simeq 10°\Delta\phi \simeq 15°$	± 1 unit	NONE

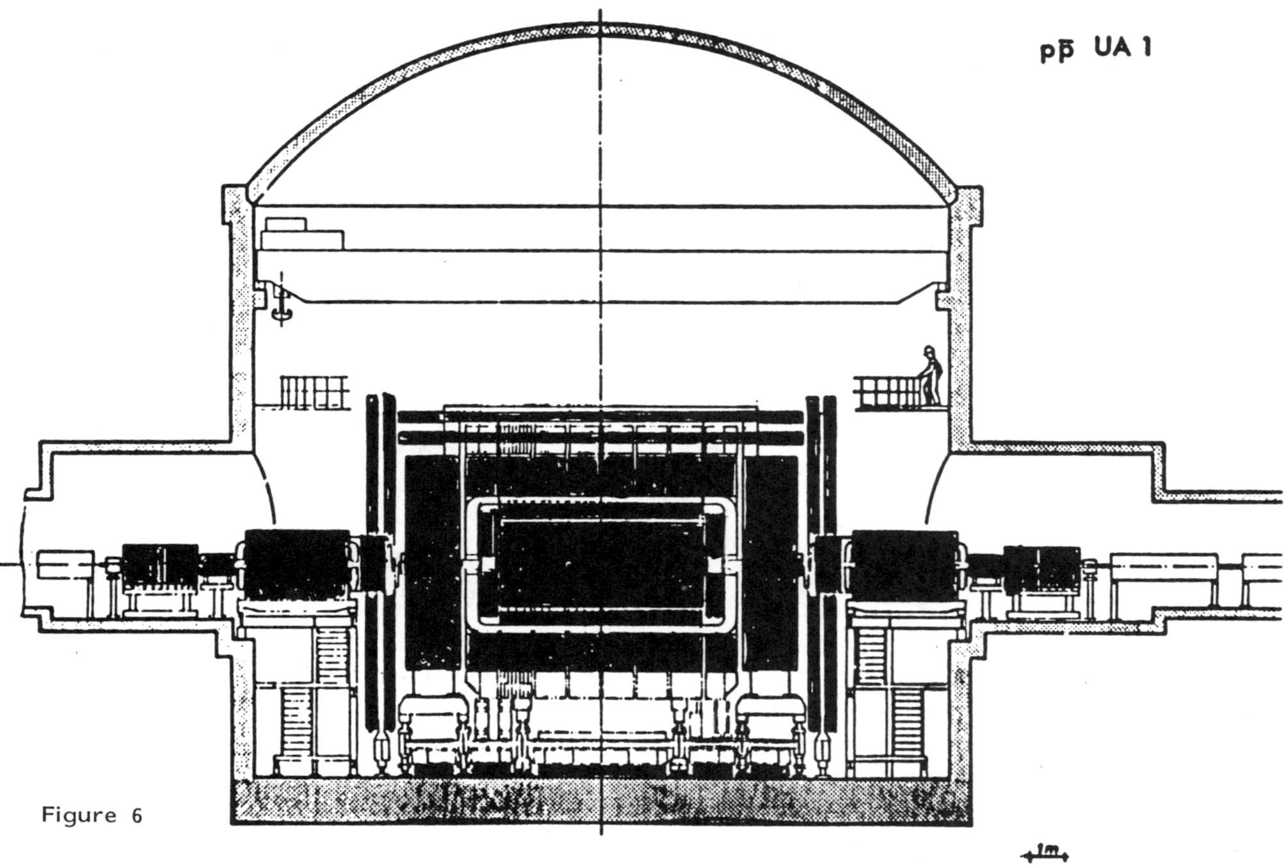

Figure 6

UA2: Figure 7 shows the UA2 detector. It has no central magnetic
field but has a torroidal spectrometer in the angular region
20° < θ < 37.5° with a field of ~0.38 T. The central coordinate
device is based on a drift chamber and multiwire proportional chamber.

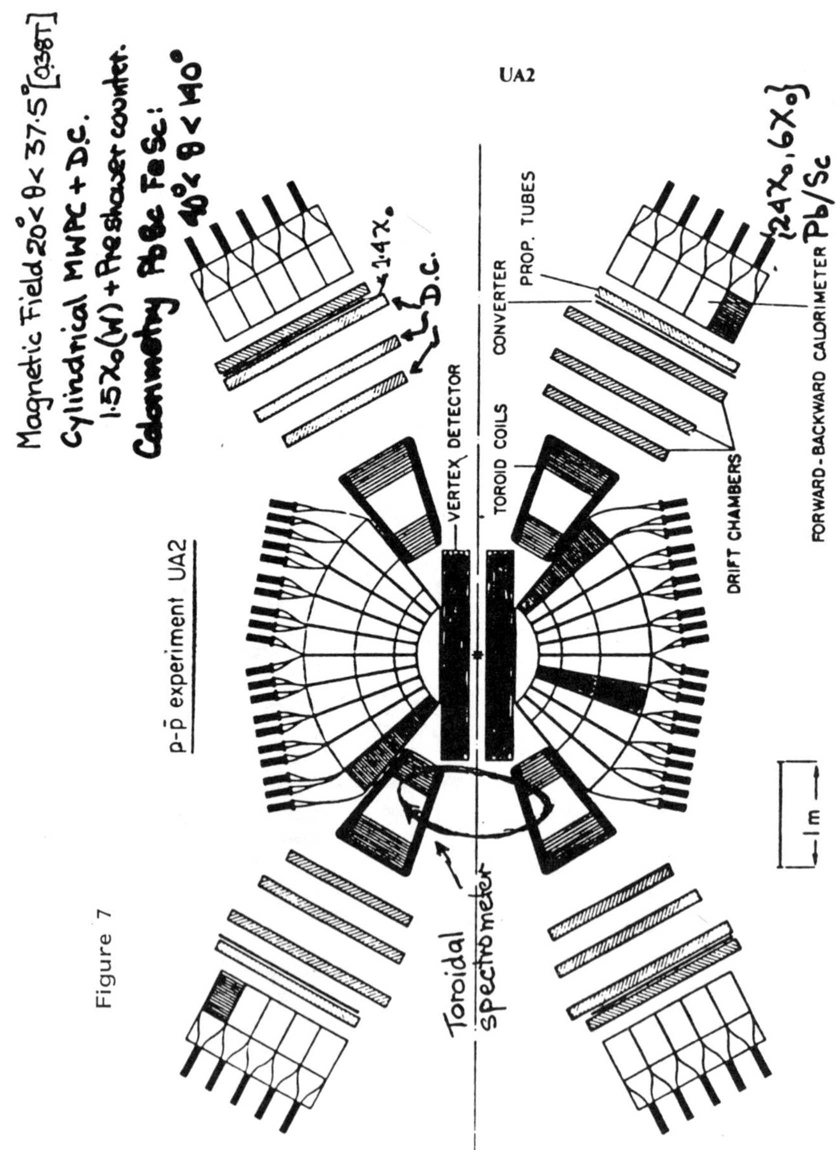

Figure 7

The central calorimetry, electromagnetic (lead scintillator) and hadronic (iron scintillator), covers the region 170° < θ < 40°, leaving the ends free of hadronic coverage. The localization of the shower in the electromagnetic calorimeter is provided by a preshower counter consisting of a 1.5 radiation length piece of lead backed up by proportional chambers giving ~3 mm precision. The calorimeter has good granularity achieved by the pointing geometry.

CDF (FNAL): The CDF detector is shown in Figure 8. The central tracking devics sits inside the superconducting coil which gives a 1.5 T solenoidal field. Outside the coil are electromagnetic and hadronic calorimeters. At forward angles the experiment uses gas calorimetry in an attempt to get free of the radiation damage problems with scintillators, now well established by UA1 and UA2. There is large angle muon detection, plus the interesting feature of a forward spectrometer for muons, based on toroids of magnetized iron. Like UA2, and unlike UA1, the calorimetry has tower geometry which points at the interaction point.

Figure 8

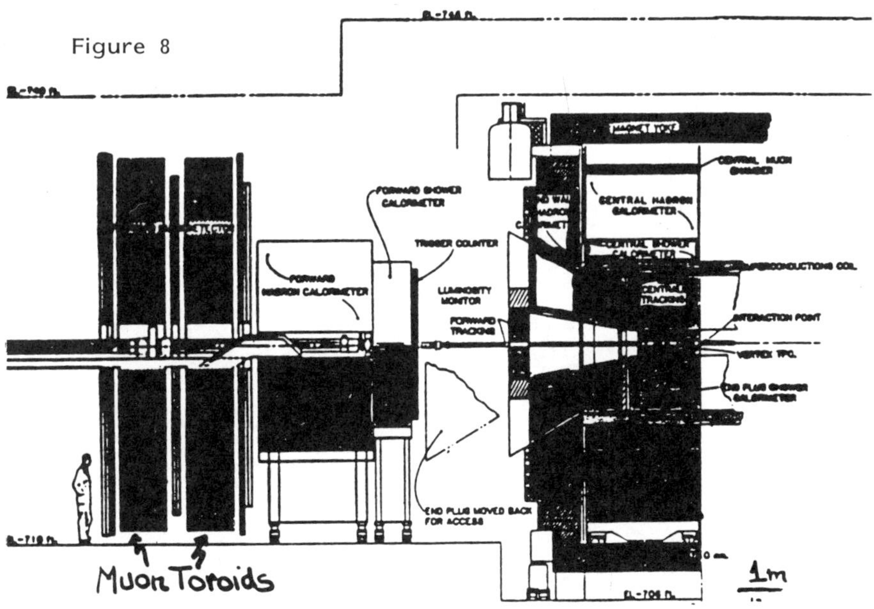

What have we learned about Detector requirements at a Hadron Collider?

1. These lessons have come about from running in a relatively low luminosity environment $\pounds < 10^{30}$ where a great deal still depends on a good track reconstruction of the event. The detector must have the ability to trigger selectively at several levels; e.g., using the UA1 experiment.

Level zero – is derived from scintillation counter hodoscopes upstream and downstream. The trigger cross-section is probably ~35 mb (UA1 at $\sqrt{s}$ = 630 GeV) resulting in a trigger rate of ~36.5 kHz at luminosity of ~10^{30} cm^{-2}sec^{-1}. This rate must be reduced to a tape writing speed ~1 Hz.

Level 1 – is dependent on the calorimetry and decisions are made between bunch crossings: typically ~3.6 μs (7.2 for 3 bunch operation at SPS). This trigger selects, for example, electron candidates E_T>10 GeV, jets E_T>25 GeV, ΣE_T over central calorimetry >80 GeV, and if the calorimetry is good enough, "missing E_T>17 GeV. The "working" thresholds depend on luminosity.

Muon Trigger: an external track – derived form the muon chambers but not drift times – which points towards the vertex, ±15 mradians.

These two 1st level triggers must reduce the high rate to ~10 – 20 Hz, which can be handled by the "local intelligence"; e.g., 168E – emulators in the UA1 experiment.

Level 2 – 5 × 168E emulators running 25,000 lines of Fortran program. At this level there is full calorimeter reconstruction using off-line calibration constants, cluster algorithms, and geometrical corrections. Jets are found, isolation criteria applied to electron candidates, and missing energy is validated, for example. Tracks are reconstructed in selected regions of the central detector and matching sought with the reconstructed muon track. These emulators can then flag the

events as "rejected", "normal", or of "special interest",
achieving a final tape writing of ~1 - 2 Hz for events which
typically contain ~120 Kbytes of data. Without this Level 2
selection with full "off-line" reconstructural ability, the
data rates at high luminosity would be prohibitive and one
could only resort to raising E_T thresholds all round at Level
1 - resulting in a serious loss of physics. The dead-time for
the full chain is kept typically to $\leq 10\%$.

2. **Lepton signatures** The charge leptons e and μ serve to flag
the effects of the weak interaction at the hadron colliders. The W
and Z bosons were both discovered via lepton decays $W \rightarrow e\nu$ and $Z \rightarrow e^+e^-$;
in fact, neither UA1 nor UA2 have reconstructed a boson mass from
their jet-jet decays. The e and μ signature also signals the semi-
leptonic decay of heavy quarks; here one has to deal with leptons in
or near jets and the muon detection provides the better opportunity.
If the detector has both e and μ capabilities a lot can be learned
from the complementarity of examining both channels with their vastly
different systematic problems.

3. **Neutrino detection** The weak decays yielding e or μ also produce a neutrino in the detector. For efficient detection of missing
transverse energy the detector should be as hermetic as possible. As
an example of the power of being able to "detect" neutrinos, Figure 9
compares the W detection in the eν channel in the UA1 and UA2
detectors[10]). Figure 9a shows the inclusive electron spectrum of the
two experiments with both detectors turning in their "best efforts"
at electron identification. The discovery of the W would still be in
debate if this were the only evidence. Figure 9b shows the inclusive
ν spectrum for events which already have an "electron" with $p_T > 15$ GeV.
With its better calorimetry coverage the UA1 spectrum already suggests
two families of events - the W's and background - mostly jet
fluctuations. In the UA2 spectrum these families are not separated.
When both the electron and neutrino presence is required, one then
obtains the extremely clean W signatures of Figure 9c, and in fact the
W was discovered with a handful of events.

Inclusive "e" spectrum
$P_T > 15$ GeV.
All "handles" used.
· Hadron leakage
· Shower profiles
· Track matching
· Isolation
$\varepsilon_{eff} \approx 75\%$:
UA1
UA2
(a)
Inclusive "ν"
for events
$P_T^e > 15$ GeV
UA1
UA2
15 GeV
15
(b)
"And "of "e" ν"
UA1
$W \to e\nu$
15 GeV
40
UA2
$W \to e\nu$:
35
(c)

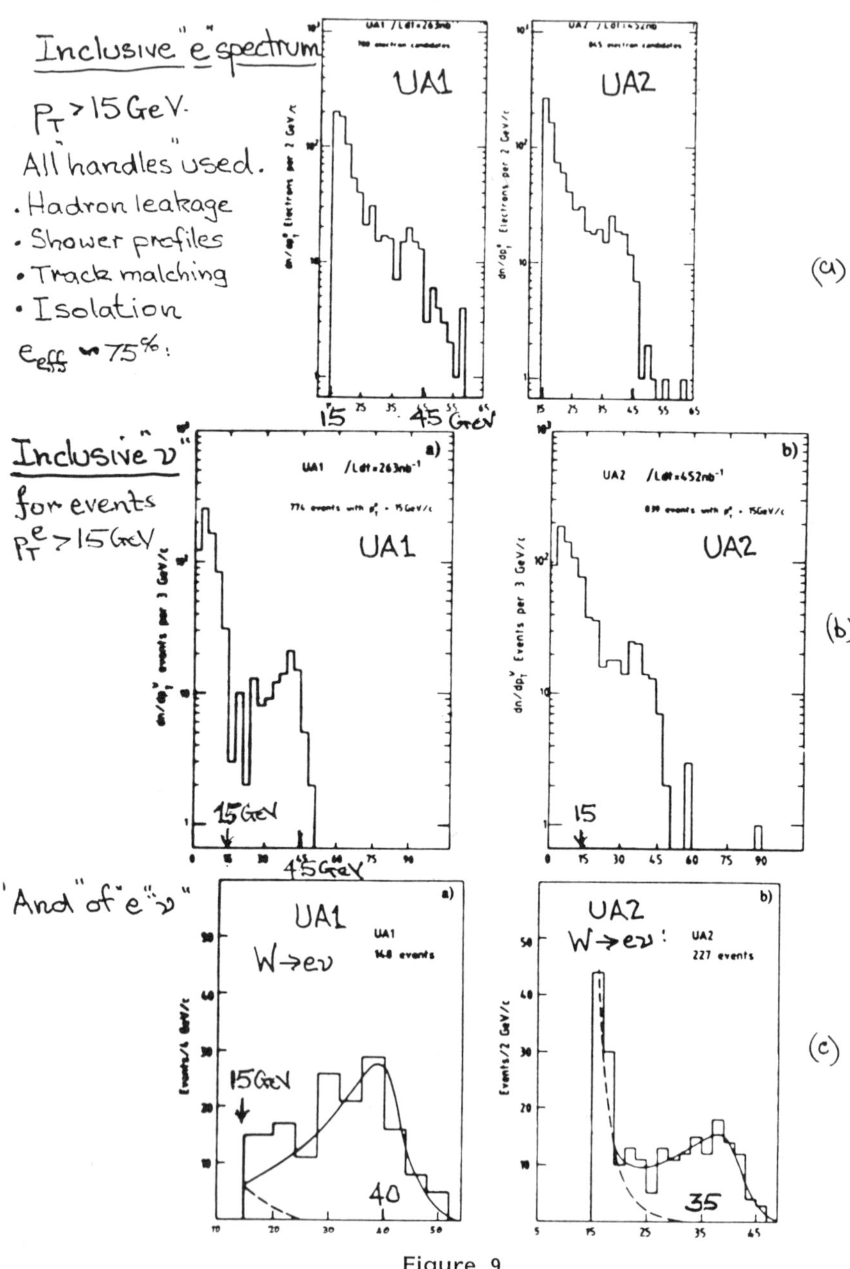

Figure 9

4. Jet measurement The measurement of the energy of jets is an essential feature in collider physics since their presence dominates the central hard collisions. The parton fragmentation can reveal extreme fluctuations, from entirely neutral (all photons from π° decays) to essentially all charged. It is therefore important that the calorimeter has the same response to hadronic and electromagnetic energy. This is certainly not the case with the lead-iron combinations. When the hadron interacts with a nucleus, energy is lost - not detected. The only way used up to now to retrieve this energy has uranium as the inactive calorimeter medium, and an active medium with good sensitivity to low energy neutrons[11]). In this way some of the fission energy of the uranium can be detected, and compensates for the unseen lost hadron energy. The ratio of electromagnetic to hadronic response is brought much closer to unity. In a jet dominated environment, a different response for the different fragmentation modes limits the precision with which the missing transverse energy can be measured, and introduces systematic biases in any attempt at "jet-jet spectroscopy". As an aside on calorimetry, one ignores at great risk a sensitivity of the active medium to radiation damage.

5. Redundancy To some extent this property is derived from the presence of a magnetic field; e.g., for electrons the momentum can be compared with the energy. If a calorimeter cell is struck by a charged track it is extremely important to know the momentum of the track, since the cell could also contain a high energy photon. Also the sign of the leptons in particular carries real information. The "hard collision" can be isolated and studied by requiring a transverse momentum cut. This is illustrated in the sequence of Figures 10a, b and c where the products of the spectator partons can be cleaned off to reveal simply the $W \rightarrow e\nu$ decay. It is equally clear that the coordinate device must be able to support and measure many extra tracks without losing or degrading the real physics.

6. Secondary vertices The microvertex detectors of e^+e^- colliders have yet to be employed successfully at the hadron colliders. Here with good spatial resolution ≤ 50 μm one is trying to see evidence for

Figure 10

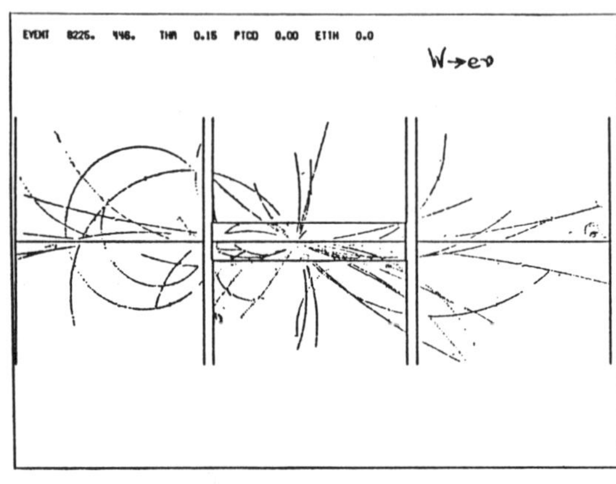

(a) Central Detector tracks only

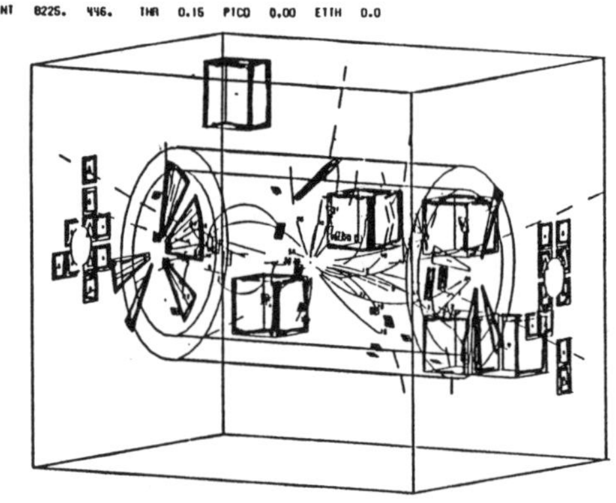

(b) Calorimetry cells added

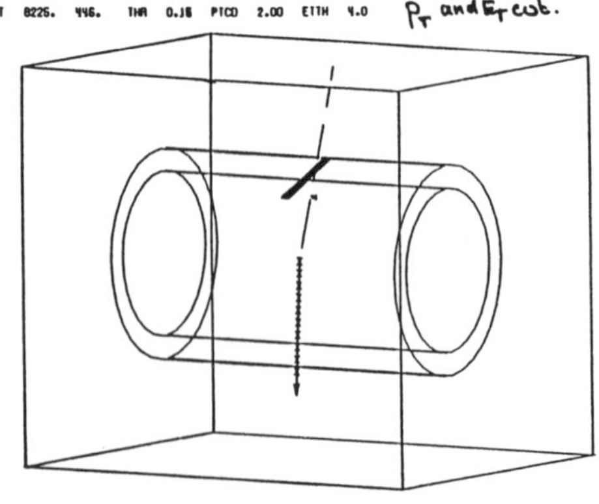

(c) A cut of $p_T >$ imposed on the tracks, and a cut of $E_T >$ imposed on the calorimeter cells – only $e\nu$ of W decay remain.

long lived ($\sim 10^{-13}$s) secondaries; e.g., τ's, charmed objects and
b-objects. The hadron collider environment will always be difficult
since the secondary vertex will be in or near a jet and the
multiplicities in the chamber will be high, >80 tracks.

The features 1 - 5 have so far been put to good use at the CERN
Collider where perhaps the major experimental discovery is that
collisions between complex objects like hadrons can indeed by revealed
as simple clean pieces of physics, almost rivaling the cleanliness of
e^+e^- collisions.

The Future

In the pre "Z° factory" era, the CERN SPS Collider will receive an
increase in luminosity by a factor of ~15. The new antiproton
accumulator (ACOL) should come onstream in September 1987, accompanied
by upgraded detectors at UA1 and UA2. The UA1 upgrade is a very
ambitious attempt to replace the central calorimeter with a warm
nonpolar liquid medium for charge collection inside depleted uranium
plates which should yield "compensation" of the hadronic energy
measurements. The UA2 upgrade is more modest, with the central
calorimetry extended to close the ends of the experiment plus the
addition of a vertex detector based on scintillating fibres.

The Tevatron Collider at FNAL should also have reached design
luminosity by 1988, making possible data samples of 10 pb^{-1}/year very
feasible in both FNAL and CERN. Currently all the UA1 physics has
been derived from 0.75 pb^{-1}. With the advent of LEP and SLC the mass
of the Z° should be measured by the accelerator physicists to ±50 MeV;
armed with this calibration point, the precise measurement of the W
mass will remain the responsibility of the hadron colliders with a
precision of ±150 MeV likely. At these levels one will access the
radiative corrections to the Standard Model.

Lecture 2 - QCD at the Collider

The emergence of jets in Hadron Collisions:

The inclusive reaction

$$a + b \to c + anything$$

can be visualized in terms of the constituents of the hadrons a, b, c as represented by Figure 1. In the box 1, $f_i^a(x)$ represents the probability that constituent i carries fraction x of the four momentum of hadron a. If hadron a is a proton and i is a quark or antiquark, then $f_i^a(x)$ can be derived from the structure functions of deep inelastic scattering. The box 2 contains the hard scattering, $2 \to 2$ process, and box 3 represents the hadronization of k, which is known in principle from the fragmentation functions of constituents as determined in e^+e^- collisions. In principle a knowledge of the contents of boxes 1 and 3 permits a study of the process in box 2.

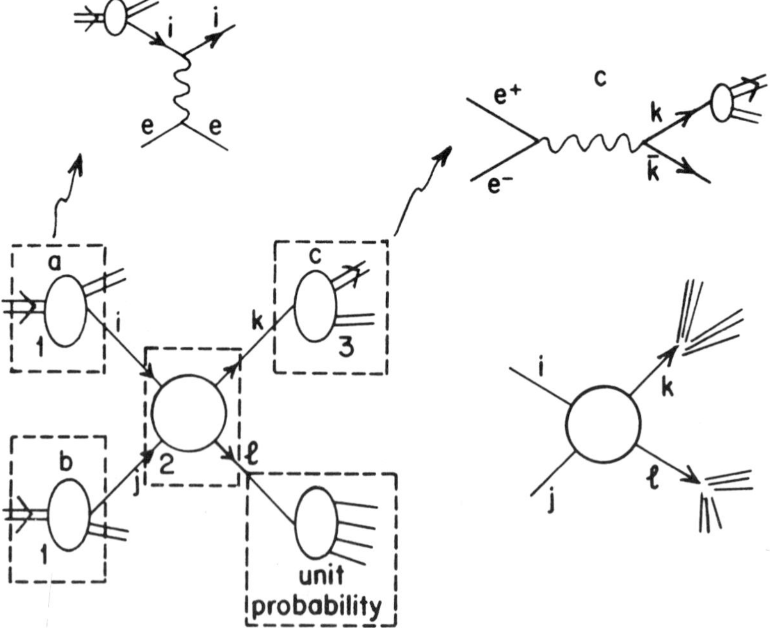

Figure 1

Some general features of the process emerge:

(a)   the $2 \to 2$ process, $i + j \to k + \ell$, is coplanar, assuming that p_T is limited in the subprocesses $a \to i$, $b \to j$, and $k \to c$;

(b)   we may expect a "jet" of particles on the "same side" as particle c, plus a jet on the away side, from the process $\ell \to$ hadrons;

(c)   in addition, we may expect some low p_T background moving along the directions of a and b, from $a \to i$ and $b \to j$.

These correlations were seen at the SPS Collider[12] using the variables ϕ – the azimuthal angle, and η – the pseudorapidity ($\eta = -\ln \tan\theta/2$). Much more spectacular evidence for jets can be observed if one examines the pattern of energy deposition in the calorimeter for high E_T events[13], again in terms of ϕ and η. Figure 2 shows a "LEGO" plot of the energy deposited per η and ϕ cell; the jets are very obvious. For jet transverse energies of $E_T > 25$ GeV the jet structure is a fairly well defined object.

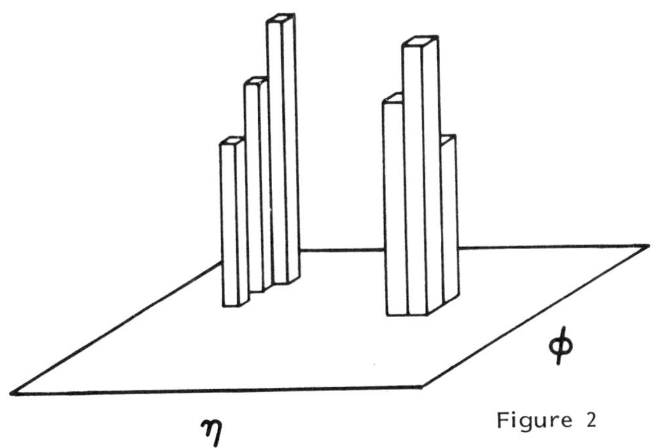

Figure 2

In order to proceed with jet studies one needs a definition of a jet, or "jet finding algorithm." In UA1 initiators are selected, usually cells in η/ϕ with $E_T > 1.5$ GeV; other cells are clustered together if the quantity $R = \sqrt{(\Delta\eta)^2 + (\Delta\phi)^2} < 1$. One is simply drawing circles of radius 1 unit in the η, ϕ plane around the initiator and summing the cells inside the circle. There are problems with any jet energy which falls outside the circle, and conversely energy from the underlying event which spills into the circle - we will not address these problems here, except to say that jet energies are corrected by Monte Carlo.

The inclusive jet cross section

At the SPS Collider the inclusive cross section for jet production is obtained from a measurement of the process

$$\bar{p} + p \to \text{"jet"} + \text{anything}.$$

The cross section may be defined as follows:

$$\left. \frac{d^2\sigma}{dE_T d\eta} \right|_{\eta=0} = \frac{\text{number of jets} \times \text{smearing correction} \times \phi \text{ correction}}{\pounds \cdot \Delta E_T \, \Delta\eta}$$

where $\pounds$ is the integrated luminosity and $\Delta\eta$ is the rapidity range (1.4). The data must be corrected for the ϕ acceptance of the apparatus, and account taken of the fact that one is measuring a rapidly varying function of E_T with a finite resolution calorimeter - the smearing correction. Figure 3 shows the UA1 data[14] at two energies compared with the QCD prediction; the theory and experiment have been adjusted in normalization (factor ~1.5) but the agreement in shape is very impressive.

The two-jet cross section:

The constituent picture of the inclusive interaction in Figure 1 is easily extended to the two-jet process

$$\bar{p} + p \to \text{jet}_1 + \text{jet}_2 + \text{"X"}$$

if one measures the jets associated with the hadronization of the final partons k and ℓ. This identification turns the jet physics

10²
10¹
10⁰
10⁻¹
10⁻²
10⁻³
10⁻⁴
dσ/dp_T dη |η = 0 (nb GeV⁻¹)
SIGMA
ET
0   20   40   60   80   100   120   140   160
p_T (GeV)
○ √s = 546 GeV
● √s = 630 GeV
QCD ---- √s = 546 GeV
—— √s = 630 GeV

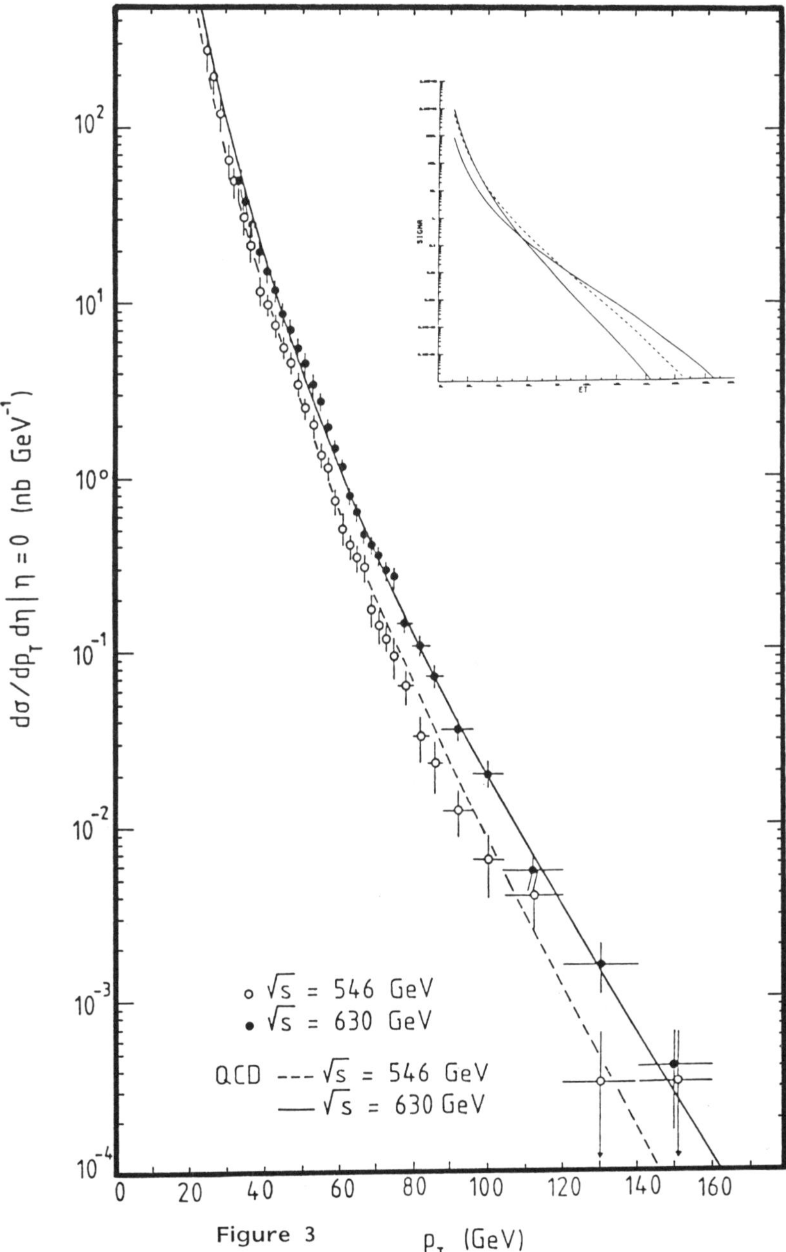

Figure 3

32

into parton physics. We can write, therefore:

$$\sigma\left(\bar{p}p \to k(jet) + \ell(jet)\right) = \sum_{ij} P(p \to i)P(\bar{p} \to j)d\sigma(ij \to k\ell)$$

$$d\sigma(\bar{p}p \to jet.jet + X) = \sum_{ij} \left[\frac{F_i(x_1)}{x_1} dx_1\right]\left[\frac{F_j(x_2)}{x_2} dx_2\right]\left[\frac{d\sigma(ij - k\ell)}{d\cos\theta^*}\right] d\cos\theta^*$$

where $\dfrac{F_i(x_1)}{x_1}$ dx_1 is the probability of finding a parton in a, of
type i, carrying a fraction of the proton's momentum between x_1 and
$x_1 + dx_1$. There are many subprocesses (ij $\to$ kℓ) when one considers that
inside the proton we might reasonably expect ($u,d,s,\bar{u},\bar{d},\bar{s},g$) partons.
Deriving the jet-jet cross section, or conversely, deriving the
constituent cross section from the jet-jet cross section, appears at
first sight hopeless, but there are some simplifications.

Suppose we consider the jet pair to be zero mass particles, produced
in the laboratory with equal and opposite p_T at angles θ_1 and θ_2. If
we define a transverse scaling variable $x_T = p_T/\sqrt{s}/2$, and recalling
$\cos\theta = \tanh\eta$, with $\eta = -\ell n \tan\theta/2$, then

$$x_1 = \frac{1}{2} x_T(e^{\eta_1} + e^{\eta_2}) \quad \text{and} \quad x_2 = \frac{1}{2} x_T(e^{-\eta_1} + e^{-\eta_2})$$

with $\hat{s} = x_1 x_2 s = M^2 = \tau s$.

At the SPS Collider, ($\sqrt{s} \approx 630$ GeV) jets with $p_T \sim 35$ GeV produced
centrally ($\eta_1 \sim \eta_2 \sim 0$) correspond to parton fractions $x_1 \sim x_2 \sim .11$.
Here the gluons in the proton will dominate the subprocess, while at
$p_T \sim 120$ GeV the central jets begin to probe quark-quark collisions.
The separate contributions from the subprocesses gg, gq, and qq to the
QCD prediction of the inclusive cross section are shown in Figure 3.

The angular dependence and the magnitude of the various sub-processes
are prescribed by QCD[15]). Figure 4 shows the scattering of quark and
antiquark of different flavours through gluon exchange. The cross
section is given by

(a)
one gluon exchange
i
j
(b)
3
3
gg → gg
3
3
4/3
qg → qg
4/3
4/3
qq̄ → qq̄
4/3
1 : 4/9 : 4/9
Main sub-process and colourfactors
+
+
+
+
Higher order terms

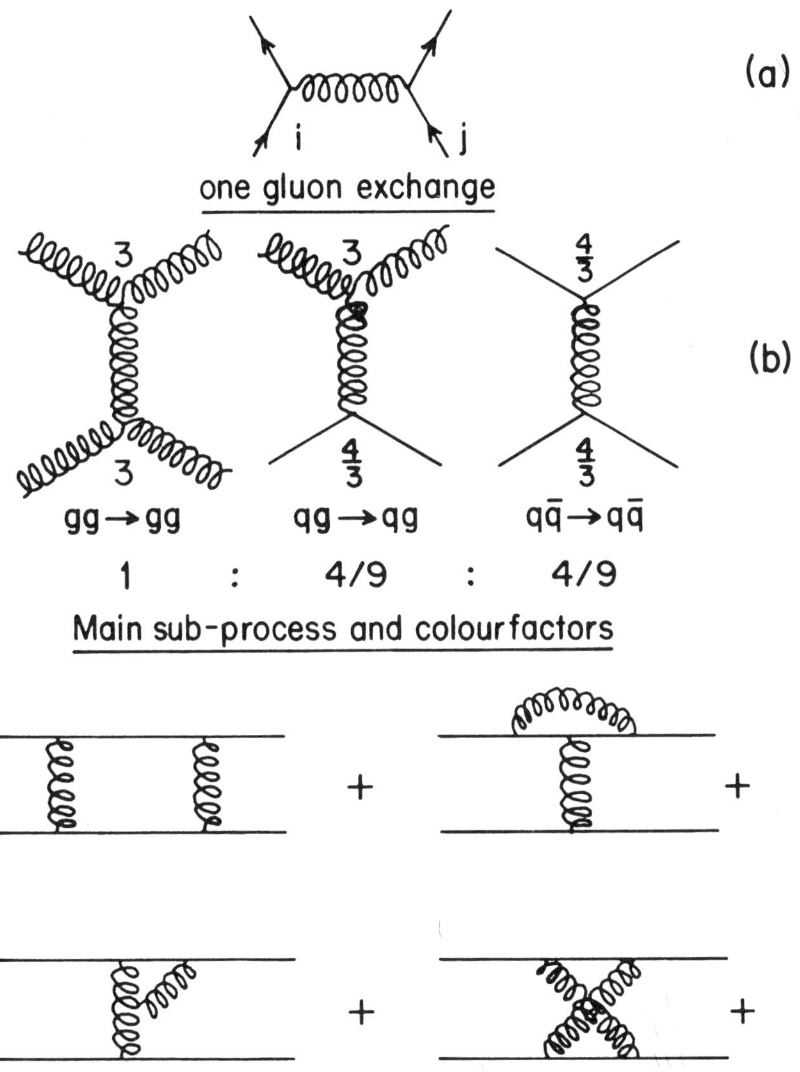

Figure 4

$$\hat{\sigma}(q_i \bar{q}_j \rightarrow q_i \bar{q}_j) = \frac{4\alpha_s^2}{9\hat{s}} \left(\frac{\hat{s}^2 + \hat{u}^2}{\hat{t}^2}\right)$$

where α_s is the strong coupling constant. When we identify the jets
with the partons, we cannot, on an event-by-event basis, determine the
parton nature of the jet, nor can we measure $\cos\theta^*$ directly, but have
to settle for $|\cos\theta^*|$ – we cannot tell "forwards from backwards".
Under these conditions the three major sub-processes depicted in
Figure 4 have experimentally indistinguishable angular distributions,
so the measured cross section becomes

$$\frac{d\sigma}{dx_1 dx_2 d\cos\theta^*} = \frac{F(x_1)}{x_1} \cdot \frac{F(x_2)}{x_2} \cdot \frac{d\sigma}{d\cos\theta}$$

where $F(x)$ is a combined gluon, and quark-antiquark, structure
function – combined with weights proportional to the colour factors of
the three sub-processes

$$F(x) = G(x) + \frac{4}{9}[Q(x) + \bar{Q}(x)].$$

The form of the angular distribution can be taken from one of the
processes; e.g., $gg \rightarrow gg$. There remains still the value of α_s,
the strong coupling constant, and since α_s is a function of Q^2,
$\alpha_s = \alpha_s(Q^2)$, the question of the effective Q^2 also arises. In the
specific $2\rightarrow2$ processes the value of Q^2 can be taken as $-\hat{t}$, but higher
order diagrams like those in Figure 4c can contribute.

To first order in QCD the strong coupling constant is given by

$$\alpha_s = \frac{12\pi}{(33 - 2n_f)\ln(Q^2/\Lambda^2)}.$$

The value of Λ sets a scale and in a sense delineates the boundary of
the non-perturbative regime. The strong interaction must bind quarks
inside a hadron at distances ≤ 1 fm; consequently one expects Λ to be
approximately the mass of a typical hadron; i.e., $0.1 < \Lambda < 1$ GeV.
For $Q^2 \gg \Lambda^2$ the exact value of Λ is not very important. At values
of $Q^2 \sim 100$ GeV^2, with $n_f = 4$ (four flavours "active"), α_s is in the
range $.2 < \alpha_s < .4$, which can be compared with the value of the fine

structure constant, $\alpha = 1/137$ of QED. Thus over distances $\sim 10^{-15}$ cm
the strong interaction is "relatively strong" but the coupling is
small enough for reliable predictions in a perturbative regime.
For values of $Q^2 \gg 100$ GeV^2 we should expect to make quantitative
comparisons with QCD.

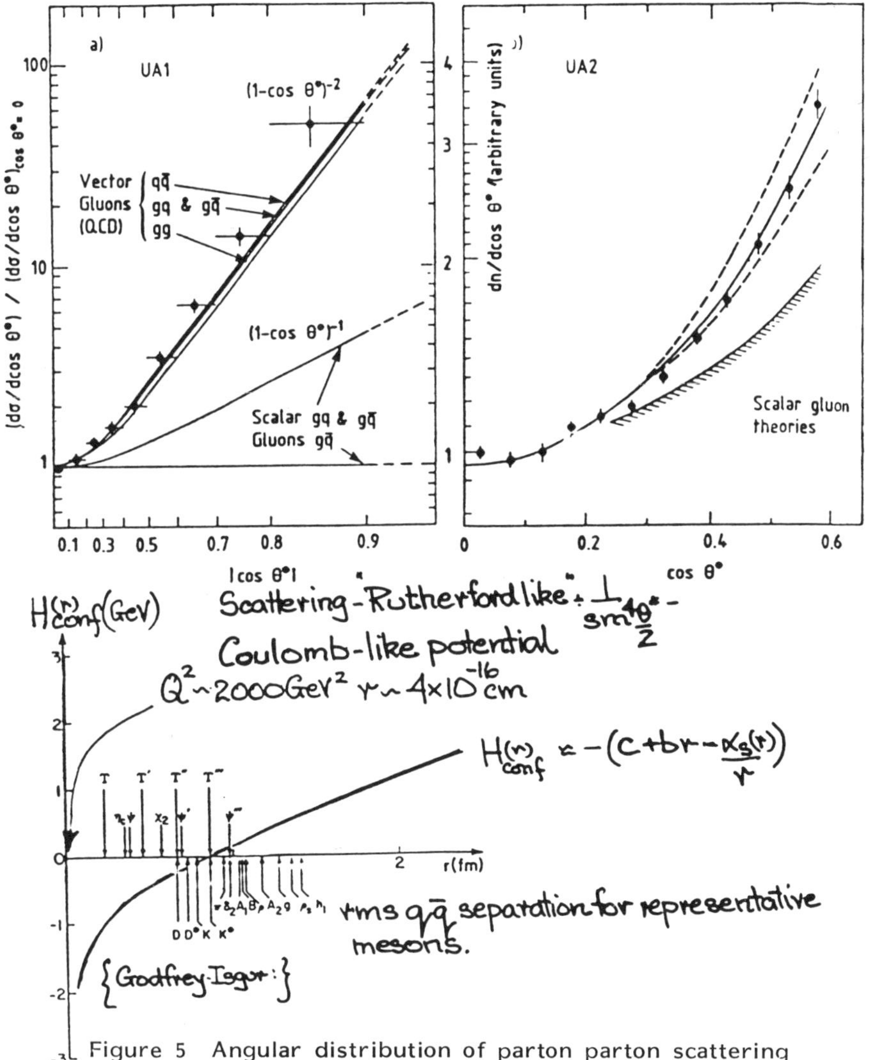

Figure 5 Angular distribution of parton parton scattering

Figure 5[16]) shows the angular distribution for the parton-parton scattering determined from the two-jet events. The experimental problem of distinguishing the different sub-processes can be appreciated from the predictions in Figure 5a. The shape of the measured angular distribution agrees with Rutherford-like scattering involving a massless vector exchange, namely the gluon. The figure also shows the rms value of the separation of q and $\bar{q}$ in various meson states superimposed on the form of the confining potential used by Godfrey and Isgur[17]) in their model of meson spectroscopy. The distinction between the non-perturbative distance scale of spectroscopy and that of perturbative QCD at the collider is dramatically evident.

If we put $S(x_1 x_2) = F(x_1)F(x_2)$, i.e. we assume factorization, we can write

$$\frac{d\sigma}{dx_1 dx_2 d \cos \theta^*} = \frac{S(x_1 x_2)}{x_1 x_2} \frac{d\sigma}{d \cos \theta^*}.$$

Experimentally on an event-by-event basis, x_1, x_2 and $\cos\theta^*$ are measured, and the shape of $\dfrac{d\sigma}{d \cos \theta^*}$ can be taken from QCD; hence one has the ingredients to derive the structure function $F(x)$.

Figure 6 shows a comparison of the combined structure function measured by UA1[18]), with that derived by the CHARM collaboration using low energy ($Q^2 \sim 20$ GeV^2) data from neutrino and electron deep inelastic scattering. The extrapolation from 20 GeV^2 to the average value of Q^2 in the UA1 experiment (~ 2000 GeV^2) is prescribed by QCD. The dotted curve shows an extrapolation taking into account the quark and antiquark components only. The difference between the two curves is due to the gluon component which is measured directly in the hadron collisions. Previously this component had only been inferred from the deep inelastic lepton scattering using a momentum sum rule for the hadrons, where one takes account of the fact that the leptons sense about 50% of the hadron's momentum, namely that fraction carried by quarks and antiquarks.

$$F(x) = G(x) + \frac{4}{9}\left(Q(x) + \bar{Q}(x)\right)$$

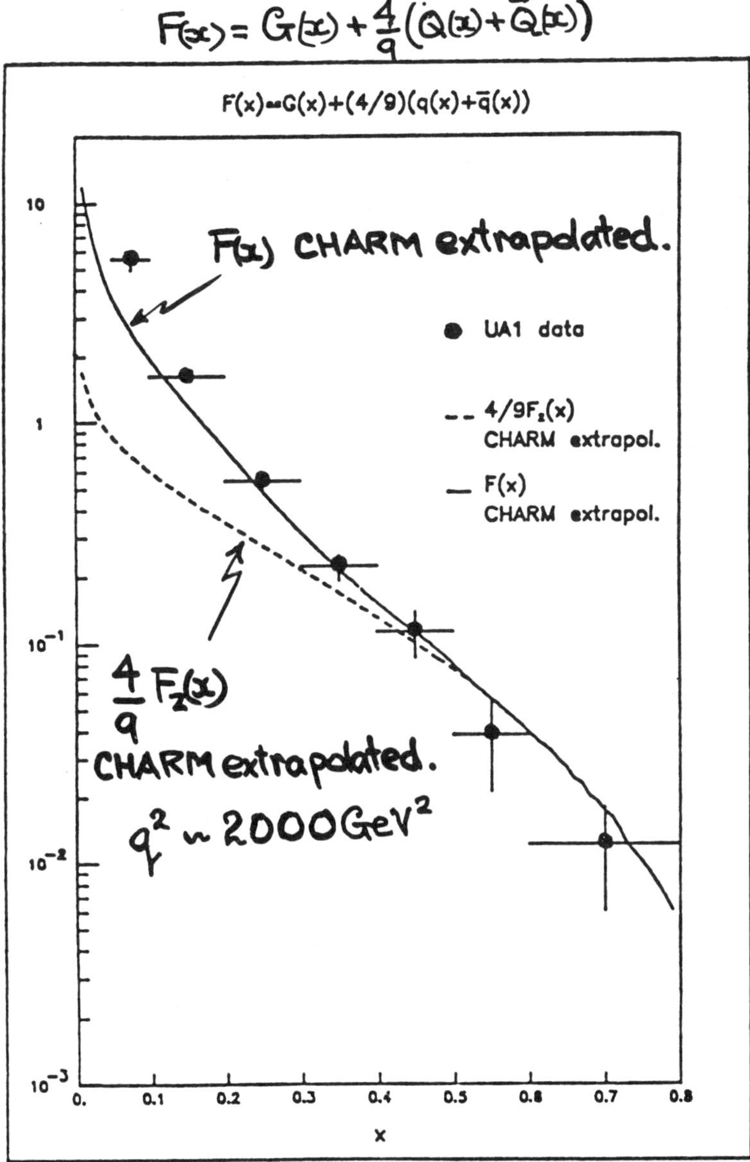

Figure 6

Evidence for scaling violation in the two-jet angular distribution

Figure 7a shows the similarity between Rutherford scattering via photon exchange and quark scattering through massless gluon exchange. The angular distribution is given by

$$\frac{d\sigma}{d\cos\theta} \propto \frac{\alpha_s^2}{s} (1 - \cos\theta)^{-2}.$$

If one defines a variable[20] χ such that $\chi = \dfrac{1 + \cos\theta}{1 - \cos\theta}$, then in the Rutherford approximation, $d\sigma/d\chi = \alpha_s^2/\hat{s}$. That is, except at small values of χ, the $d\sigma/d\chi$ plot is a constant if α_s^2 is constant, for a fixed value of $\hat{s}$. However, as one varies the scattering angle, Q^2 changes and consequently α_s changes. Also there will be effects caused by the variation of the structure function $F(x)$, since in reality one has $F(x, Q^2)$. The "non-flatness" of the data is a direct measurement of the scaling violation in QCD[19], and is well predicted by the theory.

The fragmentation of jets

The charged fragmentation function of a jet determines the manner in which the energy of the jet is shared among the charged particles. The energy of the jet, defined by the algorithm, is derived from the calorimeters. It has to be corrected by Monte Carlo for a loss of energy, due to fragments with $\Delta R > 1$, the dead regions of the detector, and the gain in energy from the underlying event. The charged tracks are measured in the central detector. The jet axis is well defined, and the momenta of the tracks are projected onto it, leading to the variable $z = p_L^{track}/E_{jet}$. The fragmentation function is defined by

$$D(z) = \frac{1}{N_{jet}} \cdot \frac{dN^{ch}}{dz}$$

and here is evaluated for tracks with $\Delta R \leq 1$ and $z > 0.1$. The data have to be corrected for two further effects arising from the finite

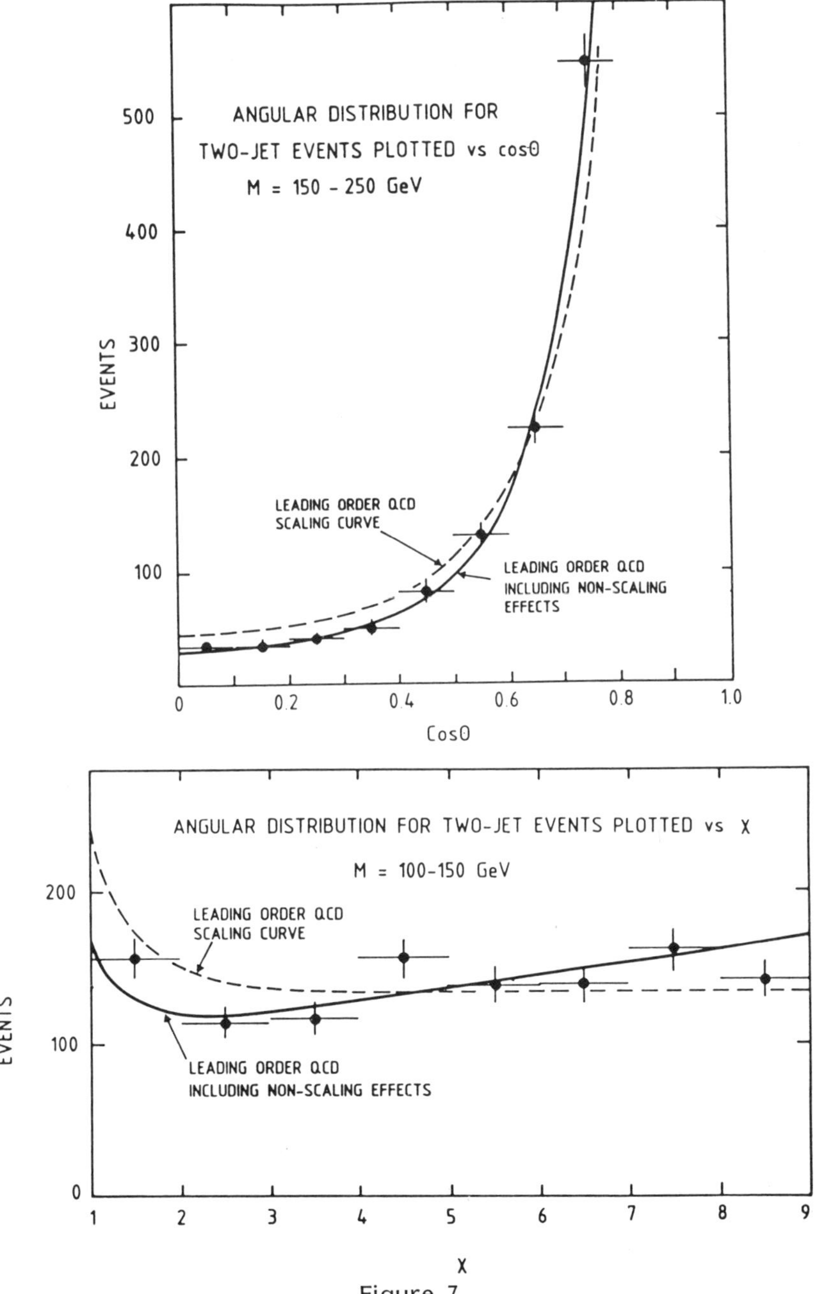

ANGULAR DISTRIBUTION FOR
TWO-JET EVENTS PLOTTED vs cosΘ
M = 150 - 250 GeV
500
400
300
200
100
EVENTS
LEADING ORDER QCD
SCALING CURVE
LEADING ORDER QCD
INCLUDING NON-SCALING
EFFECTS
0
0.2
0.4
0.6
0.8
1.0
CosΘ
ANGULAR DISTRIBUTION FOR TWO-JET EVENTS PLOTTED vs X
M = 100-150 GeV
LEADING ORDER QCD
SCALING CURVE
LEADING ORDER QCD
INCLUDING NON-SCALING EFFECTS
200
100
0
EVENTS
1
2
3
4
5
6
7
8
9
X

Figure 7

40

resolution of the detector. These amount to a "de-convolution" of the smearing due to the error in the momentum measurement in the central detector, $\{\Delta(1/p)\}$, and similarly for the jet energy as determined by the calorimeter with a resolution which varies as $\delta(E_j) \propto \sqrt{E_j}$. The resulting fragmentation function[21]) is shown in Figure 8 along with data from TASSO, e^+e^- collisions at $\sqrt{s}$ = 34 GeV, and ISR, p-p collisions at $\sqrt{s}$ = 63 GeV. The result from the SPS is clearly much softer than the other two experiments, exhibiting fewer particles at high z. This effect is mainly caused by the scaling violations brought about by the higher Q^2 of the SPS collider data, although the mix of constituents is almost certainly different. The data of TASSO and the IRS will be predominantly quark jets, while those from the SPS collider will contain a significant admixture of gluon jets.

The statistical separation of quark and gluon jets[21])

In an attempt to try to study any differences between quark and gluon jets, a statistical separation has been performed. For each two-jet event we can calculate x_1, x_2, $\hat{s}$, $\hat{t}$ and $\hat{u}$, and using the known appropriate structure functions, we calculate the probability

$$\text{Prob}(ij \to k\ell) = \frac{F_i(x_1 Q^2) \times F_j(x_2 Q^2) \times M^2(\hat{s}\hat{t}\hat{u})(ij \to k\ell)}{\underset{ijk\ell}{\Sigma} \text{ All sub-processes}}$$

where $M^2(\hat{s}\hat{t}\hat{u})$ is the relevant QCD matrix element for the sub-process $ij \to k\ell$. For each jet we can evaluate

$$\text{Prob}(\text{jet} = \text{gluon}) = \frac{\Sigma \text{ Prob}(ij \to \text{"gluon"} + \text{anything})}{\underset{ijk\ell}{\Sigma} \text{ All sub-processes}} .$$

The probabilities P(g) are shown in Figure 9. P(q) is defined as 1-P(g). The variable $\hat{s}$ is kept fixed and the range of Q^2 is limited ($1600 \leq Q^2 \leq 2600$ GeV^2) since one is looking for quark gluon differences and not effects due to scaling violations.

$$Q^2 = \frac{2\hat{s}\hat{t}\hat{u}}{\hat{s}^2 + \hat{t}^2 + \hat{u}^2}$$

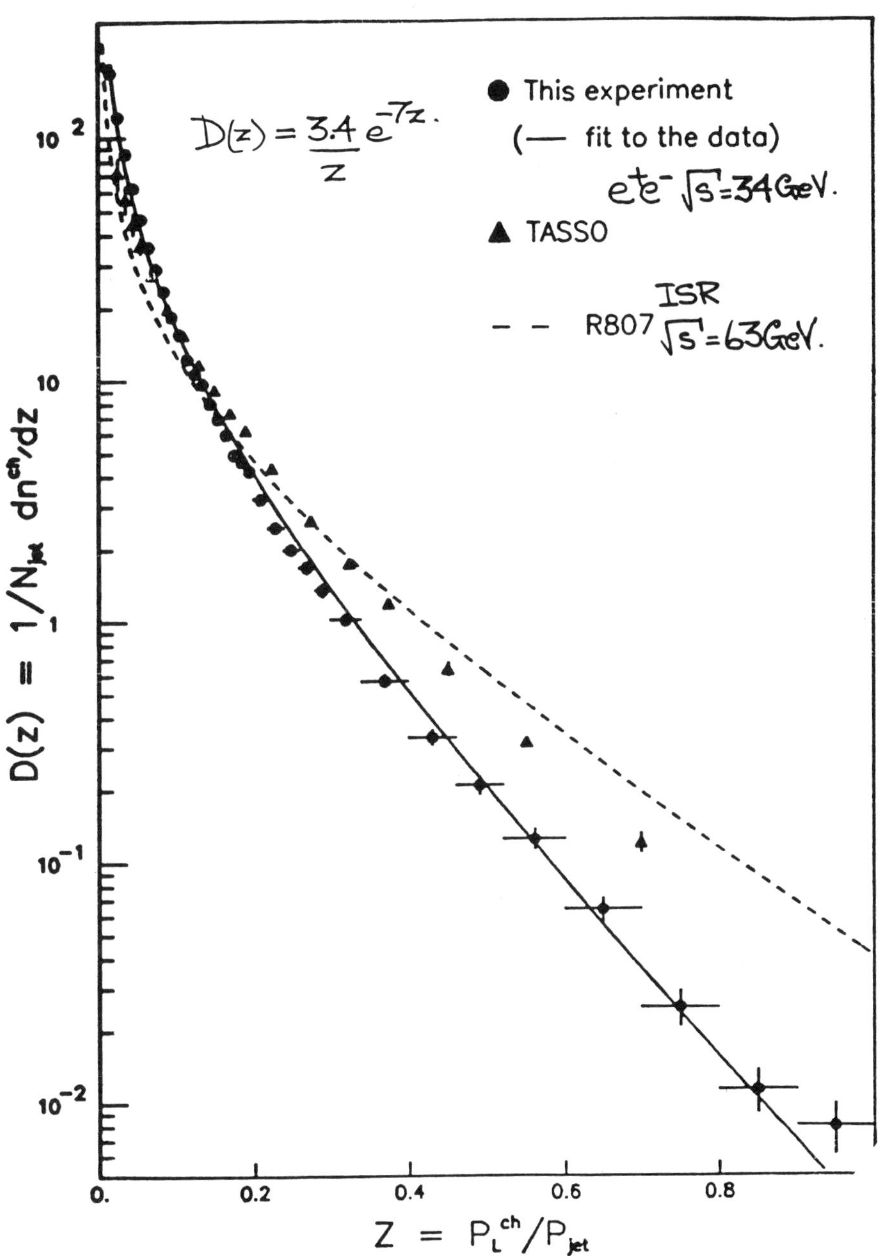

Figure 8 Charged fragmentation function

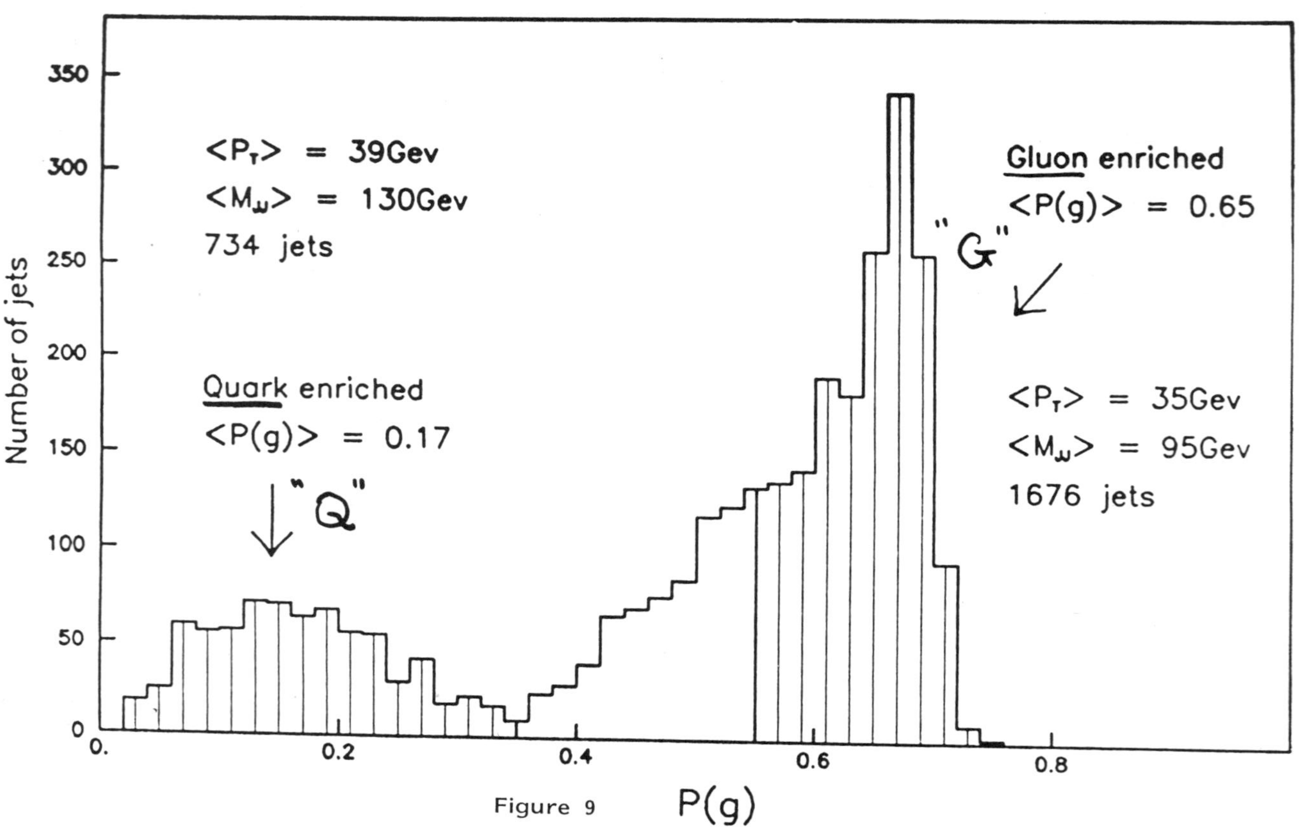

Number of jets
350
300
250
200
150
100
50
0
<P_T> = 39Gev
<M_w> = 130Gev
734 jets
Quark enriched
<P(g)> = 0.17
"Q"
Gluon enriched
<P(g)> = 0.65
"G"
<P_T> = 35Gev
<M_w> = 95Gev
1676 jets
0.
0.2
0.4
0.6
0.8
P(g)

Figure 9

If $P(g) \geq .55$ the jet is taken as "gluon" and for $P(q) = 1-P(g) \geq .65$ the jet is weighted as a "quark". Using the enriched samples defined as follows,

$$\text{"Gluon"} \quad \langle P(g) \rangle = .65 \ ,$$
$$\text{"Quark"} \quad \langle P(g) \rangle = .17 \ ,$$

the "pure" fragmentation functions $D_{Gluon}(z)$ and $D_{Quark}(z)$ can be derived from the following equations:

$$D^{"G"}(z) = 0.65 \ D_{Gluon}(z) + 0.35 \ D_{Quark}(z) \ ,$$
$$D^{"Q"}(z) = 0.17 \ D_{Gluon}(z) + 0.83 \ D_{Quark}(z) \ .$$

The ratio $D_{Gluon}(z) \ / \ D_{Quark}(z)$ thus determined is shown in Figure 10, indicating that the gluon jet appears to fragment more "softly" than the quark jet.

Scaling violation of the fragmentation function

In Figure 11 the "pure" quark and gluon fragmentation function determined as described above is plotted for various bins of z as a function of the dijet mass. Two effective mass points have been derived by UA1. The data is compared with that from PETRA, extrapolated to the SPS collider energies. Since theory does not establish clearly a correct Q^2 scale for the fragmentation function, for simplicity the data are plotted as a function of the dijet mass or $\hat{s}$. Perhaps the empirical extrapolation appears to prefer the quark data over the gluon data, but more important there is clear evidence for the scaling violation.

Average jet charge

The method used to separate gluons and quarks can be extended to separate quarks of a given flavour. If Q_{jet} is taken as

$$\langle Q_{jet} \rangle = \sum_i Q_i z_i^{1/3}$$

where i runs over all the associated charge tracks, then the results

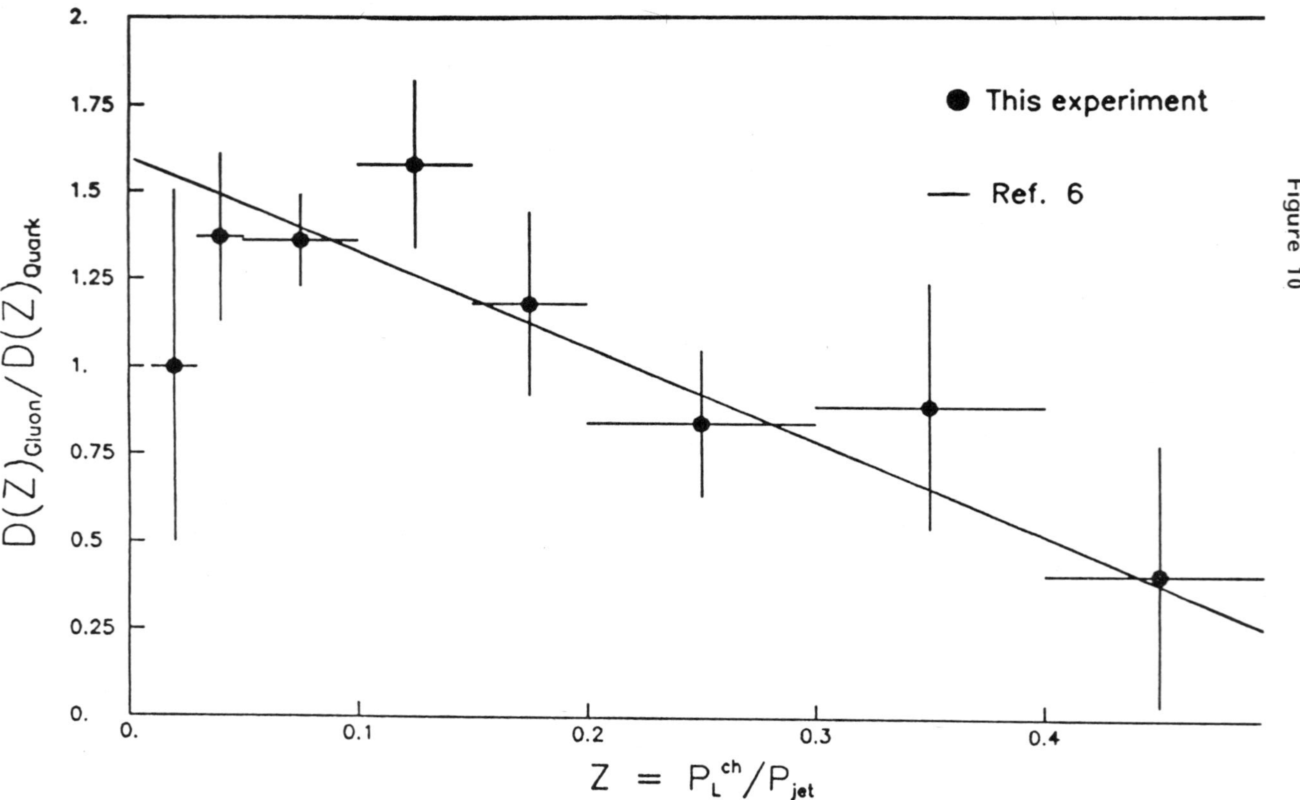
This experiment
Ref. 6
D(Z)_Gluon / D(Z)_Quark
Z = P_L^ch / P_jet
Figure 10

● ○ TASSO (see text) , ■ □ UA1 Gluon , ▲ △ UA1 Quark
D(z) = 1/N_jet dn^ch/dz
s or M²_ω(Gev)²
0.02<z<0.05
0.05<z<0.1
0.1<z<0.2
0.2<z<0.3
0.3<z<0.4
0.4<z<0.5

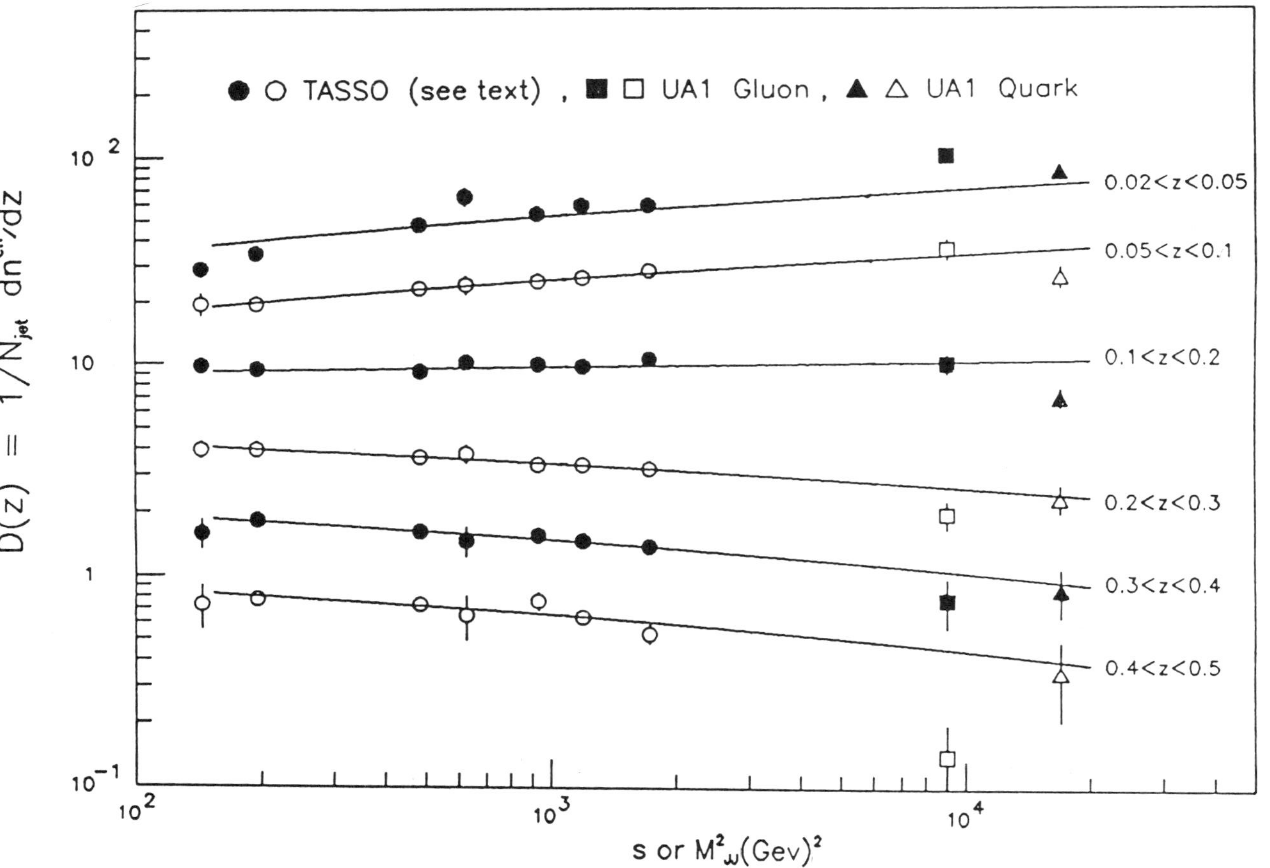

Figure 11.

are as follows:

Parton Nature	$\langle Q_{jet}\rangle$
Gluon (P(g) $\geq$ .5)	$-0.03 \pm .01$
u-Quark (P(u) $\geq$ 0.5)	$+0.15 \pm 0.03$
$\bar{u}$-Quark (P($\bar{u}$) $\geq$ 0.5)	-0.15 ± 0.03

We can conclude that the average charge for the enriched quark
(antiquark) sample is not zero, whereas the average charge of the
gluon enriched sample is compatible with zero.

Substructure of quarks

In order to look for evidence for quark substructure one needs a
framework in which to parametrize compositeness. There are many
models, and the one used in this analysis is due to Eichten et al.
The parameter Λ_c is a constant whose value sets the energy scale at
which substructure may become evident. If $\Lambda_c = \infty$, then the quarks
remain point-like; on the other hand if $\sqrt{s} > \Lambda_c$, then spectacular
signals would be seen. The angular distribution from two jet events
with a high value of the dijet mass ($250 < m_{jj} < 300$ GeV) is compared
with Eichten et al. The model predicts the form of $d\sigma/d\chi$, and
includes an amplitude coming from a four fermion point interaction
among the quarks, in addition to the usual QCD term. The data have
been fitted to $1/N \, d\sigma/d\chi$ selecting $Q^2 = P_t^2$ for the scale of Q^2. A
value of $\Lambda_c > 445$ GeV is determined at the 95% confidence level[22]).
The curves for various values of Λ_c are presented in Figure 12 along
with the data. The value of $\Lambda_c > 445$ GeV can be translated into a
distance scale and suggests that quarks remain point-like, at least
down to $\leq 4.4 \times 10^{-17}$ cm.

Figure 12

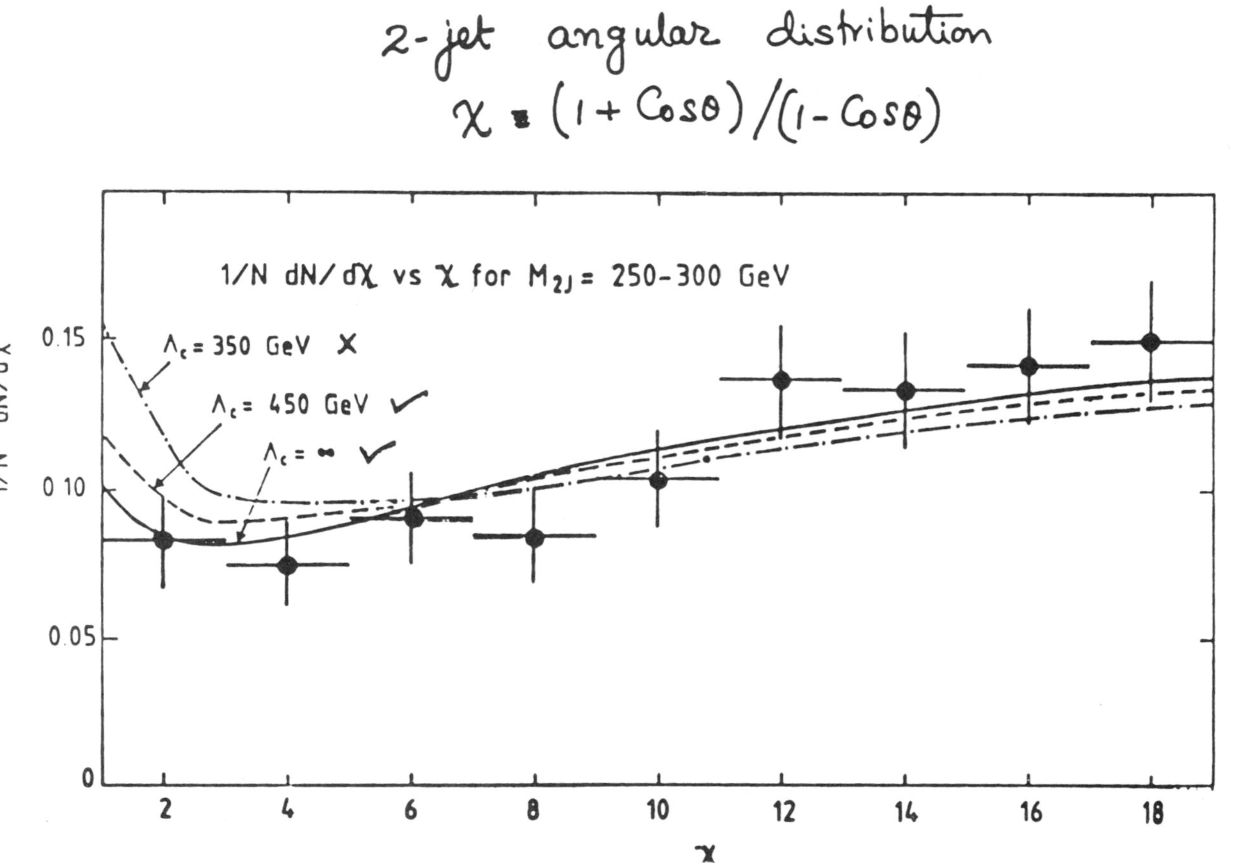

2-jet angular distribution
χ = (1+cosθ)/(1-cosθ)
1/N dN/dχ vs χ for M_{2J} = 250-300 GeV
Λc = 350 GeV X
Λc = 450 GeV
Λc = ∞
1/N dN/dχ
0.15
0.10
0.05
0
χ
2
4
6
8
10
12
14
16
18

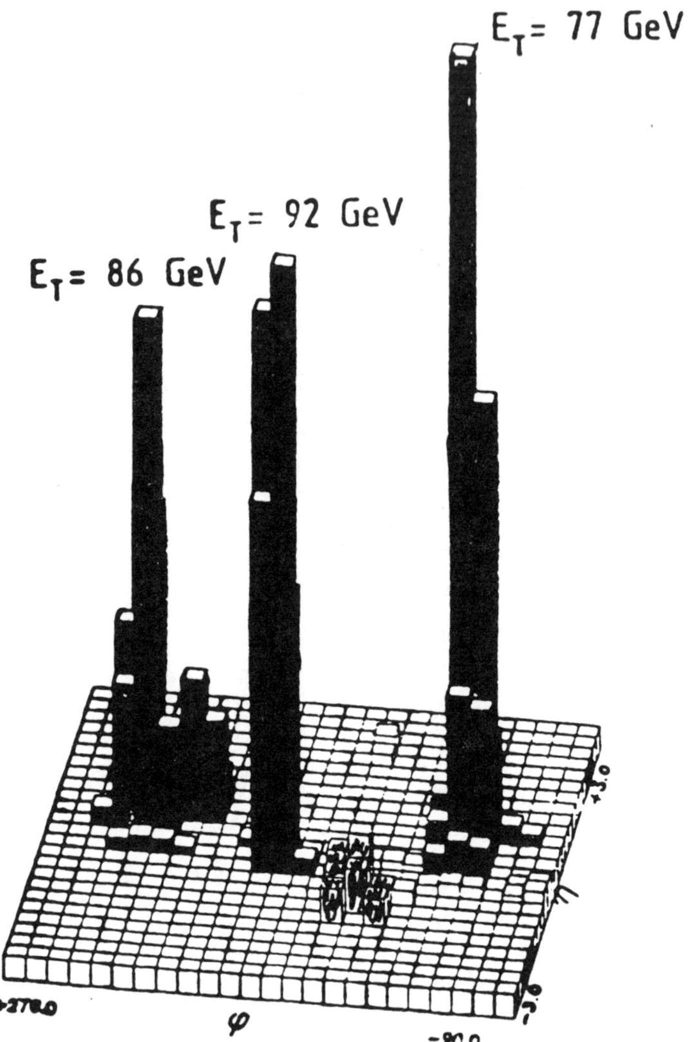

Figure 13

Three-jet events

Clean samples of three-jet events have been collected at the collider;
the LEGO plot of one of these is given in Figure 13. The value of
α_s can be measured in principle by examining the ratio of three-jet to
two-jet events. The result[23] is $\left(\dfrac{K_{3J}}{K_{2J}}\right) \alpha_s = .16 \pm .02 \pm .03$ at
$Q^2 \propto 4000$ GeV^2. The K factors come about from the inclusion of higher
order terms and are typically in the range 1.5 - 2.0. They will also
include scale breaking effects. If the effective Q^2 scales were
matched correctly for the selected two-jet and three-jet events, then
the ratio K_{3J}/K_{2J} should be near unity. This factor cannot be
eliminated until the next-to-leading-order corrections have been
calculated theoretically.

Conclusions – QCD at the Collider

(1) There is clear evidence for high p_T jets, which in itself is a measure of the success of the QCD parton picture of hadron collisons.

(2) The inclusive jet cross section has good agreement in shape and magnitude with QCD predictions. The comparison of the shape of the theory and data has been used as evidence for the existence of the three-gluon vertex.

(3) If jets are identified with partons, then the angular distribution is "Rutherford-like" providing evidence for massless vector gluon exchange. The detailed comparison of the angular distribution and theory shows clear evidence for scaling violations.

(4) The structure function, derived using the assumption of factorization, agrees well with that extrapolated in Q^2 from deep inelastic lepton scattering, and provides a direct measurement of the gluon content.

(5) A study of the charged fragmentation function reveals differences between quark and gluon jets.

(6) There is no evidence for substructure of the partons at least to a scale of $\sim 4 \times 10^{-17}$ cm.

A final global conclusion is perhaps that the collider is an excellent laboratory for testing perturbative QCD and the tests show the theory to be in good shape – quantitatively. It is clear these comparisons will become more refined in the future as higher order diagrams are calculated, with the hadron collider assuming the role of the best means of probing "short distance" phenomena because of the ready accessibility of events with very large Q^2, and the QCD framework for their analysis.

Lecture 3

Vector Boson Physics:

The main reason for the transformation of the SPS Machine to its
Collider role was the possibility of achieving hadron collisions at a
sufficiently high centre of mass energy and with enough luminosity to
produce W and Z vector bosons in detectable numbers.

The electro-weak model of Glashow-Weinberg-Salam[24] provides rather
precise prescriptions for the masses and properties of the vector
bosons. Starting from the relations between the coupling strengths

$$\frac{G_F}{\sqrt{2}} = \frac{g^2}{8m_W^2} \qquad g = e/\sin\theta_W \qquad \rho \frac{G_F}{\sqrt{2}} = \frac{g^2}{8m_Z^2 \cos^2\theta_W}$$

where θ_W is the Weinberg angle; the mass of the W boson is given by

$$m_W^2 = \left(\frac{\pi\alpha}{\sqrt{2}\,G_F}\right) \frac{1}{\sin^2\theta_W} \cdot \frac{1}{(1-\Delta r)} \;.$$

Here Δr takes account of radiative corrections[25] to $0(\alpha)$ and amounts
to $0.0696 \pm .02$. The value of $\sin^2\theta_W$, determined from neutral current
neutrino interactions is in the range $0.21 - 0.24$, and if the value of
ρ is equal to unity (the relative strength of neutral to charged weak
current coupling – a value required for models in which all Higgs
particles occur in isodoublets), then we have the relation

$$m_Z^2 = \frac{m_W^2}{\cos^2\theta_W} \qquad \text{with } \rho = 1$$

and the bosons will have masses in the following ranges: $79 < m_W <$
87 GeV/c^2 and $90 < m_Z < 95$ GeV/c^2.

The W and Z bosons are detected via their "characteristic" decays,
which so far have been restricted to purely leptonic final states.
The decays of the W$^\pm$ bosons are

$$W \rightarrow \ell\bar{\nu}_e \left\{e\bar{\nu}_e,\; \mu\bar{\nu}_\mu,\; \tau\nu_\tau\right\}$$

and $W \to q\bar{q}'$ $\{d\bar{u}, s\bar{c}, b\bar{t}$ - Cabbibo favoured decay$\}$.

The width and decay branching ratios depend on the value of $\sin^2\theta_W$ and, where appropriate, will be influenced by the mass of the top quark. However, the favoured value of $\sin^2\theta_W$ produces the values $\Gamma(W \to e\bar{\nu}_e) = 250$ MeV, $\Gamma(W \to$ hadrons$) = 2.2$ GeV, $\Gamma(W \to$ all$) = 2.95$ GeV which yield $\dfrac{e\nu}{\text{all}} \simeq 8.5\%$.

The Z° can decay into any pair of "light" fermions

$$Z \to f\bar{f}(e^+e^-), (\mu^+\mu^-), (\tau^+\tau^-), (\nu_e\bar{\nu}_e), (\bar{\nu}_\mu\nu_\mu), (\nu_\tau\bar{\nu}_\tau) \ldots$$

$$\text{and } (d\bar{d}), (u\bar{u}), (s\bar{s}), (\bar{c}c), (b\bar{b}), (t\bar{t}).$$

Again selecting a reasonable value for $\sin^2\theta_W$ and taking no account of the mass of the top quark, the widths are predicted as

$$\Gamma(Z \to e^+e^-) = 92 \text{ MeV}, \quad \Gamma(Z^\circ \to \text{hadrons}) = 2.12 \text{ GeV},$$

$$\Gamma(Z \to \text{all}) = 2.94 \text{ GeV}, \quad \Gamma(Z \to \nu_c\bar{\nu}_c) = 180 \text{ MeV}.$$

$$\therefore \quad \frac{e^+e^-}{\text{all}} \simeq 3.0\%.$$

Thus experimentally, if one uses only the leptonic decay modes involving electrons, the branching ratios impose an 8.5% and a 3.0% efficiency for the detection of the $W^\pm$ and Z° respectively. It is perhaps interesting to note that neither UA1 nor UA2 have so far produced any convincing evidence of the W or Z in "jet-jet mass" plots where the jets are the products of quark fragmentation.

The production of Vector Bosons:

In order to produce a W in a hadronic collision, the centre of mass energy of the constituent collision must have the appropriate energy; i.e., $\hat{s} = M_W^2 = x_1 x_2 s$. For $M_W \simeq 80$ GeV and $\sqrt{s} = 540$ GeV, we can have $x_1 \sim x_2 \sim 0.15$. The model for the Intermediate Vector Boson production is the Drell-Yan[26]) process shown in Figure 1a with the dominant basic quark subprocesses

$$u + \bar{d} \to W^+, \quad d + \bar{u} \to W^-, \quad u + \bar{u} \to Z^\circ, \quad d + \bar{d} \to Z^\circ$$

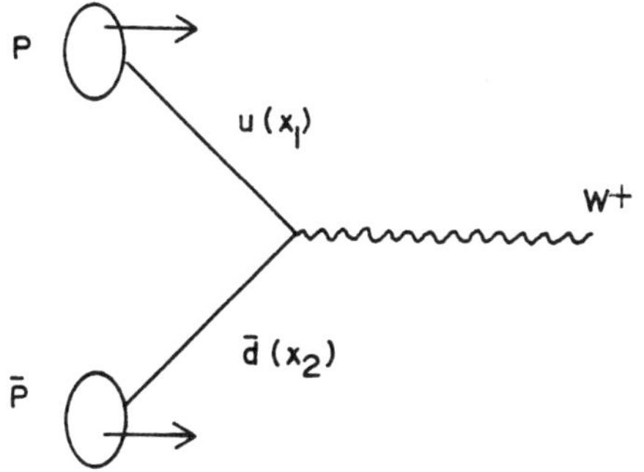

Figure 1 (a) Drell-Yan for W

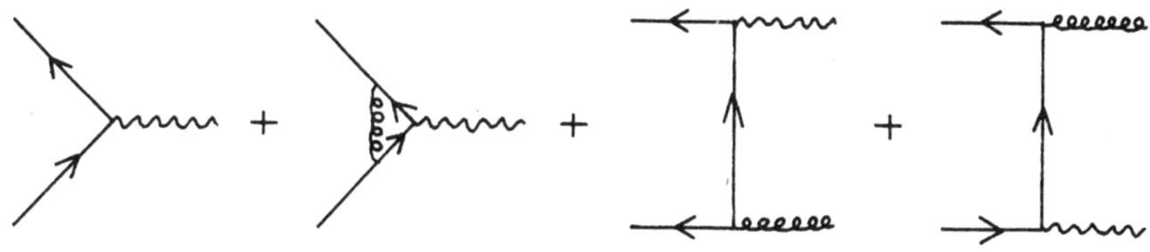

INITIAL GLUON STATES

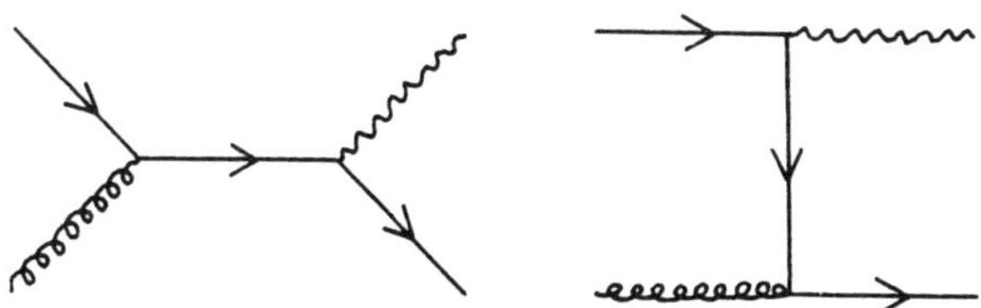

Figure 1 (b) QCD - additional terms

for proton antiproton collisions. The cross-section for W^+ production will clearly involve a term of the form

$$\sigma W^+(x_1 x_2) \sim U(x_1)D(x_2) + D(x_1)U(x_2)$$

where $U(x)$ and $D(x)$ represent the x distributions of the u and d quarks in the proton.

In addition to the leading order term of the Drell-Yan process of Figure 1a, the strong interaction will modify any cross section estimates through the emission of gluons, governed by QCD. The additional diagrams of the kind shown in Figure 1b lead to a gluon or a quark being produced in the final state in addition to the vector boson. The presence of these QCD terms to $0(\alpha_s)$ will increase the cross section by a factor $K(\sim 1.3 - 1.5)$. The precise value of K depends on the scale of Q^2 appropriate for the W production.

The production cross sections for W and Z bosons as a function of $\sqrt{s}$ of the hadron collision are shown in Figure 2, taken from the work of Altarelli et al.[27] At $\sqrt{s}$ of 540 GeV $\sigma_0^{W^+ + W^-} = 4.2 \, ^{+1.3}_{-0.6}$ nb and $\sigma_0^{Z^0} = 1.3 \, ^{+0.4}_{-0.2}$ nb; the errors are theoretical, and terms $0(\alpha_s)$ represent $\sim 30\%$ of the cross section. A quantity less affected by theoretical errors is the ratio $\dfrac{\sigma_0^{W^+ + W^-}}{\sigma_0^{Z^0}} = 3.3 \pm 0.2$.

Experimentally the higher order QCD terms lead to a long tail in the transverse momentum distribution of the bosons, with a 6-10% probability of a W with a transverse momentum greater than 25 GeV/c. Hadronic jets should also be observed in association with a high p_T boson. These events afford good quantitative tests of perturbative QCD.

Experimental Signatures of W and Z bosons

The dominant decays of the bosons will produce pairs of hadronic jets in the final state; however, these will be buried in a QCD continuum which is at least ten times larger. Nonetheless with good statistics, and calorimetry which responds independently of the fluctuations in

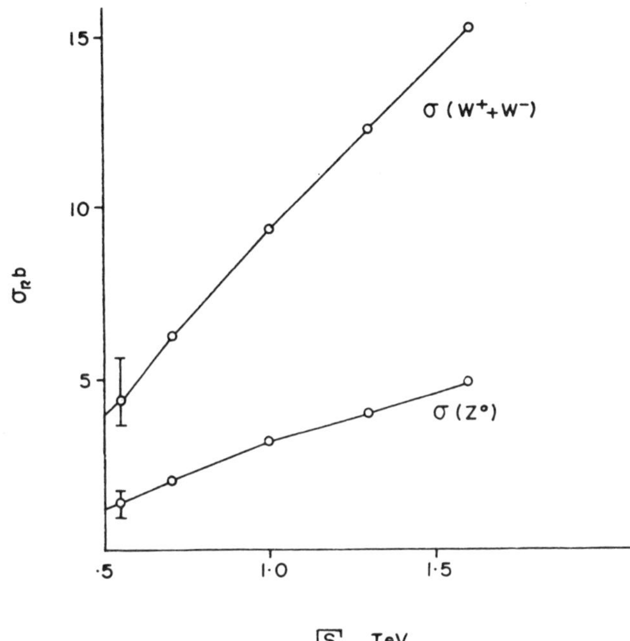

Figure 2. W and Z° cross sections of Altarelli et al.

jet fragmentation, it should be possible to see the boson structures.
The absence of a signal in UA1 and UA2 is more a comment on the
calorimetry in these experiments than the current statistics. The
discoveries of the J/ψ and T suggest that the clean identification and
measurement of pairs of charged leptons should produce a very clean
signal for the Z°. This has also proved to be the case for the $W^{\pm}$[28]),
provided the neutrino in the final state can be identified and
measured using the technique of "missing transverse energy". The
final state involving a "$\mu\nu$"[29]) pair remains more problematical
because the μ also "leaks" energy from the detector unless its
momentum is well measured. Figure 3 shows schematically, in the
plane transverse to the colliding beams, the principle of W and Z
detection through their leptonic modes of decay. Good electron and
muon identification are essential ingredients, and are best achieved
with isolated high p_T leptons.

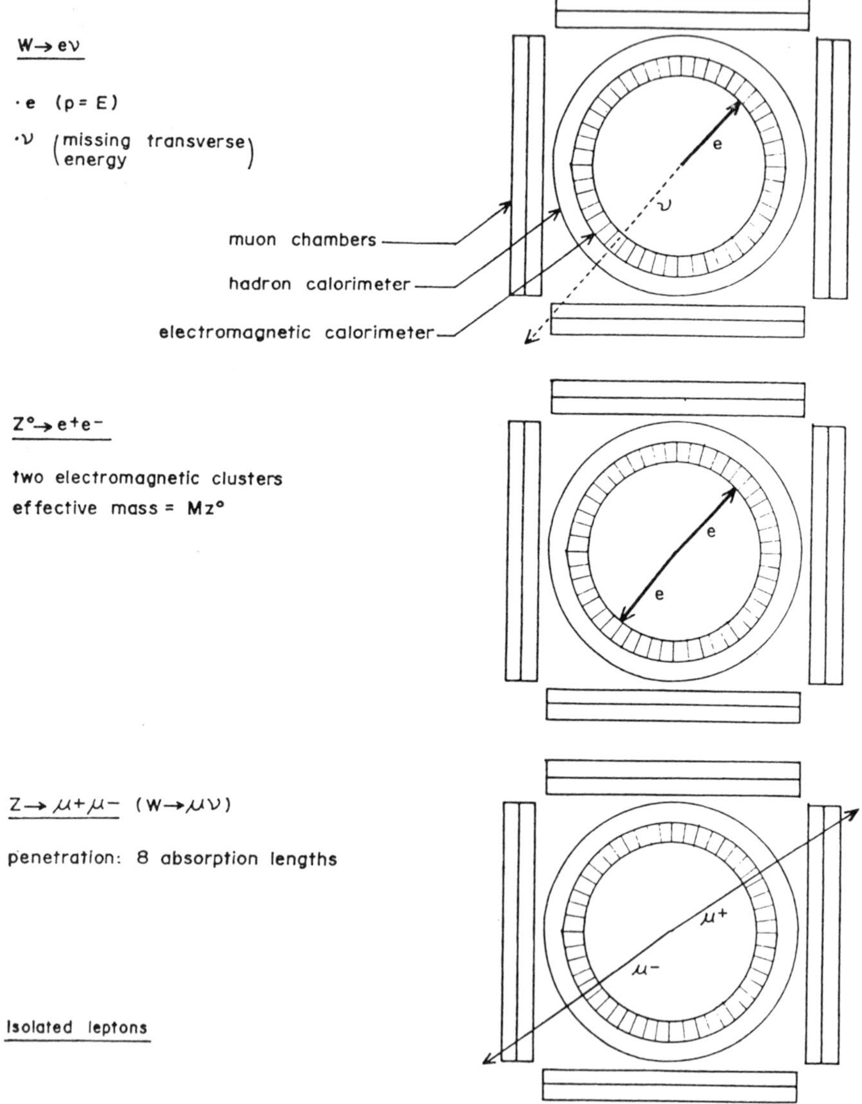

Figure 3. Schematic Boson detection - lepton decays transverse to boson

Masses of W and Z

The transverse momentum distribution of the leptons from the decay of
W bosons should reveal a Jacobian peak at $p_T^{\ell} \simeq M_W/2$ with a sharp drop
on the high p_T side. In reality the production of W's with a p_T
distribution determined by QCD smears out this peak. In order to
avoid feeding in knowledge of the QCD smearing, use can be made of the
fact that the transverse momentum of the W is determined once the p_T^e
and p_T^{ν} are measured. Fitting the electron spectrum, corrected for
p_T^W, is equivalent to fitting the "transverse mass"[30]

$$M_T^2(e\nu) = (p_T^e + p_T^{\nu})^2 - (\vec{p}_{\pi}^{\,e} + \vec{p}_T^{\,\nu})^2$$

$$M^2(e\nu) - M_T^2(e\nu) \geq 0 \qquad \{M_T(e\nu) = M(e\nu) \text{ if } p_L^W = 0\}.$$

The distribution of $M_T(e\nu)$ has an invariant end point and a shape,
predictable up to corrections $\langle p_T^{W\,2}/M_W^2 \rangle \sim 1\%$, in contrast to the p_T^e
spectrum which suffers corrections $\sim \langle p_T^W/M_W \rangle \sim 10\%$. The stability and
strength of the transverse mass distribution is exhibited in Figure 4,
where effects due to widely different values of p_T^W and Γ_W are
illustrated. The UA1 result[31] for the "enhanced transverse mass"
distribution is shown in Figure 5; a cut of $p_T^e > 30$ GeV/c and
$p_T^{\nu} > $ GeV/c has been applied. The resulting distribution is entirely
free from background and can be related to the "full" transverse mass
distribution in a straightforward manner. This particular result
yields the following values for the mass and width of the W,

$$M_W = 83.5^{+1.1}_{-1.0} \text{ GeV/c}^2, \quad \Gamma_W = 2.7^{+1.4}_{-1.5} \text{ GeV/c}^2 \{\Gamma_W \leq 6.5 \text{ GeV/c}^2 \text{ at } 90\% \text{ CL}\};$$

a systematic error of ± 2.7 GeV/c^2 must be added to uncertainty in M_W.

Figure 6 shows the effective mass distribution for pairs of high p_T,
isolated, electrons. The resulting mass and width values for the Z°
are $M_Z = 93.0 \pm 1.4$ GeV/c^2, $\Gamma_Z = 3.9^{+2.3}_{-1.5}$ GeV/c^2 $\{\Gamma_Z \leq 8.3$ GeV/c^2 at
90% CL$\}$, with an additional systematic uncertainty of ± 3.0 GeV/c^2 in
the mass of the Z°.

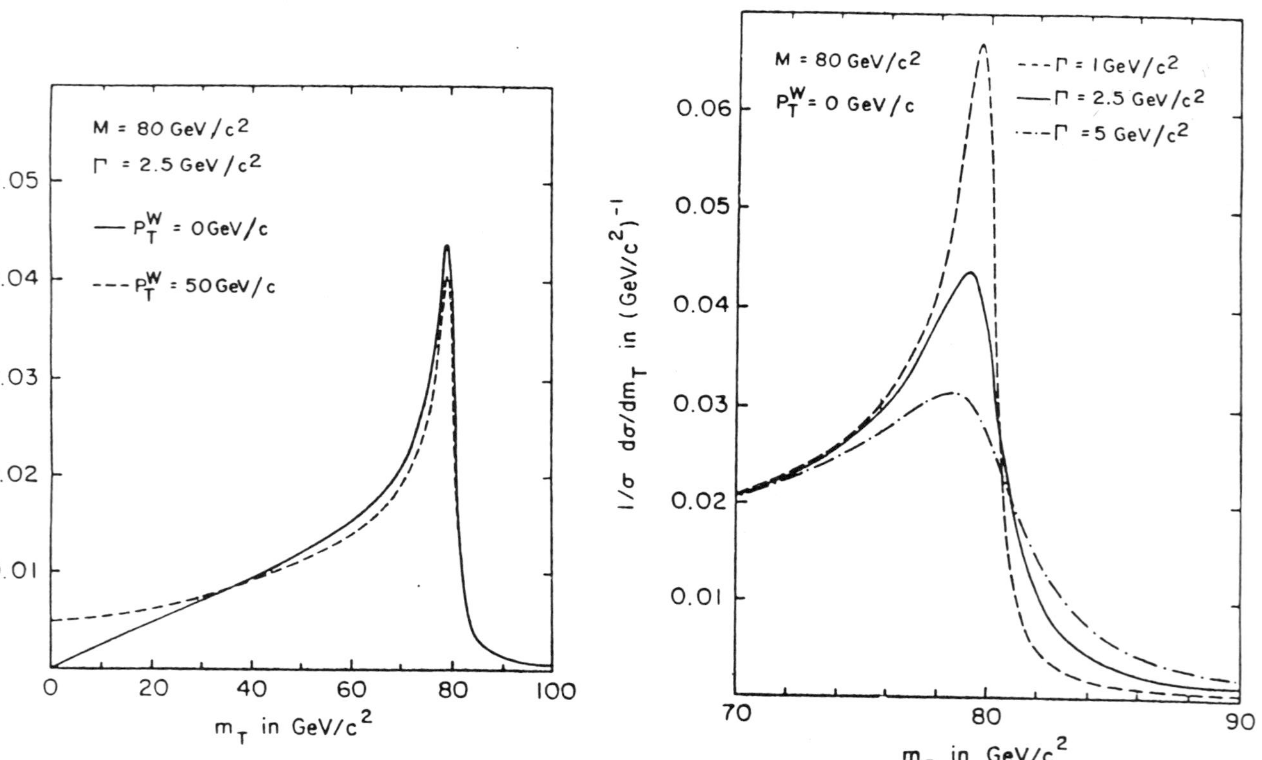

Figure 4.   Transverse mass distributions

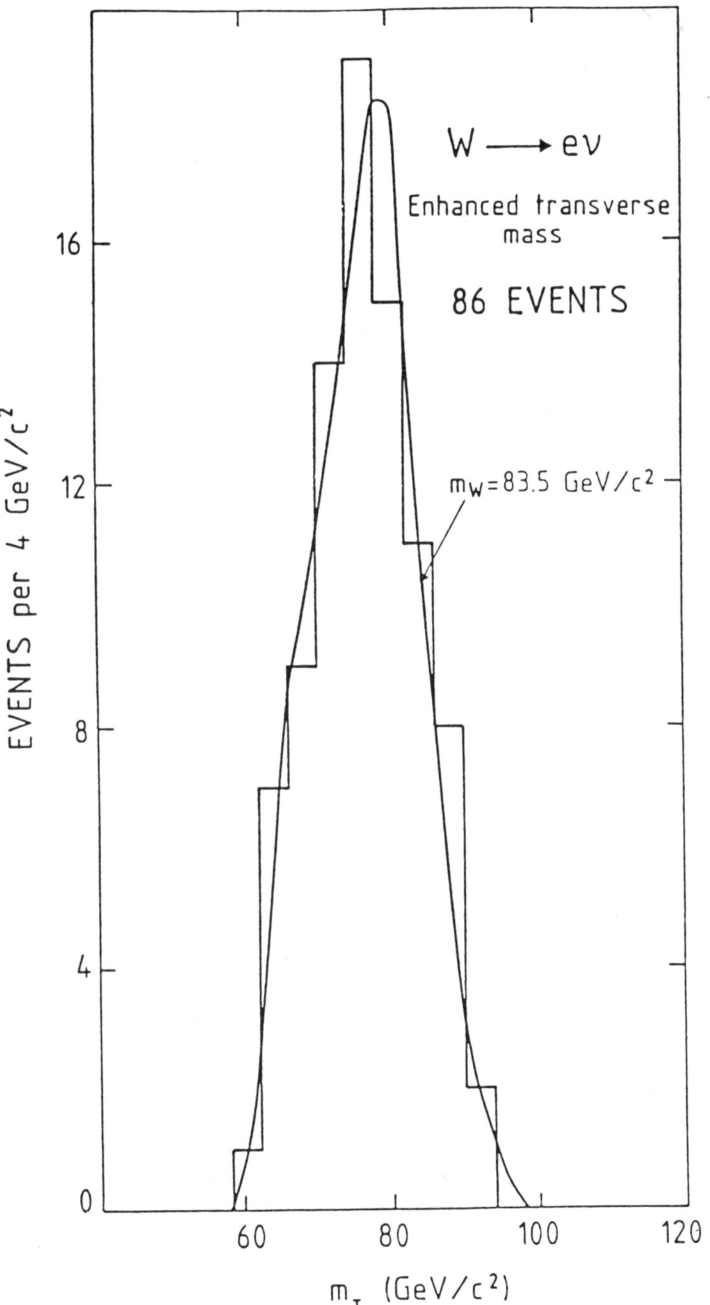

Figure 5.   Enhanced transverse mass for W – UA1

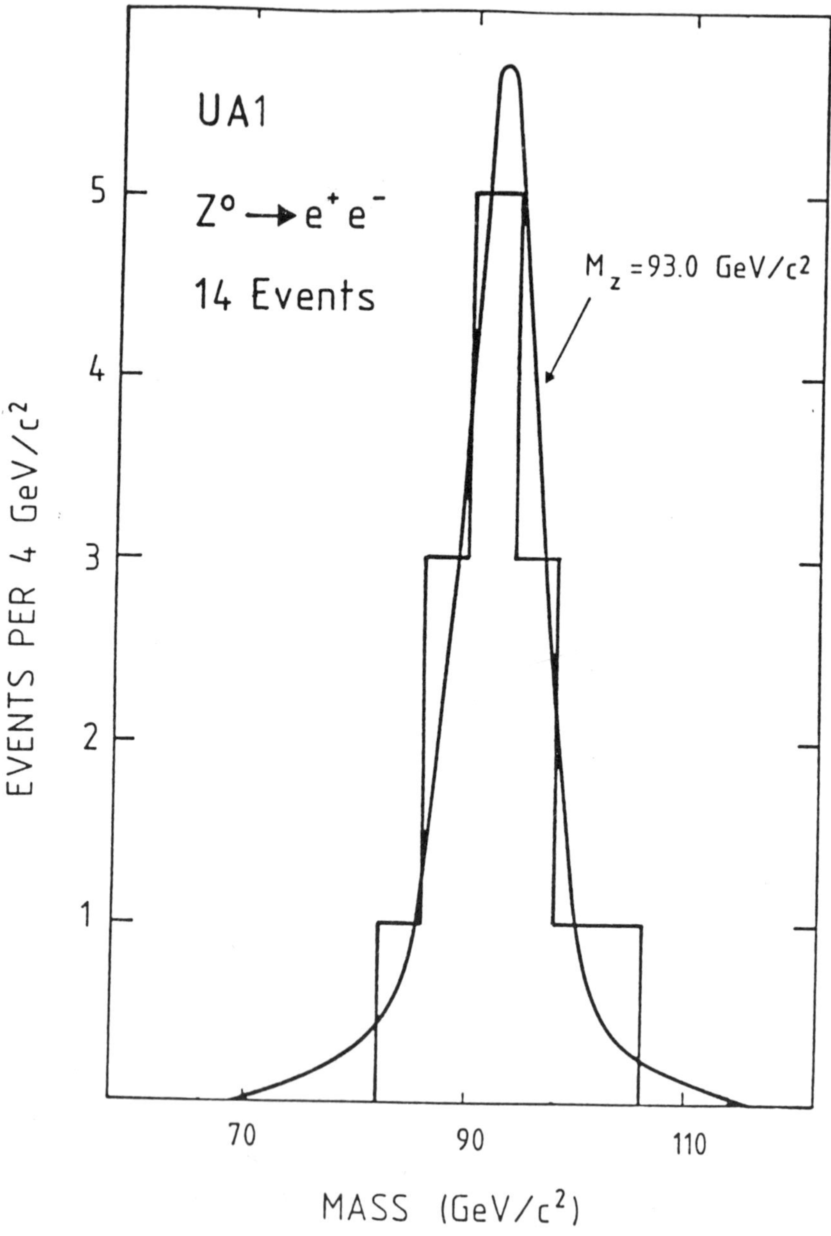

Figure 6. $Z^{\circ} \rightarrow e^+ e^-$

The UA1 detector has good muon detection capabilities over a large solid angle and has observed and measured the decays $W \rightarrow \mu\nu$ and $Z^\circ \rightarrow \mu^+\mu^-$.[35] The results are shown in Figure 7. The integrated luminosities and number of W and Z events determined thus far in the UA1 experiment are given in Table I.

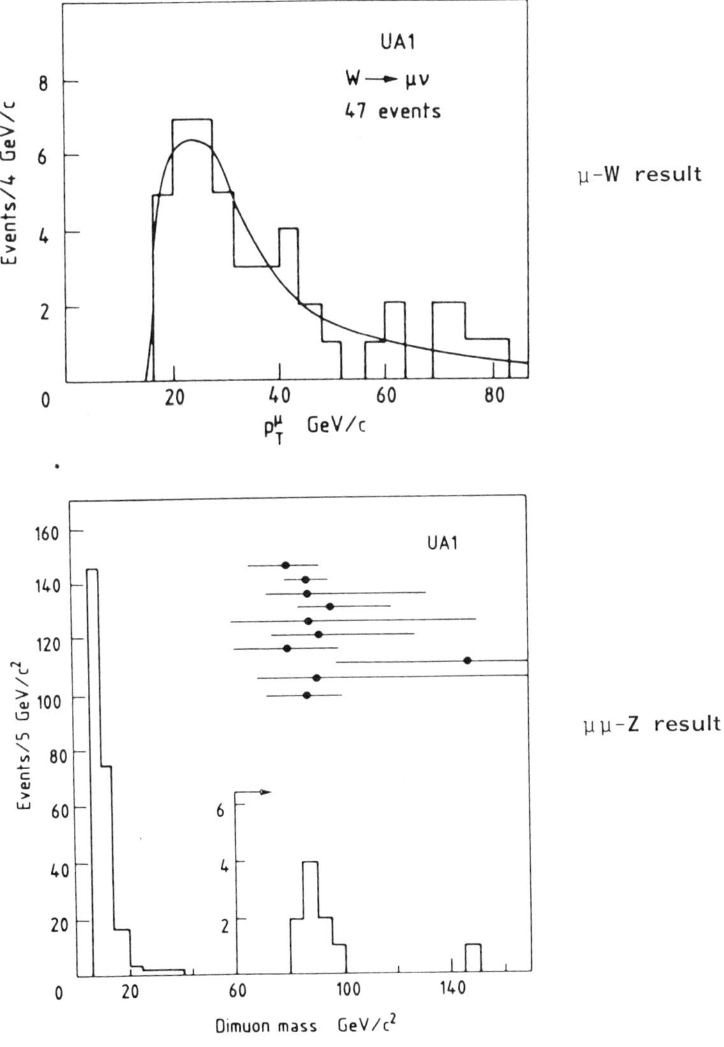

Figure 7.

Table I. Number of W and Z events in the UA1 experiment

	$\sqrt{s}$ GeV	$\int \mathcal{L}dt$ nb^{-1}	$W \to e\nu$	$W \to \mu\nu$	$Z \to e^+e^-$	$Z \to \mu^+\mu^-$
1982/3	546	136	52	14	4	5
1984	630	263	172	47	18	10
1985	630	330	218	60	22	9

The Helicity of the W boson

If the W boson discovered at the SPS Collider is indeed the "quantum
of the weak field" it should be left-handed. This is a consequence
of the (V-A) nature of the weak interction; i.e., neutrinos are left-
handed. Figure 8a shows the production of a polarized W^+ through the
fusion of $\bar{d}$ and u quarks, the produced W^+ is longitudinally polarized
with helicity h = ±1. The subsequent leptonic decay configurations,
determined by helicity conservation, are given in Figure 8b. The
example chosen is for muon decays; the arguments are equally
applicable for electron final states. The resulting asymmetry in the
angular distribution of the decay products is driven by the simple
rule that "the anti-lepton prefers to follow the direction of the
anti-quark from the anti-proton".

In reference to Figure 8c, the angular distribution of the leptons is
predicted by (V-A) to have the form $(1 + \cos\theta_e^*)^2$ with the average
value of $\cos\theta_e^*$ given by

$$\langle \cos\theta_e^* \rangle = \frac{\langle\lambda\rangle\langle\mu\rangle}{J(J+1)}$$

where $\langle\mu\rangle$ is the helicity of the $(u\bar{d})$ system, $\langle\lambda\rangle$ that of the $(e\bar{\nu})$
system, and J the spin of the W. The (V-A) interaction implies
$\langle\lambda\rangle = \langle\mu\rangle = -1$ and a value of J = 1 gives $\langle\cos\theta^*\rangle = 0.5$. The UA1
result is shown in Figure 8c for electrons or positrons whose sign is
unambiguously determined. The value of $\langle\cos\theta^*\rangle = 0.43 \pm .07$ provides
proof of the vector nature of the W; however, strictly speaking, a
choice of sign of $\langle\lambda\rangle$ or $\langle\mu\rangle = \pm 1$ cannot be separated without an
additional polarization measurement.

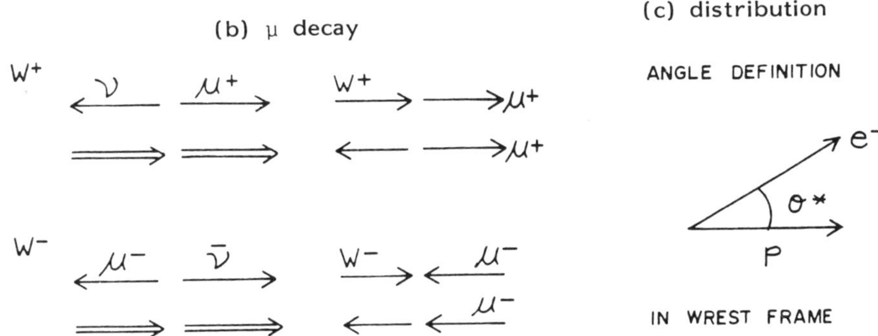

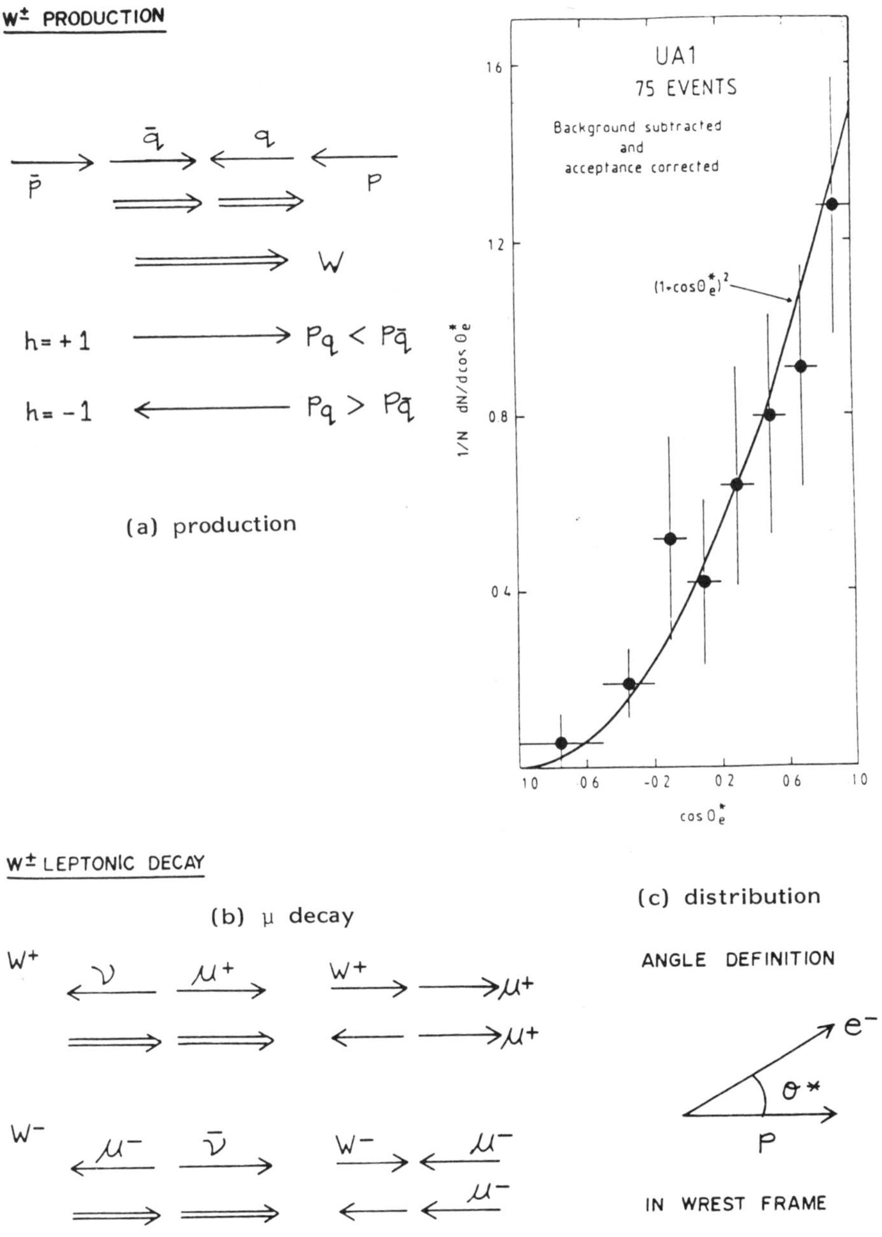

Figure 8. W Helicity figure

The Production Properties of the W

The successful model for W production, as pointed out earlier, is based on the Drell-Yan mechanism with quark-antiquark annihilation, modified by QCD, since the partons are coloured and one or more gluons may be radiated. A quantitative comparison of theory and experiment is governed mainly by the Q^2 scale - namely, selecting events in which perturbative QCD is applicable.

The longitudinal momentum distribution of the W's reflects the structure functions of the annihilating partons, predominantly u and d quarks. Experimentally, only the transverse component of the neutrino momentum is measured, and consequently imposing the W mass on the electron-neutrino system yields two solutions for the momentum of the W. In about one-third of the cases one solution for x_w ($x_w = 2p_L^W/\sqrt{s}$) is nonphysical; and in another third the ambiguity can be removed by using the forward calorimetry in the detector to impose energy and momentum conservation on the whole event. Use can be made of the relations

$$x_p x_{\bar{p}} = m_W^2/s \text{ (energy conservation)}$$

$$x_w = |x_p - x_{\bar{p}}| \text{ (momentum conservation)}$$

to determine the parton distributions in the proton and antiproton sampled by the W production. Using "signed" events, i.e.

$$W^- \rightarrow \bar{u}(\bar{p}) \text{ and } d(p)$$
$$W^+ \rightarrow u(p) \text{ and } \bar{d}(\bar{p})$$

the u and d quark distributions for the proton can be derived. These are shown[33] in Figure 9 along with the structure functions of Eichten et al.[34] with $\Lambda = 0.2$ GeV.

The measured values of p_T^e and p_T^ν can be used to determine the transverse momentum of the W. The resulting spectrum of p_T^W is shown in Figure 10, which exhibits a "tail" extending up to ~ 40 GeV/c. It should be noted that for $p_T^W > 20$ GeV/c every event has at least one jet reconstructed. The predictions from the work of Altarelli et al.[27] is superimposed on the data and provides an excellent description of the "high p_T tail".

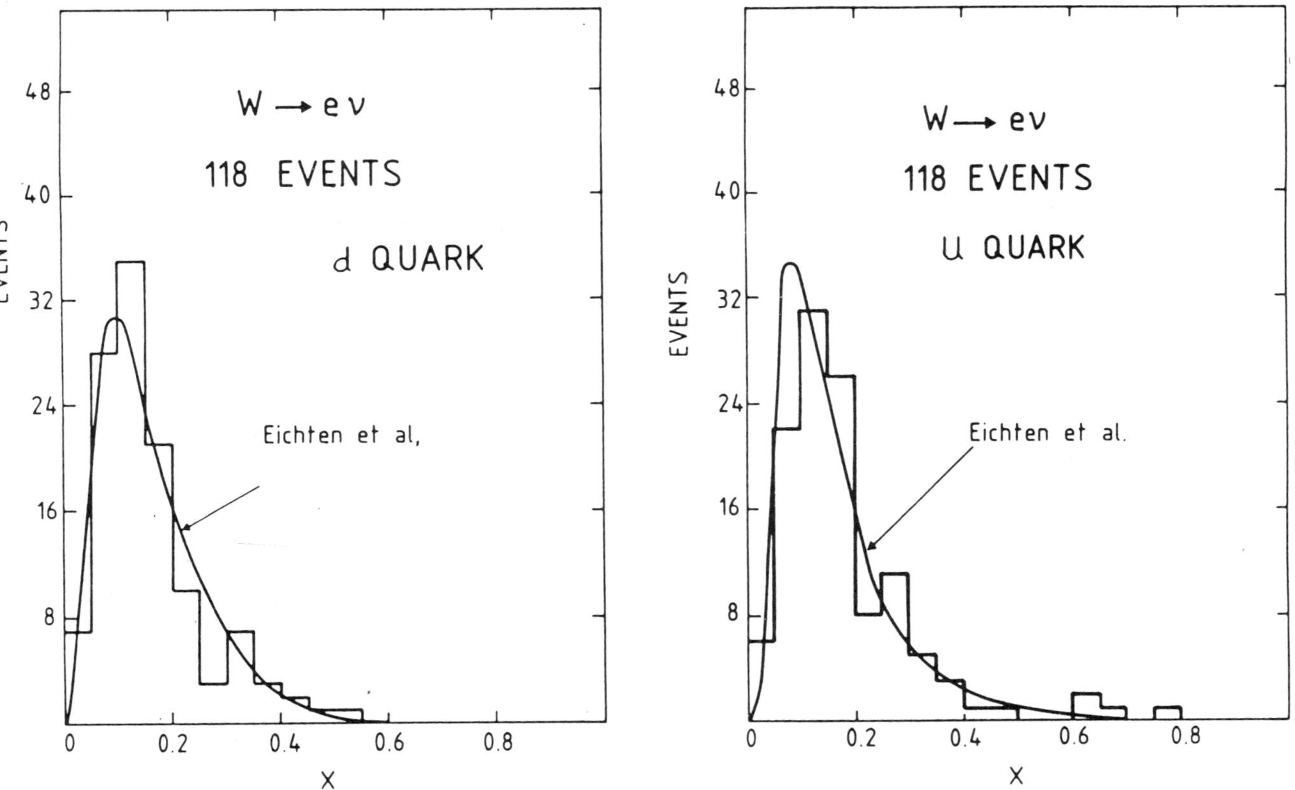

Figure 9.   u and d structure functions

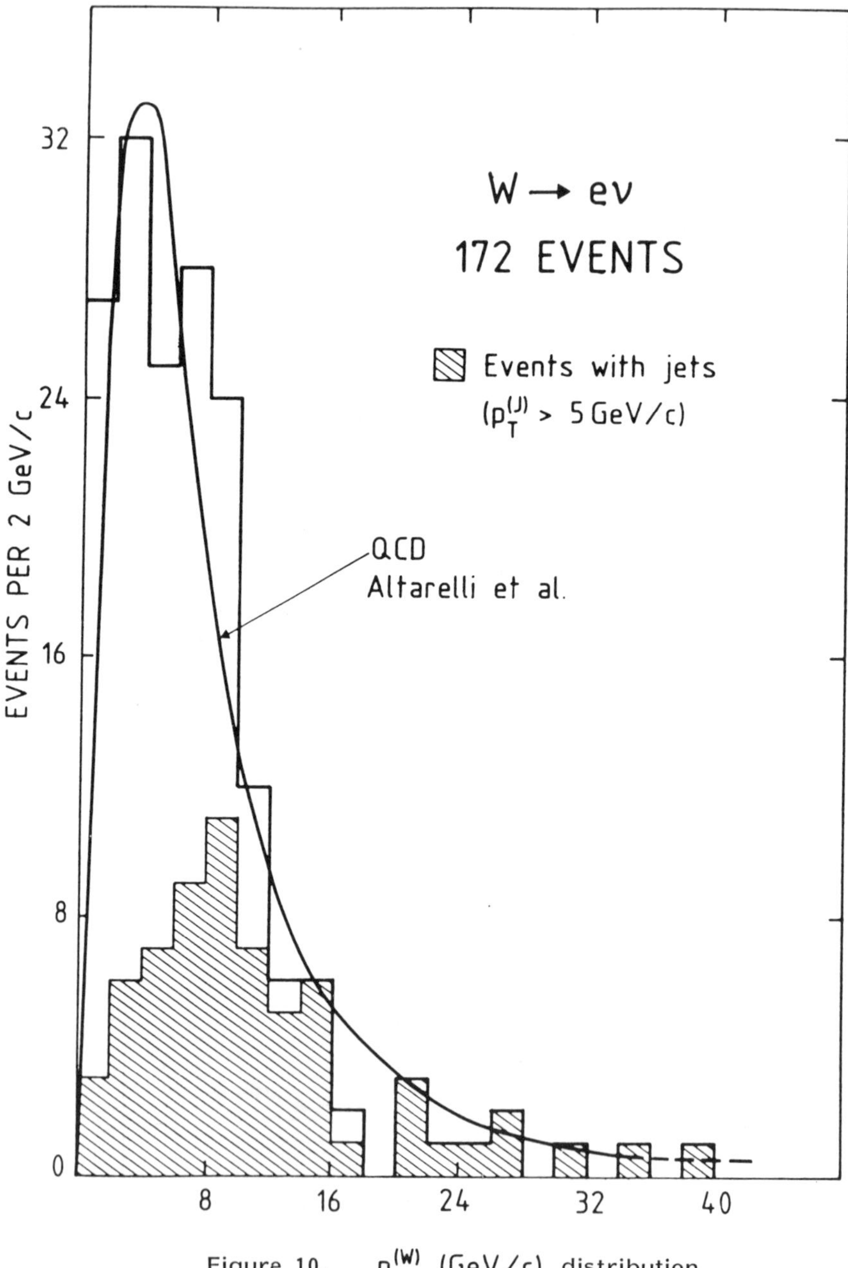

Figure 10. $p_T^{(W)}$ (GeV/c) distribution

In order to make a better comparison with the theory, events are selected in which the transverse momentum of the jet is greater than 20 GeV/c. This has two advantages: the jets are more clearly defined in the detector, and optimistically the perturbative regime of QCD is more appropriate. At $\sqrt{s}$ = 630 GeV, 6 events have one jet with p_T^{jet} > 20 GeV/c. Correcting for W and jet efficiencies, this yields a cross-section of 0.041 ± 0.015 (± 0.006) nb to be compared with a QCD prediction of 0.016 ± 0.002 nb, confirming that the jet activity produced in association with W is in good agreement with the expectations from initial state gluon radiation. This is further confirmed by an examination of the angular distribution of the jets. Figure 11 shows the distribution of jets versus $|\cos\theta^*|$, where θ^* is the angle between the jet and the average beam direction in the rest frame of the W and the jet(s). The shape is well fitted by the QCD expectation for initial state bremsstrahlung, basically $(1 - |\cos\theta^*|)^{-1}$.

Table II presents a comparison between the production cross sections for W and Z as determined in UA1 and UA2[10]) with the "QCD-based" predictions of Altarelli et al. As can be seen, the Drell-Yan model modified by QCD represents an excellent description of boson production.

Table II

		$\sqrt{s}$ = 540 GeV	$\sqrt{s}$ = 630 GeV
$W \rightarrow e\nu$:			
$(\sigma \cdot B)$nb	UA1	0.55 ± .08 ± .09	0.63 ± .05 ± .09
	UA2	0.50 ± .09 ± .05	0.53 ± .06 ± .05
Theory: nb		$0.36\ ^{+\ 0.11}_{-\ 0.05}$	$0.45\ ^{+\ 0.14}_{-\ 0.08}$
$Z \rightarrow ee$:			
$(\sigma \cdot B)$pb	UA1	41 ± 20 ± 6	85 ± 23 ± 13
	UA2	110 ± 39 ± 9	52 ± 19 ± 4
Theory: pb		$42\ ^{+\ 13}_{-\ 6}$	$51\ ^{+\ 16}_{-\ 10}$

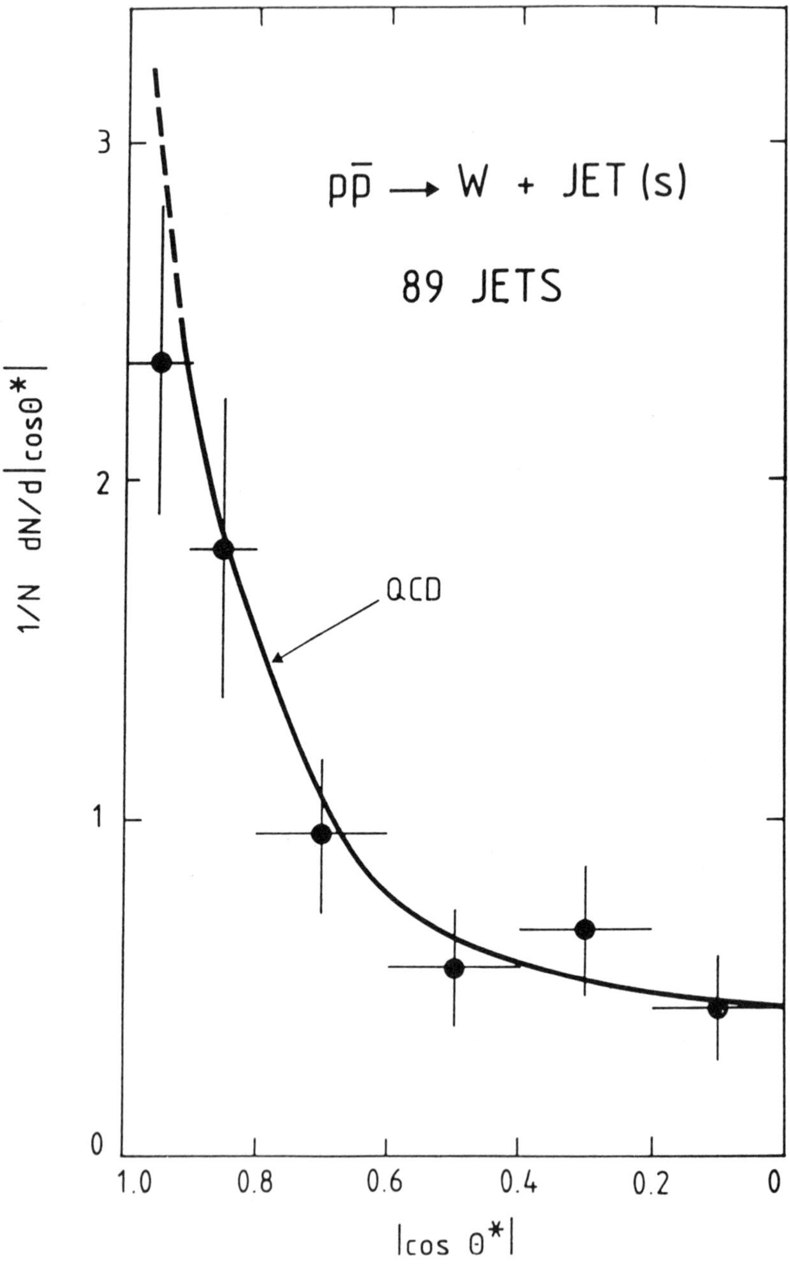

Figure 11. QCD jet - bremsstrahlung distribution

A Comparison of the masses of the W and Z bosons within the Standard Model

The values of the masses of the W and Z bosons determined by the UA1 and UA2 experiments are as follows[10]:

$$M_W: \quad UA1 = 83.5 \; {}^{+\,1.1}_{-\,1.0} \; (\pm 2.7) \; \text{GeV}/c^2,$$
$$UA2 = 81.2 \pm 1.1 \; (\pm 1.3) \; \text{GeV}/c^2,$$

$$M_Z: \quad UA1 = 93.0 \pm 1.4 \; (\pm 3.0) \; \text{GeV}/c^2,$$
$$UA2 = 92.5 \pm 1.3 \; (\pm 1.5) \; \text{GeV}/c^2.$$

The parameters of the $SU(2)_L \times U(1)$ electroweak model are related by

$$e_0 = g_0 \sin\theta_W^0, \qquad \rho_0 = \frac{M_{W_0}^2}{M_{Z_0}^2 \cos^2\theta_W^0} = 1 \quad \text{for a Higgs isodoublet.}$$

Here one has "bare" constants and the relationship for $\sin^2\theta_W^0$ can be written as

$$\sin^2\theta_W^0 = \frac{e_0^2}{g_0^2} = 1 - \frac{M_{W_0}^2}{M_{Z_0}^2}.$$

The quantities measured physically will depend on the "distance scale"; i.e.,

$$e_0 = e - \delta e, \quad g_0 = g - \delta g \; \ldots \; \text{and}$$

$$\frac{e^2}{g} = \left(1 - \frac{m_W^2}{m_Z^2}\right)(1 - \Delta r) = \sin^2\theta_W (1 - \Delta r)$$

where Δr represents the effects of the one-loop radiative corrections and has been computed to be equal to 0.0696 ± 0.002. This value for Δr depends on the masses of the top quark and the Higgs; the above value corresponds to $m_t \sim 40 \; \text{GeV}/c^2$ and $m_H \sim m_{Z^0}$.

The predictions for the W and Z masses are:

$$M_W = \left\{\frac{\pi\alpha}{\sqrt{2}\; G_F \sin^2\theta_W (1-\Delta r)}\right\}^{1/2} \qquad M_Z = \frac{M_W}{\cos\theta_W}$$

$$\left\{ \frac{\pi\alpha}{\sqrt{2}\ G_F(1-\Delta r)} \right\}^{1/2} = \frac{37.2804 \pm 0.0003}{(1-\Delta r)^{1/2}}\ \text{GeV/c}^2 = \frac{A}{(1-\Delta r)^{1/2}} = B$$

$$M_W = \frac{38.65 \pm 0.04}{\sin\theta_W}\ \text{GeV/c}^2 \qquad\qquad M_Z = \frac{77.30 \pm 0.08}{\sin 2\theta_W}\ \text{GeV/c}^2$$

The mass values of the W and Z can be used to derive values of $\sin^2\theta_W$ from the UA1 and UA2 experiments[10]).

$$\text{UA1:}\quad \sin^2\theta_W = 0.216\ {}^{+\ .004}_{-\ .005}\ (\pm\ .014)$$

$$\text{UA2:}\quad \sin^2\theta_W = 0.226 \pm .005\ (\pm\ .008)$$

A less precise value but free from the systematic errors in the mass, determinations can be derived from the ratio M_W/M_Z. The weighted mean of the UA1 and UA2 results is $\sin^2\theta_W = .212 \pm .022$. The value of $\sin^2\theta_W$ determined from the low energy ν data, and radiatively corrected, is $0.220 \pm .008$. A value can also be determined for ρ from the expression

$$\rho = M_W^2\ /\ [M_Z^2(1-B^2\ /\ M_W^2)]$$

recalling that in the minimal Standard Model $\rho = 1$.

The results are
$$\rho(\text{UA1}) = 1.028 \pm 0.037 \pm 0.019$$
$$\rho(\text{UA2}) = 0.996 \pm 0.033 \pm 0.009$$

which can be compared with the value obtained from low energy compilations; i.e., $\rho = 1.02 \pm 0.02$. Clearly from these comparisons the minimal Standard Model is an excellent description of the data at the current level of precision of the two experiments.

The comparison can be displayed by plotting $M_Z - M_W$ against M_Z; this is shown in Figure 12[10]). The ellipses represent the 68% confidence contours of UA1 and UA2, with the "freedom of movement" allowed by systematic errors indicated by the error bars on the centres. The shaded region represents the allowed area bordered by the low energy $\sin^2\theta_W$ and ρ measurements; the importance of the radiative corrections is also shown. While the mass values provide excellent confirmation of the minimal model, the present level of precision permits no firm comment to be made on the need for radiative corrections.

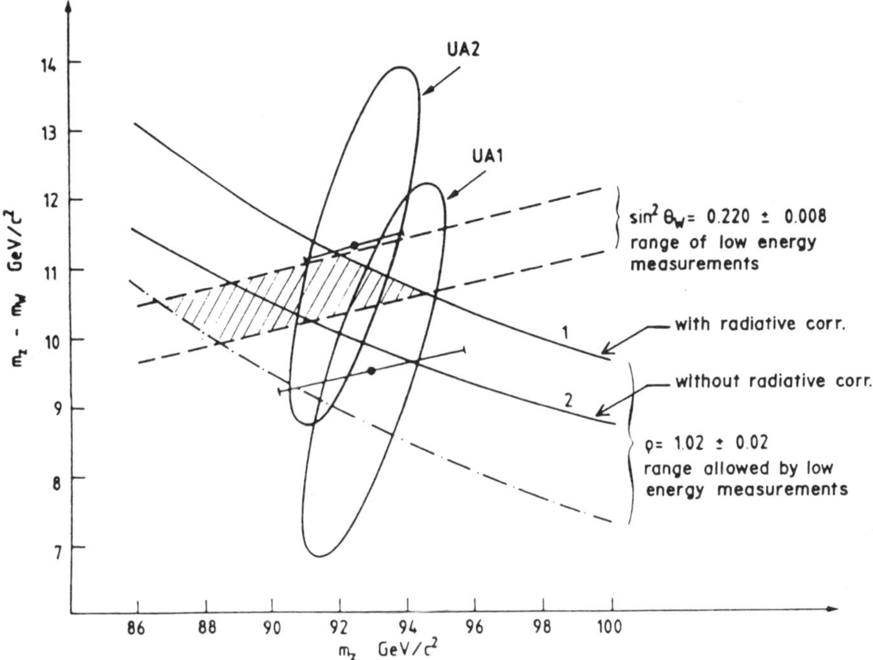

Figure 12. $m_z - m_w$ versus m_w – Standard model plot

A Limit on the number of neutrino species in the Standard Model

The Z^0 can decay into any pair of fermions $f\bar{f}$ provided that $m_f < m_z^0/2$.
However, in the event of additional families of massive fermions, it
still seems a reasonable assumption that the associated neutrinos
would be sufficiently light such that the decay $Z^0 \rightarrow \nu_f\bar{\nu}_f$ is allowed
and Γ_z would be increased by ~ 180 MeV. This kind of precision on Γ_z
is not yet achieved at the collider. However one can begin to limit
the number of neutrino species by considering the following
experimental ratio[36])

$$R = \frac{\sigma^{W^++W^-} \cdot B \; (W \rightarrow e\nu)}{\sigma^Z \cdot B \; (Z \rightarrow ee)} = \frac{\text{number of e's from W decay}}{\text{number of e's from Z decay}}$$

corrected for the known detector efficiences. We can rewrite

$$R = \frac{\Gamma(W \to e\nu)}{\Gamma(Z \to ee)} \cdot \frac{\Gamma_Z}{\Gamma_W} \cdot \frac{\sigma^{W^+ + W^-}}{\sigma^Z}$$

where $\Gamma(W \to e\nu)$ and $\Gamma(Z \to ee)$ can be calculated accurately. The ratio of the production cross sections can be taken from Altarelli et al.[27] and is estimated to be 3.3 ± 0.2. The value of Γ_Z is affected by the neutrino decay of an extra generation, adding 180 MeV. It is assumed that Γ_W remains unchanged; i.e., it is assumed that any new lepton L, associated with another generation, has a mass such that $m_L > m_W$. With these assumptions[37]

$$R = (2.715)(3.3\pm0.2)\frac{\Gamma_Z(\text{standard}) + (N_\nu-3)\Gamma_Z \to \nu\bar\nu}{\Gamma_W(\text{standard})} \,.$$

Inserting values for Γ_Z = 2.83 GeV and Γ_W = 2.82 GeV

$$N_\nu = (1.73 \pm 0.10)R - 12.55.$$

Figure 13 shows the plot of R against N_ν along with the results of UA1 and UA2.

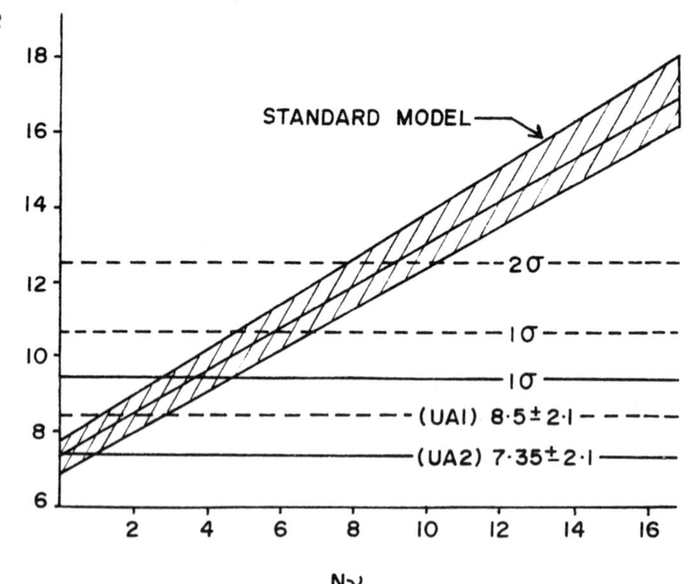

Figure 13.   R versus N_ν plot

$$R(UA1) = 8.5 \pm 2.1 \qquad R(UA2) = 7.35 \pm 2.1$$

These results[10] translate into $N_\nu \leq 10(90\% \text{ CL})$ for UA1, whereas UA2 prefers to quote a limit on additional neutrino species $\Delta N_\nu \leq 2.6 \pm 1.7$. The results clearly leave room for additional generations, but nonetheless begin to place interesting limits on N_ν.

Production of Heavy Flavours at the Collider

Heavy quarks are produced at the Collider in $Q\bar{Q}$ pairs. This can occur via the strong interaction through gluon fusion

$$gg \to Q\bar{Q}$$

or via the electromagnetic interaction by the Drell-Yan mechanism

$$q\bar{q} \to \gamma^* \to Q\bar{Q}.$$

Typical cross section predictions are in the range

$$\sigma(t\bar{t})\{m_t \sim 30 \text{ GeV}\} \sim 1 \text{ nb}$$

$$\sigma(b\bar{b})\{p_T^b > 15 \text{ GeV}\} \sim 21 \text{ nb for } m_p \sim 5 \text{ GeV}.$$

For massive quarks, still kinematically feasible as decay products, the decay of vector bosons becomes a competitive process; e.g. $W \to t\bar{b}$; at $m_t \sim 30$ GeV this branching ratio is $\sim 20\%$ and since the $W^\pm$ production cross section is ~ 5 nb one is beginning to achieve an effective cross section ~ 1 nb for the production of t or $\bar{t}$ quarks.

Experimentally the signature of the heavy quark has to be sufficiently distinctive within the detector to distinguish it above the background of QCD derived jets. For example the QCD cross sections have the following typical values

$$\frac{d\sigma}{dE_t d\eta}\Big|_{\eta=0} \sim 15 \text{nb/GeV} \{35 < p_T < 40 \text{ GeV}\} \qquad \frac{d\sigma}{dm_{jj}} \sim 5 \text{ nb/GeV} \{70 < m_{jj} < 90 \text{ GeV}\}$$

for the "inclusive jet" and "two jet" processes. The feature used to distinguish the heavy flavours is the $\sim 10\%$ branching ratio for semi-leptonic decays of the t, b and c quarks. The process is shown in

Figure 14, and leads to high p_T leptons in, or closely associated with jets. The experimental situation is easier in the case of muons, since there is the possibility, although not easy, of using the calorimeters to filter out the muons associated to the jets. The

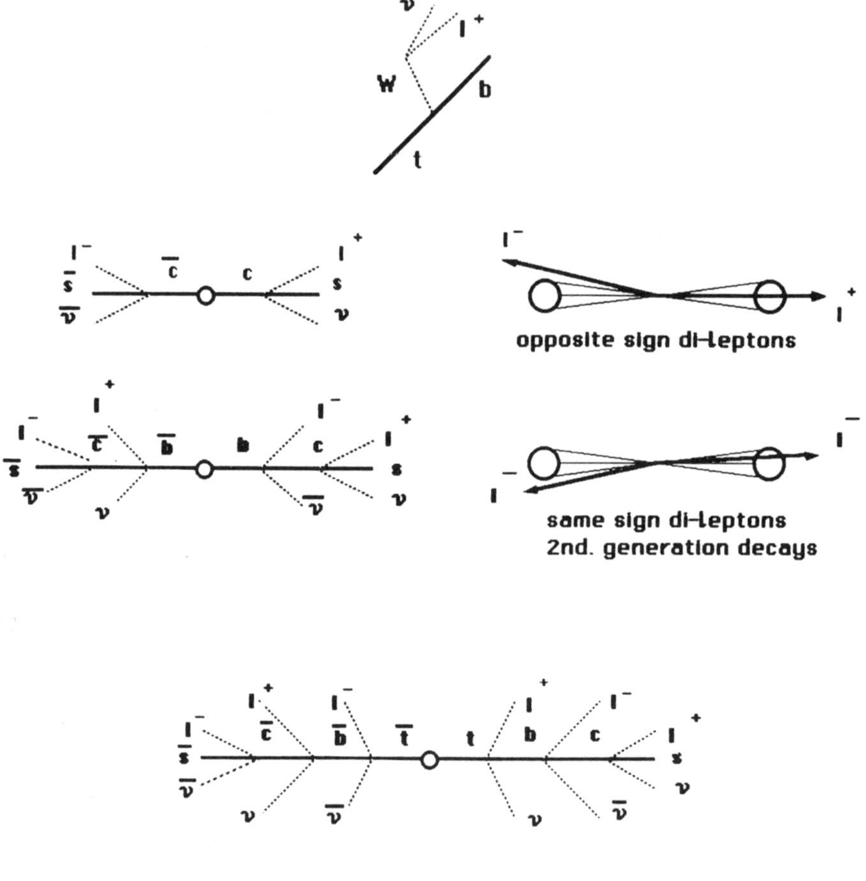

Figure 14. Heavy quark decays

detection and identification of electrons using the characteristic electromagnetic profile of the energy deposition in the calorimeter limits the electron detection to well isolated situations. An examination of Figure 14 leads to the following experimental configurations.

(1) from first generation decays $\{Q\bar{Q}\}$ – opposite sign leptons in or close to jets, i.e. non-isolated "back to back" leptons.

(2) from second generation decays – like or opposite sign leptons in or close to jets, i.e. non-isolated "back to back" or "same side" leptons.

(3) the final state of $t\bar{b}$ resulting from W decay has some special features which make it particularly attractive from an experimental point of view – these will be discussed later.

Opposite sign, isolated pairs of leptons, can of course be produced via the Drell-Yan process and the decay of vector mesons.

In the UA1 detector a muon must have a momentum ~ 3 GeV/c in order to penetrate the calorimeters and register in the muon chambers. A pair of muons with transverse momentum ~ 3 GeV/c, in a "back to back" configuration, corresponds to a mass ~ 6 GeV/c^2 for the dimuon system. This sets the approximate threshold for the dimuon "spectroscopy".[38] An integrated luminosity of ~ 380 nb^{-1} has produced 222 dimuon candidates. There is a large gap in the mass spectrum from ~ 40 GeV/c^2 to a cluser of ten events with an effective mass ~ 90 GeV/c^2; these correspond to $Z^0 \rightarrow \mu^+\mu^-$ decays.

The calorimetry is used to try to classify the remaining 221 dimuon events into heavy-flavour and Drell-Yan candidates. The transverse energy is summed in the calorimeter cells in a cone around the muon direction, a value of E_t < 4 GeV in a cone $(\Delta\eta^2 + \Delta\phi^2)^{1/2}$ = .7, where η is pseudo rapidity and ϕ the azimuthal angle in radians, is used to define isolated muons. The signs and effective mass of the dimuon system are also used to classify events. A further "dimuon isolation" criterion is invoked, where the transverse energy sum around both muons is formed.

The events break down into the following groups:

	unlike sign	like sign
	150	62
isolated	44	7
non-isolated	106	55

The isolated dimuons appear to be well described by a Drell-Yan continuum, with some evidence for structure around the region of the upsilon, plus a high mass tail (up to ~ 25 GeV) from $b\bar{b}$ production. In order to try to understand the non-isolated candidates, extensive use is made of comparisons with the EUROJET Monte Carlo. The main features of the data are consistent with the production and subsequent decays of heavy flavours. There is some suggestion that one should also include $2 \to 3$ QCD processes, i.e. reactions of the form $gg \to Q\bar{Q}g$.

The decay $W \to t\bar{b}$ as a source of heavy flavour

The topology of the decay $W \to t\bar{b}$ provides some very good experimental constraints for a search for the "top" quark. The W bosons are produced predominantly along the beam axis and hence the $t\bar{b}$ quarks will be almost "back-to-back" in the plane transverse to the beams. This leads to a "Jacobian peak" for the $\bar{b}$ jet as illustrated in Figure 15a[39]). The total momentum of this $\bar{b}$ quark far exceeds the mass of the $\bar{b}$ quark and hence one should expect a well defined jet. The subsequent decay of the t-quark, $t \to b\ell\nu$ results in a lepton at large angles to the axis defined by p_T^b, plus of course there will be missing transverse energy. In addition to these topological features shown diagramatically in Figure 15b, there exist the additional mass constraints that $M_W \simeq m(\ell\nu \text{ jet1 jet2})$ while $m_t \simeq m(\ell\nu \text{ jet2})$.

In 1984 UA1 had three electron and three muon events which appeared to fit the features of $W \to t\bar{b}$, suggesting a top mass ~ 40 GeV/c^2.[40]) Additional running produced further candidates but in numbers too large to be explained wholly by $W \to t\bar{b}$. This situation is not yet

resolved experimentally. It is perhaps worth making the point that
the analysis, expecially in the case of electrons, is made more
difficult in a detector optimized to look for isolated electrons from
parent masses ~ 80 - 90 GeV; i.e., W and Z bosons. At low p_T and in
association with jets, the identification and measurement of electrons
is difficult and plagued by QCD backgrounds.

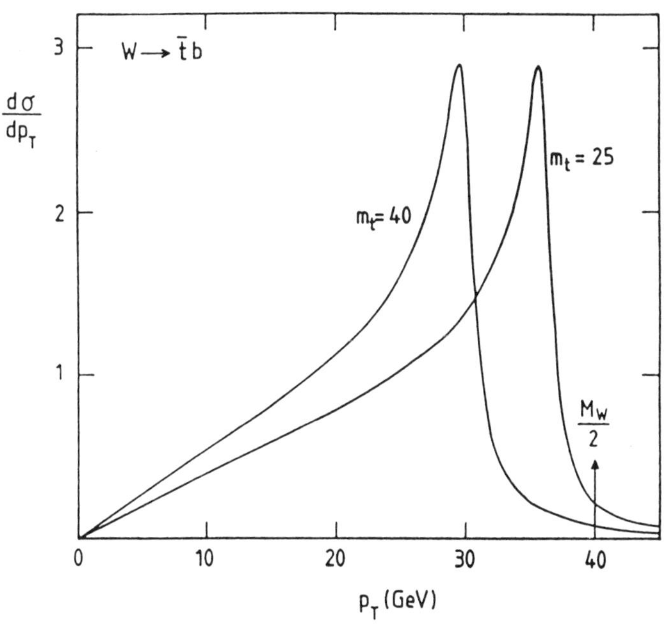

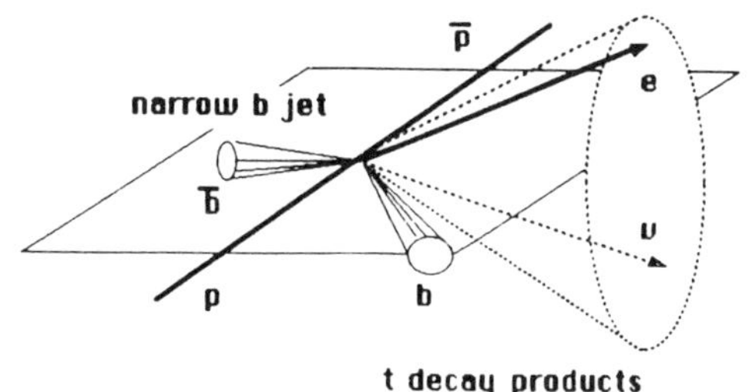

Figure 15. top

The Longterm Future - Super Hadron Colliders

The outstanding "technical" successes of the CERN SPS Collider came in two main areas:

(1) Accelerator physics - the functioning of the SPS as a bunched hadron collider proved the feasibility, previously doubted, of achieving high centre of mass energies using colliding hadrons with useful luminosities.

(2) Experimental physics - the UA1 and UA2 experiments have proved that by applying cuts in transverse energy one can examine the physics of the constituent collisons, without having signals buried in a confusing background from the "unused" constituents of the colliding hadrons. The fact that the detector does not sit in the true centre of mass of the constituent collision is not an insurmountable problem once it is realized that the optimization of the detector for a particular "parent mass" in effect selects the centre of mass form the broad band constituent collisions. Thus, provided the detector is not confused by extra tracks and energy, hadron collisions can be made simple, perhaps not to the level of e^+e^- collisions, but at least to a level not really anticipated before the SPS Collider data.

These two facts have led to proposals that in the future the highest energy collisions may well be realized in hadron colliders. In the U.S.A. the Superconduction Super Collider (SSC) is becoming a firm proposal with a reference design. The table on the following page lists a few of the parameters which are important to the experimenter.

The experimental challenge presented by a luminosity of $10^{33}\mathrm{cm}^{-2}\mathrm{sec}^{-1}$, a bunch separation of 15 nsec, and more than one event per crossing is daunting and a very long extrapolation from current experience. The physics drive behind the high energy is a belief that the Higgs particle must appear before 1 TeV. Detecting and identifying a heavy Higgs is far from trivial and places a huge burden on the calorimetry

Beam Energy	20 TeV	i.e. $\sqrt{s}$ = 40 TeV
Circumference	92 km	(6.6 T magnets)
Luminosity	10^{33}cm^{-2}sec^{-1}	
Bunch Separation	5 m	15 nsec
Average number of reactions/crossing	2.0	(σ_T ~ 120 mb)
β values	0.5 m	(10 m)
Interaction regions	± 20 m	(± 100 m)

Table III. Some SSC Parameters

in the detector. For example, at $\sqrt{s}$ = 40 TeV, a fairly modest Higgs
of mass ~ 200 GeV is produced with a cross section of ~ 50 pb.[34]) The
predominant decay is mostly into W^+W^- pairs. At this energy and with
a luminosity of ~ 10^{33}cm^{-2}sec^{-1}, about 100 $W^\pm$ are produced per sec,
along with a great deal of QCD derived jets. The key to success rests
with the ability of the calorimetry to select interesting events, and
to reconstruct well defined "jet-jet" masses - a task not yet achieved
in any calorimeter in much more congenial environments that the
hostile one at SSC. It remains an open question as to whether SSC
will be built; the price tag is very considerable: ~ 6 bn$.

There also exists a scheme to use the real estate of the LEP tunnel
at CERN for the site of a Large Hadron Collider (LHC). The centre of
mass energy is proposed in the range 10 - 20 TeV depending on the
achievable superconducting guide field.

It is believed that SSC or LHC could be realized as an accelerator
project and the design goals achieved. However, it is by no means as
clear that the existing detector techniques could be extrapoalted in a
working fashion to match the problems presented by the physics goals
of these machines; this is particularly true in the crucial area of
calorimetry, and remains the principal experimental challenge of
physics at the Super Hadron Colliders.

REFERENCES

1) Close, F.E. An Introduction to Quarks and Partons.
Academic Press (1979) p. 306.

2) Bjorken, J.D. The Tevatron Phase I Project.
December 1979 FNAL.

3) Gareyte, J. The SPS $\bar{p}$p Collider.
Proceedings of CERN Accelerator School, CERN 84-15, p. 291.
The Staff of the CERN Proton Antiproton Project.
Phys. Lett. 107B (1981) 306.

4) Rubbia, C., McIntyre, P., and Cline, D. Proc. Int. Neutrino
Conf., Aachen 1976 (Vieweg Braunschweig, 1977) p. 683.

5) Möhl, D. Stochastic cooling for beginners.
Proceedings of CERN Accelerator School, CERN 84-15, p. 97.

6) Van der Meer, S. Proceedings of Workshop on Producing High
Luminosity High Energy Proton-Antiproton Collisions.
Berkeley, 27-31 May 1978, p. 73.

7) Wilson, E.J.N. Proton Synchrotron Accelerator Theory.
CERN 77-07, p. 4.

8) Evans, L.R. The Beam-Beam Interaction.
Proceedings of CERN Accelerator School, CERN 84-15, p. 319.

9) Major Detectors in Elementary Particle Physics.
Editors G. Gidal et al. LBL-91 Supplement.

10) DiLella, L. Physics from the $\bar{p}$p Collider. CERN-EP/85-184.
Raporteurs talk given at the Int. Conf. on High Energy Physics,
Bari, 18-24 July 1985.

11) Akesson et al. Nucl. Instr. and Meth. A241 (1985) 17.
Leroy, C. et al. CERN-EP/86-66. To be published in NIM.

12) Arnison, G. et al. (UA1) Phys. Lett. 118B (1982) 173.

13) Repellin, J. P. et al. (UA2) Proceedings of 21st Int. Conf. on
High Energy Physics, Paris, 26-31 July 1982.

14) Arnison, G. et al. (UA1) Phys. Lett. 172B (1986) 461.

15) Combridge, B.L. et al. Phys. Lett. 70B (1977) 234.

16) DiLella, L. Jet Production in Hadronic Collisions.
CERN-EP/85-13. To be published in Ann. Rev. Nucl. Sci.

17) Godfrey, S. and Isgur, N. Mesons with Chromodynamics.
Thesis, University of Toronto, December 1983.

18) Arnison, G. et al. (UA1) Phys. Lett. 132B (1983) 214,
and 163B (1984) 294.

19) Arnison, G. et al. (UA1) CERN-EP/85-98.

20) Combridge, B.L. and Maxwell, C.J. Nucl Phys. B239 (1984) 428.

21) Arnison, G. et al. (UA1) CERN-EP/86-55.
Submitted to Nucl. Phys. B.

22) Arnison, G. et al. (UA1) CERN-EP/86-92.
Submitted to Phys. Lett.

23) Arnison, G. et al. (UA1) Phys. Lett. 158B (1985) 494.

24) Weinberg, S. Phys. Rev. Lett. 19 (1967) 1264.
Glashow, S.L. Nucl. Phys. 22 (1961) 579.
Salam, A. Proc. 8th Nobel Symposium,, Aspenasgaden, Almqvist,
Wiskell, Stockholm (1968)

25) Review by Bég, M.A.B. and Sirlin, A. Phys. Rep. 88 (1982) 1.

26) Drell, S.D. and Yan, T.M. Ann. Phys. (NY) 66 (1971) 578.

27) Altarelli, G., Ellis, R.K., Greco, M., and Martinelli, G.
Nucl. Phys. B246 (1984) 12 and Z. Phys. C27 (1985) 617.

28) Arnison, G. et al. (UA1)
Phys. Lett. 122B (1983) 103 and 129B (1983) 273.
Banner, N. et al. (UA2) Phys. Lett. 122B (1983) 476.

29) Arnison, G. et al. (UA1) Phys. Lett. 134B (1984) 469.

30) Smith, J. et al. Preprint. ITP-SB-83-11. NIK HEF-H/83-4.

31) Arnison, G. et al. (UA1) Phys. Lett. 16B (1986) 484.

32) Jacob, M. Nuovo Cimento 9 (1958) 826.

33) Arnison, G. et al. (UA1) Nuovo Cimento Lett. 44 (1985) 1.

34) Eichten, E. et al. Rev. Mod. Phys. 56 (1984) 579.

35) Arnison, G. et al. (UA1) Phys Lett. 147B (1984) 241.

36) Cabibbo, N. Proceedings of Third Topical Workshop on
Proton-Antiproton Collider Physics.
Rome, 12-14 January 1983. CERN 83-04.

37) Deshpande, N.G. et al. Phys. Rev. Lett. 54 (1985) 1757.

38) Arnison, G. et al. (UA1) Phys. Lett. 155B (1985) 442.

39) Barger, V. et al. Preprint. RL-83-015.
Phillips, R.J.N. Preprint. RAL-84-033.

40) Arnison, G. et al. (UA1) Phys. Lett. 147B (1984) 493.

FUTURE FRONTIERS FOR e^+e^- COLLISIONS: PHYSICS OF SLC AND LEP[*]

JONATHAN M. DORFAN

Stanford Linear Accelerator Center
Stanford University, Stanford, California 94305

1. INTRODUCTION

This report summarizes three lectures given at the Lake Louise Winter Institute on New Frontiers in Particle Physics. The audience comprised about 25% graduate students in high energy particle physics, 20% low and medium energy physicists and the remainder Ph.D's involved in high energy particle physics. I was asked to set the level of the lectures such that it would be of benefit to the two former groups. This report reflects strongly the level at which I lectured and I have made no attempt to broaden the scope for this writeup. The lectures are based on two previous lecture series which I gave[1] and I have borrowed liberally and directly from these reports.

The theme of the Winter School at Lake Louise was the future frontiers in particle physics. My job was to present the e^+e^- frontier, which I chose to define as the physics which will be done at the SLC and LEP. I placed considerably more emphasis on the Z^0 physics than the higher energy running – although the important high energy tests are covered. For a more complete discussion of the high energy LEP physics program the reader is referred to reference 2. John Ellis's lectures at this school cover the search for Higgs and SUSY particles in great detail. They should be considered as supplementary to what is discussed here – in particular his discussion of SUSY searches is much more thorough.

The outline of these lectures is as follows. We begin with a brief historical review of the contribution to particle physics of e^+e^- interactions and the advantages of this laboratory. This is followed by a discussion of the LEP and SLC machines and the reasons for developing linear colliders. A brief overview of the Standard Model and some essential formalism for the process $e^+e^- \rightarrow f\bar{f}$ are presented, followed by a discussion of detectors. Next we look at how tests of the Standard Model and physics beyond the Standard Model can be made running at the Z^0. Finally, LEP physics at energies above the Z^0 is discussed.

---

[*] Work supported by the Department of Energy, contract DE − AC03 − 76SF00515.

2. HISTORICAL PERSPECTIVE

The next frontier in e^+e^- interactions is "just around the corner" with the impending commissioning of the Z^0 "factories" SLC at SLAC (fall 1986) and LEP at CERN (fall 1988). These machines will provide copious ($\sim 10^6$ per year) production of Z^0's and we will explore here why this frontier promises to be so exciting. Following the physics of the Z^0, the next planned e^+e^- frontier will be provided by upgrading the energy of the LEP machine. By adding conventional RF, LEP can be pushed to a center-of-mass energy ($E_{c.m.}$) of $\lesssim 170$ GeV and by adding superconducting RF, upwards of 200 GeV is possible. To go substantially beyond $E_{c.m.} = 200$ GeV with e^+e^- interactions will require building a large-scale linear collider. It is too early to propose such a machine. We need to see how well the prototype for this machine (the SLC) works and what we learn in the next few years about where the next interesting frontier lies. There are many technical problems with building high luminosity, high energy linear colliders. These technical problems require many years of study before a sensible design will emerge.

During the past 15 years, e^+e^- colliding beam facilities have served as important frontiers in our quest for a better understanding of the forces of nature. Many discoveries and considerable elucidation have come from the e^+e^- colliding beam machines. A short, and by no means complete, history is illustrative. During the early to mid 1970's the e^+e^- colliding beam machines CEA, FRASCATI, SPEAR and DORIS provided a dazzling array of discoveries and an impressive list of measurements. These include: 1) the measurement of $R = \sigma_{\text{hadrons}}/\sigma_{\mu^+\mu^-}$ which provided verification that quarks come in three colors, 2) the discovery of the charm quark which verified the GIM mechanism and provided a (shortlived) equality between the number of quarks and leptons, 3) the discovery of the τ lepton which indicated the presence of a 3rd generation, 4) the discovery of jets which added more credence to the notion of spin 1/2 quarks, 5) the discovery of open charm, both charmed mesons and baryons, and 6) the study of $c\bar{c}$ spectroscopy and the discovery of the χ_c states, which provided important tests of potential models. Following the discovery of the b quark in fixed target experiments, the CESR and DORIS II machines were able to contribute: 1) careful studies of the $b\bar{b}$ spectroscopy including the discovery of the higher Υ resonances and the χ_b states, 2) the discovery of bottom mesons, 3) an impressive set of measurements on weak decays which provide valuable input for the structure of the weak mixing matrix, 4) the absence of flavor changing neutral currents which imply that the t quark must exist as a partner for the b quark (the B_L quark cannot exist as a singlet), and 5) a precision measurement of $\Lambda_{\overline{MS}}$ which is an important test of QCD. Studies during the 1980's at PETRA, DORIS II, CESR, PEP and SPEAR have provided 1) clear evidence of 3 jet events which provides strong evidence of the validity of QCD and is often quoted as the "discovery of the gluon", 2) verification of the Electro-Weak Theory from measurements of $\mu^+\mu^-$ and $\tau^+\tau^-$ charge asymmetries, 3) vast body of tests of QCD including clear evidence for scaling violations in particle momentum distributions, measurements of $\alpha_s(\Lambda_{\overline{MS}})$ and first indications that, as predicted by QCD, quarks and gluons fragment differently, 4) first measurements of the τ (known now to 10%) lifetime which provides an important test of τ/μ universality, 5) first measurement of the (surprisingly long, ~ 1 picosecond) b quark lifetime which provides important constraints for the weak mixing matrix, 6) measurements of the D^o, D^+ and F lifetimes, 7) most comprehensive study of quark fragmentation which provides considerable input for model building and non-perturbative QCD, 8) vast number of (negative) searches for light techni-pions, charged Higgs, SUSY "anythings", top quark, monojets, 4th generation quarks and leptons..., and 9) discovery of unexpected states $\theta(1420)$, $i(1640)$ and $\xi(2200)$ with implications that some of these could be "glueballs" (particles made up of two gluons in a color singlet).

This catalog is impressive and is testament to the fact that e^+e^- colliders provide an ideal frontier environment. This is mainly because they offer:

1. Clean production of natures' fundamental building blocks – the quarks and leptons. This is typically in the form of pair production of the fermion and anti-fermion.

2. All the center-of-mass energy is available for the production of these fundamental building blocks. This can be contrasted with pp (or $\bar{p}p$) collisions where hard collisions between two constituents happen at an (unknown) energy typically $\lesssim 1/6$ of the available energy. The 4 quarks not involved in the collision provide debris which render study of the hard constituent collision very difficult – the pp environment is not "clean".

3. Because the e^+e^- environment is so clean, there is an enormous potential for discoveries and in-depth study of phenomena as was justified in the previous paragraph.

At the same time it should be acknowledged that, to move beyond the planned frontiers at SLC and LEP, one encounters major technical problems with the e^+e^- colliders. Firstly, as $E_{c.m.}$ is raised, the cross section is dropping (like $E_{c.m.}^{-2}$) and correspondingly higher luminosities are required. Whereas the pp technology can be scaled up to $\gtrsim 1$ TeV effective collision energy, the scaling laws for the e^+e^- colliders for that energy range are extremely demanding and enumerable technical problems are still to be solved. The present technology cannot be scaled up – new accelerator technologies are needed.

If we consider our present theoretical understanding of nature, where does it indicate that we should be planning our next frontier? Our present understanding of the three basic particle interactions – strong, weak and electromagnetic – is in terms of the so-called Standard Model of $SU(2) \wedge U(1) \wedge SU(3)_{\text{color}}$. This model has been enormously successful at explaining all data at presently available energies. The absence of significant conflicts between the Standard Model and experiment is most impressive. To be sure, there are some essential ingredients of Standard Model as yet unobserved – the top quark and the neutral Higgs. But their experimental absence in no way threatens the model. With this "happy" situation, why do we need a new frontier? Perhaps a sharper way to state this would be to ask "what do we expect to learn from experiments at a higher energy frontier?"

There are many reasons to push to higher energies, three of which are:

1. Most theorists believe that, despite the clear success of the Standard Model, there are compelling reasons to believe that it is none other than a very good low energy approximation to the ultimate theory. Another mass scale will be encountered at $\lesssim 1$ TeV. Some of the reasons cited are the gauge hierarchy problem (incredibly fine tuning is required to make the Standard Model work), the left-right asymmetry of the Standard Model, the presence of too many parameters, the absence of any understanding of the quark and lepton mass spectrum, the presence of generations, etc.

2. History has taught us that experimental knowledge is always limited by the energy available at the last frontier crossed – new frontiers almost always bring with them new input. There are many instances of successful low energy approximations to the real world – Newtonian mechanics and the Fermi four-point theory of the Weak Interactions – to name but two.

3. Curiosity! No "self-respecting" experimentalist will accept the constraints of the current theories. Higher energy machines are the only way to know for sure what nature has in store – in the end, the **ultimate test** lies with the experimental data.

So these then are some of the reasons to push to the next frontier. We will first examine what we can learn from the Z^0 frontier ($E_{c.m.} = 93$ GeV). The Z^0 does not offer us a very substantial gain in energy relative to PETRA which has run at 44 GeV. However, as we will discuss, there are some special features of this Weak Interaction laboratory which give it an experimental reach far greater than the relatively small energy increase.

3. THE Z^0 MACHINES

To study most of the physics covered by these lectures we will require large numbers ($\gtrsim 10^5$) of Z^0 decays. We will also want to have an environment in which there is a minimal loss of decay channels arising from trigger and/or analysis techniques. The $\bar{p}p$ machines will provide valuable information about the Z^0, but the number of events will be sparse and all the Z^0 decay channels are not analyzable. As of now the UA1 and UA2 detector groups have less than 100 identified Z^0 events all of which are in the decay channel $Z^0 \rightarrow e^+e^-$ or $\mu^+\mu^-$. These two decay modes represent only about 6% of all Z^0 decays.

The high energy physics community is constructing two machines capable of providing $\simeq 10^6$ Z^0's per year. The LEP machine is under construction at CERN and the SLC machine is being built at SLAC. These two Z^0 "factories" are quite different machines and they offer different experimental possibilities. The LEP machine is a conventional e^+e^- storage ring – a scaled up version of PETRA at DESY and PEP at SLAC. It uses well-understood, proven technology and should perform close to its design specifications shortly after beam turn-on. The SLC (Stanford Linear Collider) uses an entirely new concept in accelerator technology and, in that sense, is a less certain path to high luminosity. However as we will see, the SLC is a pioneering effort in the area of linear colliders which provide the only affordable means to TeV e^+e^- colliding beam physics. The SLC will serve both as a prototype for future high energy colliding linacs and as a copious source of Z^0's. How do the two approaches differ?

3.1 CONVENTIONAL e^+e^- STORAGE RINGS

In a conventional e^+e^- storage ring one or more bunches of electrons and positrons are stored, travelling in opposite directions, in a magnetic guide field. Collisions occur at fixed points around the ring (so-called interaction regions) and there are $2n_b$ collision points possible where n_b is number of bunches. The particle detectors are placed in the interaction regions. The magnetic guide field comprises a) dipoles which provide the restoring force for a closed $e^\pm$ orbit b) quadrupoles for focussing the e^+ and e^- beams and c) sextuples to remove or reduce chromatic abberations in the magnetic focussing system.

Storage rings suffer substantial energy loss from synchrotron radiation. An electron of energy E_{beam}, travelling in a circle of radius R, loses an amount

$$\Delta E = 88.5 \times 10^{-6} \ (\text{meter GeV}^{-3}) \ \frac{E_{\text{beam}}^4}{R} \qquad (1)$$

of energy per revolution. An $e^\pm$ at PEP loses about 10 MeV/revolution for E_{beam} = 14.5 GeV. This power must be restored by RF cavities placed at strategic points around the storage ring. At full current (40 mamps) the PEP machine requires 6 MW of RF power. It is important to notice the E_{beam}^4 dependence in the synchrotron radiation loss – there is a <u>substantial</u> penalty paid as one raises the beam energy of a storage ring.

The rate for a process with cross section σ is

$$\text{rate} = \mathcal{L}\,\sigma$$

where $\mathcal{L}$ is the luminosity measured typically in units of $\text{cm}^{-2}\ \text{sec}^{-1}$. For the collision of an e^+ and e^- bunch, the luminosity is given by

$$\mathcal{L} = \frac{N^+ N^- f}{A} \tag{2}$$

where $N^\pm$ is the number of $e^\pm$/bunch, f is the collision frequency and A is the area of the larger of the two beams. Typical luminosities for existing storage rings are $\simeq 10^{31}\ \text{cm}^{-2}\ \text{sec}^{-1}$. The luminosity does not grow without bound; as one adds increasing amounts of $e^\pm$ to the beams, the continuous passage of one beam through the other causes one or both of the beams to grow, thereby reducing the luminosity. It is the cumulative effect of many small perturbations that causes the beam-beam interaction to limit the luminosity. In addition one can only tolerate as much beam current as one has RF power to suitably restore the energy lost to synchrotron radiation.

Typical beam sizes in a storage ring are

$$\sigma_x \approx 500\ \mu\text{m}$$

$$\sigma_y \simeq 50\ \mu\text{m}$$

and
$$\sigma_z \simeq 2\ \text{cm}$$

where x is the coordinate in the direction of the dipole magnet field (*i.e.* horizontal), y is vertical and z is measured along the beam direction. The beam size is limited by the synchrotron radiation damping and excitation, again a process resulting from the multiple revolution nature of the machine.

What limits the center of mass energy ($E_{\text{c.m.}} = 2E_{\text{beam}}$) achievable with storage rings? It turns out that the economics of very high energy storage rings is very unfavorable. We can write the equation for the cost (C) of a storage ring as

$$C = \alpha R + \beta\,\frac{E_{\text{beam}}^4}{R} \tag{3}$$

where α and β are constants and R is the radius of the machine. The first term in the cost equation arises from elements needed to build the ring – tunnels, vacuum system, ring magnets etc. The second term comes from the RF system (see equation (1)). Let us suppose that we minimize the cost as a function of radius R. Differentiating and setting $dC/dR = 0$ one finds

$$R = (\beta/\alpha)^{1/2} \, E_{\text{beam}}^2$$

$$C = 2(\beta/\alpha)^{1/2} \, E_{\text{beam}}^2 \ .$$

Hence the cost of the construction of a storage ring scales like E_{beam}^2 as does the radius (real estate). Lets look at some concrete examples starting with the LEP machine as a guide. The first phase of LEP will be a 50×50 GeV machine with conventional RF, circumference $= 27$ km and $C = \$500$ M. Suppose we scaled this up to a 500×500 GeV machine: circumference $\rightarrow 2700$ km and $C \rightarrow \$50,000$ M! Clearly such a machine is prohibitively expensive. Can one improve the situation by using superconducting RF? The second (superconducting) phase of LEP will be a 100×100 GeV machine at a cost of about \$700 M. Hence using superconducting RF our 500×500 GeV machine will have parameters circumference $= 675$ km and $C = \$17,500$ M – still far too costly! So clearly we need a different technology to pursue e^+e^- physics in the TeV energy range. This brings us to option 3.2.

3.2   The Linear e^+e^- Collider

In a linear collider machine one envisages two linear accelerators firing beams of electrons and positrons at each other. Following the collision, the beams are discarded. The detector is placed at the collision point. In such a machine the cost will scale like E_{beam}; $C = \alpha' E_{\text{beam}}$ and one gets away from the E_{beam}^2 scaling law of the storage ring. If one started building machines from scratch (no existing accelerator facilities) the constants α, β and α' are such that the cost of the linear collider and a storage ring are equal at roughly $E_{\text{c.m.}} \approx 150-200$ GeV. Above this energy range the linear collider becomes increasingly more economical. How does one achieve useful luminosities in a linear collider? The luminosity is given by equation (2). For LEP $f \approx 50,000$ while for a linac (SLAC) $f \approx 200$. Typically N^+N^- will be larger for the storage ring than for linear colliders, but not by much. The only way then to get a linear collider luminosity comparable to a storage ring is to reduce the beam size A in the collider by about 10^5 relative to the beam size in the storage ring. As discussed earlier the beam size in a storage ring is limited by the synchrotron radiation losses. The colliding linac does not suffer from this problem – the beam size is limited by the emittance of the linac beam. The emittance can be controlled to yield beam sizes on the order of 10 $(\mu\text{m})^2$. Hence, in principle, the reduction in frequency and bunch particle number density can be largely offset by the reduction in beam size and a colliding linac luminosity of $\approx 10^{31}$ should be possible.

The dynamics of the beam-beam interaction is very different in colliding linacs than in a storage ring. This problem is discussed fully in reference 3. The major difference comes about from the fact that the charge density in the colliding linacs is considerably (several orders of magnitude) higher than in a storage ring. The maximum current which can be collided in the colliding linac machine will still be limited by the beam-beam interaction. However the nature of the beam-beam interaction is very different in the colliding linac. The collision of the two high current density beams is very disruptive and tends to blow the beams apart. For sufficiently high currents (charge density) the passage of the one beam through the other causes a reduction (focusing) of beam size prior to the destructive disruption of the beams. This so-called "pinch" effect therefore enhances the luminosity in a linear collider system. Figure 1 (taken from reference 3) shows the pinch effect graphically. Four "snapshots" of the beam profiles in x, transverse to the beam, and z, along the beams, are shown. The upper two snapshots are taken as the e^+ and e^- beams approach each other. The third snapshot shows dramatically how the beam size has been squeezed down and in the forth the beams have passed "through" each other and are beginning to "explode." In a machine like the SLC, the "pinch" effect is expected to produce a factor of ~ 6 increase in luminosity. As stated before, linear colliders are an untested technology – the problems of producing and colliding micron size beams are by no means solved. However they will receive their first real test with the commissioning of the SLC.

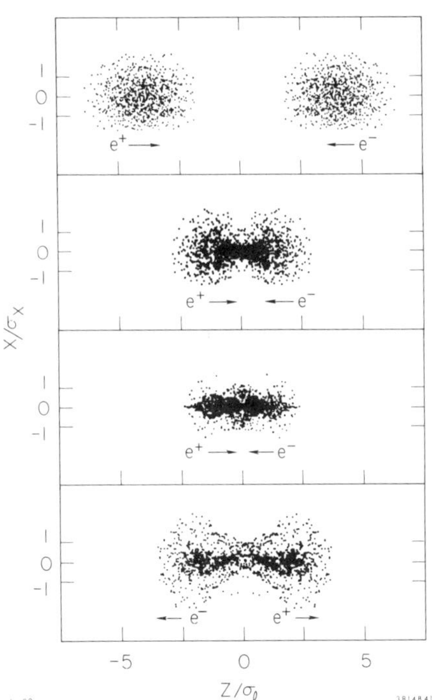

Fig. 1. Side view of the collision of oppositely charged beams showing the "pinch" effect. The coordinate z is measured along the beam direction, x is transverse to the beam direction.

The LEP Machine

LEP will be a conventional e^+e^- storage ring and a comprehensive description can be found in reference 4. The ring is being built at CERN and will have a circumference of 27 km. In its first incarnation (LEPI) it will achieve a maximum collision energy of 100 GeV and a luminosity of 10^{31} cm^{-2} sec^{-1}. 16 MW of conventional RF will be required for LEPI and the machine is expected to deliver collisions in late 1988. The initial outlay for LEP will be $500 M and it will have eight experimental halls – four of which will be instrumented at the beginning. The initial detectors go by the names of LEP3, OPAL, DELPHI and ALEPH. Typically these detectors will cost $50 M to build.

The LEPI machine will be upgraded to $E_{c.m.} \sim 170$ GeV by the addition of 80 MW of RF power and then to $E_{c.m.} \gtrsim 200$ GeV using superconducting RF cavities. The time scale for these upgrades is not yet known. Since the LEPI machine relies on conventional techniques, design performance should be reached soon after the first collisions.

The SLC Machine

The SLC machine is being built at SLAC and is slated to deliver colliding beams at the Z^0 in late 1986. The design luminosity of the machine is 6×10^{30} cm^{-2} sec^{-1} and the maximum energy at turn-on will be 100 GeV. A complete description of the SLC can be found in reference 5. However, since the SLC is not a conventional e^+e^- storage ring, we provide here a short description of the machine referring to figure 2. The existing linac will be upgraded to

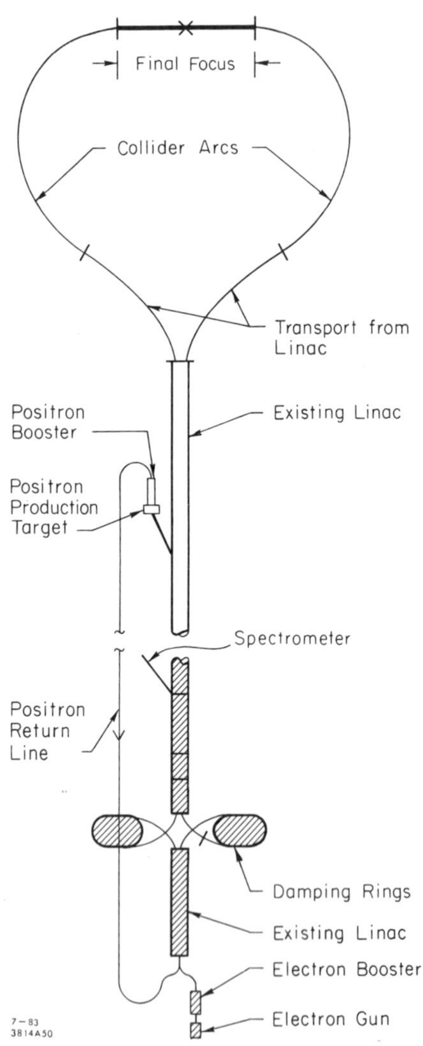

Fig. 2. Schematic of the SLC machine.

50 GeV using an extension of the SLED ideas which enabled SLAC to raise the linac energy from 22 GeV to 34 GeV. An electron bunch is diverted out of the linac and collided with a target to produce positrons. These positrons are then fed back into the front end of the accelerator. Following passage through damping rings, which provide cooling for the electron and positron bunches, a bunch of positrons immediately followed in the next linac bucket by a bunch of electrons, is transported down the accelerator to the colliding arcs. The positrons and electrons are switched to different arcs and are brought into collision by an elaborate system of optics, termed the final focus. Following the collision, the beams are dumped. So unlike a storage ring, the SLC operates as a single pass collider. The repetition rate of the linac is 180 Hz, many orders of magnitude less than that of typical storage rings. To produce a usable luminosity, this slow collision rate will be compensated for using an intense electron gun capable of producing 7×10^{10} electrons per bunch and by designing the final focus optics such that the transverse dimensions of the colliding beams are a few microns. The expected luminosity as a function of collision energy is shown in figure 3. The SLC is optimized to run at the Z^0. However, the luminosity remains good down to energies of 60 GeV. If toponium is beyond the reach of the TRISTAN machine,[6] the SLC could be used to study toponium. The energy spread of the SLC machine will be about 0.2% at full luminosity and about 0.1% at a somewhat reduced luminosity. This can be compared with LEP which will have an energy spread of 0.1%.

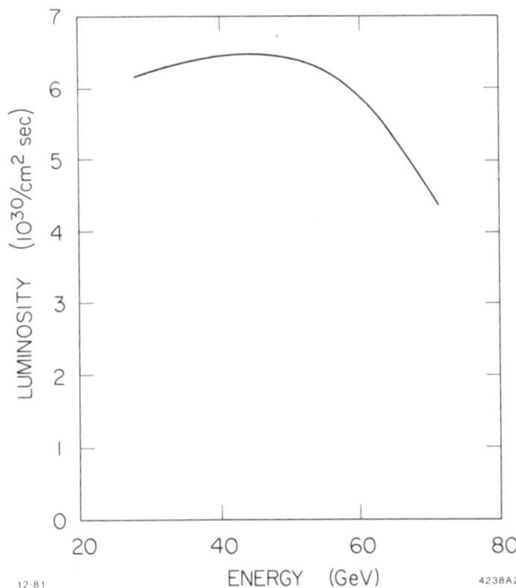

Fig. 3. The luminosity of the SLC as a function of $E_{c.m.}$.

Another feature of the SLC is the promise of longitudinally polarized beams, which, as we shall discuss, is a powerful tool for the study of Z^0 physics.[7] Polarized electrons are produced by shining circularly polarized laser light on a gallium arsinide cathode. Such an electron gun exists and has been successfully tested. Polarized electrons have already been transported down the linac and simulations of transport through the SLC arcs indicate that the transmission efficiency for the polarized electrons is $\geq$ 80%. The sign of the laser polarization can be reversed on a linac pulse by pulse basis yielding successive beam pulses of opposite polarizations. Hence it appears that, with very high probability, beams with polarizations of $\geq 50\%$ will be available at the SLC. Beam polarization at LEP is much less certain. The problems of producing and retaining the longitudinal polarization are many and no good solutions exist at this time. (See reference 4, page 132.)

The SLC machine has the distinct disadvantage of having only one interaction region. Because of the newness of the technology, it will take a considerable time and machine physics effort to reach design luminosity. The first detector for the SLC will be an upgraded MARK II detector. This detector has had six months of testing at PEP and is being installed at the SLC. A LEP competitive detector, the SLD, is also being built. This detector is scheduled to begin physics running at the SLC in late 1989 and once it has been checked out it will replace the MARK II.

What about the event rate at the Z^0? As we will see later, the cross section running on the Z^0 is about 50 nb. However initial state radiation reduces this to a usable cross section of about 40 nb. Assuming an average luminosity of 1.5×10^{30} cm^{-2} sec^{-1}, one finds an event rate of 5200 Z^0/day! Assuming 200 days for physics one has an event rate of 10^6 Z^0/year. During these lectures we will use this as a benchmark for calculating rates. Realistically during its first full year, the SLC might achieve 10^5 Z^0's, corresponding to an average luminosity of 1.5×10^{29} cm^{-2} sec^{-1}.

4. THE STANDARD MODEL AND ITS APPLICATION TO $e^+e^- \rightarrow Z^0 \rightarrow f\bar{f}$

For most of these lectures we will assume the Standard Model. During the last lecture we will look beyond the Standard Model at which time we will develop whatever formalism we need. The goal of this section is not to be complete or detailed – but merely to build a foundation from which we can extract useful experimental tests at the Z^0 .

The Standard Model is characterized by the gauge group

$$SU(3)_{\text{color}} \wedge SU(2) \wedge U(1) \quad .$$

Leptons are pointlike particles which couple to the gauge bosons of $SU(2)$ through their weak charge and to the photon of $U(1)$ through their electric charge. There are six leptons e, μ, τ, and their zero mass partners ν_e, ν_μ, and ν_τ. There are six quarks u, d, s, c, b and t which carry color and there are three color states for each quark. Leptons have no color charge and are therefore "blind" to the strong interaction.

The left handed fermions are arranged in weak iso-doublets

$$\begin{pmatrix} \nu_e \\ e \end{pmatrix}_L \qquad \begin{pmatrix} \nu_\mu \\ \mu \end{pmatrix}_L \qquad \begin{pmatrix} \nu_\tau \\ \tau \end{pmatrix}_L \qquad \begin{matrix} T_3 = 1/2 \\ -1/2 \end{matrix}$$

$$\begin{pmatrix} u \\ d' \end{pmatrix}_L \qquad \begin{pmatrix} c \\ s' \end{pmatrix}_L \qquad \begin{pmatrix} t \\ b' \end{pmatrix}_L \qquad \begin{matrix} T_3 = 1/2 \\ -1/2 \end{matrix}$$

where T_3 is the 3rd component of the weak charge. The primes on the quarks indicate that flavor conservation in the quark sector is not perfect. This generation mixing can be summarized by the elements of the Kobayashi-Maskawa matrix – the most familiar component being the Cabibbo angle which tells us that the d quark has a $\sim 5\%$ strange quark admixture. More succinctly – in the quark sector the weak eigenstates are related by a rotation matrix to the mass eigenstates. There are no analogous flavor changing currents in the neutral sector. We notice in passing the peculiarity of the three generations, the ν_e, e, u and d' being the members of the lightest generation. The Standard Model does not explain why nature chooses to replicate itself in this peculiar manner.

Right handed fermions appear in singlets, u_R, d_R $\dots t_R$, e_R, μ_R, τ_R and, since the ν's are massless, there are no right handed ν's. $T_3 = 0$ for all right handed fermions.

There are nine massless bosons in the Standard Model – 8 gluons and the photon. There are 3 massive vector bosons W^+, W^- and Z^0 and, in the minimal model with one Higgs doublet, there is one neutral scalar, H^0. Gluons carry color (unlike photons which don't carry charge) and hence $SU(3)_{\text{color}}$ is non-abelian. Since gluons carry color they can couple to other gluons. The polarization of the QCD vacuum by virtual quark and gluon pairs results in an <u>anti-screening</u> of color charge. This can be contrasted with the <u>screening</u> of electric charge by virtual e^+e^- pairs in QED. This anti-screening leads to the notion of confinement of quarks and the decrease of the strong coupling constant α_s, with increasing q^2. Free quarks should not be seen and this notion will be tested at the Z^0 although not discussed further in these lectures.

The Standard Model does not predict masses for the fundamental particles. The $W^\pm$, Z^0 masses are given in terms of the parameter $\sin^2 \theta_W$:

$$M_W^2 = \frac{\pi \alpha}{\sqrt{2} G_F} \left(\frac{1}{\sin^2 \theta_W} \right)$$

$$M_{Z^0}^2 = \frac{M_W^2}{\cos^2 \theta_W} = \frac{\pi \alpha}{\sqrt{2} G_F} \left(\frac{1}{\sin^2 \theta_W \cos^2 \theta_W} \right)$$

where α is the fine structure constant and G_F is the Fermi coupling constant. The H^0 mass is expected to fall in the range $7.5 \lesssim M_{H^0} \lesssim 10^3$ GeV. This however is of no consolation to the experimentalist searching for the H^0. The presence of the neutral Higgs is crucial to the success of the Standard Model.

The electroweak interactions of all the gauge fields are specified by the Model and are determined by e, the electric charge, and one free parameter θ_W. Spinors couple to the photon field with strength e and to the Z^0 with strength

$$-\frac{e}{\sin\theta_W\cos\theta_W}\left(T_3^{R/L} - Q\sin^2\theta_W\right) = 2\sqrt{2}\left(\frac{M_Z^2 G_F}{\sqrt{2}}\right)^{1/2}\left(T_3^{R/L} - Q\sin^2\theta_W\right)$$

where R/L indicates left and right couplings and Q is the charge of the fermion.

Aside from Higgs and fermion masses, the Electro-Weak theory is totally specified if we know α, G_F and M_Z. α and G_F are extremely accurately (better than one part in 10^5) known, M_Z is known only to about 3%. A precise measurement of M_Z will constrain considerably the Standard Model.

For almost all the physics discussed in these lectures, we are interested in the basic process $e^+e^- \to f\bar{f}$ where the symbol f signifies a fundamental fermion, either a quark or a lepton. There are two processes which contribute to the cross section as shown in figure 4, namely $e^+e^- \to \gamma \to f\bar{f}$ and $e^+e^- \to Z^0 \to f\bar{f}$. The Standard Model specifies all the couplings and hence the cross section for these processes can be calculated. If θ is the fermion polar angle, the differential cross section has the form[8]

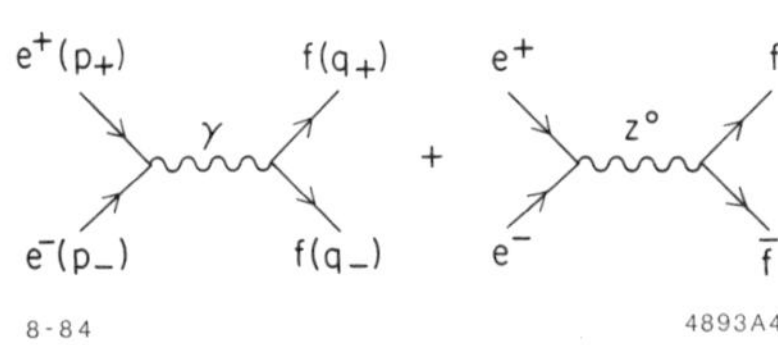

Fig. 4. The basic $e^+e^- \to \gamma$, $Z^0 \to f\bar{f}$ process.

$$\frac{d\sigma_{f\bar{f}}}{d\cos\theta} = \frac{\pi\alpha^2 Q_f^2 D}{2S}(1+\cos^2\theta) - \frac{\alpha Q_f D G_F M_z^2(S-M_z^2)}{8\sqrt{2}[(S-M_z^2)^2 + M_z^2\Gamma_z^2]}$$

$$\left[(R_e+L_e)(R_f+L_f)(1+\cos^2\theta) + 2(R_e-L_e)(R_f-L_f)\cos\theta\right]$$

$$+ \frac{DG_F^2 M_z^4 S}{64\pi[(S-M_z^2)^2 + M_z^2\Gamma_z^2]}$$

$$\left[(R_e^2+L_e^2)(R_f^2+L_f^2)(1+\cos^2\theta) + 2(R_e^2-L_e^2)(R_f^2-L_f^2)\cos\theta\right]$$

$$\tag{4}$$

where Q_f is the fermion charge, $S = E_{c.m.}^2$, M_Z the mass of the Z^0 and D takes into account the number of color degrees of freedom. For $f \equiv$ quark, $D = 3$, otherwise $D = 1$. The left and right handed weak coupling constants are given by

$$L_f = T_3^f - Q_f \sin^2 \theta_W$$

$$R_f = -Q_f \sin^2 \theta_W \quad .$$

The three terms in the cross section are the purely electromagnetic contribution, the interference between the weak and electromagnetic diagrams and the purely weak contribution. Notice that a) the interference term disappears at $\sqrt{S} = M_Z$ as it should b) the first term is just the point QED differential cross section and c) at $\sqrt{S} = M_Z$ the purely weak term dominates.

It is illustrative to integrate over $\cos\theta$ and plot the cross section as a function of $E_{c.m.} = \sqrt{S}$. This is shown in figure 5. One sees that below the region of the Z^0 mass the purely electromagnetic cross section dominates as is reflected by the $E_{c.m.}^{-2}$ behavior. On the Z^0 pole however the weak cross section dominates, providing 10^3 times more particle production than the electromagnetic process. This is part of the magic of running at the Z^0 – the Z^0 provides an enormous enhancement in event rate over running in the continuum (*i.e.* off resonance). Studying e^+e^- interactions at ~ 93 GeV in the absence of the Z^0 with the SLC or LEP ($\mathcal{L}_{\text{peak}} \sim 5 \times 10^{30}$) would be extremely painful if not in many cases impossible. The presence of the Z^0 however renders these relatively low luminosity machines capable of very high event rates.

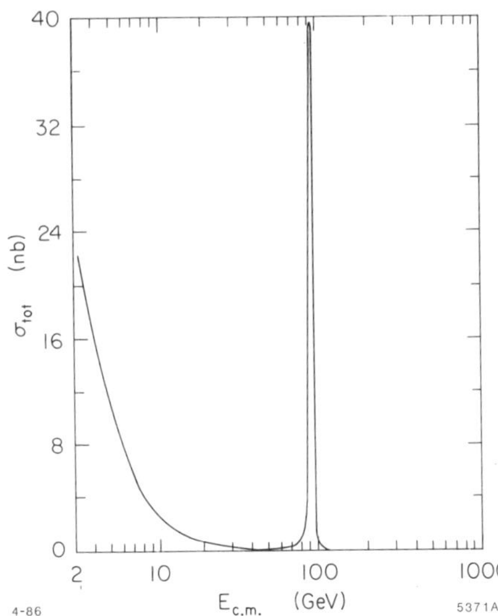

Fig. 5. The cross section for $e^+e^- \to \gamma$, $Z^0 \to f\bar{f}$ as calculated in the Standard Model. The $c\bar{c}$ and $b\bar{b}$ threshold behavior are omitted from the plot.

Let us now return to $d\sigma/d\cos\theta$ and consider running at $S = M_Z^2$, namely on the Z^0 pole. Changing notation to axial and vector coupling constants

$$a = \frac{1}{2}(L - R) \quad \text{and} \quad v = \frac{1}{2}(L + R)$$

one finds $L = a + v$, $R = v - a$, $L^2 + R^2 = 2(a^2 + v^2)$, $L^2 - R^2 = 4av$ and

$$\frac{d\sigma_{f\bar{f}}}{d\cos\theta} = \frac{DG_F^2 M_Z^4}{16\pi\Gamma_Z^2} \left[(a_e^2 + v_e^2)(a_f^2 + v_f^2)(1 + \cos^2\theta) + 8a_e v_e a_f v_f \cos\theta \right] \quad . \tag{5}$$

It is useful to tabulate the couplings and the sum of their squares. Assuming $\sin^2\theta_W = 0.22$ (which we will do throughout for convenience) we find the values in table I.

TABLE I

	Q	T_3	a	v	$a^2 + v^2$
e, μ, τ	-1	$-1/2$	$-1/2$	$-.06$	$.2536$
ν_e, ν_μ, ν_τ	0	$1/2$	$1/2$	$1/2$	$1/2$
d, s, b	$-1/3$	$-1/2$	$-1/2$	$-.35$	$.375$
u, c, t	$+2/3$	$+1/2$	$1/2$	$.21$	$.29$

We turn our attention back to equation (5). The term linear in $\cos\theta$ contributes a front-back asymmetry, A_{F-B}. $A_{F-B} \propto v_e v_f$ which, for charged leptons, is a very small number. However a measurement of A_{F-B} for charged leptons has great sensitivity to $\sin^2\theta$ as we will see later in this section. Since $\int_0^\pi \cos\theta\, d\theta = 0$ the term linear in $\cos\theta$ does not contribute to the total cross section.

Integrating the term in $(1 + \cos^2\theta)$ yields the total cross section for producing a final state $f\bar{f}$ at the Z^0 :

$$\sigma_{f\bar{f}} = \frac{DG_F^2 M_Z^4}{6\pi\Gamma_Z^2} (v_e^2 + a_e^2)(v_f^2 + a_f^2) \ .$$

We omit here the derivation of Γ_Z but note that

$$\Gamma_Z = \frac{G_F M_Z^3}{24\sqrt{2}\pi} \sum_i (v_i^2 + a_i^2) D_i \tag{6}$$

where i ranges over all fundamental fermions and D_i is the color factor (3 for quarks, 1 for leptons). We can obtain σ_{point}, which is the lepton point QED cross section, from the first term in equation (4):

$$\sigma_{\text{point}} = \frac{\pi\alpha^2}{2S} \int (1 + \cos^2\theta) \, d\cos\theta$$

$$= \frac{4\pi\alpha^2}{3S} \simeq \frac{87 \text{ nb}}{S} \quad .$$

Hence we can write

$$R_{f\bar{f}} = \frac{\sigma_{f\bar{f}}}{\sigma_{\text{point}}} = \frac{D(a_f^2 + v_f^2)(a_e^2 + v_e^2)}{16\alpha^2(1 - 2x_W + 8x_W^2/3)^2} \quad . \tag{7}$$

Assuming 6 quarks, ignoring the finite t mass and setting $x_W = \sin^2\theta_W = .22$, one finds at the Z^0 the R values in table II. Also shown in table II are the branching fraction for each process. Hence under these assumptions $R_{Z^0} \approx 5200$ and $B(Z^0 \rightarrow \text{hadrons}) \simeq 72\%$. This value of R_{Z^0} has not been corrected for initial state radiation effects which has the effect of lowering the peak cross section with a compensating "radiative tail" on the high side of the resonance. These radiative effects are discussed more fully in reference 9 – we quote here the approximate result. For a narrow resonance with peak cross section σ_0 the actual cross section, after the inclusion of radiative effects, is

$$\sigma_{\text{peak}} \approx \left(\frac{\Gamma}{m}\right)^t \sigma_0 + \left(\delta_0 + \frac{13}{2}t\right)\sigma_0$$

TABLE II

CHANNEL $(f\bar{f})$	$R_{f\bar{f}}$	$\Gamma_{f\bar{f}}/\Gamma_{Z^0}$ (%)
each $\nu\bar{\nu}$	313	6.1
$\mu^+\mu^-$, $\tau^+\tau^-$, e^+e^{-*}	159	3.1
$u\bar{u}$, $c\bar{c}$, $t\bar{t}$	550	10.6
$d\bar{d}$, $s\bar{s}$, $b\bar{b}$	704	13.6

* We have ignored t channel diagrams which are only important at small values of θ.

where $t = 2\alpha/\pi \; (\ell n \; S/m_e^2 - 1)$ is the so called equivalent radiator and $\delta_0 = 2\alpha/\pi \; (\pi^2/6 - 17/36) \simeq 0.005$. At $\sqrt{S} = M_Z$, $t = .11$ and

$$\sigma_{\text{peak}} \approx 0.8\sigma_0 \; .$$

Hence the radiatively corrected R is approximately 4200 on the Z^0 .

We return now to the problem of how to incorporate the effects of large masses ($wrt \; \sqrt{S}$) for the final state fermion in equation (4). In a general way we can write

$$\frac{d\sigma_{f\bar{f}}}{d\theta} = f(\beta_f, \theta) \; \sigma(m_f = 0)$$

and

$$\sigma_{f\bar{f}} = f(\beta_f)\sigma(m_f = 0)$$

where β_f is the fermion velocity and m_f is the fermion mass. For <u>vector</u> couplings

$$f(\beta_f, \theta) = \frac{3}{16\pi} \; \beta_f[(1 + \cos^2\theta) + (1 - \beta_f^2)\sin^2\theta]$$

and

$$f(\beta_f) = \frac{1}{2} \; \beta_f(3 - \beta_f^2) \; .$$

For <u>axial-vector</u> couplings

$$f(\beta_f, \theta) = \frac{3}{16\pi} \; \beta_f^3(1 + \cos^2\theta)$$

and

$$f(\beta_f) = \beta_f^3 \; .$$

Therefore for the t quark with velocity β_t, the correct form of the contribution to the Z^0 width is (see equation (6))

$$\Gamma(Z^0 \to t\bar{t}) = \frac{G_F M_Z^3}{8\sqrt{2}\pi} \left(v_t^2 \; \frac{1}{2} \; \beta_t(3 - \beta_t^2) + a_t^2\beta_t^3 \right) \; .$$

Figure 6 shows the suppression of $t\bar{t}$ relative to a full strength (light) charge two-thirds quark as a function of the t quark mass. Since we know from PETRA that $M_t \gtrsim 23$ GeV/c^2 the $t\bar{t}$ final state at the Z^0 is suppressed at least to 0.7 of the $u\bar{u}$ rate.

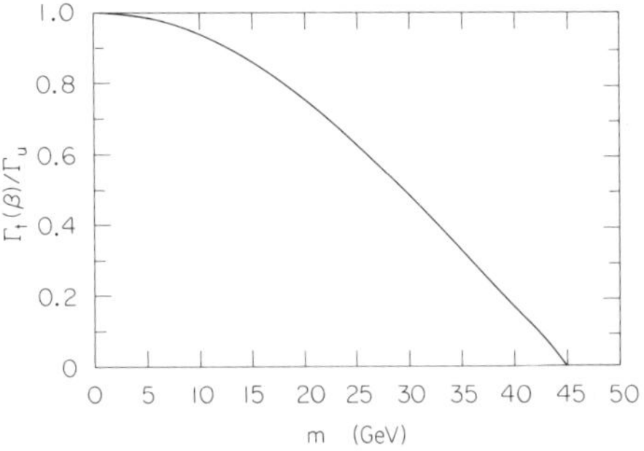

Fig. 6. The suppression factor of $t\bar{t}$ decays
of the Z^0 as a function of M_t.

We return now to the forward backward charge asymmetry (A_{F-B}) discussed earlier. Consider for the moment the concrete example of the final state $\mu^+\mu^-$ as applied to our master formula (4). We will get a contribution to A^μ_{F-B} from terms linear in $z = \cos\theta$.

$$A^\mu_{F-B} = \frac{\int_0^1 \frac{d\sigma}{dz}dz - \int_{-1}^0 \frac{d\sigma}{dz}dz}{\int_0^1 \frac{d\sigma}{dz}dz + \int_{-1}^0 \frac{d\sigma}{dz}dz} = \frac{N^F_{\mu-} - N^B_{\mu-}}{N^F_{\mu-} + N^B_{\mu-}}$$

where $N^F_{\mu-}(N^B_{\mu-})$ is the number of μ^- in the forward (backward) hemisphere relative to the incoming e^- direction. On the Z^0 pole

$$A^\mu_{F-B} = \frac{3a_e a_\mu v_e v_\mu}{(v_e^2 + a_e^2)(v_\mu^2 + a_\mu^2)} \simeq 4.3\% \quad . \tag{8}$$

A^μ_{F-B} is proportional to $v_e v_\mu$ which makes it small but very sensitive to $\sin^2\theta_W$. Rewriting the couplings in terms of $\sin^2\theta_W$ we find at the Z^0

$$A^\mu_{F-B} = \frac{3(1 - 4x_W)^2}{4(1 - 4x_W + 8x_W^2)^2} \, , \quad x_W = \sin^2\theta_W$$

and

$$\frac{1}{15} \frac{dA^\mu_{F-B}}{A^\mu_{F-B}} \simeq \frac{d\sin^2\theta}{\sin^2\theta} \quad . \tag{9}$$

Hence the statement that a measurement of A^μ_{F-B} provides substantial sensitivity to $\sin^2\theta_W$.

From equations (7) and (8) applied to $e^+e^- \to Z^0 \to e^+e^-$ one finds

$$A^e_{F-B} = \frac{3a_e^2 v_e^2}{(a_e^2 + v_e^2)^2}$$

and

$$R^{e^-e^+} \propto (v_e^2 + a_e^2)^2 \ .$$

From these two equations one can determine a_e and v_e but not their relative sign. Is it possible to measure the relative sign? The answer is yes, as long as one can measure the fermion polarization of one of the charged $f\bar{f}$ final states. It turns out that the only practical final state for a polarization measurement is $\tau^+\tau^-$. Since parity is violated in the neutral current interaction, even in the absence of $e^\pm$ beam polarization, the Z^0 is produced polarized as are its decay products. The polarization P is given by

$$P^{f\bar{f}} = \frac{\sigma_R - \sigma_L}{\sigma_R + \sigma_L} = \frac{R_f^2 - L_f^2}{R_L^2 + L_f^2} = -\frac{2a_f v_f}{(a_f^2 + v_f^2)} \ .$$

Now the ratio

$$A^f_{F-B}/P^{f\bar{f}} = -3a_e v_e/(a_e^2 + v_e^2)$$

is independent of the final state fermion couplings and measures directly the relative sign of v_e and a_e. By measuring $R^{e^+e^-}$, A^e_{F-B} and the τ polarization one finds a_e and v_e and their relative sign. Then from $R^{\mu^+\mu^-}$, A^μ_{F-B}, $R^{\tau^+\tau^-}$, A^τ_{F-B} one can obtain the μ and τ axial and vector couplings. In this way the universality of the weak interactions is checked. In addition each measurement of a vector coupling provides a measurement of $\sin^2\theta_W$.

There remains one important issue which I have avoided but which is important to raise at this juncture. We have calculated the cross section for $e^+e^- \to Z^0 \to f\bar{f}$ in lowest order. However there are important weak radiative effects arising from one-loop diagrams such as are shown in figure 7. In order to compare the measurements at the Z^0 with the Standard Model, we must take these effects into account. To give you a feeling for their magnitude, inclusion of these one-loop diagrams changes (raises) the Z^0 mass by about $3\frac{1}{2}\%$. Happily these effects have been calcualted by several authors[10] and there is good agreement on their size. The exact

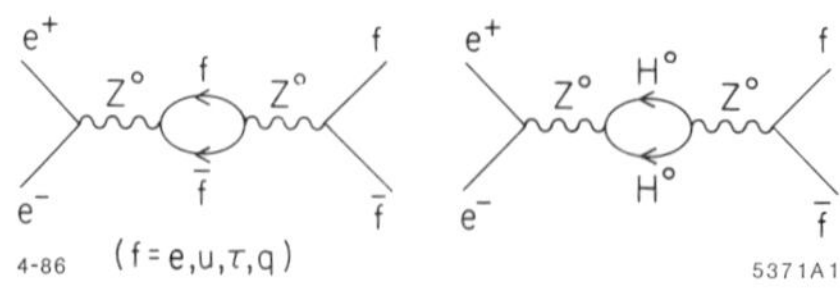

Fig. 7. One-loop weak radiative corrections to the process $e^+e^- \to Z^0 \to f\bar{f}$.

impact of those effects on an experimental measurement will depend crucially on the cuts which are applied to the data. Hence application of a Monte Carlo simulation, as opposed to analytical calculations, is the most reasonable way to extract the physics. Such Monte Carlo programs – with both weak and electromagnetic radiative effects – are being developed for both SLC and LEP. In addition, as we will see later, we will be able to check the validity of the radiative correction calculations from measurements made at the Z^0.

5. THE Z^0 ENVIRONMENT – REQUIREMENTS FOR DETECTORS

So it seems we will have two fine Z^0 "factories" – what kind of detectors do we need? The Z^0 environment has been studied in many workshops and the interested reader can find summaries of these workshops in references 11, 4 and 2. We describe here the main features of the environment, particularly as they pertain to detector design. The basic production process is shown in figure 8 where the final state particle naming convention is given in the figure caption. We now consider how these produced states decay. The final states e^+e^- , $\mu^+\mu^-$ are stable and result in opposite sign, high energy ($\sim E_b$) back to back leptons. The typical decays of the other produced states are shown schematically in figure 9. A quick glance at figure 9 and one realizes that what is needed at the Z^0 is a detector capable of a) measuring the properties of high energy jets b) measuring and tagging electrons and muons over a wide range of momenta both in isolation and in the presence of high energy jets and c) measuring the total energy and momentum in the event as an indicator of the missing energy and transverse momentum of ν's.

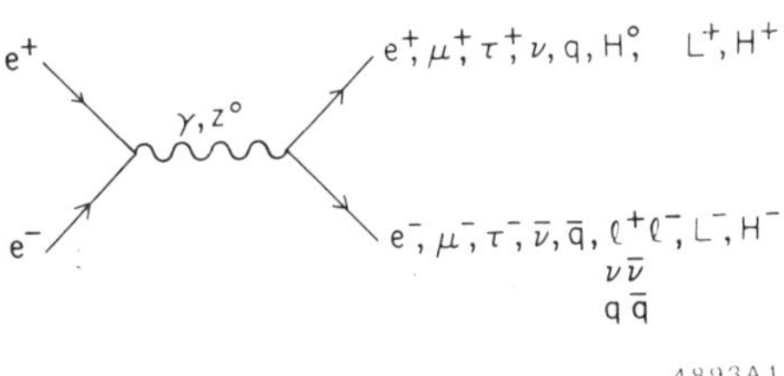

Fig. 8. The basic e^+e^- process where final states are produced via an intermediate photon or Z^0. The notation is obvious except that q stands for a quark, H^0 the neutral Higgs scalar, $\ell^\pm$ a charged lepton, $H^\pm$ a charged Higgs scalar and $L^\pm$ a (new) heavy charged lepton.

In addition there are many multi-jet and multi-lepton events which demands that the detector be uniformly instrumented over as large a solid angle as possible. Figure 10 shows the fractional momentum carried by hadrons, leptons and photons in events of the type $Z^0 \rightarrow$ hadrons. Notice the large dynamic range of the particle momenta. The detector must do an equally good job at high and low momenta. The high momentum (leading) particles carry information

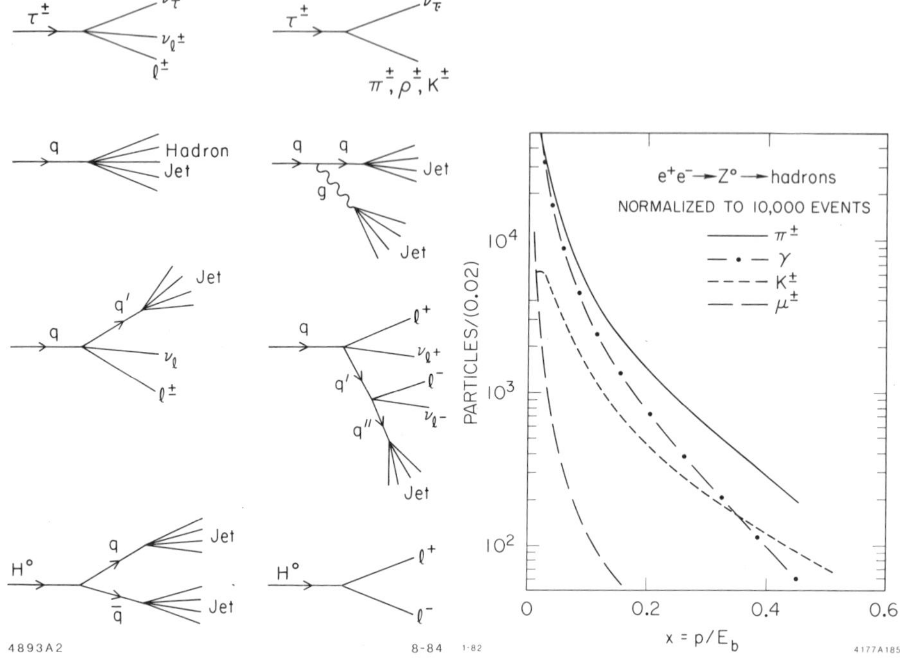

Fig. 9. Typical decays which result from the process in figure 4. The symbol g stands for a gluon.

Fig. 10. Momentum distribution for different particle species produced in the decay of $Z^0 \to$ hadrons.

about the quark flavor and the fragmentation process, while the intermediate and low energy particles provide information about the decay chains and the energy flow. Typical multiplicities for the $Z^0 \to$ hadrons are 22 charged particles and 23 photons per event – jet multiplicities on the order of 11 charged particles and 11 photons. In addition this multiplicity is highly collimated – most jets are contained in a $\lesssim 10°$ cone. Figure 11 shows the angle between various particle species and the event jet axis. The distribution peaks at $\sim 2°$ for photons and hadrons. Hence a detector will have to possess fine segmentation both in the charged tracking and the calorimetry.

Studies[11] of reconstruction of K_S^0, D^0, $D^\pm$, measurement of the invariant cross section $S \, d\sigma/dx$ (at high x), measurement of the $\tau^\pm$ polarization lead to the conclusion that a momentum resolution of $\sigma_p/p \lesssim 0.3 \, P$ (GeV/c) is needed. In order to separate leptons from hadrons cleanly will require rejection of hadrons at a level of $\geq 10^3$. This can be understood in simple terms as follows. The average charge multiplicity is 10/jet and the typical semi–leptonic branching fraction

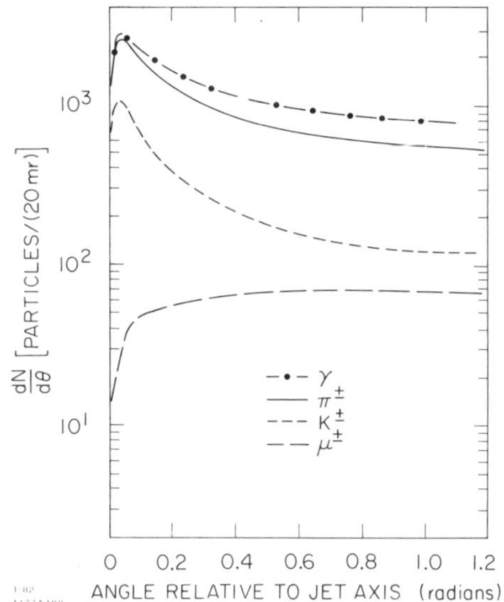

Fig. 11. Distribution of the angle with respect to the jet axis for different particle species in $Z^0 \rightarrow$ hadron events.

$(B(q \rightarrow \ell^{\pm}\nu x))$ is 10%. Hence in hadronic events one will have, on average, one $e^{\pm}$, $\mu^{\pm}$ per 100 charged hadrons. Having a signal to noise of 10 $e^{\pm}$, $\mu^{\pm}$ per 1 hadron requires a rejection of hadrons at the 10^3 level. A final requirement for a good detector in the new energy regime of the Z^0, is the ability to search for free quarks. This can be done by a) measuring the ionization of charged particles in a gas chamber (dE/dx) which measures charge directly or by b) using time of flight to look for massive particles.

It seems possible to design 4π detectors which are equal to most of the rigors of the environment described here. Although diverse in their approaches to the problems, the four LEP detectors and the SLD should do an excellent job of studying the Z^0 physics. The MARK II upgrade is a more modest approach designed to be ready for the early start of the SLC. Its main drawback is its lack of hadron calorimetry and hadronic particle identification. However as a survey detector, it will do most physics very well. For completeness a list of the detector proposals is given in reference 12 and a schematic of the upgraded MARK II detector is given in figure 12.

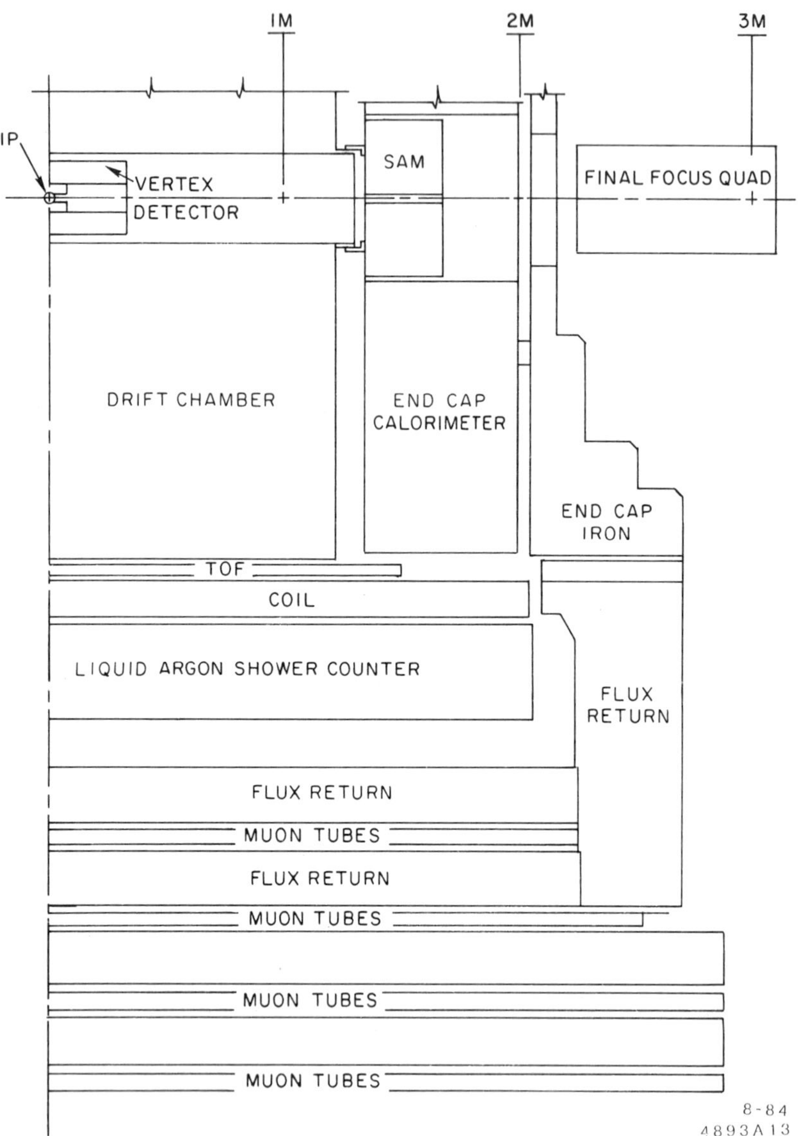

Fig. 12. A schematic of the Upgraded MARK II detector.

6. TESTING THE STANDARD MODEL AT THE Z^0

6.1 MEASUREMENTS OF THE Z^0 MASS AND WIDTH

We recall from section 4 that, with the exception of particle masses, the Electro-Weak sector of the Standard Model is entirely specified by knowing α, G_F and M_Z. α and G_F are known very precisely[13] (to better than 1 part in 10^5) and presumably then the first task at the SLC and LEP will be to make a precise measurement of M_Z. To date, the best measurement of M_Z is provided by the UA2 group[14] of $M_Z = (93.5 \pm 2.5 \pm 1.3)$ GeV.

To measure M_Z requires mapping out the resonance shape by running the machine at discrete energy settings in the neighborhood of 93 GeV. The MARK II group[15] has made a study of optimum strategies for such an energy scan. Variables used to differentiate different scans are the number of scan points, the energy step and the amount of data logged at each point. Several optimum strategies were found; an example of one has 9 scan points, each 750 MeV apart and requiring 100 nb^{-1} or 3000 equivalent Z^0 events. Even at a low average luminosity of 2×10^{28} cm^{-2} sec^{-1}, this would only take two calender months. Based on statistics alone, such a scan would yield the following errors:

$$\delta M_Z = 45 \text{ MeV}/\text{c}^2 \, , \quad \delta\Gamma_Z = 135 \text{ MeV} \quad \text{and} \quad \delta R_Z = 3.5\% \, .$$

These numbers are impressive, especially the error on the mass. What would the systematics in the mass measurement be? Initial state radiation will shift the Z^0 mass peak down by about 250 MeV/c^2 – however this can be accurately calculated to within about 10 MeV/c^2. Energy dependent errors in the luminosity measurement monitor will change the line shape. However typical errors of this type (2%) would lead to small (10 MeV/c^2) errors in the mass determination. There are other effects, however the dominant effect will be the knowledge of the e^+ and e^- beam energies. Recognizing the potential limitation, two energy measuring spectrometers have been designed[15] for the SLC e^+ and e^- beam dump areas which will be capable of measuring $E_{c.m.}$ to 0.05%. This translates into a systematic uncertainty of 45 MeV/c^2 in M_Z. For normalization $\delta M_Z = 35$ MeV/c^2 corresponds to an error in $\sin^2 \theta_W$ of 0.0002!

So it appears that with relatively little running, a few months at a very modest SLC luminosity, M_Z will be very accurately determined. Now if we believe the one-loop radiative correction calculations discussed in section 4, then we can input this accurate measurement of M_Z into the Monte Carlo simulation model and all Electro-Weak measurables and distributions should be very accurately predicted. (There will be effects arising from the unknown top and neutral Higgs masses. These will be discussed later – for almost all subsequent discussions these can safely be ignored.) In particular we can use the Monte Carlo to predict the Z^0 line shape. How do the predicted width and peak cross section agree with data? New physics (including here the t quark) will change both of these quantities from the model predictions. Table III summarizes the changes in Γ_Z for various hypotheses. The combined measurements of Γ_Z and R_Z provided by the scan would have sensitivity to the presence of several of these alternatives. The exact nature of the anomalous width and cross section would not be revealed of course; we consider later how each one of those alternatives and others can be distinguished by looking at the event topologies.

TABLE III

Channel	Change in Γ_Z MeV
$\nu\bar{\nu}$	160
L^+L^-, $M_L = 30$ GeV/c^2	36
$q\bar{q}$, $M_q = 30$ GeV/C^2, $Q = -1/3$	211
$q\bar{q}$, $M_q = 40$ GeV/C^2, $Q = -1/3$	109
$q\bar{q}$, $M_q = 30$ GeV/C^2, $Q = 2/3$	144

Another way to measure Γ_Z is to recall from section 4 that for any $f\bar{f}$ final state:

$$\sigma_{f\bar{f}} = \frac{DG_F^2 M_Z^4}{6\pi\,\Gamma_Z^2}\left(v_e^2 + a_e^2\right)\left(v_f^2 + a_f^2\right) \quad .$$

Hence for each $f\bar{f}$ state one gets

$$\frac{d\Gamma_Z}{\Gamma_Z} = \frac{1}{2}\frac{d\sigma_{f\bar{f}}}{\sigma_{f\bar{f}}} \quad .$$

Given an accurate measurement of M_Z, the major experimental uncertainty will come from the luminosity measurement which can safely be done to $< 5\%$. This would yield an error $\delta\Gamma_Z < 70$ MeV per $f\bar{f}$ species. With a sample of 20,000 Z^0's the statistical error of both $\mu^+\mu^-$ and e^+e^- would be significantly less than that coming from the luminosity and, by adding these two channels, a 50 MeV measurement of the width would be possible. Referring again to table III, this would be a powerful indicator of new physics!

In summary then, with relatively little data, one can measure Γ_Z in two independent ways each one of which will yield an error $\lesssim 100$ MeV, these errors each being significantly smaller than the contribution of one ν species.

6.2 CHARGED LEPTON FINAL STATES

We now turn our attention to what can be learnt from specific topologies. Much can be learnt from studying the charged lepton final states. To do this we must isolate events of the type $Z^0 \to e^+e^-$, $\mu^+\mu^-$ and $\tau^+\tau^-$. This is a rather simple experimental task and is routinely done at PEP and PETRA. The experimental problems are even easier at the Z^0. For the e^+e^- and $\mu^+\mu^-$ final states one requires two opposite sign, charged particles which are "back-to-back" and carry the full beam energy. Electrons are trivially distinguished from muons using a rudimentary electromagnetic shower counter. To measure that the tracks have opposite sign requires only a modest momentum precision of $\sigma_p/p^2 \approx 1\%$ – all the LEP and SLC detectors will do far better than this. These channels have high rates ($B(Z^0 \to \ell^+\ell^-) = 3\%$) and there are no background problems. To identify the $\tau^+\tau^-$ final state one will probably require a topology in which the one τ decays to a single charged prong ($B(\tau \to 1$ charged prong $= 84\%)$) and the other τ decays to three charged prongs and any number of neutrals ($B(\tau \to 3$ charged prongs $= 16\%)$). This gives a very clean $\tau^+\tau^-$ sample at a rate of 3% $\times (2 \times 0.84 \times 0.16) \approx 1\%$. Hence the $\tau^+\tau^-$ final state will contribute information with a statistical weight of $\approx \sqrt{3}$ less than $\mu^+\mu^-$ or e^+e^- .

Consider now our canonical 10^6 produced Z^0's which will provide 30,000 $\mu^+\mu^-$ events. The asymmetry measurement will suffer a statistical error of $(\sqrt{30,000})^{-1} \simeq .005$. Hence (see equations (8) and (9))

$$\frac{\delta A}{A} = \frac{.005}{.043} = .12$$

and

$$\frac{\delta(\sin^2\theta)}{\sin^2\theta} = \frac{1}{15}\,(.12) = .008 \quad .$$

Assuming $\sin^2\theta_W = 0.22$, $\delta(\sin^2\theta_W) = .0017$! (Notice this is about an order of magnitude better than measurements from ν interactions or the polarized e^-d experiment.) The measurement error for the coupling constants is obtained after laborious propagation of errors which we omit here but are found in reference 11b page 28:

$$\delta(v/a) \simeq 0.008 \quad \text{for} \quad e^+e^-, \mu^+\mu^- \quad .$$

The measurements for the $\tau^+\tau^-$ channel will be less precise by about $\sqrt{3}$ as discussed above.

Notice now that we have a measurement of M_Z ($\sin^2\theta_W$) and at least two more measurements of $\sin^2\theta_W$. Hence we are able to check the validity of the weak radiative correction calculations which, until now, we have assumed to be correct. Their effect was to modify $\sin^2\theta_W$ by 7%.

108

The measurement of R^{ee}, $R^{\mu\mu}$ and $R^{\tau\tau}$ amounts to counting the number
of events in each category, making a correction for inefficiencies and normaliz-
ing to the luminosity. Typically these measurements can be done to $\approx 3\%$, the
main limitation arising from the normalization. As we mentioned earlier, the
final ingredient needed to measure the couplings is the measurement of the τ
polarization. This is best done using the decay $\tau \to \pi\nu$, although the leptonic
decays $\tau \to e\nu\nu$ and $\tau \to \mu\nu\nu$ are also useful. With modest particle identification the different τ modes can be identified. The measurable sensitive to the τ
polarization is the $\pi^{\pm}$ or $\ell^{\pm}$ momentum spectrum. For a τ of polarization P_τ the
fractional momentum of the π in the decay $\tau \to \pi\nu$ is given by[16]

$$\frac{dN_\pi}{dx_\pi} = 1 + P_\tau(2x_\pi - 1)$$

where $x_\pi = 2E_\pi/E_{c.m.}$. The average value of x_π is

$$\langle x_\pi \rangle = (3 + P_\tau)/6$$

and hence a measurement of $\langle x_\pi \rangle$ yields P_τ. Likewise[16] for $\tau \to \ell\nu\bar{\nu}$.

$$\frac{dN_\ell}{dx_\ell} = \frac{1}{3}\left[5 - 9x_\ell^2 + 4x_\ell^3 + P_\tau(1 - 9x_\ell^2 + 8_\ell^3)\right]$$

and

$$\langle x_\ell \rangle = (7 - P_\tau)/20 \ .$$

So for the $\tau \to \pi\nu$ measurement we can select two prong and four prong events
as shown in figure 13. One must now ensure that the single π's are indeed
π's. This involves making sure that the track is neither a muon nor an electron. The separation of pions from muons and electrons in such a low multiplicity environment is easy particularly for momenta above 1 GeV/c. All the
LEP and SLC detectors will be able
to make a good separation. In addition making a good determination of
$\langle x \rangle$ requires a momentum precision
of $d\sigma_p/p^2 \lesssim 0.5\%$. The experimental details of the measurement are
discussed in great detail in reference
11b page 103 and we will borrow
liberally from that discussion. We
should remind ourselves that $P_\tau =$
$-2a_\tau v_\tau/(a_\tau^2 + v_\tau^2) = f(\sin^2\theta_W)$ so

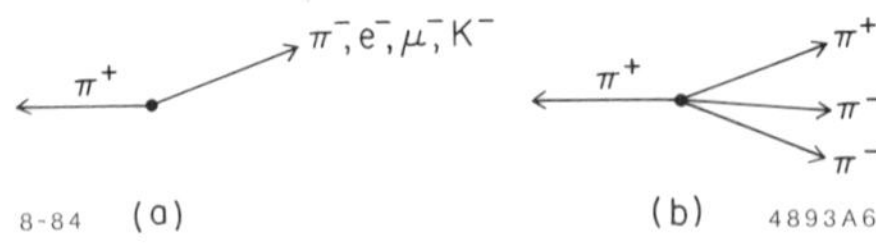

Fig. 13. Event topologies which could
be used to study the decay $\tau^{\pm} \to \pi^{\pm}\nu_\tau$
via the production process $Z^0 \to \tau^+\tau^-$.
In (a) the τ^- decays to a single charged
prong, whereas in (b) the τ^- is envisaged
as decaying to three charged pions.

that a measurement of P_τ is also a measurement of $\sin^2\theta_W$. In particular if $\sin^2\theta_W = 1/4$, $P_\tau \equiv 0$. Figure 14 shows the predicted dN/dx spectra for different values of $\sin^2\theta_W$. The simulation discussed in reference 11b used 10^6 Z^0's and a detector with parameters similar to the typical SLC/LEP detector. Figure 15 shows the simulated experimental momentum spectra for the decay pion and lepton where $\sin^2\theta_W$ has been set to 0.23. From these spectra the average values obtained are

$$\langle x_\pi \rangle = 0.483 \pm 0.006$$

$$\langle x_\ell \rangle = 0.359 \pm 0.003 \ .$$

The solid lines on the figure correspond to the theoretical curves from figure 14 and demonstrate how the finite momentum resolution ($\sigma_p/p^2 = 0.5\%$ for this simulation) distorts the high x end of the spectrum.

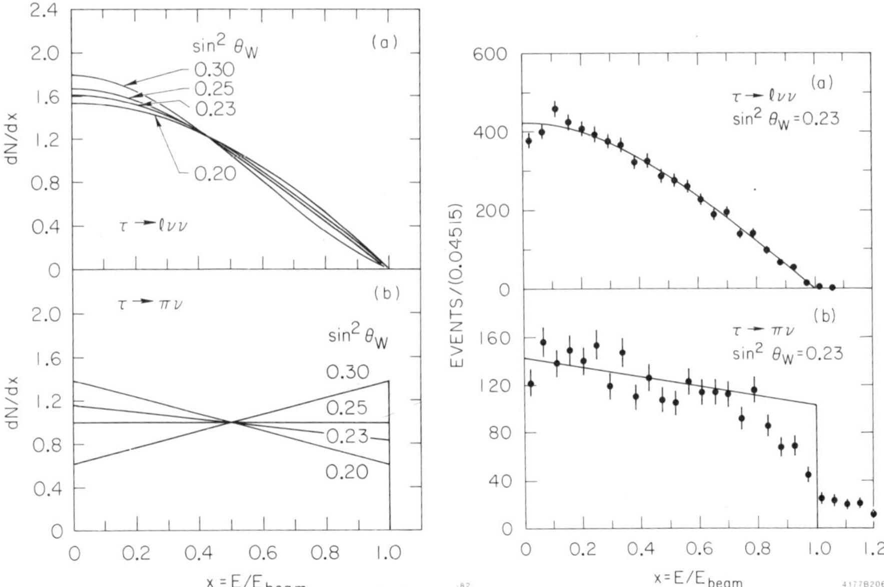

Fig. 14. Momentum spectra for τ decay products for different values of $\sin^2\theta_W$ (a) for charged leptons arising from $\tau \rightarrow \ell\nu\bar\nu$ and (b) for pions arising from $\tau \rightarrow \pi\nu$.

Fig. 15. The solid lines are taken from figure 14 with $\sin^2\theta_W = 0.23$. The data points result from a computer simulation which includes realistic detector components.

For the decay channel $\tau \to \pi\nu$, $P_\tau = 6\langle x \rangle - 3$ and therefore from this simulation

$$P_\tau \approx 0.11$$

$$\delta P_\tau = 6\delta\langle x \rangle = 0.036 \ .$$

Also from the previous discussion we have

$$A_{F-B}^\tau \approx 0.043 \pm 0.008 \ .$$

We obtain the relative sign of a_e and v_e from the ratio

$$\frac{A_{F-B}^\tau}{P_\tau} = -\frac{3a_e v_e}{(a_e^2 + v_e^2)} \ .$$

Clearly our ability to tell the relative sign is limited by the P_τ measurement and hence for this toy experiment one would determine the relative sign of v_e and a_e to $\gtrsim 3\sigma$ from the decay $\tau \to \pi\nu$. Additional statistical power would come from the decay modes $\tau \to e\nu\bar{\nu}$ and $\tau \to \mu\nu\bar{\nu}$.

From $R^{e^+e^-}$ and $A_{F-B}^{e^+e^-}$ one obtains a_e and v_e, their relative sign, coming from the additional measurement of P_τ. Measuring $R^{\mu^+\mu^-}$, $R^{\tau^+\tau^-}$, $A_{F-B}^{\mu^+\mu^-}$ and $A_{F-B}^{\tau^+\tau^-}$ will then provide the μ^- and τ^- vector and axial vector couplings. Having measured all the couplings provides further checks on the Standard Model and in particular of the universality of the Weak Interactions.

Now life becomes much easier if one has a longitudinally polarized electron (or positron) beam. As we discussed in section 2, the SLC is expected to have a longitudinally polarized e^- beam with polarization $P_{e^-} \gtrsim 50\%$. In addition, on a pulse by pulse basis, the sign of the polarization can be switched from left to right. Now one can do a very simple experiment namely to measure the total cross section for left polarized electrons (σ_L) and that for right polarized electrons (σ_R). These cross sections will not be equal and we can form an asymmetry

$$A_{L-R} = \frac{\sigma_L - \sigma_R}{\sigma_L + \sigma_R} = -2P_{e^-} \frac{a_e v_e}{(a_e^2 + v_e^2)} \ .$$

Recognize that A_{L-R} immediately gives the relative sign of v_e and a_e thus obviating the need for the τ polarization measurement. In addition this is a very simple experiment to perform and, unlike the measurement of A_{F-B}, all the Z^0 decay (except $Z^0 \to \nu\bar{\nu}$ of course) events can be used and statistics are no problem at all. For $P_{e^-} = 0.5$ and $\sin^2\theta_W = 0.22$, $A_{L-R} = 12\%$ $-$ three times larger than A_{F-B}^μ. So it will be much easier to measure A_{L-R} on the Z^0 peak than A_{F-B}^f.

The error in A_{L-R} is dominated by the measurement error in P_{e^-}. How do we measure P_{e^-}? Two polarimeters will be built[17] for the SLC one of which is designed to achieve a 1% measurement of P_{e^-}. If 1% were actually achieved, one would find from

$$\frac{\delta A_{L-R}}{A_{L-R}} = \frac{\delta P_{e^-}}{\delta P_{e^-}} \simeq \frac{7.3\delta(\sin^2\theta_W)}{\sin^2\theta_W}$$

that $\delta(\sin^2\theta_W) = 0.0003$. If we consider a more conservative starting error of $\delta P_{e^-}/P_{e^-} \sim 5\%$, we find $\delta(\sin^2\theta_W) \approx 0.0015$. Notice that to get similar precision for A_{F-B} would take 10 times more running! One begins to see why polarized e^- beam at the SLC is a powerful tool. Further examples will follow in the later sections.

6.3 SEARCHING FOR THE TOP QUARK

How about searching for the particles which should be present in the minimal Standard Model but have not been observed – namely the top quark and the neutral Higgs boson. We discussed already that we might have an indication from the width measurement that top was being produced at the Z^0. To make sure that we have a new, heavy quark being produced at the Z^0 is quite simple, requiring rather little data, as long as the quark mass is $\lesssim 40$ GeV/c^2. Establishing the quark charge (2/3 or $-1/3$) is more of a challenge but, with enough data, this too seems possible.

As we saw in figure 6, the production of $t\bar{t}$ falls with increasing mass. However the experimental problems of isolating $t\bar{t}$ events become easier with increasing mass. Hence, as we shall see, one has roughly equal sensitivity to finding top in the mass range from $M_t \doteq 25$ GeV/c^2 to 40 GeV/c^2. There are too straightforward ways to isolate $t\bar{t}$ events from hadronic events produced via the 5 light quark species. One way is to look at the shape of the events and the other is to measure the transverse momentum (P_t) (measured relative to the quark axis) of leptons produced in the semi-leptonic decay of the t quark. Both methods work because the top quark is so much heavier than the 5 known quarks, which in turn have masses $\ll M_{Z^0}$.

In the former approach suppose we run at the Z^0, collect hadronic events (72% of Z^0 decays) and do a sphericity shape analysis[18] on the events. Hadronic events are very simple to isolate because of their high multiplicity and large detected energy. They will be isolated with high efficiency and no background. The sphericity analysis will provide three orthogonal axes, two of which define a plane – the event plane – which is the plane which contains most of the momentum of the detected particles. The aplanarity is a measure of the momentum out of the event plane. Because transverse momentum (P_t) is limited in the fragmentation process and because the 5 known quarks will all have high velocities in Z^0 decay, the events containing the 5 known quarks will have small aplanarity. However if a heavy quark is produced it will have a low velocity, and the same limited P_t will result in a considerably larger aplanarity.

The second approach uses the fact that the t quark will have copious ($\sim 10\%$) semileptonic decays. Again because of their heavy mass and low velocity, the leptons arising from such decays make a large angle with respect to the quark (jet) direction. This can be contrasted with leptons arising from the 5 known quarks. Rather than measuring the decay angle, we choose to use the transverse momentum relative to the quark direction (P_t). In an experiment the sphericity (or thrust) axis is a good measure of the $q\bar{q}$ direction and momenta are usually measured relative to this axis. The muons (or electrons) coming from the $t\bar{t}$ events will have substantially larger P_t than the corresponding leptons from the lighter quarks.

The MARK II Group[19] has studied these two methods for isolating top at the SLC. The simulations were done using the LUND Monte Carlo program. Figure 16 shows the aplanarity distribution for the 5 light quarks arising from 10^4 Z^0 decays and the additional contribution which would result from the production of $t\bar{t}$ with $M_t = 40$ GeV/c^2. One sees a clear $t\bar{t}$ signal for aplanarities > 0.12. Similarly figure 17 contrasts the lepton P_t relative to the thrust axis for leptons arising from the light quarks and the $t\bar{t}$ events ($M_t = 40$ GeV/c^2). For this plot an aplanarity cut of > 0.02 has been applied. A clear $t\bar{t}$ signal is seen. The results of this study are shown in table IV from which we can see that by combining both aplanarity and lepton P_t information, 10^3 Z^0's should be enough to indicate fairly clearly that a new heavy quark ($M \lesssim 40$ GeV/c^2) was being produced at the SLC. With 10^4 events there would be little doubt. 10^3 Z^0's is 4 days of running at $\mathcal{L} = 10^{29}$ cm^{-2} sec^{-1}.

TABLE IV

	# Events Produced	Detected # Events with Aplanarity > 0.12	# e, μ with $P_t > 3$ GeV/c and Aplanarity > 0.02
Background; LUND $udscb$	7200	9	17
$t\bar{t}$; $M_t = 30$ GeV/c^2	634	48	146
$t\bar{t}$; $M_t = 35$ GeV/c^2	452	64	131
$t\bar{t}$; $M_t = 40$ GeV/c^2	265	58	88
$t\bar{t}$; $M_t = 45$ GeV/c^2	83	13	27

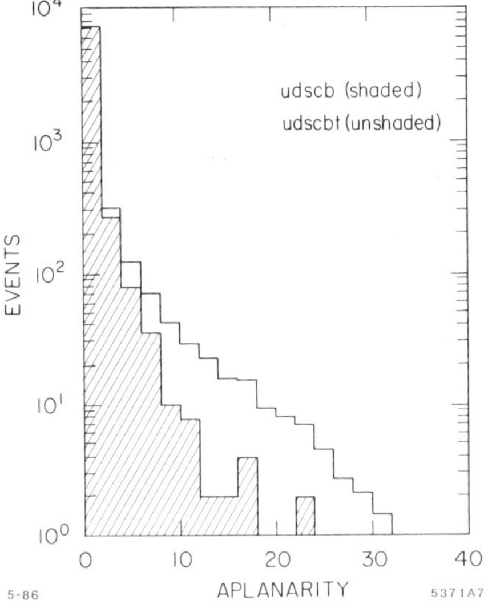

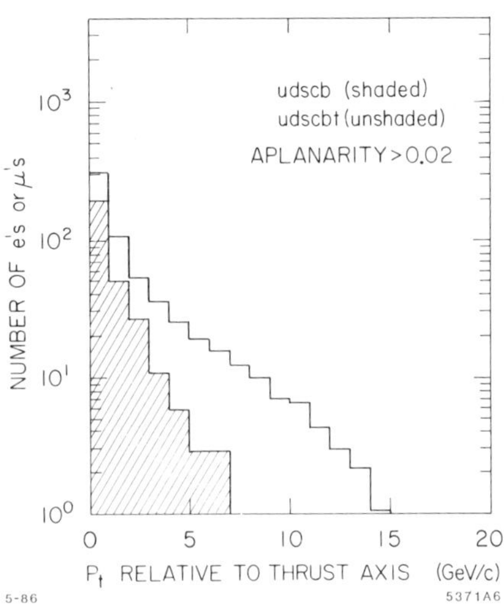

Fig. 16. The aplanarity distribution for hadronic events at the Z^0 contrasting the contribution from the five light quarks (shaded) with that coming from $Z^0 \rightarrow t\bar{t}$ where $M_t = 40$ GeV/c^2.

Fig. 17. The P_t spectrum for leptons arising from Z^0 decays to the five light quarks (shaded) contrasted with leptons arising from $Z^0 \rightarrow t\bar{t}$ with $M_t = 40$ GeV/c^2.

For studying $t\bar{t}$ events, we note that the high P_t tag will provide $\sim 10^4$ $t\bar{t}/10^6$ Z^0's, with the additional attractive feature that the sign of the lepton's electric charge distinguishes the q vs $\bar{q}$ source for the lepton. This is crucial for studying the charge asymmetry A^t_{F-B}.

How about measuring the t quark mass? A measurement with an accuracy of a few GeV will result from fitting the shapes of the aplanarity and P_t spectra. This method however suffers from the fact that it relies heavily on the input to the Monte Carlo in particular to how one models the quark fragmentation process and the higher order QCD effects. Measuring the jet mass doesn't work well because with the "fat" tt events, assignment of particles to the t and $\bar{t}$ is ambiguous and the $t\bar{t}$ production direction is not well specified, especially for large t masses. In principle the best method of determining the t quark mass is to return to figure 6, the t quark threshold curve. If one could find out where one was on the curve, one could "read off" M_t. This method is discussed fully in reference 20, briefly here.

114

Recall $\rho = \Gamma_t(\beta)/\Gamma_u$ as plotted in figure 6. Let $r = N_h/N_{\mu\mu}$ where N_h is the number of hadronic events and $N_{\mu\mu}$ is the number of $\mu^+\mu^-$ events. Then

$$\rho = 1 + r\,\frac{\Gamma_{\mu\mu}}{\Gamma_u} - \frac{\Gamma_h}{\Gamma_u}$$

where Γ_h = the hadronic width calculated in the Standard Model for <u>3</u> quark generations of <u>massless</u> colored weak isospin doublets. $\Gamma_{\mu\mu}$ and Γ_u are the partial widths for the $Z^0 \to \mu^+\mu^-$ and $Z^0 \to u\bar{u}$. The experiment is simple – measure r which provides ρ which from figure 6 provides M_t. The measurement of r does not involve a luminosity measurement and should be free of systematics at the 1-2% level. If we assume our canonical 10^6 Z^0, $N_h \approx$ 730,000, $N_{\mu^+\mu^-} \approx 31,000$ and if $\sin^2\theta_W$ is known to ≈ 0.001 then one obtains a t mass resolution shown in figure 18.

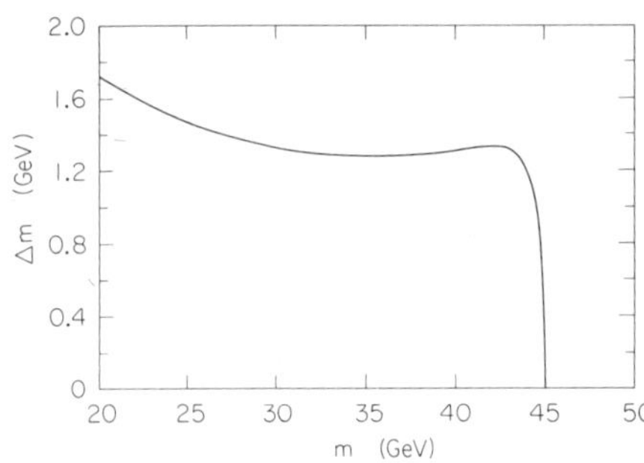

Fig. 18. The error in the determination of the top quark mass as a function of the top quark mass for a sample of 10^6 Z^0 events. No systematic errors are included in this plot.

Notice that in order to use this method one requires independent knowledge that $t\bar{t}$ events are being produced at the Z^0 and that there are no other processes which are contributing to N_h.

Suppose such a heavy quark is found. How does one know that it is indeed the $q = 2/3$ t quark as opposed to a 4th generation $q = -1/3$ b' quark? To distinguish these two possibilities requires measuring A^q_{F-B} which is 6.5% (13%) for $q = 2/3\ (-1/3)$. We will discuss this later in section 6.6.

6.4 Searching for the Neutral Higgs, H^0

At the Bonn Conference in 1981, Okun said[21] that in his mind the outstanding experimental challenge was the search for scalars. He urged experimentalists to "drop everything" and devise cunning searches for the elusive scalars. To date no search has proven successful and it is interesting to speculate how one could search for the H^0 running on the Z^0.

The H^0 will couple to the heaviest fermions available and this feature will be used in any search for the H^0. The decay rate for $H^0 \to f\bar{f}$ is given by:

$$\frac{d\Gamma}{d\Omega} = \frac{G_F M_{H^0} m_f^2}{16\pi^2 \sqrt{2}} \quad .$$

The decay rate depends on m_f^2 (m_f is the fermion mass) and is isotropic. So if $M_{H^0} < 2M_b$, the H^0 will decay mostly to $c\bar{c}$ and $\tau^+\tau^-$. If $2m_t < M_{H^0} < 2M_b$ then the H^0 will decay mostly to $b\bar{b}$. These conclusions are summarized in figure 19.

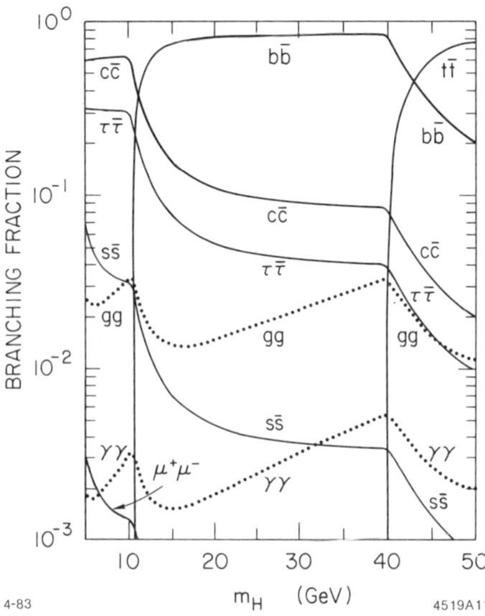

Fig. 19. Decay modes of the neutral Higgs boson as a function of its mass.

How can we search for the H^0? The process $e^+e^- \to Z^0 \to H^0 H^0$ is forbidden by spin-statistics. The process $Z^0 \to H^0\gamma$ vanishes in first order because the Z^0 and γ are "orthogonal" – in second order the rate is too small to be of any practical use. The most promising search channel seems to be $Z^0 \to H^0 Z^{0*} \to H^0\ell^+\ell^-$ (see figure 20) which was first discussed[22] by Bjorken and is also discussed in reference 23. The rate for this process is given by:

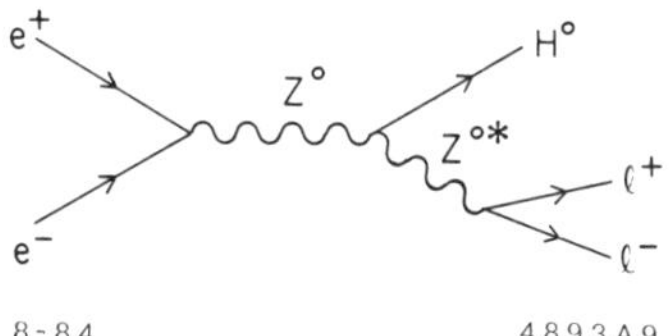

Fig. 20. The process $e^+e^- \to Z^0 \to H^0\ell^+\ell^-$.

$$\frac{1}{\Gamma(Z^0 \to \mu^+\mu^-)} \frac{d\Gamma(Z^0 \to H^0\ell^+\ell^-)}{dM_{\ell^+\ell^-}} = \frac{\alpha F}{4\pi \sin^2\theta_W \cos^2\theta_W}$$

where

$$F = \frac{10k^2 + 10\lambda^2 + 1 + (k^2 - \lambda^2)[(1 - k^2 - \lambda^2) - 4k^2\lambda^2]^{1/2}}{(1 - k^2)^2}$$

116

$$M_{\ell^+\ell^-} = \text{lepton pair mass}$$

$$k = M_L/M_{Z^0}$$

$$\text{and} \qquad \lambda = M_{H^0}/M_{Z^0} \quad .$$

This relative rate, integrated over $M_{\ell^+\ell^-}$, is plotted as a function of M_{H^0} in figure 21. Also shown for comparison is the rate for $Z^0 \to H^0\gamma$. $B(Z^0 \to \mu^+\mu^-) = 3\%$, so one sees that for $M_{H^0} \approx 20$ GeV/c^2 $B(Z^0 \to H^0\ell^+\ell^-) \approx 3 \times 10^{-5}$, a yield of 30 events for 10^6 Z^0 events. Unfortunately the rate drops off very rapidly with increasing H^0 mass and for masses above ~ 40 GeV/c^2 the measurement becomes severely rate limited.

The $H^0\ell^+\ell^-$ signal must be sought in the presence of an enormous background from $Z^0 \to$ hadrons. For $M_{H^0} \approx 20$ GeV/c^2 there are $\approx 10^4$ $Z^0 \to$ hadron events per $Z^0 \to H^0\ell^+\ell^-$ event! Luckily the event topology is very favorable and a measurement indeed seems possible. Many of the detector groups at SLC and LEP have studied the experimental problems and their conclusions are pretty uniform. We chose here the study discussed in the MARK II proposal.[12]

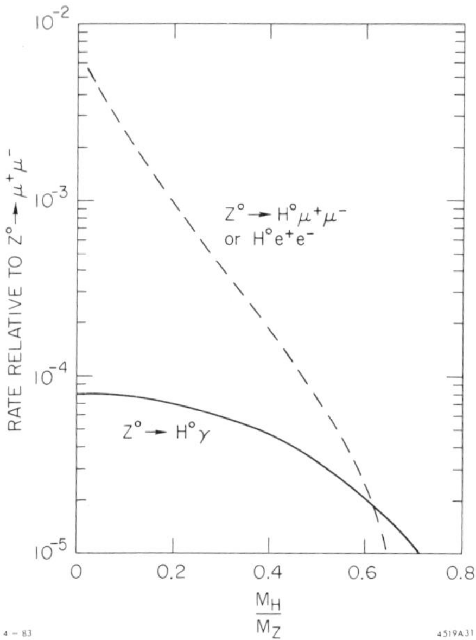

Fig. 21. The decay rate for $Z^0 \to H^0 e^+ e^-$ or $Z^0 \to H^0\mu^+\mu^-$ relative to $Z^0 \to \mu^+\mu^-$ which has a branching fraction of 3%.

The favorable topology arises from the fact that most of the energy in the process $Z^0 \to H^0\ell^+\ell^-$ goes to the virtual Z^0 and hence the two leptons which result from the decay of the virtual Z^0 have very high invariant mass and momenta. The H^0 is produced with a fairly small fraction of the available energy and will decay mostly into two quark jets. In addition there is very little correlation between the H^0 direction and the e^+ or e^- direction and in most events the $e^\pm$ will be well separated from the H^0 decay products. The topology is schematically shown in figure 22.

The main source of background comes from the process $Z^0 \rightarrow t\bar{t}$ where both the t and $\bar{t}$ decay semi-leptonically. However requiring the angle between sphericity axis of the hadronic system (all particles except the ℓ^+ and ℓ^-) and the leptons to be $\gtrsim$ 200 mrad virtually eliminates this background for $M_{H^0} \lesssim$ 40 GeV/c^2. This cut loses very little signal ($\approx$ 6%) because there is virtually no correlation between the direction of the leptons and the hadronic sphericity axis.

The mass of the hadronic system (the H^0) is obtained from the missing mass recoiling against the lepton pair. The experiment can be done with either a e^+e^- or $\mu^+\mu^-$ lepton pair providing that the energy resolution of the leptons is sufficiently good to see a peak in the missing mass. The missing mass recoiling against the e^+e^- ($\mu^+\mu^-$) pair is shown in figure 23 (24) for the MARK II simulation for Higgs masses of 10, 25 and 35 GeV/c^2. Clear signals are seen. The main issue for this measurement will be statistics. For Higgs masses of 10, 20, 30 and 40 GeV/c^2 one expects to have 180, 50, 25 and 10 events produced in the $H^0e^+e^- + H^0\mu^+\mu^-$ channels per 10^6 Z^0 events.

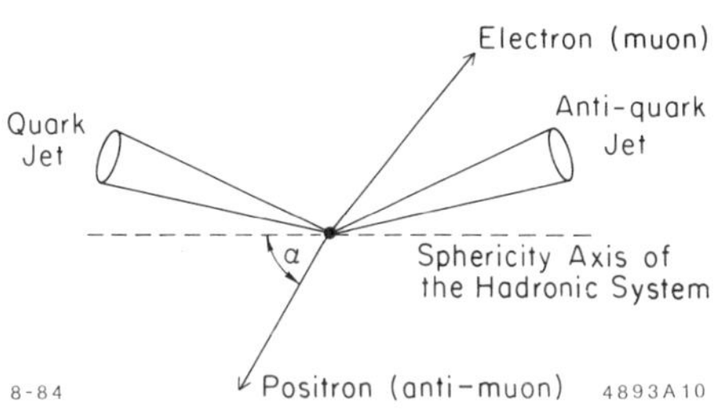

Fig. 22. A schematic representation of the topology of the $Z^0 \rightarrow H^0\ell^+\ell^-$ events.

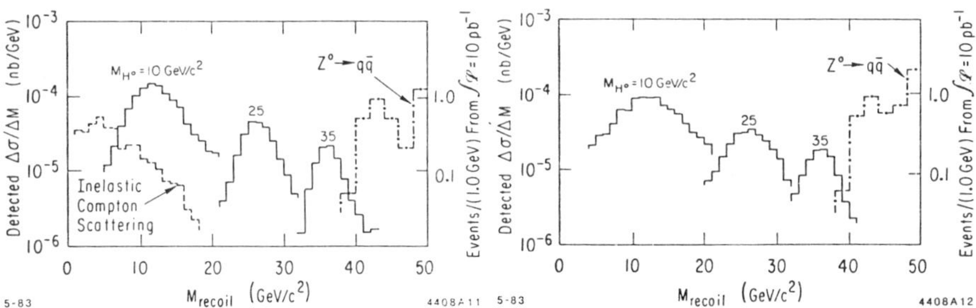

Fig. 23. The Higgs signal from $Z^0 \rightarrow H^0e^+e^-$. The expected backgrounds are also shown.

Fig. 24. The Higgs signal from $Z^0 \rightarrow H^0\mu^+\mu^-$. The expected backgrounds are also shown.

Assuming the search was successful and we found a peak in the recoil mass spectrum how do we know that we have discovered the Higgs scalar? We would have to verify that it decayed isotropically and that the couplings favored the heaviest fermion pair available.

We can measure the decay angular distribution as follows. First we would reconstruct the two jet directions from the particles associated with the jets. From the ℓ^+ and ℓ^- momenta we can reconstruct $\vec{P}_{H^0}$. Knowing M_{H^0} and $\vec{P}_{H^0}$, we can transform the jet directions into the H^0 center of mass and plot the decay angular distribution. (This method will work as long as we can make the assumption that the decay angular distribution is symmetric about $\theta^* = 90°$. This is because we don't know how to distinguish the jet from the anti jet (θ^* from $\pi - \theta^*$) and hence by plotting both we are assuming a symmetric decay distribution). Realistically the major problem with this procedure will be the limited statistics. Optimistically one might have ≈ 50 events to play with.

Now how about measuring if the coupling is proportional to m_f^2? Here the procedure would depend on M_{H^0}. Suppose, as is likely, that $M_{H^0} > 10$ GeV/c^2 in which case $H^0 \to b\bar{b}$ almost exclusively (see figure 19). We will see in the next section that using a vertex detector one can expect to tag events containing two b jets with an efficiency $\gtrsim 50\%$ and this with very little contamination from c jets. This can be done because the b quark has a long measured ($\sim$ psec) lifetime. So one would subject the $H^0\ell^+\ell^-$ candidate events to this test and if indeed half ($=$ tag efficiency) the events were tagged as having a b jet, one would feel fairly confident that the H^0 decayed predominantly to $b\bar{b}$. If $M_{H^0} < 10$ GeV/c^2 the obvious signal to look for would be $H^0 \to \tau^+\tau^-$.

To summarize the H^0 search then, it is probable that if $M_{H^0} \lesssim 40$ GeV/c^2 it can be found at the Z^0 . We will require a machine with excellent luminosity – $\langle \mathcal{L} \rangle > 10^{30}$ cm^{-2} sec^{-1} – and a detector with good electromagnetic calorimetry and/or momentum resolution. All the LEP and SLC detectors appear capable of doing this measurement. With sufficient statistics ($\gtrsim 50$ events) the H^0 decay angular distribution and coupling can probably be inferred.

6.5   What Will We Learn From $Z^0 \to$ Hadrons?

An obvious question is can we learn anything form $Z^0 \to$ hadrons which cannot be obtained from PETRA ($E_{c.m.} \lesssim 46$ GeV) and PEP ($E_{c.m.} \lesssim 36$ GeV)? The answer is yes and probably the main reason is that the Z^0 offers a very large statistical advantage over the PEP and PETRA machines. At present the largest PEP/PETRA hadronic dataset is the MARK II which has 100,000 hadronic events at a PEP energy of $E_{c.m.} = 29$ GeV. It has taken three years to accumulate this data and the present performance of PEP is that a good PEP year is worth 60,000 hadronic events. Contrast this with the expectation that a good SLC/LEP year will yield $\approx 180 \times 10^4$ hadronic events or 30 times as much as PEP. So there will be a considerable improvement in statistics. We now examine some of the physics which will be covered.

<u>QCD Tests</u>

As discussed by the authors in reference 24 (and probably many others) the QCD corrections to the Z^0 hadronic final states are exactly those calculated for lower energy e^+e^- interactions. In particular one recovers the familiar Sterman-Weinberg formula. All the usual low energy tools like sphericity, thrust, etc. are equally useful at the Z^0. The familiar 3 jet Dalitz plot distributions for $e^+e^- \rightarrow Z^0 \rightarrow qqg$ are the same as for the continuum:

$$\frac{d^2\Gamma(Z^0 \rightarrow 3 \text{ jets})}{dx_1 dx_2} = \Gamma(Z^0 \rightarrow \text{hadrons}) \frac{2\alpha_s(M_Z)}{3\pi} \frac{(x_1^2 + x_2^2)}{(1 - x_1)(1 - x_2)}$$

where x_i ($i = 1, 2, 3$) are the fractional parton energies ($x_i = 2E_i/E_{c.m.}$) and $\sum_i x_i = 2$. We can study the three jet events at the Z^0 in much the same way as we study them at PETRA and PEP. These studies will probably be easier at the Z^0 because the jet cone angles will be $\sim$ 2-3 times smaller ($\approx 1/E_{\text{jet}}$) than at PEP or PETRA. Hence the problems of which particles belong to which jet should be easier. This will provide more reliable measurements of x_i, quark and gluon jet multiplicities and jet directions. In addition the efficiency for finding well reconstructed 3 jet events should be higher than at lower energies. And of course there will be a copious supply of 3 jet events. Simulations have shown that about 60% of the produced three jet events are cleanly reconstructed which would yield about 5×10^4 reconstructed 3 jet events/10^6 produced Z^0's. By contrast the MARK II has about 5×10^3 reconstructed 3 jet events and many of the PETRA results at 34 GeV have been published on $\lesssim$ 1000 3 jet events. Back to what we will learn.

We will try to measure α_s, a task which has been difficult at lower e^+e^- energies.[25] Part of the problem with the lower energy measurements has been understanding the QCD corrections and removing the model dependence. Combining the new data at $E_{c.m.} = M_{Z^0}$ with the low energy data will allow one to measure some of these effects which are now parametrized in a variety of models. α_s would be measured using the same techniques as at lower energies (see reference 25 as an example) namely studying the Dalitz plot distributions, or measuring the ratio of 3 jet to 2 jet events, or measuring particle energy correlations, event shapes, etc. I expect the model dependent problems encountered at lower energies will be much improved at the Z^0. However new model dependent effects may prove troublesome, an example of which is the appearance, at higher energies, of many soft gluons. I would not speculate with confidence that α_s will be more easily measured at the Z^0, but in all probability things will be better. With many reconstructed jets at energies hitherto not available in e^+e^-, more information will be gained on the parton fragmentation process. In particular the question of whether quarks and gluons fragment differently can be studied.

It has been argued in many places (see reference 26 for but a few) that for highly perturbative parton regimes (high energy partons) gluon jets should be considerably broader than quark jets. This is an important test because it arises from the gluon self-coupling which relates directly to the non-abelian nature of QCD. The difference in the fragmentation of quarks and gluons comes about from the fact (see figure 25) that the color charge at the triple gluon vertex is 9/4ths larger than at the quark-quark-gluon vertex. The ratio of the cone angle δ (a la Sterman and Weinberg) of a gluon and a quark jet is given roughly by

$$\delta_g(E) \approx \delta_q(E)^{4/9}$$

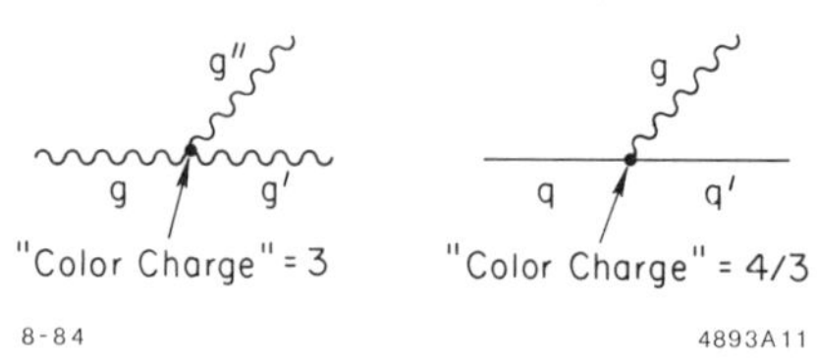

Fig. 25. The contrasting "strengths" of the triple gluon vertex and the quark-quark-gluon vertex.

where δ is measured in radians. The cone angle δ is such that most ($\gtrsim 90\%$) of the parton energy is contained in the cone. At the Z^0 one expects $\delta_q \simeq 10°$ which would imply a gluon jet of the same energy would have $\delta_g \approx 27°$. Such large differences should be seen easily and the Z^0 3 jet events should provide a meaningful test of differences in quark and gluon jets.

Flavor Tagging

We have already seen earlier in this section that $t\bar{t}$ events can easily be tagged using event shape parameters or high P_t leptons. The advantage of the high P_t tag is that the sign of the lepton charge flags which jet is t and which is the $\bar{t}$. The importance of this will become apparent soon. Studies by the MARK II Upgrade Group[19] have shown that the high P_t lepton tag has high efficiency for selecting $t\bar{t}$ events and in addition backgrounds (from $b\bar{b}$ mainly) are small. Requiring a high P_t lepton they find $\approx 10^4$ tagged $t\bar{t}/10^6$ Z^0's with a background of $< 10\%$.

How about tagging $b\bar{b}$ events? The B meson appears[27] to have a lifetime on the order of 1 psec. At the Z^0, they will travel $\gamma\beta c\tau \approx 3$ mm on average before they decay. The decay particles of the B meson, when extrapolated back towards the primary vertex, will appear to "miss" the primary vertex (see figure 26).

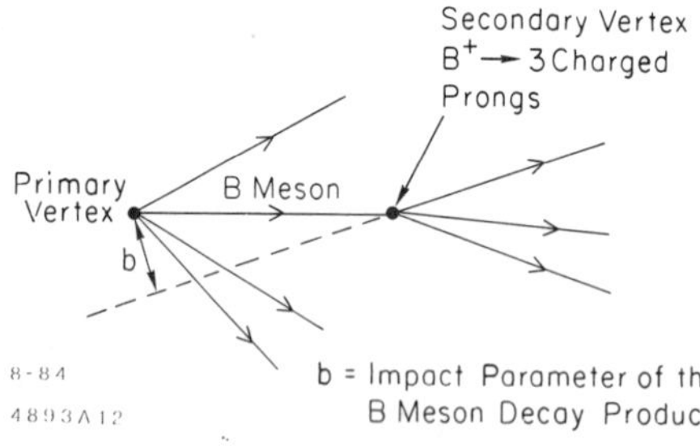

Fig. 26. The production and subsequent decay of a B meson indicating the primary vertex, secondary vertex and the impact parameter b of one of the B decay tracks.

The amount by which they "miss" is called the impact parameter, b. Large impact parameter tracks will signal the decay of a long lived particle. From simulations one finds that for $\tau_B = 10^{-12}$ secs typical tracks from B meson decay in $Z^0 \to b\bar{b}$ events have $b \gtrsim 200\ \mu$. This can be contrasted with expected measurement errors of 50-100 μ. In a study done by the MARK II Upgrade Group,[28] efficiencies of $\geq 50\%$ were found for tagging events of the type $Z^0 \to b\bar{b}$. The technique used was to require at least 3 tracks in a jet with $\geq 3\sigma_b$ where σ_b was the error in the measurement of the track's impact parameter. Multiple scattering in the apparatus walls can cause tracks to have large impact parameters and hence provide bogus tagging information. Requiring three tracks with a substantial impact parameter alleviates this problem. In addition the invariant mass of the three large impact parameter tracks was required to be > 1.95 GeV/c^2 which eliminates almost all background from D decays. The tagged $b\bar{b}$ events sample was found to have $< 10\%$ background from non $b\bar{b}$ events.

Using this efficiency as prototypical, one would expect 6.8×10^4 tagged $b\bar{b}$ events/$10^6\ Z^0$ events. If in addition one required an electron or a muon to distinguish quark and antiquark b jets, one would have a tagging efficiency of about $8 \times 10^3\ b\bar{b}/10^6\ Z^0$ events.

So it seems as if one will be able to tag b and t jets at the SLC and LEP with impressive event yields. What physics can be done? Clearly the fragmentation process, both longitudinal and transverse, for heavy quarks can be studied. Jet multiplicity can be studied. Comparisons with low energy data will provide additional information on the fragmentation process.

The B lifetime will be measured with better precision and better statistics than at PEP. Current τ_B measurements rely on $\lesssim 1000$ events which affects not only the statistical error in τ_B but also limits the ability of the experiments to understand their systematics. Presently the systematics are limiting the measurements at the $\sim 25\%$ level. We can use the tagged $b\bar{b}$ events to measure the B meson lifetime and divide the events into two jets. The one jet will provide the b jet tag as discussed above. The other jet can be used in an unbiased way to measure the B lifetime. Estimates from simulations done by the MARK II Upgrade Group indicate that using the same method employed at PEP[27] a systematic error of $\sim 5\%$ should be achieved for τ_B.

With tagged $b\bar{b}$ and $t\bar{t}$ events we could measure the charged 2/3rds and $-1/3$rd quark couplings. Recall (section 4) that if we measure

$$R_q = \frac{\sigma_{q\bar{q}}}{\sigma_{\text{point}}} \propto \left(a_q^2 + v_q^2\right)$$

and

$$A_{F-B}^q = \frac{3a_e v_e a_q v_q}{\left(a_q^2 + v_q^2\right)\left(a_e^2 + v_e^2\right)}$$

we can obtain a_q and v_q. (We are assuming a_e, v_e are measured as discussed earlier in this section.) With a tagged sample of $b\bar{b}$ and $t\bar{t}$ we can make these measurements. For the forward–backward asymmetry we need to distinguish q from $\bar{q}$, so we will have to use events with an electron or muon. Even with this restriction the statistical errors in the measurement of the couplings will be $\lesssim 2\%$ for 10^6 Z^0 events. The R_q measurement requires an accurate measurement of luminosity which will be possible at the $\leq 5\%$ level. In order to determine the quark direction one will use the thrust axis. At the Z^0 this will be well determined (except for $t\bar{t}$ as $M_t \to 40$ GeV/c^2) and should not effect the quality of the measurement of A^q_{F-B}. It will be very important to have good detector coverage at small θ angles. The solid angle (25% of 4π) for which $\theta < 40°$ contains as much asymmetry information as the remaining 75% of 4π. Based on those considerations, I would expect one could measure the b couplings to $\lesssim 10\%$. A^t_{F-B} is a little bit trickier, especially as $M_t \to 40$ GeV/c^2. In this case it becomes harder to define the thrust axis and the correlation between the sign of the lepton charge and the parent (t or $\bar{t}$) becomes weaker. Nonetheless in the MARK II simulation[19] of A^t_{F-B} for $M_t = 40$ GeV/c^2, a 4σ effect was seen for an equivalent sample of 3×10^5 produced Z^0's. These measurements of A^b_{F-B} and A^t_{F-B} will constitute important tests of the Standard Model in the quark sector.

We return now to the question posed earlier. Suppose we discover a new heavy quark at the SLC or LEP. How do we know its charge? Is it the t or a b'? The key is the difference in the couplings. Simple substitution using the values in table I gives

$$A^t_{F-B} = 6.5\%$$

and

$$A^b_{F-B} = 13\% \quad .$$

The simulation discussed above provides a $\sim 4\sigma$ differentiation for 3×10^5 produced $\dot{Z}^0$'s.

The tagged $b\bar{b}$ sample will be an excellent place to search for mixing in the neutral B meson system. The method would be to search for hadronic events which contain leptons of the same sign in opposite hemispheres. This would result from the production of $B^0\bar{B}^0$ where, through mixing, the B^0 ($\bar{B}^0$) evolves to a $\bar{B}^0$ (B^0). Both B mesons are then required to decay semi-leptonically.

Very little mixing is expected[29] to occur in the B_u mesons. However maximal (100%) mixing could be possible in the B_s mesons. We can make a crude estimate of the number of same sign dileptons we would have for 10^6 Z^0's. Assuming that the mixing is maximal in the B_s system, that the fraction of B_s produced is 15%

and that $B(b \to \ell x) = 10\%$ one would have:

$$\# \text{ same sign } e^{\pm}e^{\pm} = (.1)^2 \times (.15) \times 7 \times 10^4$$
$$\simeq 100 \quad .$$

Here 10^6 Z^0's will provide ~ 400 same sign dilepton (ee, $\mu\mu$ and $e\mu$) events. The backgrounds for these events should be small because the hadron rejection capabilities of the SLC and LEP detectors will be excellent. Charm backgrounds are all but eliminated by the $b\bar{b}$ tag as explained above. The SLC and LEP could well be the best hunting ground in the near future for $B^0 - \bar{B}^0$ mixing effects.

7. Z^0 PHYSICS BEYOND THE STANDARD MODEL

Possibly to most interesting physics at the Z^0 will be the surprises. Certainly we all hope so! How about some "predictably" surprises – namely things which spoil the tidy Standard Model predictions?

7.1 NON-MINIMAL HIGGS SCHEME, SEARCHING FOR CHARGED AND NEUTRAL HIGGS

We have, until now, considered the minimal Higgs structure of one Higgs doublet. However there is nothing in the Standard Model which prevents us from having more than one Higgs doublet. With two Higgs doublets (8 fields) one gives up three of these fields to produces masses for the $W^{\pm}$ and Z^0 leaving 5 physical Higgs particles. They are:

$$\begin{array}{ll} \text{Two neutral scalars} & H_1^0, \ H_2^0 \\ \text{One pseudoscalar} & h^0 \text{ (the axion in some models)} \\ \text{and two charged pseudoscalars} & H^+, \ H^- \quad . \end{array}$$

For decay purposes the usual scalar rule applies – couplings are largest for the heaviest fermion decay products permissable.

The search for the two neutral scalars proceeds exactly as discussed earlier for H^0 except that now the lower 7.5 GeV/c^2 bound on the H^0 is removed. Looking for H^0's below 7.5 GeV/c^2 has the advantage of increasing production rate (see figure 21). However at masses of a few GeV/c^2 two photon backgrounds become a major nuisance.

What about searching for the $H^{\pm}$. The only process available at the Z^0 is $e^+e^- \to Z^0 \to H^+H^-$ which has a rate of $\Gamma(Z^0 \to H^+H^-) = \frac{1}{4}\beta_{H\pm}^3(3\%) \simeq 10^{-2}\beta_{H\pm}^3$, where $\beta_{H\pm}$ is the charged Higgs' velocity. From PETRA measurements we know that $M_{H\pm} > 15$ GeV/c which means that the dominant decay mode for the $H^\pm$ will be $H^\pm \to b\bar{c}$. (There will also be a small fraction of $H^\pm \to \tau^\pm \nu_\tau$.) So most of the events arising from $Z^0 \to H^+ H^-$ would contain four jets, two in each back-to-back hemisphere. The major background comes from QCD 4 jet events of the type $qqgg$ which occur at a rate $\sim \alpha^2 \times 0.72 \simeq 1.5 \times 10^{-2}$. The major handle one has in rejection of this background is that the signal has 4 long-lived quarks whereas at test half the background has only two long lived quarks. Hopefully a good vertex detector will provide the necessary background rejection.

The $H^\pm$ mass would have to be obtained from the di-jet masses in the two hemispheres. Presumably one would have sensitivity up to charged Higgs masses of $\lesssim 40$ GeV/c^2.

7.2 THE GENERATION PUZZLE – SEARCHING FOR NEW GENERATIONS

The discovery of the τ and the b quark has led to a very beautiful symmetry between the quark and lepton sectors. Nature at present appears to have three generations of both quarks and leptons. While this symmetry is indeed attractive, we are led to an obvious question – why three generations? Why not five or ten? We readily understand the need for one generation – our very being is dependent on it. But more than one generation seems superfluous and it is interesting to speculate on why nature chose to replicate itself in this strange way.

The distinguishing generation element is mass – successive generations have higher masses. A perfectly defensible reason why we see three generations then is that the energy of our machines is not sufficient to yield the next generation(s). The prospect of higher energy machines implies more quarks and leptons. We may go to our theoretical friends and ask them where we need to look; where will the next generation appear? The answer is that none of the current theories understands the generation puzzle and no mass predictions exist.

The prospect of a factor of $\gtrsim 2$ in available energy, plus the large rate makes the Z^0 a good place to look for new generations. How do we search for new generations? There are three obvious possibilities:

a) Search for a new charged lepton, $L^\pm$,

b) Search for a new $Q = -1/3$ quark, and

c) Search for more ν's.

We do not include searches for $Q = 2/3$ quarks because if such a quark were found, it would satisfy our need for the top quark. Consider the search for $L^\pm$. The $W^\pm$ which mediate the decays of $L^\pm$ is __democratic__ with respect to fermion coupling strengths. Allowing for three quark colors we have

$$B(L^\pm \to \ell^\pm \nu\nu) = \frac{1}{12} = 8\%$$

and

$$B(L^\pm \to \text{hadrons}) = 76\% .$$

(These numbers will be modified slightly by QCD corrections but, for the argument being made here, these small modifications are unimportant.) We will therefore be able to use the standard low multiplicity searches for $L^\pm$ which will be pair produced with $B(Z^0 \to L^+L^-) = 3\% \ \beta_{L\pm}(3 - \beta_{L\pm}^2)/2$ where β_L is the heavy lepton velocity. It would be like searching for the τ all over again. From PETRA we know that $M_{L\pm} > 18$ GeV/c^2 and hence $\beta_{L\pm} \leq 0.92$. As an example one might search for acoplanar $e^\mp \mu^\pm$ events. For $M_{L\pm} = 40$ GeV/c^2 ($\beta_L = .4$), there would be 200 $e^\pm \mu^\mp$ events/10^6 Z^0 events. In the channel acoplanar ex or μx there would be 2.5×10^3 events/10^6 Z^0 events. τ pair production is not a serious background because, while the charged particle multiplicity topology is the same, the decay products of the (light) τ are entirely back-to-back (*i.e.* coplanar) unlike those coming from the $L^\pm$. So it is easy to search for $L^\pm$ at the Z^0 as long as $M_{L\pm} < M_{Z^0}/2$.

We have already discussed how to find a new heavy quark and how the charge asymmetry A_{F-B} can be used to separate charge $-1/3$ and charge $2/3$ quarks. If a new heavy quark were found with a charge $-1/3$ this would signal the presence of a new generation.

Finally we discuss the search for additional ν's. We derived in section 4 $\Gamma(Z^0 \to \nu\bar\nu) \simeq 170$ MeV and decided in section 6.1 that we could possibly measure Γ_{Z^0} with a precision $\lesssim 100$ MeV. But how do we know that the additional width comes from a 4th ν? Is there a way to count the number of ν species?

The answer is yes, but not by running on the Z^0, but rather by running above the Z^0 and observing the radiative transition

$$e^+e^- \to \gamma Z^0$$
$$ \hookrightarrow \nu\bar\nu .$$

One can chose an $E_{c.m.}$ such that the mass recoiling against the photon is M_{Z^0}. In this way one gets an enhanced event rate. This measurement is discussed by Barbiellini et al.[30] and the theoretical background can be found in reference 31. The Feynman diagrams are shown in figure 27. It turns out that the $W^\pm$ exchange diagrams are very small (see figure 28) and can be ignored. If this is done then

$$\frac{d^2\sigma}{dxdy} \simeq \frac{G_F \alpha S f(x,y)(v_\nu^2 + a_\nu^2)N_\nu}{\{[1 - S(1-x)/M_Z^2]^2 + \Gamma_Z^2/M_Z^2\}}$$

where

$$f(x,y) = \frac{(1-x)[(1-x/2)^2 + x^2 y^2/4]}{6\pi^2 x(1-y^2)}$$

and

$$x = 2E_\gamma/E_{c.m.} \quad , \quad y = \cos\theta_\gamma$$

$$N_\nu = \# \text{ of } \nu \text{ species} \quad .$$

A measurement of $\sigma_{\nu\bar\nu} = \int \frac{d^2\sigma}{dxdy} dxdy$ measures <u>directly</u> the number of neutrino species in the world! Each new species adds about 33% to $\sigma_{\nu\bar\nu}$. We need to chose $E_{c.m.}$ sufficiently high so that the backgrounds from $e^+e^- \to e^+e^-\gamma$ are sufficiently small. Choosing[30] $E_{c.m.} = 105$ GeV and integrating over $y = \cos\theta_\gamma$ in the interval $20° < \theta_\gamma < 160°$ yields the differential cross section shown in figure 28. We can now integrate over the Z^0 reflection peak, namely require experimentally that one sees a photon of energy 14 ± 2.5 GeV. The cross section so obtained is $\sigma_{\nu\bar\nu} = 0.025$ nb for $N_\nu = 3$. Each new generation will contribute $\sigma_{\nu\bar\nu} \simeq 0.008$ nb. For $\langle \mathcal{L} \rangle = 3 \times 10^{30}$ cm^{-2} sec^{-1} the event rate is 2/day/ν species. A 50 day run would yield 100 events/ν species – easily enough to measure N_ν.

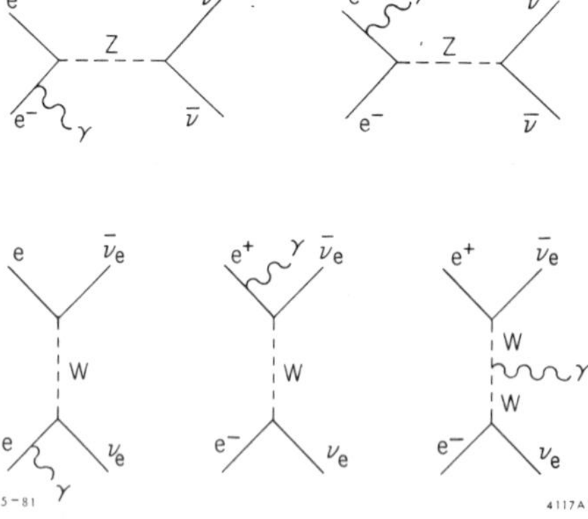

Fig. 27. Lowest order Feynman diagrams contributing to the process $e^+e^- \to \gamma\nu\bar\nu$.

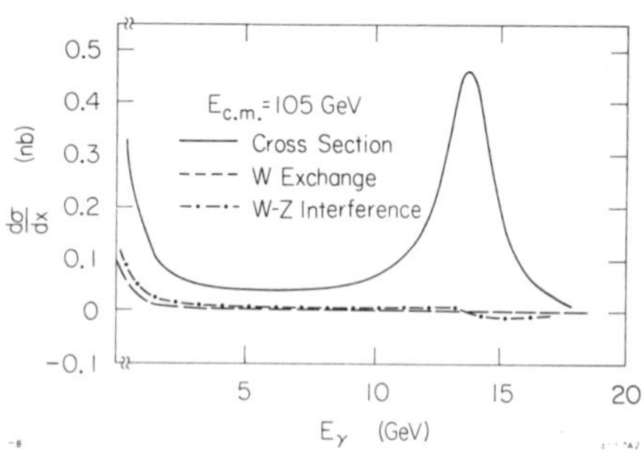

Fig. 28. The differential cross section $d\sigma/dx$ ($x = 2E_\gamma/E_{c.m.}$) is shown as a function of E_γ for the process $e^+e^- \to \gamma\nu\bar\nu$. The calculation assumes $E_{c.m.} = 105$ GeV.

The experimental signal is very simple – one hard ($E_\gamma = (14 \pm 2.5)$ GeV) photon in the angular range $20° < \theta < 160°$ and nothing else. What about backgrounds from QED processes like $e^+e^- \to e^+e^-\gamma$ and $e^+e^- \to 3\gamma$? The former background is potentially much larger. Suppose we observe a 14 GeV photon from the $e^+e^-\gamma$ process at $\theta = 20°$. The P_t of this photon will be balanced by the $e^\pm$ which radiated it. To a reasonable approximation we can write

$$\tan\theta_{e\pm} \simeq \theta_{e\pm} = \frac{E_\gamma \sin 20°}{E_{e\pm}} \simeq 6° \quad .$$

In other words there is a <u>minimum</u> angle, θ_{min}, beyond which one must see an electron (or positron) if one sees the 14 GeV photon with $20° < \theta < 160°$. The real kinematics and cross section appear in figure 29. With a veto for the $e^\pm$ down to $\theta = 6°$, the signal to noise would be 40:1.

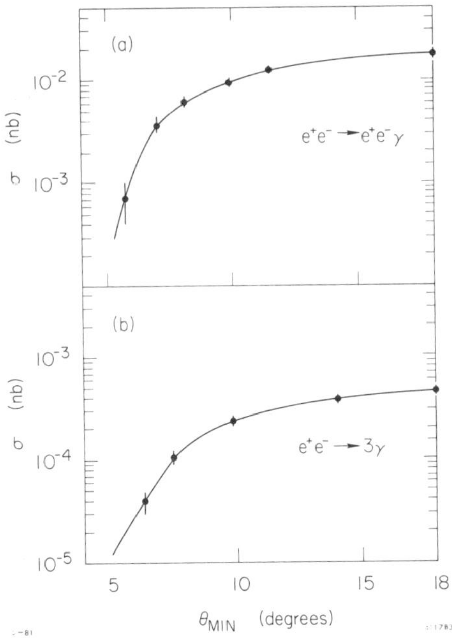

Fig. 29. Estimates of the backgrounds from the processes $e^+e^- \to e^+e^-\gamma$ and $e^+e^- \to 3\gamma$ as a function of θ_{min} for $|y| \le 0.94$ and $E_\gamma = 14.5 \pm 2.5$ GeV.

The detector required for this experiment is very simple. Electromagnetic shower counters down to within 6° of the beamline and a charged particle tracker (no magnet needed) for $20° < \theta < 160°$ to ensure that the γ is not an electron. All the SLC and LEP detectors are equipped to do this experiment.

There is nothing magic about the $E_{c.m.}$ chosen in the example of reference 30. In a recent study, the MARK II Group have decided that they are sufficiently well instrumented to run at about 5 GeV above the Z^0 peak.

This ν counting experiment is not that difficult to do experimentally. The main limitation will come from the need for a high luminosity machine. Without an average luminosity of $\gtrsim 10^{30}$ cm^{-2} sec^{-1} the search will be rate limited. There is also an important theoretical point to be made. One doesn't only count N_ν but rather contributions from all neutral stable, particles which couple weakly to the Z^0. If SUSY is correct, then

$Z^0 \to \bar{\nu}_s \nu_s$ will contribute to this rate. We will see later tht this potential contribution is, in all likelihood, considerably less than the contribution from one ν species.

7.3 SUPERSYMMETRY

Since John Ellis's lectures at this School were devoted to this topic, I provide here a sketchy outline of the value of the Z^0 machines for testing SUSY. Also reference 32 by Kane and Haber is an excellent primer for the uninitiated. There are many models and many decay schemes and what I write down here is presumably true in some model(s). But this doesn't mean that it is correct $-$ *i.e.* SUSY doesn't demand it, rather the model does.

Production cross sections for the partners of the normal fermions are characteristic of scalars namely

$$R_{\tilde{s}\tilde{s}} = \frac{1}{4} R_{f\bar{f}} \quad .$$

Here $\tilde{s}$ indicates a SUSY scalar whose normal partner is denoted by f. However there are two SUSY partners for each normal fermion so in reality

$$R_{\tilde{s}\tilde{s}} = \frac{1}{2} R_{f\bar{f}}$$

and

$$\frac{d\sigma_{\tilde{s}\tilde{s}}}{d\cos\theta} = \frac{1}{2}\beta^3 \sin^2\theta \sigma_{f\bar{f}} \quad .$$

So, if $M_{\tilde{s}} < M_Z/2$, SUSY scalars could add considerably to the width of the Z^0. As we said previously, if Γ_Z is too wide there could be many reasons for it. One would have to search for each possibility separately.

Scalar leptons with $M_{\tilde{\ell}} < M_Z/2$ will be copiously produced and $B(Z^0 \to \tilde{\ell}^+\tilde{\ell}^-) = 1\frac{1}{2}\% \; \beta^3$ where β is the scalar lepton velocity. Presumably $\tilde{\ell}^\pm \to \ell^\pm \tilde{\gamma}$ and, assuming the $\tilde{\gamma}$ is stable, one gets a very distinctive signature namely events at the Z^0 which have two high energy leptons (e^+e^- , $\mu^+\mu^-$ or $\tau^+\tau^-$) with large missing P_t and energy. The presence of a stable light particle ($\tilde{\gamma}$) in the decay chains of all the SUSY particles implies that SUSY events are characterized by missing P_t and energy. This is a key element in the search for SUSY signatures.

For scalar quarks $B(Z^0 \to \tilde{u}\tilde{\bar{u}}) = 6.6\% \; \beta^3$, $B(Z^0 \to \tilde{d}\tilde{\bar{d}}) = 5.3\% \; \beta^3$ where β is the quark velocity. The scalar quark will decay to a quark and a gluino or $\tilde{\gamma}$ and hence one has events with two jets which are not back to back but have substantial missing P_t and energy. Again this is a distinctive signature and, provided β is not too small, there is copious production.

For scalar neutrinos $B(Z^0 \to \tilde{\nu}\,\tilde{\bar{\nu}}) = 3\%\,\beta^3$ where β is the $\tilde{\nu}$ velocity. In order to discuss this channel further requires a decay scheme for the $\tilde{\nu}$. The schemes are complicated by the fact that one has no idea of the scalar electron, scalar $\nu, \ldots$ masses. Certainly a prominent decay mode will be $\tilde{\nu} \to \nu\tilde{\gamma}$ which is an invisible mode which could have a branching fraction $\simeq 0.6$. There are also multiple charged particle modes possible as known in figure 30 taken from Barnett et al.[33] How much will $Z^0 \to \tilde{\nu}\,\tilde{\bar{\nu}}$ contribute to the ν counting experiment? The contribution per SUSY species relative to a ν species will be

$$N_{\tilde{\nu}}/N_\nu = B^2(\tilde{\nu} \to \nu\tilde{\gamma}) \frac{\Gamma(Z^0 \to \tilde{\nu}\,\tilde{\bar{\nu}})}{\Gamma(Z^0 \to \nu\bar{\nu})} \approx 0.2$$

where I have used $B(\tilde{\nu} \to \nu\tilde{\gamma}) \simeq 0.4$. In all likelihood then, it will be hard to see a scalar ν species in the neutrino counting experiment. However the possibility that scalar neutrinos exist could place systematic limits on how well one would measure N_ν.

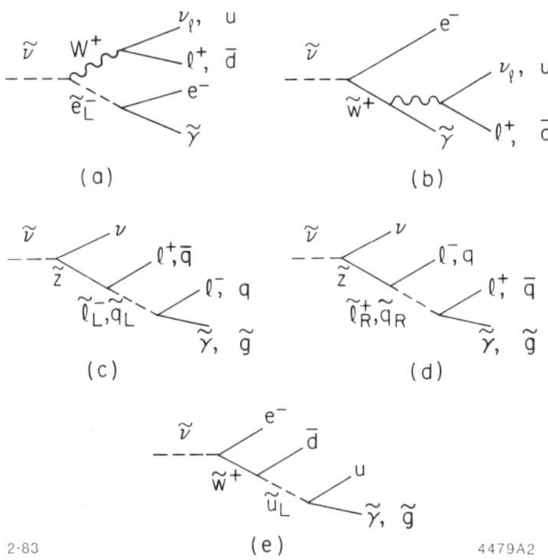

Fig. 30. Possible decay modes for $\tilde{\nu} \to$ multiple charged particles taken from the model of Barnett *et al.*, reference 33.

The multicharge decays shown in figure 30 could generate some spectacular events at Z^0. The topology

$$Z^0 \to \tilde{\nu}\,\tilde{\bar{\nu}}$$

would yield an electron and positron in one hemisphere of the detector and nothing else! Even if $B(\tilde{\nu} \to \nu_e e^+ e^- \tilde{\gamma}) \approx 10^{-3}$, 10^6 Z^0's would yield ~ 40 such events! Another interesting topology would be

$$Z^0 \to \tilde{\nu} \, \tilde{\bar{\nu}}$$
$$\llcorner \to e^- q\bar{q}\tilde{g}$$
$$\llcorner \to \tilde{\gamma}\nu$$

yielding an electron, two quark jets and a gluino in one hemisphere and nothing visible recoiling against them. The electron energy is expected to be large ($\langle P_e \rangle \approx$ 8 GeV) making them easy to detect. Certainly, if SUSY is correct, there is a chance that we could see some spectacular events at the Z^0 .

What about the charginos $\omega^\pm$, $h^\pm$ which are the spin 1/2 partners of the $W^\pm$ and $H^\pm$. Since they couple weakly, these particles look like heavy leptons $L^\pm$ discussed earlier. They decay via

$$h^\pm, w^\pm \to W^\pm \tilde{\gamma}$$
$$\llcorner \to \ell^\pm \nu \text{ or } q\bar{q} \quad .$$

The decay will be the same as $L^\pm \to W^\pm \nu$ except for effects arising from large $\tilde{\gamma}$ mass. How are they distinguished from $L^\pm$? Consider for the moment the unmixed case for which the weak couplings are

$$v = (T_{3L} + T_{3R} - 2Q\sin^2\theta_W)$$

$$a = T_{3L} - T_{3R}$$

with

$$T_{3L} = T_{3R} = \pm 1 \qquad \text{for } w^\pm$$

$$T_{3L} = T_{3R} = \pm 1/2 \quad \text{for } h^\pm .$$

Hence $v_{W^\pm} = 1.56$, $v_{h^\pm} = 0.56$ and $a_{w^\pm} = a_{h^\pm} = 0$. So for this unmixed case

$$\frac{\Gamma(Z^0 \to w^+ w^-)}{\Gamma(Z^0 \to \tau^+ \tau^-)} = 9.6$$

and

$$\frac{\Gamma(Z^0 \to h^+ h^-)}{\Gamma(Z^0 \to \tau^+ \tau^-)} = 1.2 \quad .$$

Hence charginos could add as much as $\sim 33\%$ to Γ_Z. The search for $w^\pm$, $h^\pm$ proceeds exactly as the search for $L^\pm$ discussed earlier.

How does one distinguish the $L^\pm$ from the $w^\pm$, $h^\pm$? Their weak interactions are very different! The charge asymmetry is

$$A_{F-B} = \frac{3 v_e a_e v_f a_f}{(v_e^2 + a_e^2)(v_f^2 + a_f^2)}$$

$$= 4.3\% \text{ for } L^\pm$$

$$= 0 \text{ for } w^\pm,\ h^\pm \text{ in unmixed cases.}$$

We have considered the simplest unmixed case. Suppose the $w^\pm$ and $h^\pm$ are maximally mixed in states $\tilde{w}_1$ and $\tilde{w}_2$. Then

$$\frac{\Gamma(Z^0 \to \tilde{w}_1^+ \tilde{w}_1^-)}{\Gamma(Z^0 \to \tau^+ \tau^-)} = \frac{\Gamma(Z^0 \to \tilde{w}_2^+ \tilde{w}_2^-)}{\Gamma(Z^0 \to \tau^+ \tau^-)} = 5.5$$

and

$$A^{\tilde{w}_1}_{F-B} = 14\%$$

$$A^{\tilde{w}_2}_{F-B} = -14\% \quad .$$

Of course we have no guidance from the theory as to what level of mixing, if any, there is.

7.4 MODELS WITH EXTENDED ELECTRO-WEAK GAUGE GROUPS: $SU(2) \wedge U(1) \wedge \mathcal{G}$

At low Q^2, the effective Hamiltonian of the Standard Model for charged current interactions is given by

$$H_{cc} = \frac{4 \mathcal{G}_F}{\sqrt{2}} J_\mu^+ J_\mu^-$$

and for neutral current interactions by

$$H_{NC} = -\frac{e^2}{q^2} J_{em}^2 + \frac{4 \mathcal{G}_F}{\sqrt{2}} \left\{ \left(J^3 - \sin^2 \theta_W J_{em} \right)^2 \right\}$$

where (J^+, J^-, J^3) form the weak isospin current. The success of the Standard Model at low energies is testament to the validity of this current-current description. Neither neutrino scattering experiments nor the e-d parity violation

132

experiment probe the electromagnetic part of the weak current. Hence we can add to H_{NC}

$$H'_{NC} = -\frac{e^2}{q^2}\, J^2_{em} + \frac{4\mathcal{G}_F}{\sqrt{2}}\left\{\left(J^3 - \sin^2\theta_W\, J_{em}\right)^2 + C J^2_{em}\right\}$$

without conflicting any of the low energy experimental data. The only low energy bound for C is $C < 4$ which comes from measurements of the anomalous magnetic moment of the muon.

The lepton pair charge asymmetry measurements at PETRA and PEP probe the term $C J^2_{em}$ via the interference of the weak and electromagnetic propagators. However these measurements do not place stringent limits on C. The freedom implied by the addition of the term $C J^2_{em}$ permits models which extend the Electro-Weak gauge group to $SU(2) \wedge U(1) \wedge \mathcal{G}$. Examples of such models are $SU(2) \wedge U(1) \wedge U(1)'$ (references 34,37), $SU(2) \wedge U(1) \wedge SU(2)'$ (reference 35) and $SU(2)_L \wedge SU(2)_R \wedge U(1)$ (references 36,37). The common feature of these models is the presence of two Z^0's with $M_{Z_1} < M_{Z^0} < M_{Z_2}$ where M_{Z^0} is the mass of the Z^0 in the Standard Model. The reason why one gets two energy levels is that $C J^2_{em}$ can be thought of as a perturbation and this perturbation splits the single energy level (M_{Z^0}). We discuss briefly some models for which $\mathcal{G} = U(1)$ and $SU(2)$, to illustrate how the formalism works.

In the model discussed in reference 34 in which $\mathcal{G} = U(1)$, all fermions transform under $SU(2) \wedge U(1)$ in the usual manner and they are invariant under $U(1)'$. The spontaneous symmetry breaking is achieved using a pair of complex Higgs fields, ϕ_1 and ϕ_2. The Higgs field ϕ_1 follows the Standard Model prescription. However, ϕ_2, which is invariant under $SU(2)$, has non-trivial transformations under $U(1) \wedge U(1)'$. Hence we recover the same $W^\pm$ structure as in the Standard Model. However ϕ_2 gives rise to an additional heavy neutral boson which can be associated with $U(1)'$. We thus obtain

<u>for $U(1)$</u>

$$(e^-, \nu_e) \begin{pmatrix} V_0 & 0 \\ 0 & V_0 \end{pmatrix} \begin{pmatrix} e^- \\ \nu_e \end{pmatrix}$$

<u>for $SU(2)$</u>

$$(e^-, \nu_e) \begin{pmatrix} W^0 & W^- \\ W^+ & W^0 \end{pmatrix} \begin{pmatrix} e^- \\ \nu_e \end{pmatrix}$$

<u>Standard Model</u>: $SU(2) \wedge U(1)$

$$(\gamma, Z^0) = \begin{pmatrix} \cos\theta & -\sin\theta_W \\ \sin\theta_W & \cos\theta_W \end{pmatrix} \begin{pmatrix} V^0 \\ W^0 \end{pmatrix};$$

one parameter, $\sin^2\theta_W$

and for $U(1)'$

$$(e^-, \nu_e) \begin{pmatrix} V_0' & 0 \\ 0 & V_0' \end{pmatrix} \begin{pmatrix} e^- \\ \nu_e \end{pmatrix}$$

$\underline{SU(2) \wedge U(1) \wedge U(1)'}$

$$(\gamma, Z_1, Z_2) = (3 \times 3) \begin{pmatrix} V_0 \\ V_0' \\ W_0 \end{pmatrix};$$

three parameters .

The three parameters of $SU(2) \wedge U(1) \wedge U(1)'$ can be taken as $\sin^2 \theta_W$, M_1 and M_2. The parameter C is given by

$$C\,(\mathcal{G} = U(1)) = \cos^4 \theta_W \left(\frac{M_{Z^0}^2}{M_1^2} - 1 \right) \left(\frac{M_{Z^0}^2}{M_1^2} - 1 \right)$$

so as either $M_{1,2} \to M_{Z^0}$ we recover the Standard Model with $C = 0$.

In the model with $\mathcal{G} = SU(2)^{35}$ one goes through a similar procedure but in this instance ϕ_2 has non-trivial transformations under $SU(2) \wedge SU(2)'$. Again one obtains two neutral heavy bosons. In this model

$$C\,(\mathcal{G} = SU(2)) = \sin^2 \theta_W \left(\frac{M_{Z^0}^2}{M_1^2} - 1 \right) \left(\frac{M_{Z^0}^2}{M_1^2} - 1 \right)$$

and again $C = 0$ as $M_{1,2} \to M_{Z^0}$.

How does one test for these extended Electro–Weak gauge groups at the Z^0 energy range? Inspection of the modified H'_{NC} will convince you that in essence the extra $C J_{em}^2$ term has the effect of modifying the value of what we measure for $\sin^2 \theta_W$. As we discussed previously, the most sensitive measure of $\sin^2 \theta_W$ comes from looking at the left–right asymmetry A_{LR}. Shown in figure 31 is A_{LR} as a function of $E_{c.m.}$ for the Standard Model and for $SU(2) \wedge U(1) \wedge U(1)'$ in which $\sin^2 \theta_W$ is taken to be 0.22. Running at the Z^0 pole one would easily see the deviation from the Standard Model. However one will have to run at a substantially higher $E_{c.m.}$ in order to get sensitivity to M_2. Although not discussed here in detail, we show the sensitivity of A_{LR} to tests for the model $SU(2)_L \wedge SU(2)_R \wedge U(1)$.[36] Figure 32 shows A_{LR} as a function of $E_{c.m.}$ for the Standard Model and the left–right symmetric model.[7] In this case running at the nominal Z^0 mass can distinguish the extended model from the Standard Model and yield the mass of the second heavy boson. This is shown more explicitly in figure 32b.

Another example comes from superstring models which add a $U(1)$ group to the Standard gauge structure thereby producing as second Z^0-like particle. Tests of models of this type using polarized electrons at the Z^0 are discussed in great detail in reference 37. As an illustration we show one figure from these

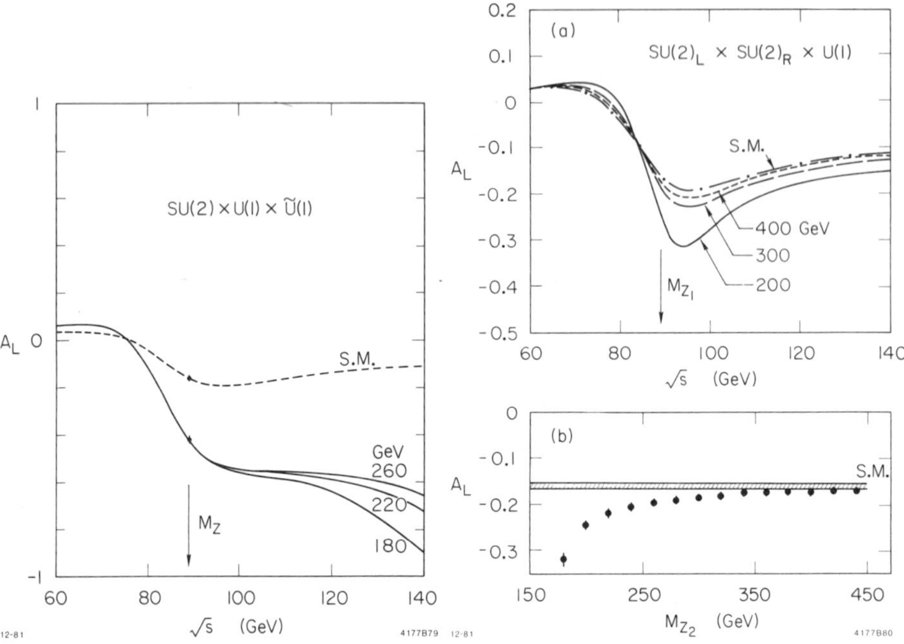

Fig. 31. A_{LR} as a function of $\sqrt{s}$ calculated in the Standard Model (S.M.) and in a $SU(2) \times U(1) \times \widetilde{U}(1)$ model for different values of the mass of the second neutral boson. The two data points indicate the statistical accuracy expected for 3×10^4 Z^0 decays.

Fig. 32. A_{LR} calculated in the Standard Model (S.M.) and in the $SU(2)_L \times SU(2)_R \times U(1)$ model; (a) as a function of $\sqrt{s}$ and for different values of the mass of the second neutral boson; (b) at $\sqrt{s} = M_Z$, comparison of the A_{LR} value in the Standard Model and in the $SU(2)_L \times SU(2)_R \times U(1)$ model as a function of the mass of the second neutral boson for 3×10^4 Z^0 decays.

authors (figure 33) which shows A_{LR} as a function of $M_{Z'}/M_Z$ for two parameter choices in a generalized superstring type model. For comparison the Standard Model value for A_{LR} is obtained from the figure by letting $M_{Z'}/M_Z \to \infty$. Considerable sensitivity to high mass Z''s would be obtained – for the high (low) resolution polarimeters A_{LR} will be measured to 1% (5%) at the SLC.

From the above examples, it is clear that polarized beams at the SLC provide an important tool for searching for Nature's correct gauge group.

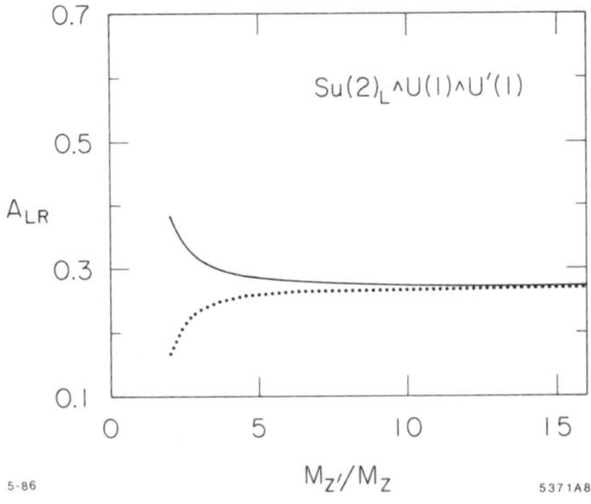

Fig. 33. A_{LR} on the Z^0 resonance as a function of $M_{Z'}/M_Z$ for two parameter choices of a superstring model as discussed in reference 37. The Standard Model value of A_{LR} is obtained by letting $M_{Z'}/M_Z \to \infty$.

8. PHYSICS ABOVE THE Z^0, THE LEPII AND LEPIII FRONTIER

There is considerable physics which can be studied using the LEPII machine ($E_{c.m.} \leq 170$ GeV) and the LEPIII machine ($E_{c.m.} \lesssim 250$ GeV). This program is thoroughly discussed in reference 2. In these lectures, given the constraints of time, just a few topics were chosen for discussion. Many of them relate to the kinds of tests which were discussed for Z^0 running.

8.1 SEARCHES FOR NEW CHARGED FERMIONS

For all new fermion searches, we were limited at the Z^0 to masses $< M_{Z^0}/2$. At LEPII the limiting masses for these searches are $E_{c.m.}/2$. Hence LEPIII provides considerable improvement in search domains. However we must now consider that the production cross-sections are much smaller and in order to perform these searches will require good luminosity and considerable running time.

The most basic measurement one can perform is a measurement of $R = \sigma_{\text{hadrons}}/\sigma_{\text{point}}$ as discussed in section 4. Using formula (4) in section 4, one can calculate R for $S > M_Z^2$. In the limit $S \gg M_Z^2$ ($E_{c.m.} > 200$ GeV) R is independent of S and for 5 quarks has a value of ~ 7 for $\sin^2 \theta_W = 0.22$. For $S < 200$ GeV one still sees the effects of the Z pole – at the highest LEPII energy of 170 GeV $R = 10.4$, at 150 GeV $R = 12.2$. Can we use a measurement of R to search for new quarks at LEPII and LEPIII? The increase in R from a t quark at $E_{c.m.} = 170$ GeV is about 2.4 or 25% of R (5 quarks), where I have ignored finite quark mass effects (*i.e.* threshold effects). The increase for a b type quark is 1.72 or 17% of R (5 quarks). Typical systematic errors are at the $\sim 3\%$ level for R measurements, so the measurements will not be systematics limited. However in order to establish a 4σ effect in R will take good average luminosity.

Assuming $\langle \mathcal{L} \rangle = 5 \times 10^{30}$ cm^{-2} sec^{-1}, one finds that $R = 10.4$ ($E_{c.m.} = 170$ GeV) corresponds to 15 hadronic events/day. Hence to see a 4σ effect for $t\bar{t}$ ($b'\bar{b}'$) production would take 17 (25) days.

Presumably these heavy quark events will have an event shape which is easily distinguishable from the light quark decays as we discussed for the Z^0 searches. A search based on event shapes would also suffice to establish the presence of a new heavy quark.

A new heavy lepton will show up in the low multiplicity events which would not, of course, satisfy the requirements for hadronic events. Production will be via $e^+e^- \to L^+L^-$. The simplest way to look for a new heavy lepton will be to search for non-coplanar ex, μx and $e\mu$ events. The same event types will be produced by $\tau^+\tau^-$ events – however because the τ is so light compared to $E_{c.m.}$ the events would be coplanar. We recall from section 7.2 that $B(L^+ \to e, \mu) = 8\%$ and therefore:

$$\# \; ex \; \text{events/day} = 0.2 \; \beta_L(3 - \beta)L^2)/2$$

where β_L is the heavy lepton velocity and I have assumed $\langle \mathcal{L} \rangle = 5 \times 10^{30}$ cm^{-1} sec^{-1} and $E_{c.m.} = 170$ GeV. Adding both the ex and μx events and having a little patience, one would be able to search for a new heavy lepton.

From these too examples one sees how crucial the peak luminosity for LEPII,III will be. Without $\mathcal{L} \sim 10^{31}$ cm^{-2} sec^{-1} useful physics will be hard to come by. But given good luminosity one will be able to search for new quarks and charged leptons up to masses $\lesssim E_{\text{beam}}$.

8.2   SEARCHING FOR H^0, $H^\pm$

What about searching for H^0? As discussed in John Ellis's lectures at the School, the best search procedure will be via the process

$$e^+e^- \to Z^{0*} \to Z^0 H^0$$

as depicted in figure 34. The total cross section for this process is given by (see reference 2, volume II)

$$\sigma(e^+e^- \to Z^0 H^0) = \frac{\pi\alpha^2 P_Z[3M_Z^2 + P_Z^2][1 + (1 - 4\sin^2\theta_W)^2]}{24\sin^2\theta_W \cos^4\theta_W \sqrt{S}\,(S - M_Z^2)^2}$$

where P_Z is the Z^0 momentum

$$P_{Z^0} = \frac{1}{2}\left[S - 2M_Z^2 - 2M_H^2 + S^{-1}\left(M_Z^2 - M_H^2\right)^2\right]^{1/2} \quad .$$

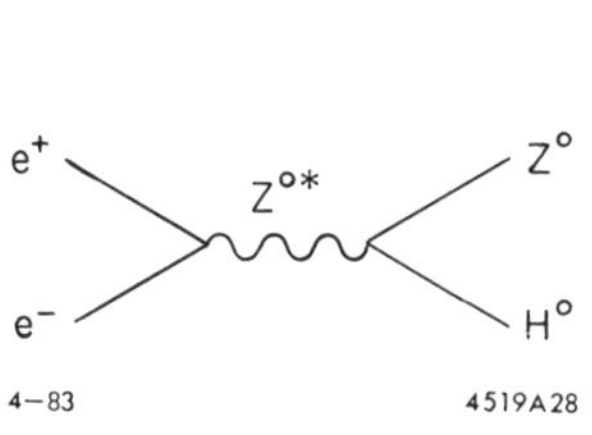

Fig. 34. The process $e^+e^- \to Z^{0*} \to Z^0 H^0$.

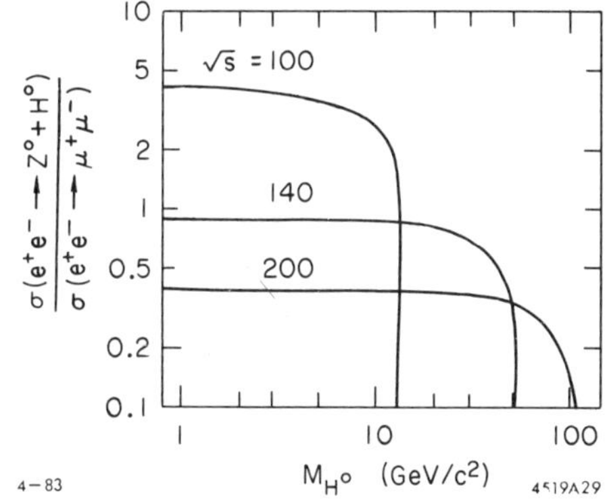

Fig. 35. The rate for the process $e^+e^- \to Z^0 H^0$ is shown as a function of $E_{c.m.}$ and M_{H^0}. The rate is normalized to the point cross section which has the value $86.8/E_{c.m.}^2$ nb.

This cross section is plotted in figure 35 as a function of M_{H^0} for three choices of $E_{c.m.}$. For a given $E_{c.m.}$, the cross section is relatively constant as a function of M_{H^0}, falling rapidly as the kinematic limit of $(E_{c.m.} - M_{Z^0})$ is reached. For $\langle \mathcal{L} \rangle = 5 \times 10^{30}$ cm^{-2} sec^{-1} the event rate is ~ 0.8 event/day at $E_{c.m.}$ of 170 GeV with a maximum search mass of $\lesssim 70$ GeV/c^2. So in a year one would have ~ 300 events in total. The corresponding event rate at $E_{c.m.} = 140$ GeV would be ~ 800 with a maximum search mass of 40 GeV/c^2. We should recall here that LEPI and SLC will have a maximum search mass at $E_{c.m.} = M_{Z'}$ of $\lesssim 40$ GeV. So LEPII,III has a sizeable advantage in search mass. However the rates are low. Realistically to search for these events one would have to require that the Z^0 decay to e^+e^- or $\mu^+\mu^-$. One has the Z^0 constraint plus two identified leptons which will provide a clean signal. The H^0 mass would result from calculating

138

the mass recoiling against the Z^0 decay products. Requiring the leptonic decay channels will cost a factor of 16 in rate. With larger $E_{c.m.}$ this will cause severe statistics problems – in that case one would have to try to augment these clean channels with $Z^0 \rightarrow \nu\bar{\nu}$, $\tau^+\tau^-$ and $q\bar{q}$.

How about searching for $H^\pm$ at LEPII,III? The production cross section for $e^+e^- \rightarrow H^+H^-$ is given by

$$\sigma_{H^+H^-} = \frac{1}{4}\,\beta_{H^\pm}^3\,\sigma_{\mu^+\mu^-}$$

where $\sigma_{\mu^+\mu^-}$ is not the point cross section but also has a contribution coming from the Z^0 propagator. (To get an exact value of this cross section one can evaluate equation (4) in section 4.) $\sigma_{\mu^+\mu^-}$ has the value of ~ 4 pb at $E_{c.m.} = 170$ GeV and ~ 1.5 pb at $E_{c.m.} = 250$ GeV. So roughly speaking H^+H^- production is about 5% of the continuum $e^+e^- \rightarrow$ hadrons. To detect the presence of the H^+H^- will require reconstructing four jet events. The QCD background will be present at a level of roughly half the H^+H^- signal (for $\beta = 1$). The two sources can be distinguished by their production angular distribution – $\sin^2\theta$ for the H^+H^- $(1 + \alpha\cos\theta + \cos^2\theta)$ for the QCD events. Event rates will be ~ 400 H^+H^- events per year for $\langle\mathcal{L}\rangle = 5 \times 10^{30}$ cm^{-2} sec^{-1} and $E_{c.m.} = 170$ GeV. With sufficient running searches for $H^\pm$ up to energies $\lesssim E_b$ should be possible at LEPII,III.

8.3 $e^+e^- \rightarrow W^+W^-$

LEPII and III are by far the best place to study W^+W^- production. The three lowest order diagrams with contribute to this process are shown in figure 36 and the resulting cross section as a function of E_b is shown in figure 37. We see that once above threshold the cross-section is roughly constant in the LEPII,III energy range and is also large $(\sim 3000$ W^+W^- per year for $\langle\mathcal{L}\rangle = 5 \times 10^{30}$ cm^{-2} sec^{-1}). Besides providing a valuable laboratory for studying $W^\pm$ decays, LEPII also provides the only place to study to gauge couplings γW^+W^- and $Z^0 W^+W^-$, thus making a unique contribution to testing the Standard Model. The form of the cross section shown in figure 37 is crucially determined by cancellations amount the three competing lowest order diagrams (figure 36). In particular, the ν exchange diagram would grow without bound as $E_{c.m.}$ increases unless moderated by the two other graphs. Hence a measure of the total W^+W^-

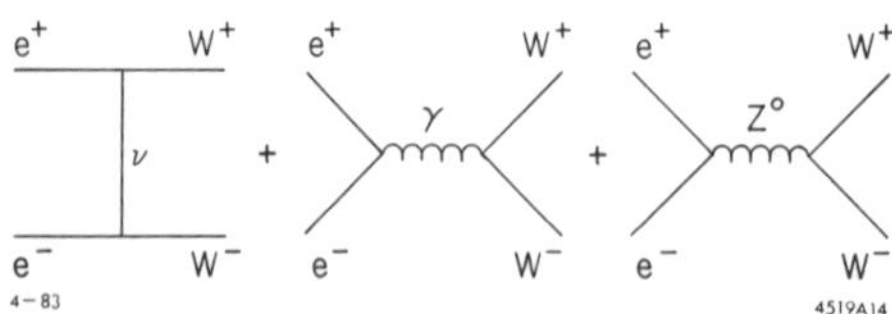

Fig. 36. The three processes which contribute to $e^+e^- \rightarrow W^+W^-$.

cross section provides a crucial test of the Standard Model. A precise check will require knowledge of weak radiative corrections. However SLC and LEPI should provide an excellent understanding of these effects and, in all likelihood, we will know how to include these effects.

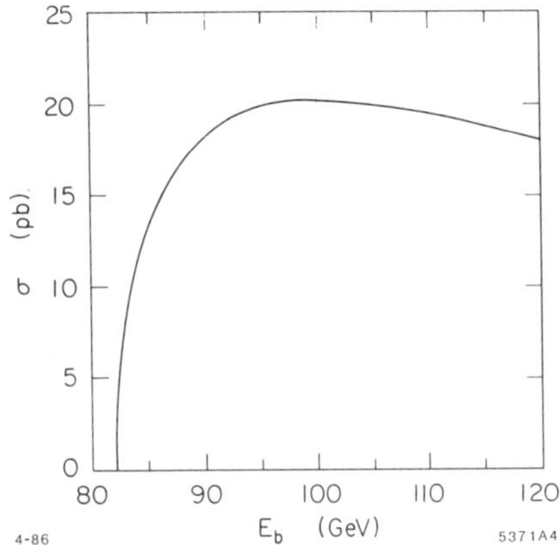

Fig. 37. The cross section for $e^+e^- \to W^+W^-$ as a function of beam energy. LEPII has a maximum beam energy of 85 GeV.

What do the events look like. The $W^\pm$ coupling strength is democratic with respect to fermions and hence for $N_G = 3$ generations

$$\frac{B(W \to \ell\nu)}{B(W \to \text{all})} \simeq \frac{1}{4N_G} = 8\%$$

and

$$\frac{B(W \to q'\bar{q})}{B(W \to \text{all})} \simeq 25\% \, |V_{qq'}|^2$$

where $V_{qq'}$ is the Kobayashi-Maskawa weak mixing matrix element and where for $q = t$ finite mass effects are important. For $W \to tb$ with $M_t = 40$ GeV/c^2 $B(W \to \bar{t}q) \simeq 18\% \, |V_{tq}|^2$. Hence W^+W^- decays predominantly to 4 jets (76%) with little background from $e^+e^- \to 4$ jets where we can estimate the cross section to be $\alpha_s^2 R\sigma_{\text{point}} \approx 0.6$ pb at $E_{c.m.} = 170$ GeV. For an average luminosity of 5×10^{30} cm^{-2} sec^{-1} one would get six 4 jet events and one $\ell^\pm q\bar{q}'$ per day from W^+W^- production. Both of these provide clean topologies for the study of $W^\pm$.

How well could we measure the $W^\pm$ mass? The $p\bar{p}$ measurements are limited to ~ 2 GeV errors because of the inability of the calorimeters to reconstruct all the jet energy. This problem can be overcome at LEPII,III because one can use E_b as a constraint for $E_{W\pm}$ thereby removing this limitation. As an example figure 38 shows the di-jet mass (2 entrees per events) reconstructed in a LEP study (reference 2, volume 2) for $E_b = 100$ GeV. From this study an error of $\lesssim 150$ MeV/c^2 is obtained for the W mass. Other methods which could be used are a fit to the threshold curve (figure 37) or a study of the endpoint of the electron momentum spectrum arising from $W \rightarrow e\nu$ decays. With high probability then, an order of magnitude improvement in our knowledge of M_W should be achieved at LEPII.

As a final example of what can be achieved at LEPII,III we remind ourselves that whenever we increase $E_{c.m.}$ we are able to probe increasingly smaller characteristic fermion sizes. If fermions were not point-like, their production via $e^+e^- \rightarrow f\bar{f}$ would be characterized by an intrinsic mass scale Λ such that $\langle r \rangle \sim 1/\Lambda$. These scales introduce so-called contact terms which modify the production cross-sections predicted by the Standard Model. The simplest case to consider is Bhabha scattering $e^+e^- \rightarrow e^+e^-$. Studies by the LEP groups[2] indicate that limits on compositeness down to $\langle r \rangle \sim 10^{-18}$ cm should be possible for electrons.

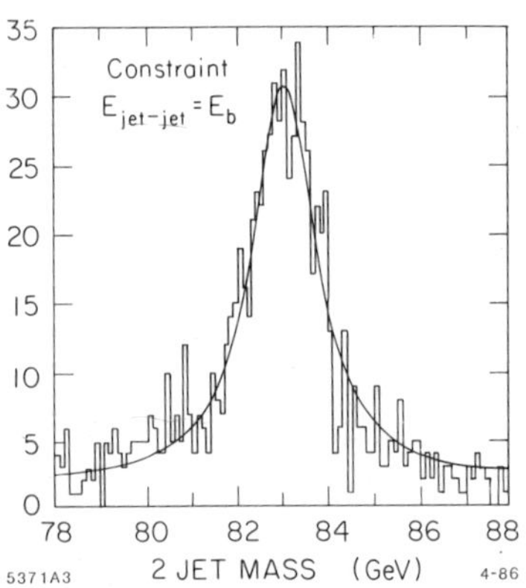

Fig. 38. Di-jet mass for events simulated as $e^+e^- \rightarrow W^+W^-$. A clear $W^\pm$ peak is seen. This simulation was done for the LEP study (see reference 2).

Because of the constraints of time, all the physics of LEPII and LEPIII have not been covered here. But enough examples have been chosen to indicate that this will be a rich and fruitful physics frontier.

9. CONCLUSIONS

As outlined in section 2, e^+e^- interactions have served the physics community well, providing many crucial discoveries and a large body of rich experimental data. The future for e^+e^- interactions with the imminent turn-on of SLC and LEP promises to continue this tradition with the very real possibility of new discoveries and the certain program of diverse and important measurements.

ACKNOWLEDGEMENTS

I spent a most enjoyable week at Lake Louise for what was their inaugural Winter School. They are to be congratulated on a fine School. I learnt much from my co-lecturers and enjoyed enormously the warm hospitality of our Canadian hosts. I thank the organizers for including me in their program. I also wish to thank the staff at SLAC who prepared this document despite my constant tardiness.

REFERENCES

1. (a) e^+e^- *Interactions at Very High Energy: Searching Beyond the Standard Model*, SLAC-PUB-3093 and Proceedings of the 1982 SLAC Summer Institute on Particle Physics. (b) Z^0 *Decay Modes – Experimental Measurements*, SLAC-PUB-3407 and Proceedings of the 1984 Theoretical Advanced Study Institute on Elementary Particle Physics.

2. Physics at LEP, edited by John Ellis and Roberto Peccei, CERN86-02, Volumes 1 and 2.

3. R. Hollebeek, Nucl. Instrum. Methods **184**, 333 (1981).

4. ECFA 81/54, General Meeting on LEP, Villars-sur-Ollon, Switzerland, June 1981.

5. *SLAC Linear Collider Conceptual Design Report*, SLAC-299 (June 1980); B. Richter, SLAC-PUB-2854 (1981).

6. S. Ozaki, Proceedings of the 1981 International Symposium on Lepton and Photon Interactions at High Energies, Bonn, 1981, p. 935.

7. C.Y. Prescott, SLAC-PUB-2854 (1981); Proceedings of the SLC Workshop, SLAC-247 (March 1982).

8. For a derivation of this cross section and most of the formulae quoted in this chapter see C. Quigg, *Gauge Theories of the Strong, Weak and Electromagnetic Interactions*.

9. J.D. Jackson and D. Scharre, Nucl. Instrum. Methods **128**, 13 (1975).

10. W. Marciano and A. Sirlin, Phys. Rev. **D22**, 2695 (1980); M. Veltman, Phys. Lett. **91B**, 95 (1980).

11. (a) CERN Yellow Reports 76-18 (1976) and 79-01 (1979). (b) Proceedings of the SLC Workshop on Experimental use of the SLC; SLAC-247 (1982). (c) Proceedings of the Cornell Z^0 Theory Workshop; CLNS 81-485.

12. LEP Detectors:
OPAL CERN/LPC/83-4
L3 CERN-Proposal-LEP-L3
ALEPH CERN/LPC/83-2
DELPHI DELPHI-83-66/1
SLC Detectors:
MARK II SLAC-PUB-3561/CALT-68-1015
SLD SLD Design Report.

13. Particle Data Group, Rev. Mod. Phys. **56** (1984).

14. UA2 Collaboration, Z. Phys. **C30**, 1 (1986).

15. MARK II Collaboration, *Proposal for Precise Beam Energy Measurement at the SLC.*

16. Y.-S. Tsai, Phys. Rev. **D4**, 2821 (1971).

17. H. Ogren *et al.*, *Polarization at the SLC: A Study by the SLC Polarization Group.*

18. G. Hanson *et al.*, Phys. Rev. Lett. **35**, 1609 (1975).

19. G. Hanson, Data taken from interim report at the MARK II Collaboration Asilomar Workshop, March 1986.

20. G. Tarnopolsky, SLAC-PUB-2842 (1981).

21. L.B. Okun, Proceedings of the 1981 International Symposium on Lepton and Photon Interactions at High Energy, Bonn, 1981, p. 1018.

22. J.D. Bjorken, SLAC-198 (1976).

23. J. Finjord, Physica Scripta, Vol. **21**, 143 (1980).

24. W. Marciano *et al.*, Phys. Rev. Lett. **43**, 22 (1979); D. Albert *et al.*, Nucl. Phys. **B166**, 460 (1980).

25. See for instance J.M. Dorfan, Proceedings of the 1983 International Symposium on Lepton and Photon Interactions at High Energies, Cornell University, p. 686.

26. M.B. Einhorn and B.G. Weeks, Nucl. Phys. **B146**, 445 (1978); K. Shizuya and S.-H.H. Tye, Phys. Rev. Lett. **41**, 787 (1978).

27. For a summary see for instance G. Baranko, Proceedings of the International Europhysics Conference on High Energy Physics, Bari, Italy, July 1985, p. 226.

28. K. Hayes, *B Tagging at the SLC*, MARK II/SLC NOTE #73.

29. See for instance, Ling-Lie Chau, Brookhaven Print-85-0555, Invited talk at the Symposium on High Energy e^+e^- Interactions, Nashville, Tenn., April 1984 and APS Spring Meeting, Washing, D.C., April 1984.

30. G. Barbiellini *et al.*, Phys. Lett. **106B**, 414 (1981).

31. E. Ma and J. Okada, Phys. Rev. Lett. **41**, 287 (1978); K. Gaemers *et al.*, Phys. Rev. **D19**, 1605 (1979).

32. G. Kane and H. Haber, UM-HE-TH83-17.

33. M. Barnett *et al.*, SLAC-PUB-3224 (September 1983).

34. E.H. De Groot *et al.*, Z. Phys. **C5**, 127 (1980). This reference discusses the general question of $SU(2) \wedge U(1) \wedge G$.

35. V. Barger *et al.*, Phys. Rev. **D25**, 1384 (1982).

36. A. De Rujula *et al.*, Annals of Phys. **109**, 242,258 (1977).

37. M. Cvetič and B.W. Lynn, SLAC-PUB-3900 (March 1986).

HERA: PHYSICS, MACHINE AND EXPERIMENTS

Günter Wolf

Deutsches Elektronen-Synchrotron, DESY, Hamburg, Germany

ABSTRACT

With HERA ep collisions at a c.m. energy of 314 GeV and $Q^2_{max} = 10^5$ GeV^2 will become possible. The physics opportunities, the design and status of the machine and the planned detectors are discussed.

1.   INTRODUCTION

DESY is presently constructing a new storage ring HERA that will provide collisions between 30 GeV electrons and 820 GeV protons[1]. Experimentation is scheduled to begin in 1990.

HERA offers exciting and unique physics opportunities. Electrons and quarks can be probed for substructure down to distances of a few 10^{-18} cm. A search for new mediators of the neutral and charged current is possible up to W' and Z' masses of 800 GeV. The large center of mass energy permits the detection of leptoquarks up to 180 GeV, of families of excited quarks and leptons up to 250 GeV, and squarks and sleptons can be produced with masses up to 180 GeV - provided they exist.

The following lectures present an introduction to the physics expected at HERA and describe briefly the machine and the planned experiments.

2.   HERA PHYSICS: GENERAL CONDITIONS

The large momentum transfers possible between electron and proton, $Q^2_{max} = 10^5$ GeV^2, make HERA first and foremost an electron quark collider. The relevant diagram is shown in Fig. 1. The incoming electron emits a lepton ℓ and exchanges a current j with one of the quarks of the incoming proton. This leads to the emission of a quark q' called the current quark. Depending on whether a neutral current (γ, Z^0) or a charged current (W^-) is exchanged the final state lepton is either an electron or a neutrino (Fig. 2).

The scattering process is not confined to the known currents, quarks and leptons. New currents may contribute and new quarks and leptons with masses up to the kinematic limit of 314 GeV may be produced. Any particle with electromagnetic and/or weak charge which is within the kinematic limits can be produced.

Next to electron quark scattering, current-gluon fusion (see Fig.3) will play an important role at HERA. Photon-gluon fusion depicted in Fig. 4 will presumably be the dominant process for the production of heavy Q = c,b,t... .

We shall discuss a few reactions that are archetypical for the different types of processes. More detailed studies can be found in Refs. 2, 3.

3.   ELECTRON SCATTERING AND THE PROTON STRUCTURE FUNCTIONS.

3.1   Kinematics

The diagram that describes ep scattering in lowest order is shown in Fig. 1 (taking j = γ). The event topology is illustrated in Fig. 5. The final state partons (quarks, gluons) and the "spectator" remnants from the incident proton are assumed to develop into jets of hadrons. The lepton and the current jet emerge on opposite sides of the beam axis, balancing each other in transverse momentum. The debris of the proton is emitted in a very narrow cone (of order 10 mrad) around the proton beam direction.

Apart from the total center-of-mass energy squared,

$$s = (p_p + p_e)^2 = m_e^2 + m_p^2 + 2(E_e E_p + p_e p_p) \approx 4 E_e E_p$$

E_e, E_p energies of incoming electron and proton

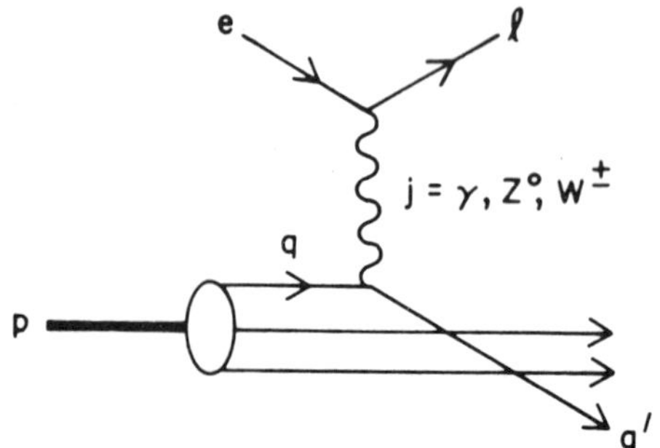

Fig. 1 Lepton proton scattering.

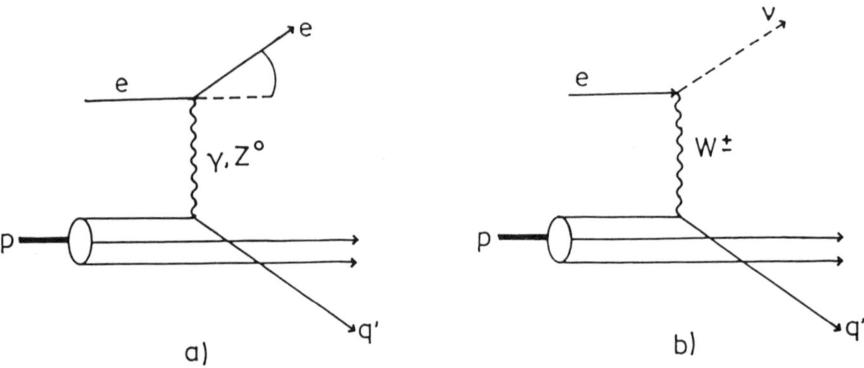

Fig. 2 Diagrams for neutral (a) and charged current (b) scattering.

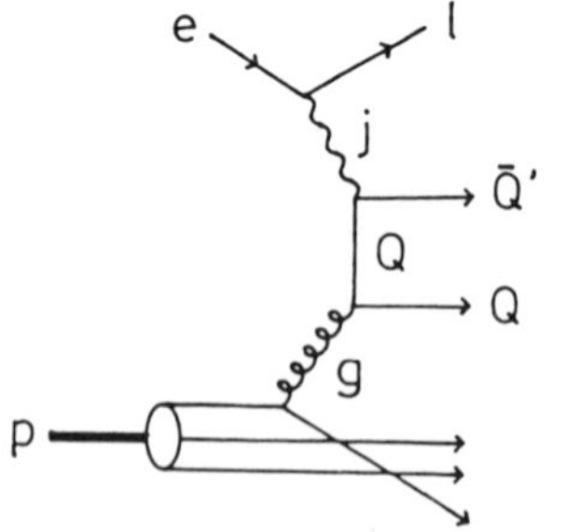

Fig. 3 Quark pair production by
current gluon fusion.

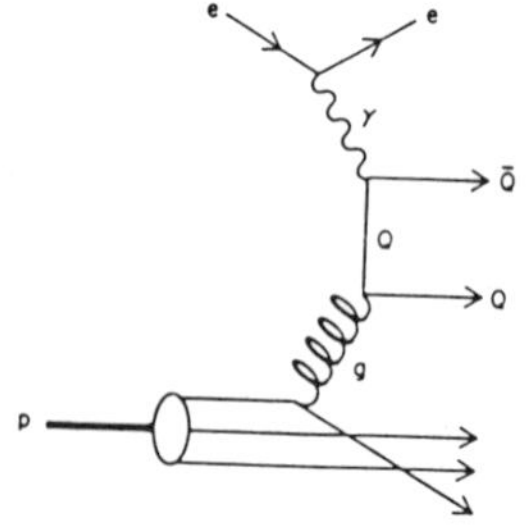

Fig. 4 Quark pair production by
photon gluon fusion.

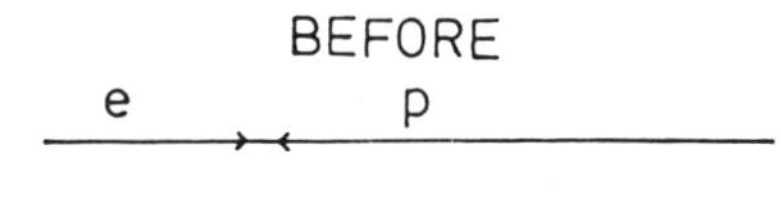

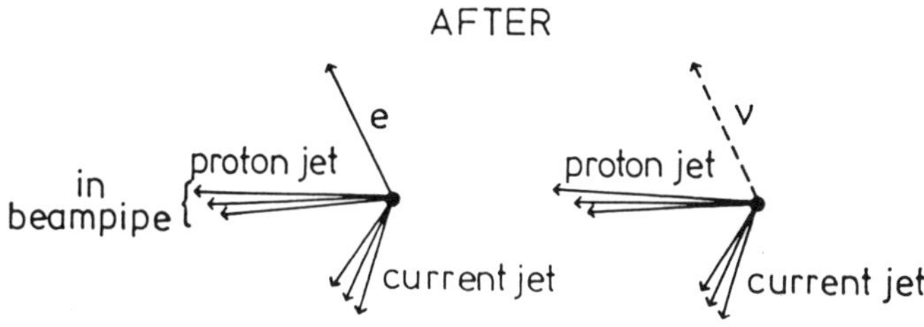

Fig. 5 Topology of deep inelastic ep scattering events.

there are two kinematic variables that describe the inclusive sactter-
ing process:

$$q^2 = (p_e - p_e')^2 = -Q^2 \qquad \text{square of the for momentum transfer}$$

$$W^2 = (q+p_p)^2 \qquad \text{square of the total mass of the final hadronic system.}$$

or equivalently

$$x = \frac{Q^2}{2(q \cdot p_p)} = \frac{Q^2}{2m_p \nu} \qquad \text{the Bjorken scaling variable.}$$

$$y = \frac{(q \cdot p_p)}{(q \cdot p_e)} = \frac{\nu}{\nu_{max}}, \qquad \text{note that } Q^2 \simeq sxy.$$

In the rest system of the incoming proton ν measures the energy transferred by the current. The maximum value which ν can reach is

$$\nu_{max} = \frac{s - (m_e + m_p)^2}{2m_p} \approx \frac{s}{2m_p} = 2E_e E_p / m_p$$

For E_e = 30 GeV, E_p = 820 GeV: ν_{max} = 52 TeV. HERA therefore is equivalent to a fixed target experiment with an incident lepton beam of
52 TeV.

The variables Q^2, x, y can be determined either from the energy E_e'
and scattering angle Θ_e of the outgoing electron, or from the energy E_j
and production angle Θ_j of the current jet.

From the electron:

$$Q^2 = 2E_e E_e' (1 + \cos\Theta_e)$$

$$x = \frac{E_e' \cos^2 \Theta_e/2}{E_p (1 - (E_e'/E_e) \sin^2\Theta_e/2)}$$

$$y = 1 - E_e'/E_e \sin^2\Theta_e/2$$

From the current jet:

$$Q^2 = \frac{E_J^2 \sin^2\Theta_J}{1 - \frac{E_J}{2E_e}(1-\cos\Theta_J)}$$

$$y = \frac{E_J}{2E_e}(1-\cos\Theta_J) = \frac{E_J}{E_e}\sin^2\Theta_J/2$$

$$x = \frac{Q^2}{sy} = \frac{E_J \cos^2\Theta_J/2}{E_p(1 - \frac{E_J}{E_e}\sin^2\Theta_J/2)}$$

The angles are always measured with respect to the proton beam. In deriving the current jet results the mass of the jet has been neglected. As an illustration of the kinematics for 30 GeV electron colliding on 820 GeV protons, Figs. 6, 7 show in the x, Q^2 plane curves for fixed energy and angle of electron and current jet.

3.2 Qualitative Behaviour of the Scattering Cross Sections

It is instructive to make the following crude guess for the cross sections of neutral current (NC) and charged current (CC) scattering. Considering only γ and W exchange, respectively (Fig. 8 a,b), they can be approximated by

$$\frac{d\sigma(\gamma p)}{dxdy} \approx \alpha^2 s \left(\frac{1}{Q^2}\right)^2 F(x,y) \qquad \text{NC}$$

$$\frac{d\sigma(Wp)}{dxdy} \approx \alpha^2 s \frac{1}{(Q^2+M_W^2)^2} F(x,y) \qquad \text{CC}$$

The W has been assumed to couple to leptons and quarks with the same coupling strength e as the γ and hence the appearence of the factor α^2 in both cases. The structure function $F(x,y)$ measures the momentum fraction x carried by the struck quark for a fixed value of y. The major difference between NC and CC cross sections stems from the different propagator contributions, Q^{-4} and $(Q^2 + M_W^2)^{-2}$, respectively, where $M_W = 82$ GeV is the W mass. Figure 8c compares the Q^2 dependence of

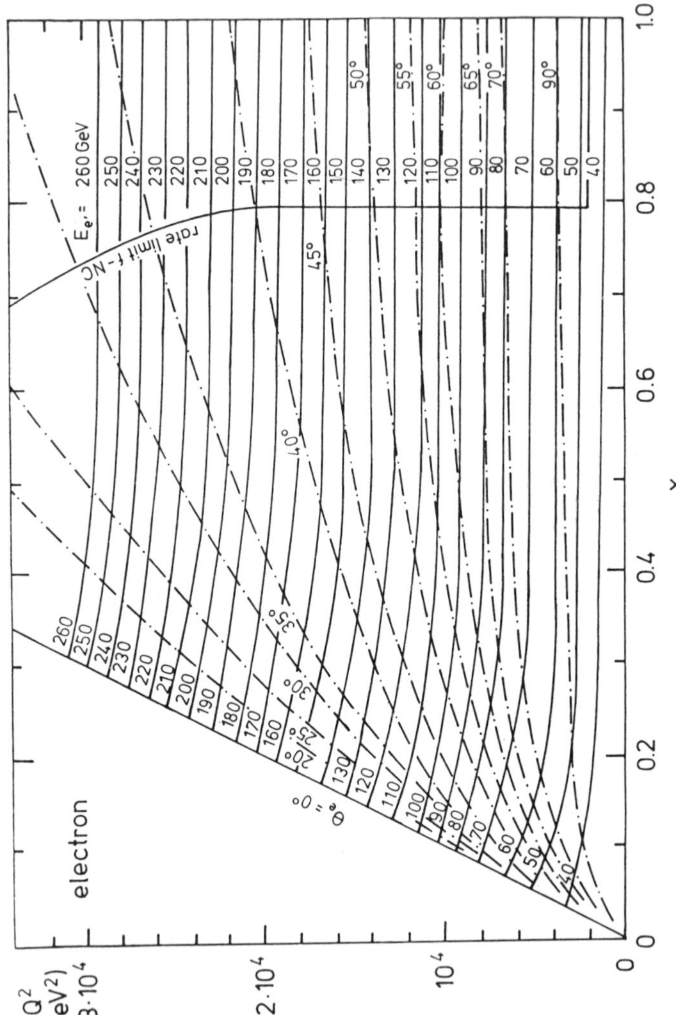

Fig. 6 30 GeV e + 820 GeV p: The x-Q^2 plane with lines of fixed electron scattering angle and energy.

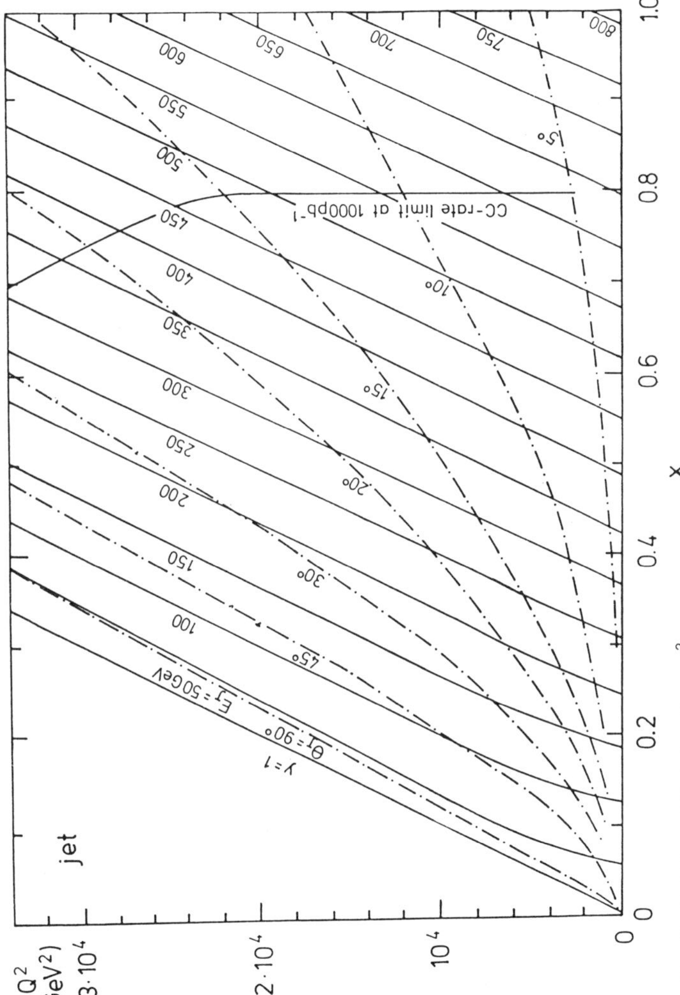

Fig. 7   30 GeV e + 820 GeV p: The x-Q^2 plane with lines of fixed current jet angle and energy.

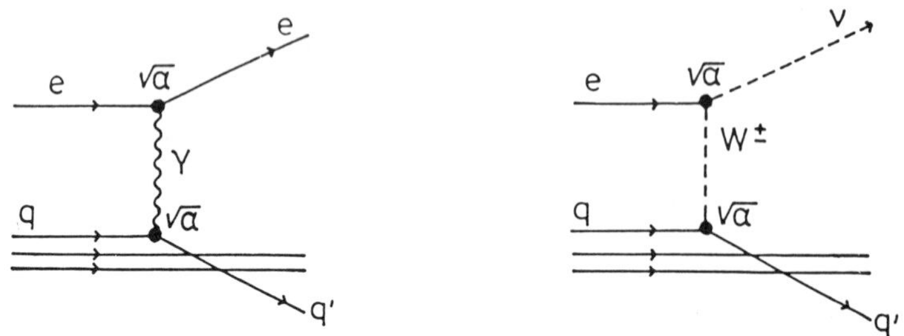

Fig. 8a,b   NC and CC scattering by γ and W exchange, respectively.

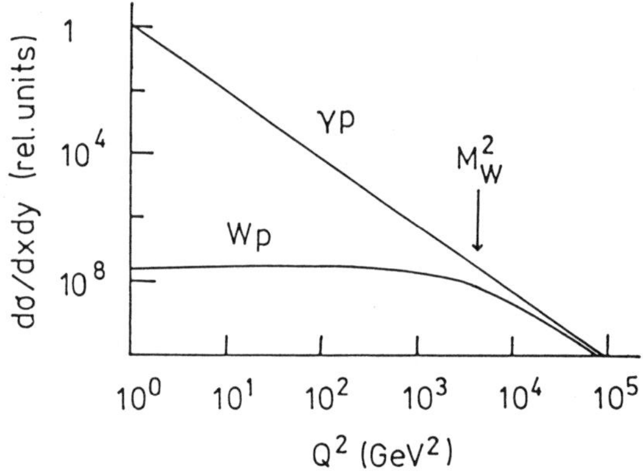

c)   Qualitative behaviour of NC scattering by γ exchange

and CC scattering by W exchange.

the two processes. Near $Q^2 = 0$ the NC cross section is $\sim 10^8$ times larger while for $Q^2 \gtrsim 10^4$ GeV^2 the weak contribution is of the same magnitude as the electromagnetic one.

We shall now present more precise expressions for the NC and CC cross sections.

3.3   Neutral Current Processes

The one photon exchange cross section can be written in terms of two dimensionless structure functions F_1, F_2 which are functions of ν, Q^2 or x,y:

$$\frac{d^2\sigma(\gamma)}{dxdy} = \frac{4\pi\,\alpha^2}{s\,x^2y^2}\left[(1 - y)F_2(x,Q^2) + xy^2\,F_1(x,Q^2)\right]$$

In the approximation that scattering on spin 1/2 partons dominates F_1 can be expressed in terms of F_2 using the Callan - Gross relation: $2xF_1 = F_2$.

This leads to

$$\frac{d^2\sigma(\gamma)}{dxdy} \approx \frac{4\pi\,\alpha^2}{s\,x^2y^2}\ \ (1 - y + \frac{y^2}{2})\ F_2(x,Q^2)$$

In the quark picture, ep scattering is described as incoherent eq scattering with x measuring the fractional momentum of the struck quark. The structure functions can then be expressed in terms of the quark distribution functions u(x), d(x),...* where u,d,... give the number of u,d,... quarks with fractional momentum between x and x + dx:

$$F_2(x) = 2xF_1(x) = x\left[q(x) + \bar{q}(x)\right]$$

where for the electromagnetic current

$$q(x) = \frac{4}{9}u(x) + \frac{1}{9}d(x) + \frac{1}{9}s(x) + \frac{4}{9}c(x)\ .$$

At Q^2 values above 10^4 GeV^2 the contribution from Z^0 exchange becomes comparable to that from photon exchange. Formally, the cross section

---

* The quark distribution functions as well as the structure functions depend on x and Q^2. For ease of writing the Q^2 dependence is not shown.

can be written as before,

$$\frac{d^2\sigma(\gamma + Z^0)}{dxdy} = \frac{4\pi\,\alpha^2}{s\,x^2 y^2}\left[(1-y)F_2(x,Q^2) + y^2 x F_1(x,Q^2)\right]$$

where now the structure functions F_1 and F_2 receive contributions from γ exchange, from Z^0 exchange and from the interference between the two processes. In terms of the quarks distribution functions the cross section reads for left (right) handed electrons allowing for left handed (L) and right handed (R) incoming electrons:[4]

$$\frac{d^2\sigma}{dxdy} = \frac{\pi\alpha^2}{sx^2 y^2}\sum_q\left\{xq(x)(A_q + (1-y)^2 B_q) + x\bar{q}(x)(B_q + (1-y)^2 A_q)\right\}$$

with

$$A_q = \left(-Q_q + g_{Lq}g_{Le}\frac{Q^2}{Q^2+M_Z^2}\right)^2 + \left(-Q_q + g_{Rq}g_{Re}\frac{Q^2}{Q^2+M_Z^2}\right)^2$$

$$B_q = \left(-Q_q + g_{Rq}g_{Le}\frac{Q^2}{Q^2+M_Z^2}\right)^2 + \left(-Q_q + g_{Lq}g_{Re}\frac{Q^2}{Q^2+M_Z^2}\right)^2$$

and

$$g_{Li} = \frac{I_L^3 - Q_i\sin^2\Theta_W}{\sin\Theta_W\cos\Theta_W} \qquad g_{Ri} = \frac{-Q_i\sin^2\Theta_W}{\sin\Theta_W\cos\Theta_W}$$

where the Q_i, $i = q,e$ measure charges of the incoming quark and lepton, I_L^3 is the weak isospin of the lepton ($I_{e^-}^3 = -1/2$, $I_{e^+}^3 = 1/2$), and Θ_W is the weak mixing angle, $\sin^2\Theta_W = 0.225 \pm 0.005 \pm 0.006$.[5]

The final expression of the cross section for left (L) or right (R) handed electron proton scattering reads[6]

$$\frac{d^2\sigma}{dxdy}\left(e^-\left\{{L\atop R}\right\} + p \to e^-\left\{{L\atop R}\right\} X\right) =$$

$$\frac{4\pi\alpha^2}{s(xy)^2}\left[Q_u^2\left(u(x) + \bar{u}(x)\right) + Q_d^2\left(d(x) + \bar{d}(x)\right)\right] x\left(1 - y + y^2/2\right)$$

$$+ \frac{\alpha}{xy}\frac{\sqrt{2}\,G_F\,M_Z^2}{sxy + M_Z^2}\left\{{1 - 2\sin^2\vartheta_W \atop -2\sin^2\vartheta_W}\right\} \cdot$$

$$\cdot\left\{\left[Q_u(1 - 4\,Q_u\sin^2\vartheta_W)\left(u(x) + \bar{u}(x)\right)\right.\right.$$

$$\left.- Q_d(1 + 4\,Q_d\sin^2\vartheta_W)(d(x) + \bar{d}(x))\right] \cdot (1 - y + y^2/2)\Big\} +$$

$$+ \frac{\alpha}{xy}\frac{\sqrt{2}\,G_F\,M_Z^2}{sxy + M_Z^2}\left\{{1 - 2\sin^2\vartheta_W \atop 2\sin^2\vartheta_W}\right\} \cdot$$

$$\cdot\left[Q_u\left(u(x) - \bar{u}(x)\right) - Q_d(d(x) - \bar{d}(x))\right] xy\left(1 - y/2\right) +$$

$$+ \frac{s}{4\pi}\frac{G_F^2\,M_Z^4}{(sxy + M_Z^2)^2}\left\{{(1 - 2\sin^2\vartheta_W)^2 \atop 4\sin^4\vartheta_W}\right\} \cdot$$

$$\cdot\left\{\left[(1 - 4\,Q_u\sin^2\vartheta_W + 8\,Q_u^2\sin^4\vartheta_W)\left(u(x) + \bar{u}(x)\right) +\right.\right.$$

$$\left.+ (1 + 4\,Q_d\sin^2\vartheta_W + 8\,Q_d^2\sin^4\vartheta_W)\left(d(x) + \bar{d}(x)\right)\right] \cdot (1 - y + y^2/2)\Big\} +$$

$$+ \frac{s}{4\pi}\frac{G_F^2\,M_Z^4}{(sxy + M_Z^2)^2}\left\{{(1 - 2\sin^2\vartheta_W)^2 \atop -4\sin^4\vartheta_W}\right\} \cdot$$

$$\cdot\left\{\left[(1 - 4\,Q_u\sin^2\vartheta_W)\left(u(x) - \bar{u}(x)\right) +\right.\right.$$

$$\left.+ (1 + 4\,Q_d\sin^2\vartheta_W)(d(x) - \bar{d}(x))\right] xy\left(1 - y/2\right)\Big\}$$

with $Q_u = 2/3$ and $Q_d = -1/3$.

Here use has been made of the prediction by the standard theory,

$$M_Z = \frac{\pi\alpha}{G_F/\sqrt{2}}\frac{1}{\sin\theta_W\,\cos\theta_W}$$

where G_F is the fermi coupling constant, $G_F = \dfrac{1.02 \cdot 10^{-5}}{M_p^2}$. To simplify the writing, only the $u,\bar{u}$ and $d,\bar{d}$ contributions are shown.

The expected event rates have been calculated in Ref. 7 using the structure functions of Ref. 8 for an integrated luminosity of 200 pb^{-1} expected after 2 years of data taking. The results are shown in Fig. 9. Note that events with $x < 0.01$ and $y < 0.01$ have been suppressed. There are $3 \cdot 10^6$ events with Q^2 values between 3 and 10^4 GeV^2. Roughly a thousand events have Q^2 values above 10^4 GeV^2 where Z^0 effects domi-nate. If we demand a minimum of a hundred events we see that neutral currents can be studied with sufficient statistics up to Q^2 values of $3 \cdot 10^4$ GeV^2. The small dark corner with $Q^2 < 200$ GeV^2, $y < 0.004$ marks the kinematical region that has so far been explored experimentally. HERA will extend the kinematic region by two orders of magnitude in either variable.

3.4   Charged Current Processes

Charged current reactions, $e^- p \rightarrow \nu X$, proceed through W^- exchange. The cross section for left handed electrons can be written in terms of three structure functions F_1, F_2 and F_3:

$$\frac{d^2\sigma}{dxdy}(e_L^- p \rightarrow \nu X) = \frac{G_F^2 s}{\pi} \frac{1}{(1 + Q^2/M_W^2)^2}$$

$$[(1 - y)F_2(x,Q^2) + y^2 x F_1(x,Q^2) + (y - \frac{y^2}{2}) x F_3(x,Q^2)]$$

The cross section for right handed electrons vanishes since the neutrino is left handed.

$$\frac{d^2\sigma}{dxdy}(e_R^- + p \rightarrow \nu + X) = 0$$

The quark-parton expression for the structure functions is:

$$F_2(x) = 2xF_1(x) = x[q(x) + \bar{q}(x)]$$

$$xF_3(x) = x[q(x) - \bar{q}(x)]$$

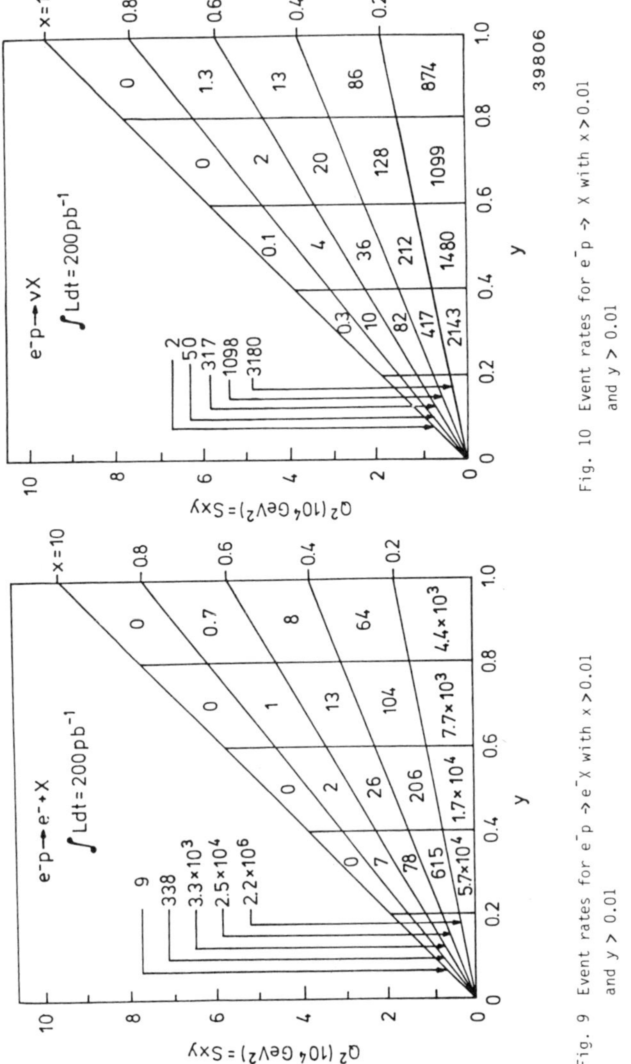

Fig. 10 Event rates for e⁻p → X with x > 0.01 and y > 0.01

Fig. 9 Event rates for e⁻p → e⁻X with x > 0.01 and y > 0.01

where

$$q(x) = u(x) + c(x) + \ldots$$

$$\bar{q}(x) = \bar{d}(x) + \bar{s}(x) + \ldots$$

This leads to

$$\frac{d^2\sigma}{dxdy}(e_L^- p \rightarrow \nu + X) = \frac{G_F^2 s}{\pi} \frac{1}{(1 + Q^2/M_W^2)^2} x\{u(x) + (1 - y)^2 \bar{d}(x)\}$$

where the higher mass quarks have been omitted. The expected event rates are shown in Fig. 10. They have been calculated[7] for the same luminosity as above. Roughly 10 000 events are produced at $Q^2 < 10^4$ GeV^2 and 1000 events at $Q^2 > 10^4$ GeV^2. Figure 10 shows that charged currents can be studied with sufficient statistics up to $\sim 4 \cdot 10^4$ GeV^2.

A precise determination of the structure functions will open the way to address several fundamental questions.

3.5.1 Test of QCD

The large Q^2 range accessible at HERA will allow a stringent test of QCD which predicts logarithmically falling structure functions. Mass corrections and higher twist contributions which affect present experiments should be negligible at HERA. Gluon radiation as illustrated in Fig. 11 leads to scale breaking of the form

$$F(x) \rightarrow \frac{F(x)}{1 + c\ell n(Q^2/\Lambda^2)}$$

The available structure function data span the Q^2 range 0-300 GeV^2; at HERA it will increase to 40 000 GeV^2. This will enhance greatly the sensitivity to power terms (in Q^2) whose presence would be at variance with QCD.

The accuracy with which one can hope to measure the QCD scale parameter is around ± 40 MeV for $\Lambda_{QCD} = 200$ MeV.

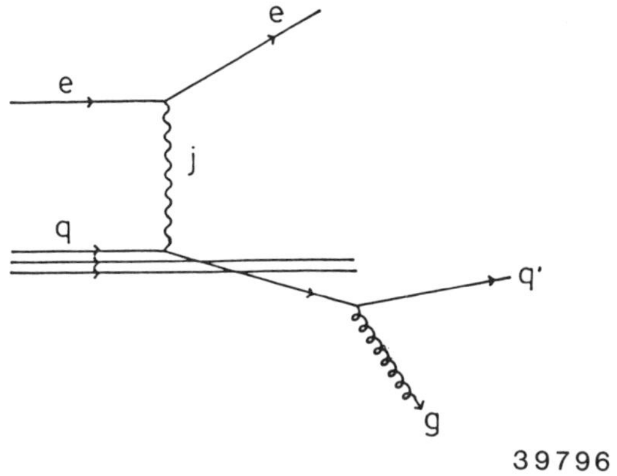

Fig. 11   eq scattering with accompanying gluon bremsstrahlung.

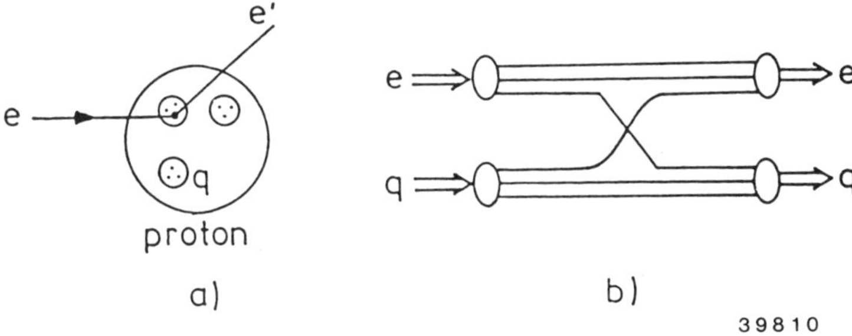

Fig. 12 a) Electron scattering on composite quarks.
       b) Composite electron and quark scattering by
          constituent interchange.

3.5.2 <u>Structure of quarks and electrons</u>

If quarks and/or electrons are extended objects the structure functions will show power law type deviations from their QCD predicted values. The sensitivity to possible structure of quarks and electrons has been estimated with the help of a model[9] which assumes quarks and electrons to be composites and to have common constituents (Fig. 12). The interchange of the constituents leads to a contact term of the form

$$\mathcal{L} \sim \frac{g^2}{\Lambda^2} \, \Sigma \, \bar{e}_\alpha \gamma^\mu e_\alpha \bar{q}_\beta \gamma_\mu q_\beta$$

where α, β denote states of a definite helicity, $\alpha = L,R$; $\beta = L,R$. The mass parameter Λ sets the compositness scale; g measures the coupling strength. For the following $g^2/4\pi = 1$ will be assumed. Figure 13 shows the ratio of the structure functions F_2 obtained with the contributions from $\gamma + Z^0$ exchange and the contact interaction, and for γ exchange alone:

$$F_2(\gamma + Z^0 + C.I.)/F_2(\gamma)$$

The ratio is shown as a function of Q^2 and x for different Λ values. At Q^2 values above 10^4 GeV^2 a contact interaction with $\Lambda = 1$ TeV leads to large deviations - factors of 5 to 10 - from the standard results. Given two years of data taking HERA will be sensitive up to Λ values of ~7 TeV corresponding to distances of $3 \cdot 10^{-18}$ cm. The present lower limits on Λ are 2-4 TeV on e,μ from $e^+e^- \rightarrow \mu^+\mu^-$ and 300 GeV for quarks from the total cross section for e^+e^- annihilation into hadrons.[10] A theoretical analysis of some composite models has indicated that the energy of HERA would still be insufficient to see structure in the quarks and electrons.[11] It suggests a lower as well as an upper limit on Λ:

$$\lambda(20 - 30)\text{TeV} < \Lambda < 250 \text{ TeV}$$

The lower limit is deduced from the limit on the decay $K^+ \rightarrow \pi^+\mu e$; λ is a number of order unity. The upper limit stems from cosmological considerations.

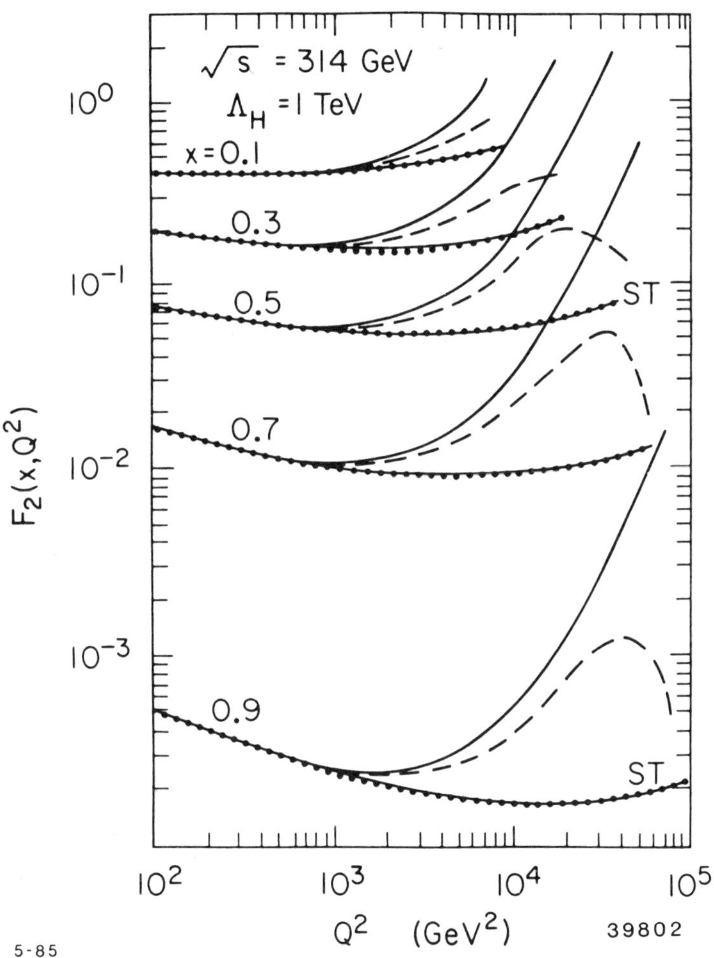

Fig. 13 The ratio $F_2(\gamma + Z^0 + C.I)/F_2(\gamma)$ for composite electron quark scattering.

3.5.3 New currents

Altarelli et al.[3] have estimated the effect of a second W_2 on the rate of charged current events. The amplitude for the exchange of the standard $W(W_1)$ can be written as

$$A(Q^2) \sim \frac{g^2}{8(Q^2 + m_N^2)} \sim \frac{G_F/\sqrt{2}}{1 + Q^2/m_W^2} \; .$$

Under the assumption that the second W couples in the same way to leptons and quarks the amplitude can be written as

$$A_{1+2}(Q^2) \sim \frac{g_1^2}{8(Q^2 + m_1^2)} + \frac{g_2^2}{8(Q^2 + m_2^2)} \; .$$

The coupling constants g_1, g_2 have to be chosen such that the low Q^2 region remains unchanged, i.e.

$$A(0) \equiv \frac{G_F}{\sqrt{2}} = A_{1+2}(0) = \frac{g_1^2}{8m_1^2} + \frac{g_2^2}{8m_2^2} \equiv \frac{g_2^2}{8m_W^2}$$

taking $m_1 = m_W$ and defining $r = g_2^2/g^2$ the ratio of the cross sections for two W's over one W is given by

$$\frac{\sigma(W_1 + W_2)}{\sigma(W_1)} = \left[1 - r \frac{m_1^2}{m_2^2} + r \frac{m_1^2}{m_2^2} \frac{1 + Q^2/m_1^2}{1 + Q^2/m_2^2} \right]$$

The ratio is shown in Fig. 14 for different W_2 masses assuming $r = 1$. For m_2 around 200 GeV large deviations from the standard theory result will be observed. The deviations become smaller as m_2 increases. With two years of data taking one will be sensitive to W_2 masses up to ~800 GeV.

A similar result holds for neutral currents and additional Z^0's.

3.5.3 Right handed currents

One of the great mysteries of weak interactions is the asymmetry between left and right handedness; there are left handed neutrinos and

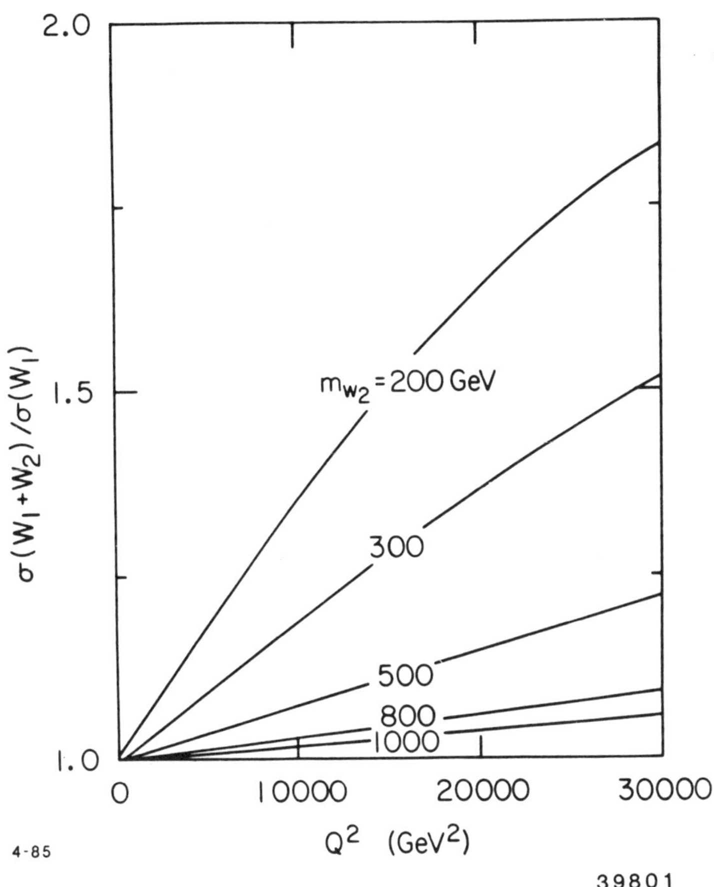

Fig. 14 The ratio of the cross sections for two and one W

left handed currents and no right handed counter parts. It has long been speculated that left-right symmetry is restored at some higher energy; namely there exist heavy right handed neutrinos N_R that couple to a right handed W_R.[12) The mass of N_R has been related to the masses of the corresponding charged lepton, m_e, and the left handed neutrino m_ν:[13) $m_{NR} \approx m_e^2/m_\nu$. For electrons this leads to the following prediction for m_{N_R}:

m_ν (eV)	0.1	1	10
m_{N_R} (GeV)	2611	261	26

Several experimental and theoretical limits have been put on the mass of W_R:

<u>β decay of μ's</u>

if ν is a Dirac spinor: $M_R > 380$ GeV   (Ref.14)

if ν is a Majorana spinor: no limit

<u>Nonleptonic decays</u> $M_R > 200$-300 GeV (Ref.15)

<u>$K_S^0 - K_L^0$ mass difference</u>

educated guess: $10 < M_R < 100$ TeV (Ref.16)

The lower limit on a right handed Z^0 is $M_{Z_R} \gtrsim 150$ GeV provided $\sin^2\Theta_W < 0.25$ (Ref.15). However, a simple Higgs structure leads to $M(Z_R)/M(W_R) \approx 1.6$. This suggests also $M(Z_R)$ to be in the region of tens of TeV (Ref. 17).

3.5.5 <u>Search for right-handed currents with longitudinally polarized electrons</u>

Longitudinally polarized electrons make HERA ideally suited to search for right handed currents. Any nonzero cross section contribution to $e_R^- p \to \nu X$ or $e_L^+ p \to \bar\nu X$ scattering signals the presence of a right handed charged current (see Sect. 3.4). In the absence of a W_R contribution:

$$\frac{d^2\sigma}{dxdy}\begin{pmatrix} e^-_L p \\ e^+_R p \\ e^-_R p \\ e^+_L p \end{pmatrix} = \frac{G_F^2 s}{\pi} \quad \frac{x}{(1 + Q^2/M_W^2)^2} \begin{pmatrix} (u+c)+(1-y)^2(\bar{d}+\bar{s}) \\ (d+s)+(1-y)^2(\bar{u}+\bar{c}) \\ 0 \\ 0 \end{pmatrix}$$

For neutral current reactions the situation is different. The photon is blind to the helicity of the electron but the Z^0 gives different contribution to e_L and e_R scattering as shown in sect. 3.3. Figure 15 shows the difference between the cross sections for $e^-_{L,R}$ and $e^+_{L,R}$ scattering. They amount to ~60 % at Q^2 values of 10^4 GeV^2.

4.   PRODUCTION OF NEW PARTICLES

4.1 Pair Production of Heavy Quarks

Photon-gluon fusion (Fig. 4) will be a strong source of heavy quarks. A discussion of the process can be found in Ref. 18. The total cross section depends on the quark mass approximately as

$$\sigma(eq \to Q\bar{Q}X) \sim M_Q^{-4}$$

Given a t quark mass of 40 GeV and a luminosity of 200 pb^{-1} the expected number of $t\bar{t}$ events is a few hundred. Top-like quarks can be searched for up to masses of 100 - 120 GeV.

The heavy quarks will essentially be photo produced ($Q^2 \approx 0$) and will be emitted in the direction of the incoming proton. Given the high mass these heavy quarks will decay into many particles (~20 - 25 for m_t = 40 GeV) which are isotropically distributed in the plane perpendicular to the beams. Thus, the production of new heavy quarks should be discernible from other processes. Figure 16 shows for illustration the production of a $t\bar{t}$ pair assuming M_t = 40 GeV.

Another possible source of Q production might be an intrinsic $Q\bar{Q}$ component in the nucleon. It has been argued for instance[19] that the $c\bar{c}$ component could be as large as 1 % and could become an important contributor once Q^2 exceeds ~4 M_c^2.

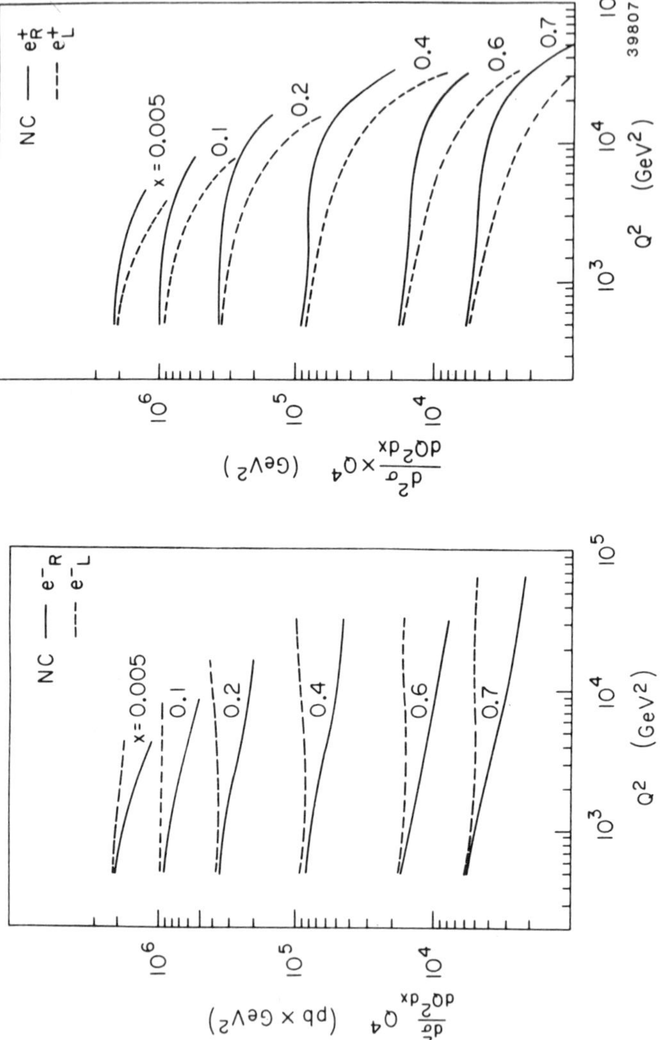

Fig. 15   The neutral current cross section for left and right handed $e^{\mp}$.

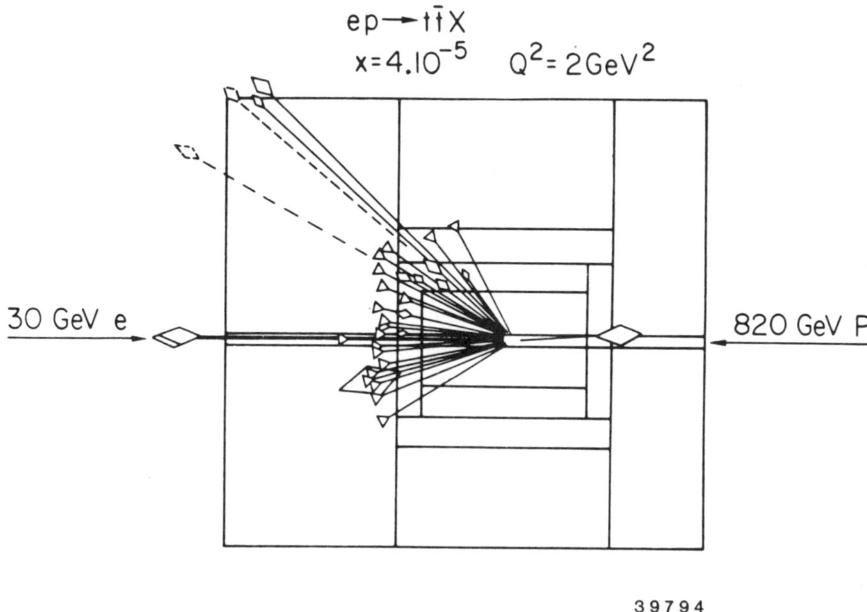

Fig. 16   Example of a tt̄ production event.

4.2   Vector Boson Production

Z^O and $W^{\pm}$ production is expected to proceed predominantly through the diagrams shown in Fig. 17a. The total cross sections are predicted to be[20]

$$\sigma(ep \to Z^O X) \approx 0.1 \text{ pb}$$

$$\sigma(ep \to W^{\pm} X) \approx 0.05 \text{ pb}$$

leading to ~20 Z^O events and ~10 $W^{\pm}$ events for 200 pb^{-1}.

In the (unlikely) case that Z^O and $W^{\pm}$ have also a strongly interacting component the time honoured pomeron exchange (Figs. 17b) may add to their production rate.

4.3   Higgs Production

Possible diagrams for production of Higgs particles are shown in Fig. 18. For a total of 200 pb^{-1} the expected yield of Higgs produced is not overwhelming[21]:

$$M_H = 10 \qquad 20 \qquad 50 \qquad 100 \text{ GeV}$$

$$\text{number of events} = 10 \qquad 10 \qquad 10 \qquad 3$$

If quarks and leptons are composites the existence of excited quarks and leptons $q^{\pm}, e^{\pm}$ appears to be natural (remember p, Δ). Their production would follow the diagrams of Fig. 19. The production rates are sufficient to search for $q^{\pm}, e^{\pm}$ up to masses of 250 GeV.[22]

4.5   New Currents, Quarks, Leptons

Suppose there exist new currents X^O, $X^{\pm}$ which link the known quarks and leptons to new quarks $Q = U, D, S, C, B, T$ and leptons L, with coupling strengths equal to that of the standard weak interactions. The cross section for the dominant associated production of D and L^O as depicted in Fig. 20 is given by

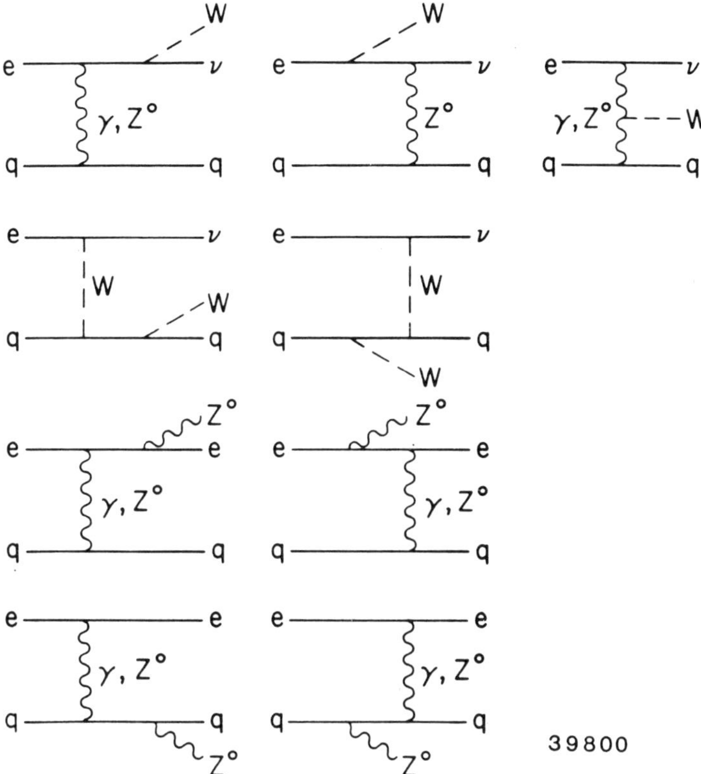

Fig. 17a   Diagrams for Z^0 and $W^\pm$ production

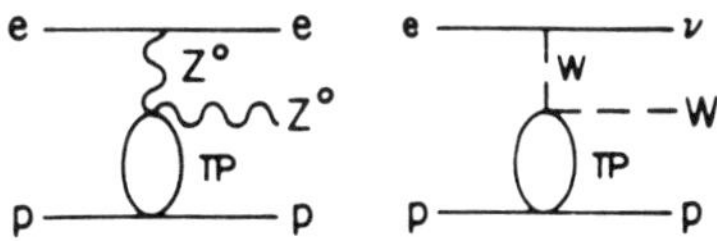

Fig. 17b   Diagrams for Z^0, $W^\pm$ production by pomeron exchange

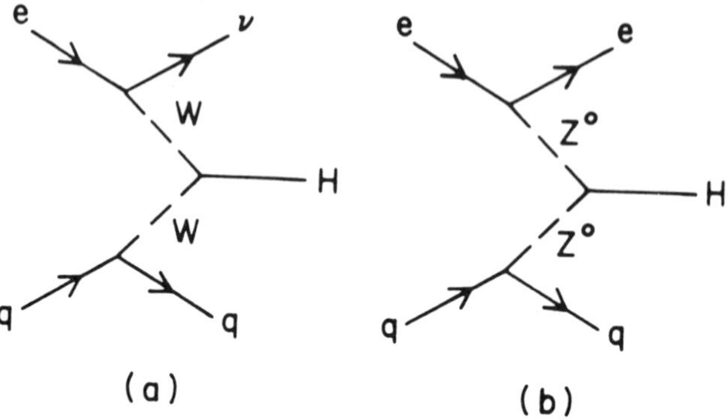

Fig. 18 Diagrams for Higgs production

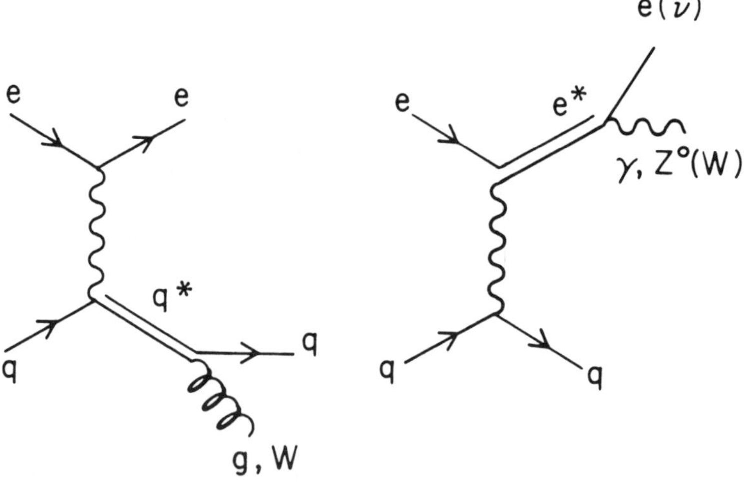

Fig. 19 Diagrams for the production of excited quarks and leptons.

$$\frac{d\sigma}{dxdy} = \frac{G_F s}{2} \left(\frac{m_W}{m_X}\right)^4 \frac{1}{(1 + \frac{sxy}{M_X^2})}$$

$$\left\{ xq(x) \left(1 - \frac{m_L^2 + m_D^2}{sx}\right) + x\bar{q}(x) \left(1 - y - \frac{m_L^2}{sx}\right)\left(1 - y - \frac{m_D^2}{sx}\right) \right\}$$

where

$$q(x) = u(x) + c(x) + \ldots$$

$$\bar{q}(x) = \bar{d}(x) + \bar{s}(x) + \ldots$$

The kinematically allowed regions for x and y are

$$\frac{(m_L + m_D)^2}{s} < x < 1$$

$$y_- < y < y_+$$

$$y_\pm = \frac{1}{2sx} \left[sx - m_D^2 - m_L^2 \pm \sqrt{(sx - m_D^2 - m_L^2)^2 - 4m_D^2 m_L^2} \right]$$

The process will lead to the event rates shown in Fig. 21 as a function of m_D for different lepton masses m_L assuming m_X = 83 GeV and 10 days of running at L = 10^{31} cm^{-2}s^{-1}; 200 pb^{-1} would probe up to $m_L + m_D \approx 220$ GeV.

4.6 Leptoquarks

Technicolour offers an alternative though not undisputed explanation for the Higgs mechanism. In this theory the Higgs is a pion-like composite: a state where a new quark and antiquark are bound together by the super strong technicolour force. Besides new quarks U,D,... new leptons E,M,... are also predicted which together lead to a new universe of particles. What is most exciting for HERA is the occurrence of leptoquarks TLQ = $\bar{u}$E,... The expected mass hierarchy is as follows:

		mass
- technipions ($U\bar{D}$,...)		0(10 GeV)
- leptoquarks ($U\bar{E}$,...)		0(160 GeV)
- technicolour Octets ($U\bar{D}$,...)		0(240 GeV)

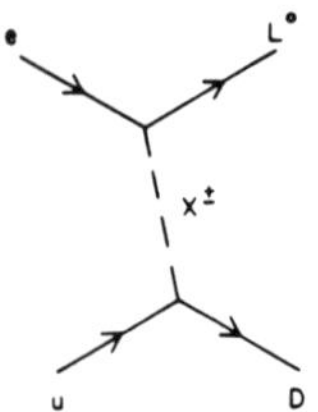

Fig. 20   Diagram for new quark and lepton production.

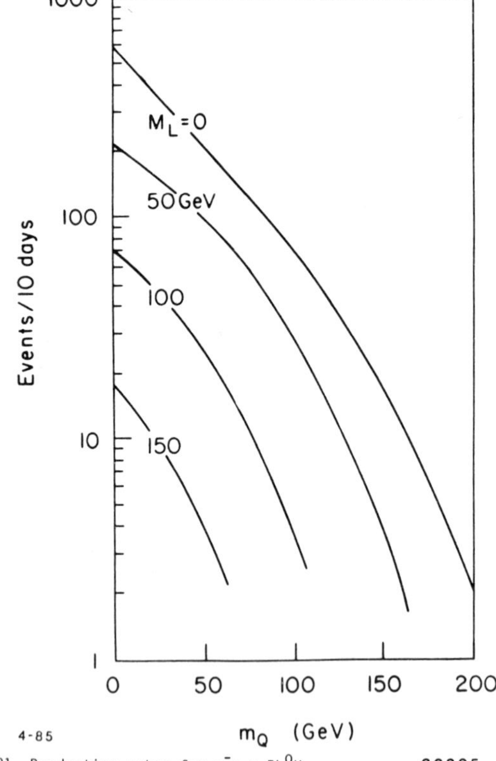

4-85

m_Q (GeV)

Fig. 21   Production rates for $e^-p \rightarrow DL^0X$.   39805

Leptoquarks would be within the reach of HERA experiments. The diagrams expected to dominate the production are given in Fig. 22. The cross section for ep $\rightarrow$ TLQX is shown in Fig. 23 as calculated by[23]. Here Θ_{et} is the mixing angle. Given a $\sin^2\Theta_{et}$ of ~0.05 leptoquarks with masses up to 180 GeV can be studied at HERA (Fig. 23).

Technicolour theories already have difficulties explaining the small $K_L^0 \rightarrow \mu^+\mu^-$ branching ratio, and no evidence has been found at PETRA or PEP for low-lying technipions.

4.7. Supersymmetric Particles

HERA is a hunter's paradise for supersymmetric perticles - if they exist in the accessible mass range. The prototype reaction is associated production of squark ($\tilde{q}$) and and slepton ($\tilde{\ell}$) by photino ($\tilde{\gamma}$), zino ($\tilde{Z}$) or wino ($\tilde{W}$) exchange (Fig. 24). The cross section has been calculated in Ref. 24. Figure 25 shows the cross section as a function of the squark mass for different slepton mass values. If we demand at least 10 events per 200 pb^{-1} then squark plus slepton production can be studied up to mass values of $m_{\tilde{q}} + m_{\tilde{e}} = 160$ GeV.

The production of scalar leptons has been studied in Ref. 20. The dependence on the $e^{\pm}$ polarization has been analyzed in Ref. 25.

5. DETECTION OF NEW PROCESSES

The mass range that is accessible to HERA for new particles was estimated in the previous sections by requiring the production of at least a few tens of these events a year. Since the yield of standard neutral and charged current events is many orders of magnitude larger one might wonder whether their detection will be as difficult as e.g. that of the top quark at the $S\bar{p}pS$ collider.

The events should be visible if the dominating background is ordinary neutral current scattering (eq $\rightarrow$ eq). This is a consequence of two facts: the kinematics in eq $\rightarrow$ eq is well defined, and, the decay of

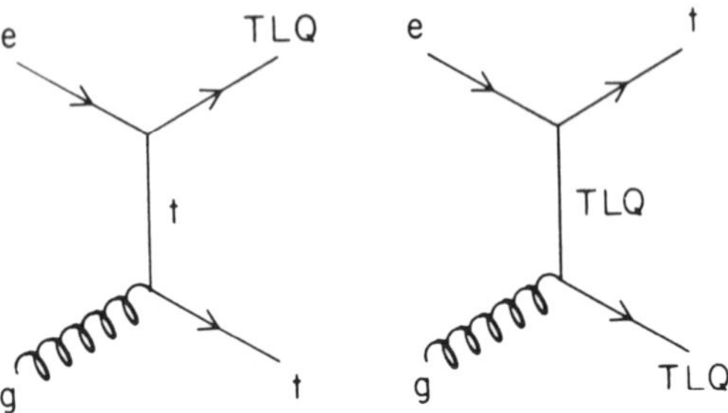

Fig. 22 Diagrams for the production of leptoquarks.

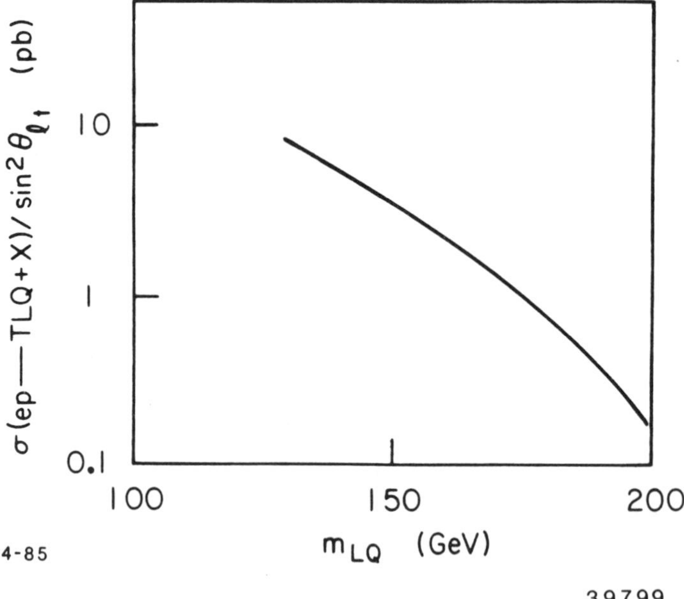

Fig. 23 Cross section for the production of leptoquarks.

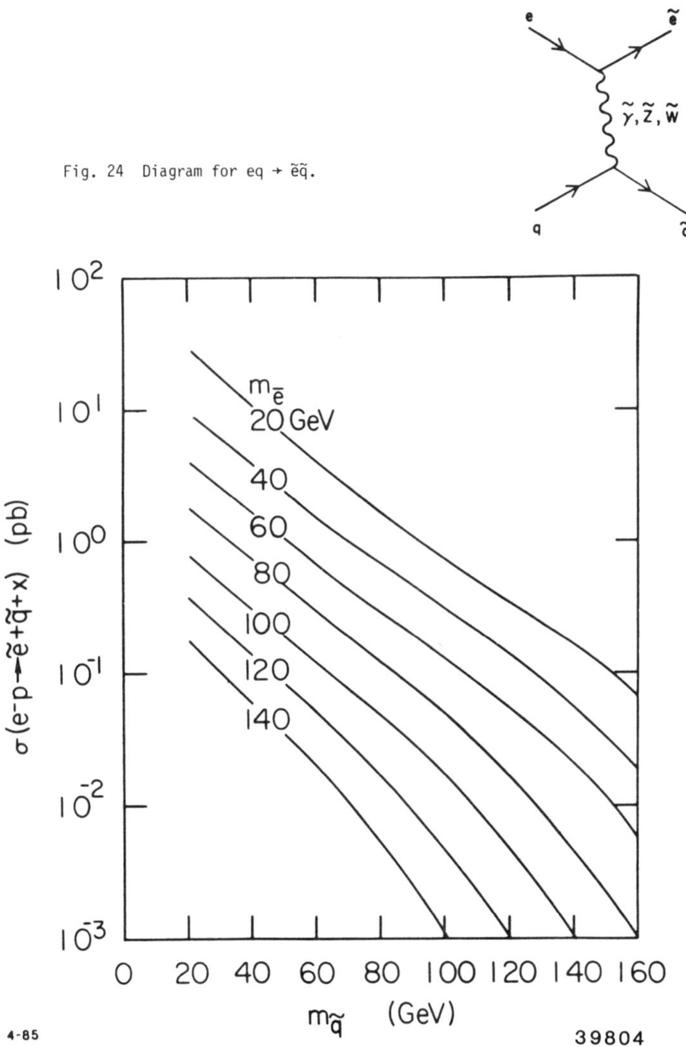

Fig. 24 Diagram for eq → ẽq̃.

Fig. 25 Cross section for the production of supersymmetric quarks and leptons.

a new heavy object will spread the emerging quark (lepton) over a large volume in phase space. For definiteness consider the supersymmetric reaction discussed before,

$$eq \rightarrow \tilde{e}\tilde{q}$$

The event rate relative to the standard neutral current reaction, $eq \rightarrow eq$, is

$$\text{for } M_{\tilde{q}} + M_{\tilde{e}} = \begin{cases} 80 \text{ GeV} \\ 180 \text{ GeV} \end{cases} \quad \frac{\sigma(\tilde{e}\tilde{q})}{\sigma(eq)} \approx \begin{cases} 10^{-3} \\ 10^{-5} \end{cases}$$

The expected decay chains for $\tilde{q}$ and $\tilde{e}$ are

$$\tilde{q} \rightarrow q + \tilde{\gamma}$$

and

$$\tilde{q} \rightarrow q + \tilde{g} \rightarrow qg\tilde{G}$$

and

$$\tilde{e} \rightarrow e + \tilde{\gamma} \text{ or } \tilde{e} \rightarrow e + \tilde{G} \ .$$

Under the assumption that $\tilde{\gamma}$ and $\tilde{G}$ are weakly interacting and leave the experiment undetected the final state consists of an electron and a quark jet as in the case of an ordinary neutral current event. The difference lies in the kinematics as seen in the plane perpendicular to the beams. This is illustrated in Fig. 26.

In the standard case the electron and the quark jet are anti-collinear and balance each other in transverse momentum. For the super-symmetric process e and q are neither anticollinear nor do they balance each other in transverse momentum. These two facts can be used to suppress the background. How efficiently that suppression can be done is demonstrated in Fig. 27. Here the acollinearity angle $\Delta\varphi$ is plotted versus Δy which measures the difference in y values computed from the current jet and the scattered electron:

$$\Delta_y = y_q - y_e$$

$$y_q = \frac{\Sigma E_i - P_{\| i}}{2E_e}$$

$$y_e = \frac{E'_e - P'_{\| e}}{2E_e}$$

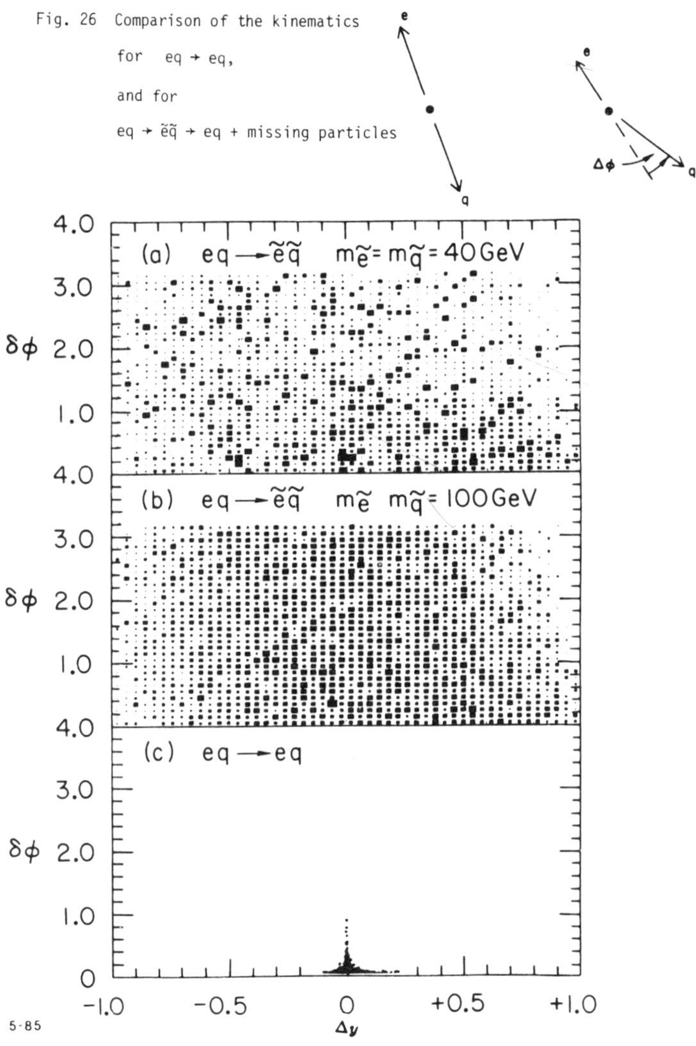

Fig. 27 Distribution of $\Delta\varphi$ versus Δy for the supersymmetric process eq → ẽq̃
with $m_{\tilde{e}} = m_{\tilde{q}} = 40$ GeV (a) and $m_{\tilde{e}} = m_{\tilde{q}} = 100$ GeV (b) and for the
standard process eq → eq (c).

178

$E_i, P_{\parallel i}$ are the energies and longitudinal momentum components of the current jet particles; $E'_e, p'_{\parallel}$ are the energy and longitudinal momentum component (measured w.r.t. the incoming proton direction) of the scattered electron.

The distributions shown in Fig. 27 have been computed by a Monte Carlo technique for a reasonably realistic detector assuming a finite hadronic energy resolution ($\Delta E/E = 60 \%/\sqrt{E}$) and a blind angular region around the beam pipe. Fig. 27(c) shows the result for $\sim 10^4$ standard neutral current events. As expected, the events cluster near $\Delta\varphi = \Delta y = 0$. In contrast, for the supersymmetric process (shown in Fig. 27(a) for $m_{\tilde{q}} = m_{\tilde{e}} = 40$ GeV and in Fig. 27(b) for $m_{\tilde{q}} = m_{\tilde{e}} = 100$ GeV). The events are spread all over the $\Delta\varphi, \Delta y$ map. Thus, supersymmetric events can be well isolated from the standard ones without significant loss in efficiency.

6. THE HERA COLLIDER

The layout of HERA is shown in Fig. 28 (Ref. 26). Two separate magnet systems guide the e^- and p beams around the 6.3 km long ring. The beams cross each other in four interaction points. The salient machine parameters are listed in Table 2.

Table 2:   Parameters of HERA

	p-ring	e-ring	units
Nominal energy	820	30	GeV
c.m. energy	314		GeV
Q^2_{max}	98.400		GeV^2
Luminosity	$1.5 \cdot 10^{31}$		$cm^2 s^{-1}$
Polarization time		25	min
Number of interaction points	4		
Crossing angle	0		
Free space for experiments	± 5.5		m
Circumference	6336		m
Bending radius	588	608	m
Magnetic field	4.65	0.165	T
Energy range	300-820	10-33	GeV
Injection energy	40	14	GeV
Circulating current	160	58	mA
Total number of particles	$2.1 \cdot 10^{13}$	$0.8 \cdot 10^{13}$	
Number of bunches	200		
Number of bunch buckets	220		
Time between crossings	96		ns
Beta function (β^*_x / β^*_y)	10/1.0	2/0.7	m
Beam size at crossing σ_x	0.27	0.26	mm
Beam size at crossing σ_y	0.08	0.070	mm
Beam size at crossing σ_z	11	0.8	cm
Energy loss / turn	$1.4 \cdot 10^{-10}$	127	MeV
Max. circumf. voltage	0.2/2.4	260	MV
Total RF power	1	13.2	MW
RF frequency	52.033/208.13	499.667	MHz
Filling time	20	15	min

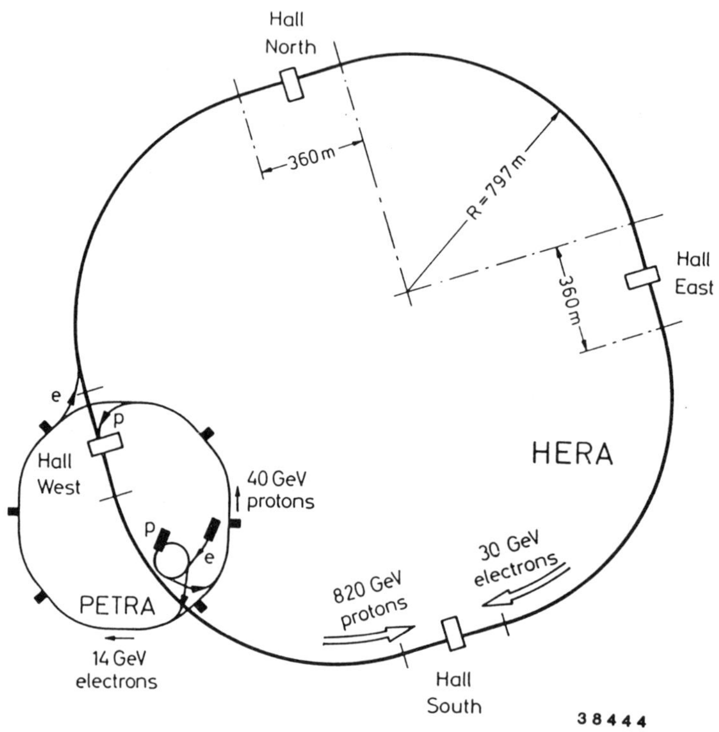

Fig. 28 Layout of HERA

6.1   Civil Engineering

The machine houses in a tunnel that is between 15 and 20 m deep underground (Fig. 29). The tunnel follows the terrain and has typically a slope of about 0.7 %. About one half of the tunnel is below the water table. Most of the tunnel runs underneath public land; only the stretch where HERA intersects PETRA is on DESY property. As a consequence, most of the installations for cryogenics, electrical and cooling power etc. are located on the DESY site and feed the rings through Hall West while the surface activities near the other halls - which are on public land - are restricted so as to keep the disturbance of the environment there at a minimum.

A cross section of the tunnel is shown in Fig. 30. It has an inner diameter of 5.2 m. The proton ring is located above the electron ring. Fig. 30 also indicates the supply lines for liquid and gaseous Helium and for cooling water.

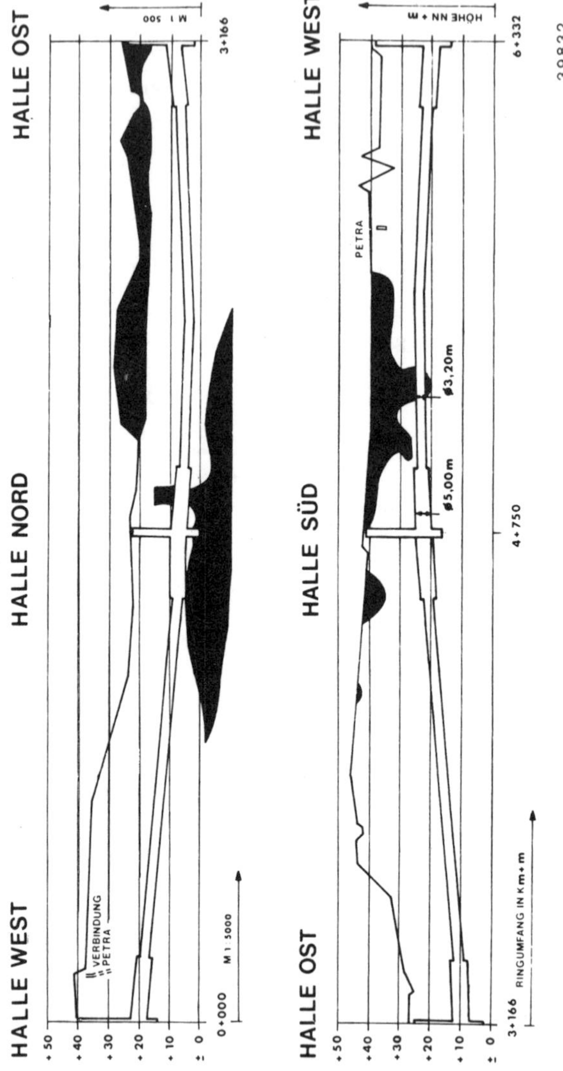

Fig. 29 The HERA tunnel.

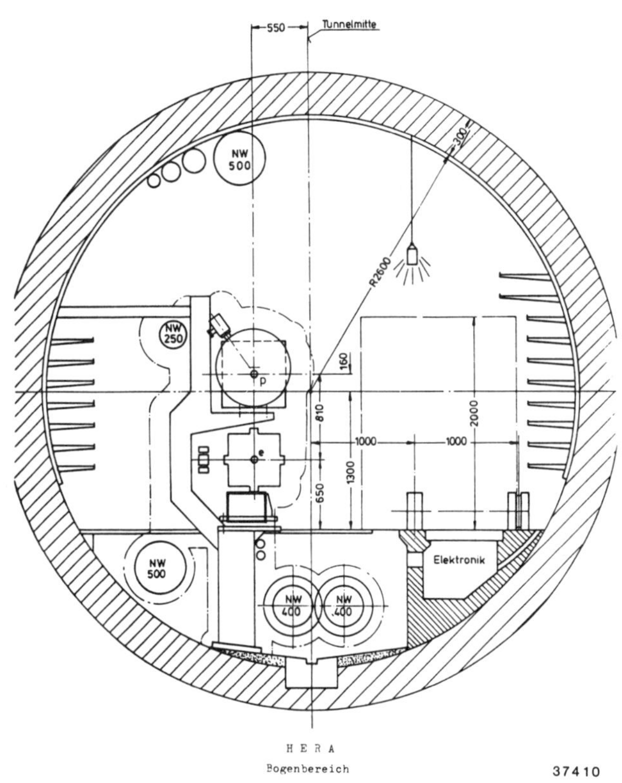

Fig. 30 Cross section of the HERA tunnel.

184

6.2   Injection System

The injection system uses the rebuilt synchrotron DESY and the storage ring PETRA (Fig. 31). Protons are accelerated to 50 MeV in a newly constructed LINAC, to 7.5 GeV in the revamped synchrotron, DESYIII, to 40 GeV in PETRA which requires bypasses around the RF cavities for the $e^{\pm}$ beams) and injected into HERA. Electrons pass from a LINAC (50 - 400 MeV) to the newly constructed synchrotron DESY II (9 GeV), to PETRA (14 GeV) and finally to HERA. The filling takes for each beam between 15 and 20 min.

6.3   Electron Ring

At 30 GeV a bending field of 0.165 T is required. This can be achieved with conventional magnets. In Fig. 32 a half cell of the magnet system is shown. Dipole, quadrupole and sextupole are premounted on one common support, the total length of a half cell being 11.8 m. Figure 32 also shows cross sections for the three different types of magnets. The dipole magnet is excited by a single conductor which carries at 30 GeV a current of 6767 A.

6.4   Proton Ring

The technological challenge of HERA lies in the superconducting magnet system for the proton beam. A bending field of 4.65 T is required to keep an 820 GeV proton beam on orbit. At this field strength iron cannot be used for field shaping and homogeneity has to be achieved by proper shaping of the superconducting coils. Figure 33a,b indicates schematically the manner in which the windings are wrapped around the beam pipe. In order to obtain a homogeneous field the current density has to follow a $\cos\varphi$ distribution, $I = I_o \cos\varphi$.

The mechanical accuracy required is quite demanding. Consider the case where the height of the left coil halves is larger by 2δ and that of the right coil halves is shorter by 2δ (Fig. 33c). If $\delta = 20$ μm, the skew quadrupole component is $a_2 = 1.7 \cdot 10^{-4}$ which is at the tolerable limit. Hence, a mechanical precision of typically 20 μm has to be kept.

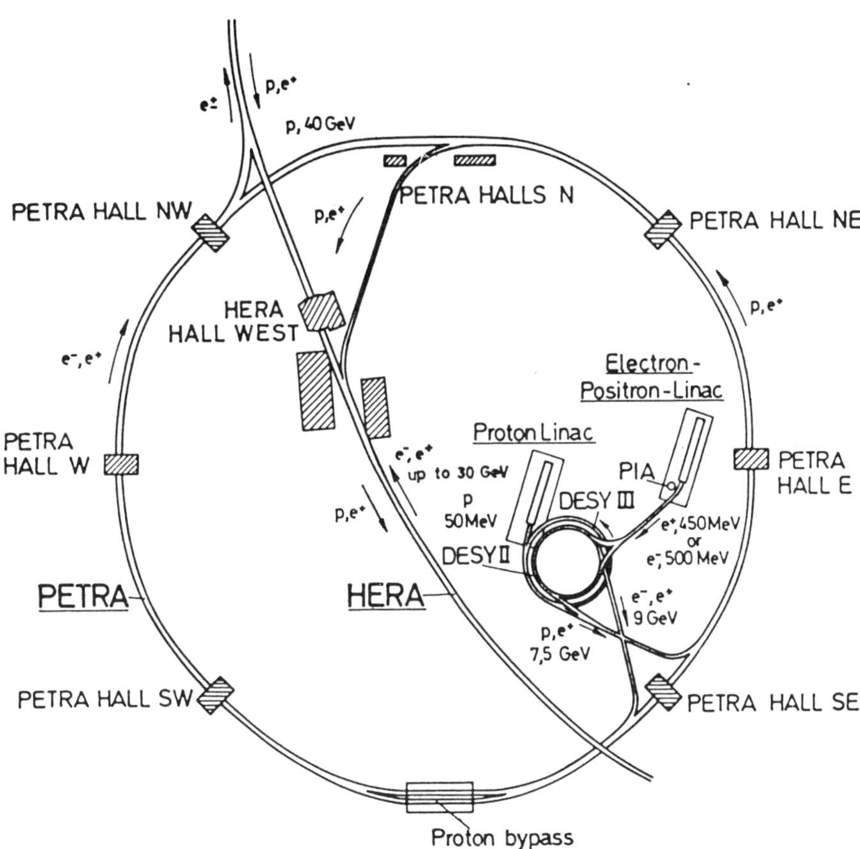

Fig. 31 The injection system for HERA

HERA: Magnetmodule for the Electron Ring

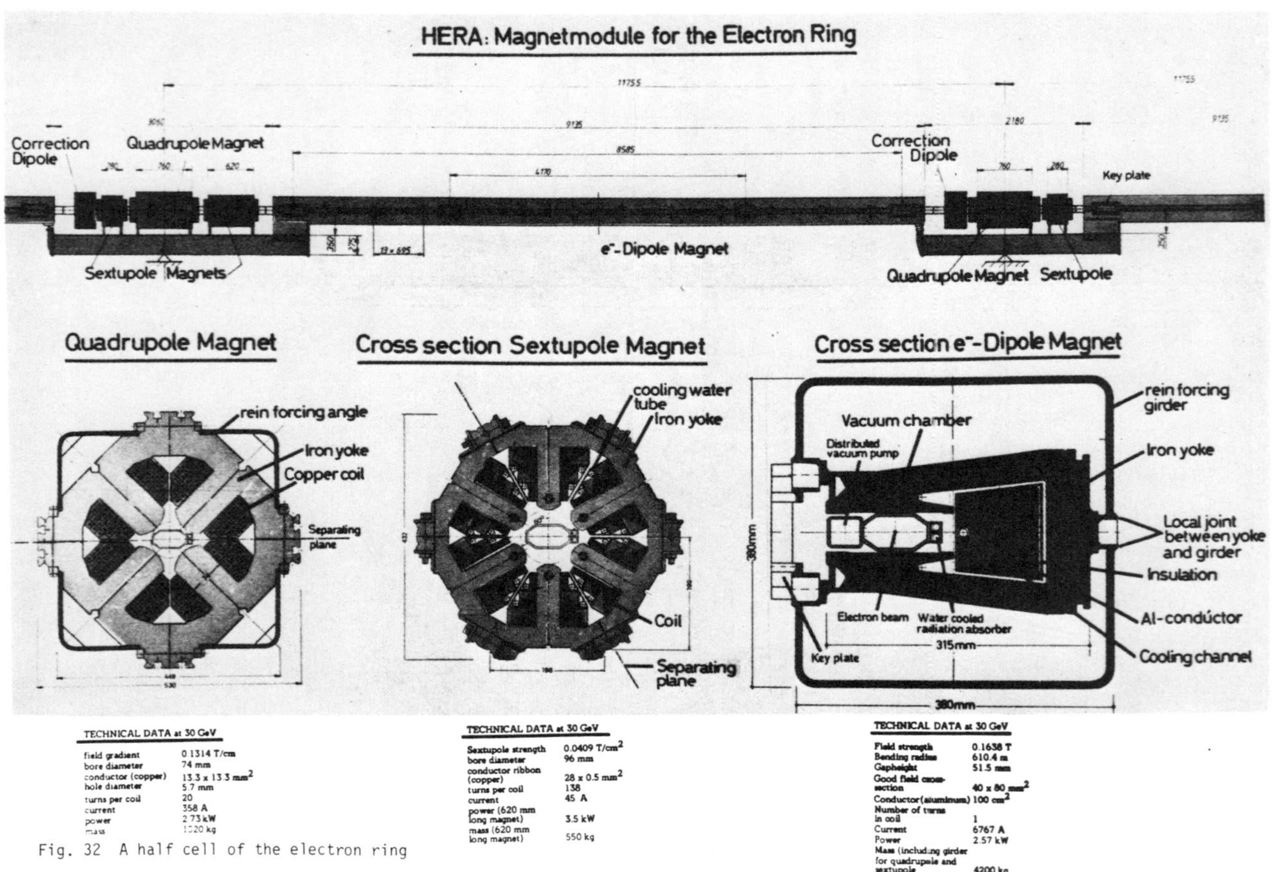

Quadrupole Magnet

TECHNICAL DATA at 30 GeV

field gradient	0.1314 T/cm
bore diameter	74 mm
conductor (copper)	13.3 x 13.3 mm^2
hole diameter	5.7 mm
turns per coil	20
current	358 A
power	2.73 kW
mass	1520 kg

Cross section Sextupole Magnet

TECHNICAL DATA at 30 GeV

Sextupole strength	0.0409 T/cm^2
bore diameter	96 mm
conductor ribbon (copper)	28 x 0.5 mm^2
turns per coil	138
current	45 A
power (620 mm long magnet)	3.5 kW
mass (620 mm long magnet)	550 kg

Cross section e⁻-Dipole Magnet

TECHNICAL DATA at 30 GeV

Field strength	0.1638 T
Bending radius	610.4 m
Gapheight	51.5 mm
Good field cross-section	40 x 80 mm^2
Conductor (aluminum)	100 cm^2
Number of turns in coil	1
Current	6767 A
Power	2.57 kW
Mass (including girder for quadrupole and sextupole)	4200 kg

Fig. 32 A half cell of the electron ring

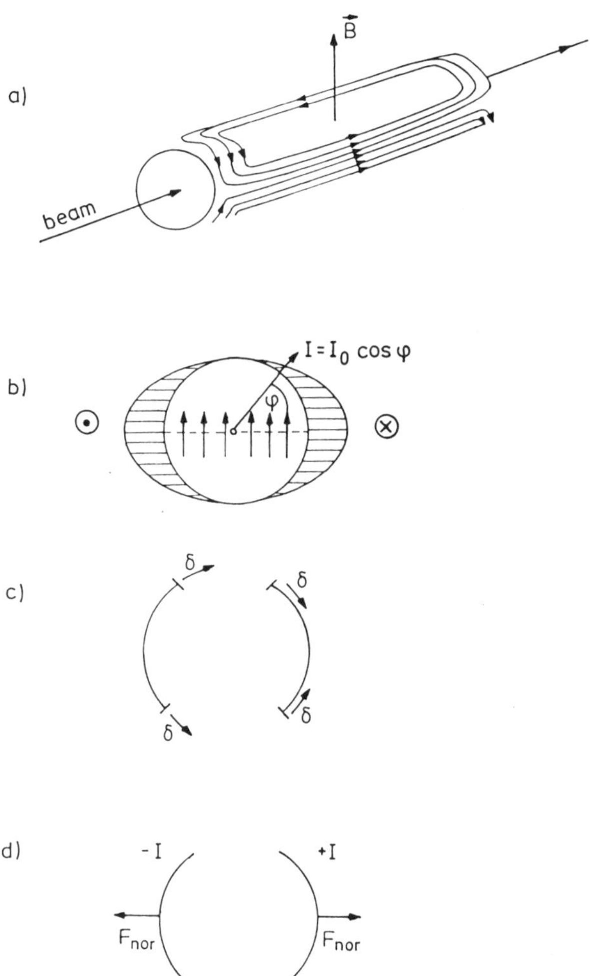

Fig. 33 The mechanics of a superconducting dipole

It has to be kept in the presence of a magnetic pressure which tries to push the two coil halves apart with a force of ~100 t per meter (Fig. 33d).

The design of the HERA magnets has started from the FNAL-TEVATRON magnets and has lead to a novel concept shown in Fig. 34. The coils are clamped by aluminum collars. The collaring is reinforced by the cold iron yoke which surrounds the collar tightly. Vacuum and superinsulation shield the iron yoke from the ambient temperature. The dipole magnet has an overall length of 9 m.

The advantages over a warm iron design are a 12 % gain in field, a smaller static heat load at 4 K and better quench protection. The disadvantage is a 5 times larger cold mass; however, the cooldown-warmup cycle takes the same amount of time in both cases, about 30 - 40 h.

A schematic drawing of a proton ring cell with the dipole, quadrupole and correction magnets is shown in Fig. 35. The length of one cell is 47.08 m.

6.5   RF System

6.5.1 <u>Electron ring</u>

The circulating electrons of energy E loose energy by synchrotron radiation:

$$\Delta E/turn = \frac{0.084\ E^4}{\rho}$$

where ρ is the bending radius in m, ρ = 608 m. For E = 30 GeV, ΔE = 112 MeV/turn. For an electron current of 58 mA the power lost by synchrotron radiation is 6.5 MW. The power loss is compensated by RF which is fed into cavities that are traversed by the electron beam.

The frequency chosen for the RF of HERA is 500 MHz, which is the same as for DESY, DORIS and PETRA (see Table 2). For operation below 30 GeV HERA will be equipped with cavities formerly used at PETRA. There is, however, a strong physics desire to run the electron beam at

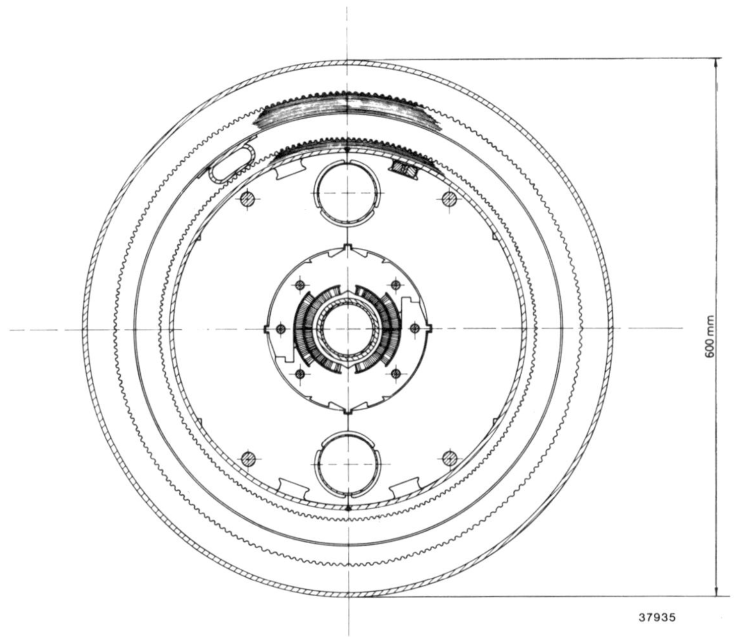

Fig. 34 Cross section of the superconducting dipole

somewhat higher energy: at E = 27 GeV the polarization time (see sect. 6.6) is 43 min compared to 12 min at E = 35 GeV.

In order to reach 35 GeV it is planned to add superconducting cavities (see Table 3). Superconducting 500 MHz cavities are under development at DESY (Fig. 36). With a 2 x 4 cell cavity in a single cryostat an accelerating field of 12 MV or a gradient of 5 MV/m is expected. This will lead to 240 kW/cryostat fed to the beam. The expected heat loss is 85 W/cryostat at 4K and 5 MV/m.

In a prototype test a 1 GHz 9 cell superconducting cavity with 2.7 MV/m has been installed in PETRA. It has been run for 5 months in normal operation with very encouraging results.[27]

Table 4:   RF for the electron ring.
A total available RF power of 7 MW is assumed.

number normal c.cavities	number super c.cavities	E (GeV)	max I (mA)	polarization time (min)
88	–	27	46	42
88	–	29	11	30
88	8	30	27	25
–	32	35	30	12

6.5.2 Proton ring

The loss of synchrotron radiation by the proton beam is negligible (remember $\Delta E \sim \text{mass}^{-4}$). For the proton beam new cavities and clystrons have to be developed. The frequency of the RF will vary from 52 MHz at injection (40 GeV) to 208 MHz at 820 GeV.

6.6   Electron Polarization

After coasting for some time the electrons will become polarized transverse to the ring plane (Fig. 37). The build-up time for polarization follows from the relation

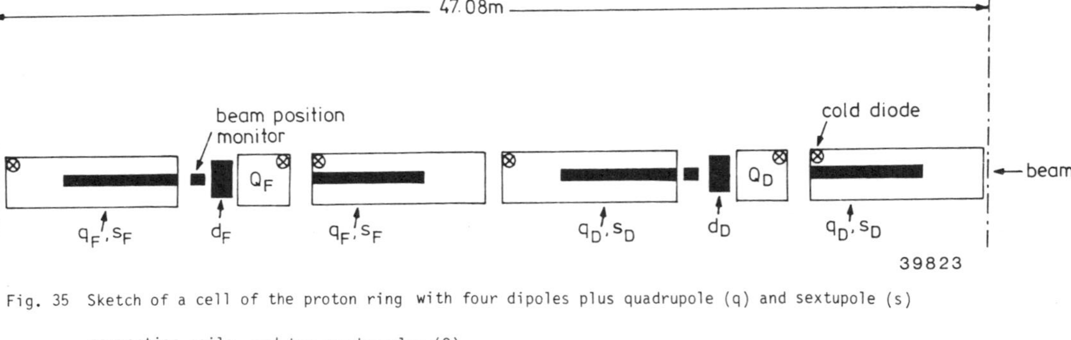

Fig. 35 Sketch of a cell of the proton ring with four dipoles plus quadrupole (q) and sextupole (s)

correction coils, and two quadrupoles (Q)

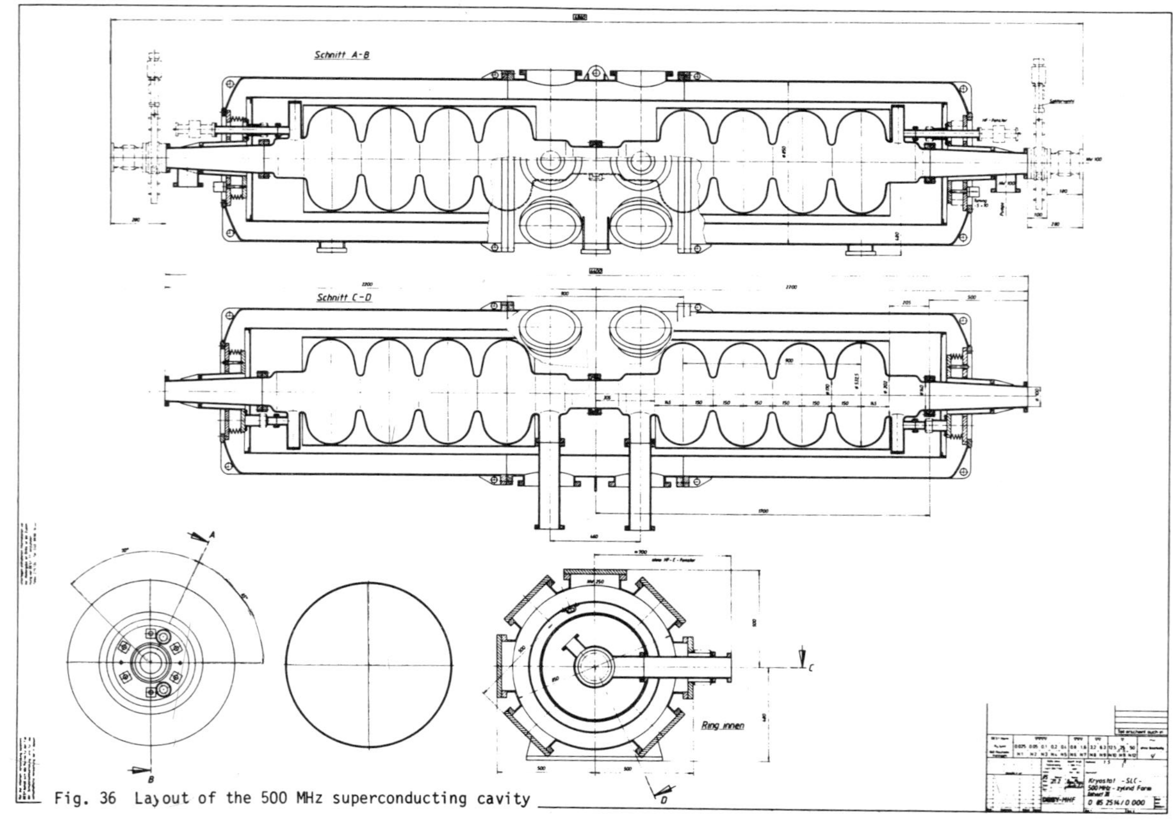

Fig. 36 Layout of the 500 MHz superconducting cavity

$$P(t) = P_0 \, (1 - e^{-t/\tau_p})$$

with
$$P_0 = 92 \, \%$$

and
$$\tau_p = \frac{98 \, \rho^2 \, R}{E^5}$$

where τ_p in s, ρ bending radius in m, R average radius in m and E in GeV.
For $E = 30$ GeV, $\tau_p = 25$ min.

For particle physics, instead of transverse polarization an electron beam of definite helicity is needed (see eg. sect. 3.5.5). This can be achieved in the following way. In the arcs the electrons remain transversely polarized. Shortly before reaching the interaction point the spin is rotated into the beam direction (or opposite to it) and as the beam leaves the interaction region the rotation is undone. Notice, that the two rotations have to be done with utmost precision since the build up time for polarization is $O(1000s)$ during which the electrons pass $2 \cdot 10^8$ times the rotators. It is clear that a tiny misalignment can destroy the polarization. The spin rotator makes use of the difference between the frequencies for spin precession (ω_{Larmor}) and revolution ($\omega_{cyclotron}$) in a magnetic field which results from the anomalous magnetic moment of the electron:

$$\omega_{Larmor} - \omega_{cyclotron} = \gamma \, \frac{g-2}{2} \, \frac{e}{m} \, B$$

where $\gamma = E/m$, m electron mass. This leads to a spin rotation angle of

$$\Theta_{rot} = 0.000159 \, \gamma\varphi$$

where φ is the angle by which the electron direction is bent in the magnet.

Numerous design studies have been made to find the optimum spin rotator (Ref. 28). A spin rotator of the siberian snake type is shown in Fig. 38; V and H define vertically and horizontally bending magnets. The total length of the rotator is about 70 m. The maximum degree of longitudinal polarization is expected to be $P_L \approx 80 \, \%$.

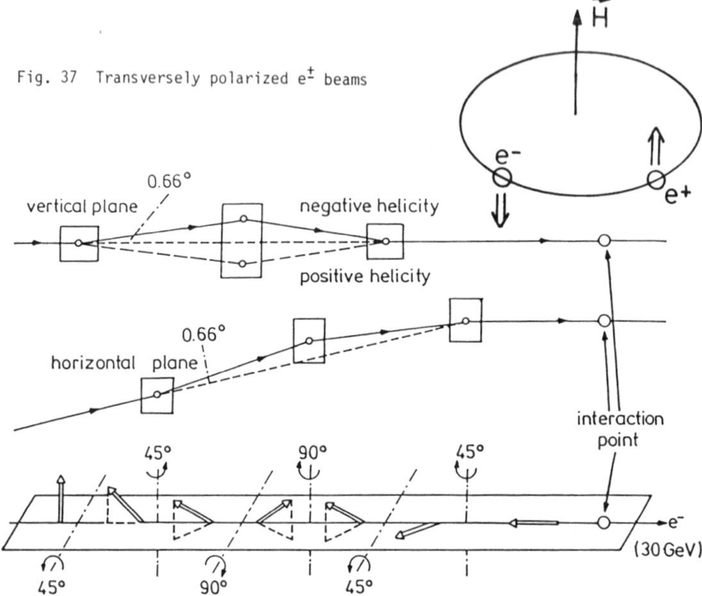

Fig. 37   Transversely polarized e$^\pm$ beams

Fig. 38   Spinrotator

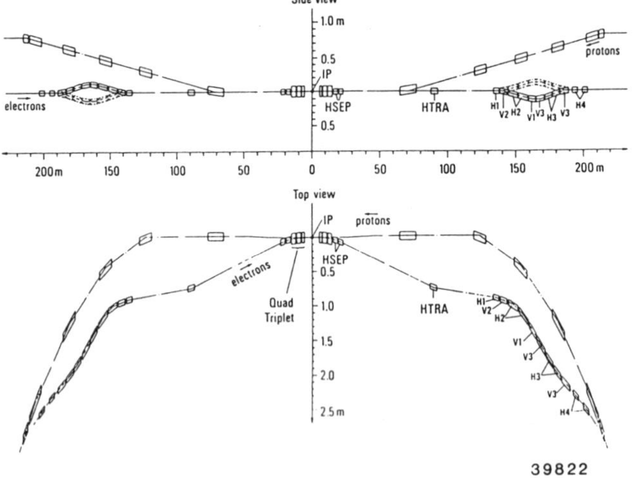

39822

6.7   Construction Schedule and Status of HERA

The schedule foreseen for the construction of HERA is given in Table 5.

Table 5:   The HERA schedule

Authorization	4/1984
Start civil engineering	4/1984
DESY II assembled	1985
Delivery of first e^- series magnets	8/1986
Start installing machine components in tunnel	10/1986
DESY II operational	end 1986

Superconducting dipoles:
    20 prototype magnets
    start series production
	
	end 1986
	2/1987

Electron injection into first quadrant	4/1987
Commissioning of proton LINAC	4/1987
Delivery of last e^- magnet	9/1987
Civil engineering completed	12/1987
Commissioning of DESY III	12/1987
Refrigeration plant operational	3/1988
Proton injection into first quadrant	1988
Delivery of last superconducting magnet	1988
Commissioning of proton ring	1989
Electron-proton physics	1990

The status of the project as of May 1986 is as follows:

<u>Buildings</u>

tunnel   : 30 % of the tunnel is done; the quadrant South Hall to West Hall is ready.

halls   : South Hall and West Hall are ready; construction of the other two halls has started.

DESY II   : is being tested.

e^- ring   : dipoles, quadrupoles, sextupoles have been ordered.

H^- LINAC   : the major components have been ordered. The ion source works within specification.

p ring   : three 9 m long dipoles have been built and tested; the quality of the field is good. The prototype quadrupole is being constructed at Saclay. Production of the preseries of the correction magnets has started in the Netherlands.

cryogenic plant: The components are being installed.

6.8   Cost and Financing

The capital costs for HERA in 1981 prices are

General: site, buildings, power	260 MDM
Electron ring	112 MDM
Proton ring	282 MDM
Sum	654 MDM

The Federal Government of Germany and the city of Hamburg share the financial burden in the ratio 9 : 1. However, 15 - 20 % of the expenditures are paid by foreign contributions. These are from

Canada	: beam transport, proton RF
China	: technical help
France	: prototype of the proton quadrupole
Israel	: currents leads for proton magnets
Italy	: half of all proton dipoles
Netherlands	: correction coils for the proton ring
Poland	: technical help and components
Switzerland	: liquid helium supply line
UK	: technical help
USA	: test superconducting cable.

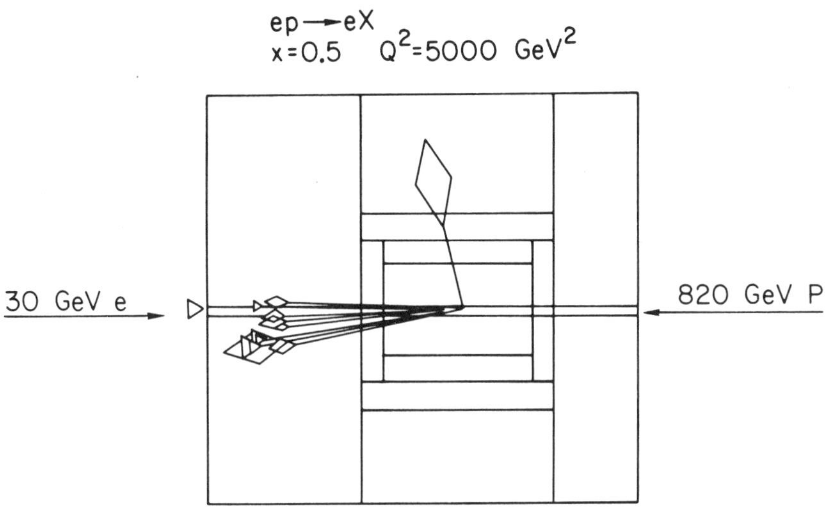

Fig. 39 Example of a neutral current event

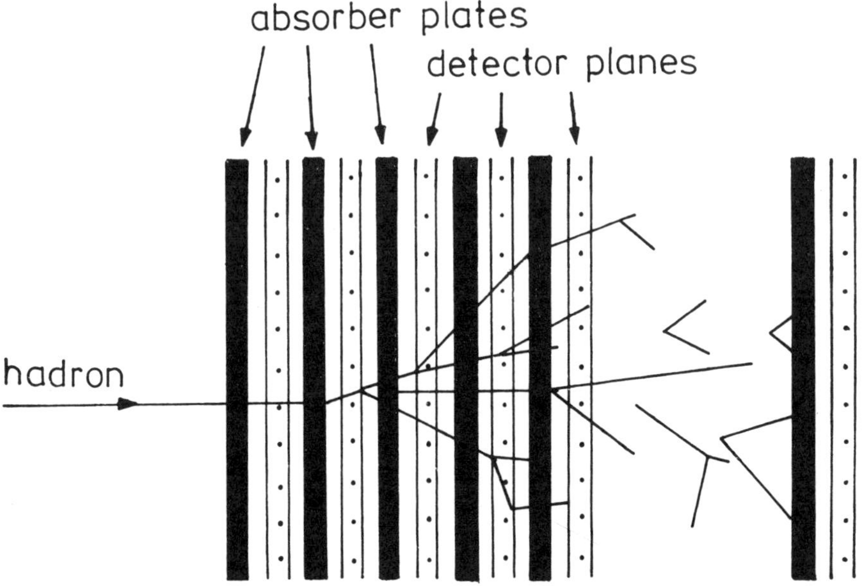

Fig. 40 Sketch of a sampling calorimeter

7.   EXPERIMENTATION AT HERA

Construction of a good detector for HERA is a challenge. Table 6 lists the requirements for several classes of physics. Particles and jets with energies up to 820 GeV have to be measured with precision. The large momentum imbalance between incident protons and electrons and the nature of space like processes send most particles into a narrow cone around the proton direction (see Fig. 39). Thus the calorimeter has to achieve simultaneously hermeticity, energy resolution and segmentation in a way not yet realized in any existing experiment. The detector has also to cope with high particle densities and with a time distance between bunch crossings of only 96 ns.

Table 6:   HERA physics and detector requirements

topic	comments	requirements
• structure of proton, quark, e	• safe prediction of standard theory; look for deviations	• inclusive measurement of e for NC, and of current jet for NC,CC
	• reactions occur at parton level large $Q^2 \leq 40\ 000$ GeV^2 accessible	→ hermetic electromagnetic and hadronic calorimeters with uniform response to photons and hadrons, best possible hadron energy resolution
	• polarized $e^{\pm}_{L,R}$ provide excellent handle on preon structure and on new currents	→ precise measurement of E_{jet}, Θ_{jet}
• new particles t, LQ, e^*, q^*, N_R, SUSY	• mass range: up to 120-200 GeV	• measurement of jets, jet topology and missing p_T
	• very clean signatures	→ calorimetry as above
	• particles produced by CC alone are unique to HERA	• measurement of isolated e, µ and on jets → tracking detectors over full solid angle; excellent e,µ identification
• photoproduction	• c.m. energies up to 250 GeV attainable	→ forward electron tagging

200

7.1 Calorimetry

Good calorimetry will be vital for HERA experiments. For this reason we shall discuss the properties of calorimeters in some detail. For reasons of economy the calorimeter has to be of the sampling type pictured in Fig. 40: it consists of absorber plates interleaved with detection layers to measure the electromagnetic and hadronic showers.

The energy resolution of a calorimeter is a combination of several contributions. Event-to-event fluctuations and sampling fluctuations usually dominate. The first contribution is called the intrinsic energy resolution.

For electromagnetic showers the ultimate resolution is determined by the sampling fluctuations of the electromagnetic cascade, which are well understood and can be described by simulation programs such as EGS (Ref. 29). For a calorimeter made of high Z material (e.g. Pb, U) of one radiation length (X_o) sampling thickness, the energy resolution is $\sigma/E = 0.15/\sqrt{E}$, E in GeV.

The performance of a hadron calorimeter depends strongly on its relative response to electrons and hadrons. A ratio e/h = 1 (compensated calorimeter) enables single hadron resolutions as good as $\sigma/E = 0.35/\sqrt{E}$ to be attained (Ref. 30) whereas only $(0.5\text{-}0.6)/\sqrt{E}$ has been reached for noncompensating calorimeters. This can be understood by considering a simple example. Assume e/h = 1.4 and hadronic showers which consist only of $\pi^{\pm}$ and π^{o}. On the average the energy carried by $\pi^{\pm}$'s, $E_{\pi^{\pm}}$, will be twice that carried by π^{o}'s, $E_{\pi^{o}}$. In individual events the $E_{\pi^{\pm}} : E_{\pi^{o}}$ ratio can fluctuate widely. This will lead to a large spread of the energy measured by the calorimeter, e.g.

$$E_{\pi^{\pm}} : E_{\pi^{o}} = 3 : 0 \qquad \text{rel. energy measured} \qquad 1.0$$
$$3 : 1 \qquad\qquad\qquad\qquad\qquad 1.13$$
$$1 : 2 \qquad\qquad\qquad\qquad\qquad 1.27$$
$$0 : 3 \qquad\qquad\qquad\qquad\qquad 1.40$$

For hadronic showers (Ref. 31), the energy is deposited through electromagnetic cascades (E_{em}), ionization by charged hadrons (E_{had}), and nuclear breakup and excitation energy (E_{nuc}), all of which have different responses in the calorimeter. As a result, the energy resolution is determined both by sampling fluctuations and by the event-to-event fluctuations in E_{em}, E_{had} and E_{nuc}, with different responses in the calorimeter.

In a sampling calorimeter, due to the different Z dependence of the individual processes involved, the response to E_{em} is generally reduced relative to the response to a minimum ionizing particle (mip). In a noncompensating calorimeter only a small fraction of E_{nuc} is detected.

Equal response to electrons and hadrons has sofar been achieved only in calorimeters made from depleted uranium (DU) and scintillator (Ref. 30).

7.2　Compensation in a DU-Scintillator Calorimeter

Five different effects determine the response to hadron showers from DU sampling calorimeters as illustrated in Fig. 41 (Ref. 7):

1. Ionization by hadrons produce energy loss in the sampling layer (dE/dx).

2. Electromagnetic showers are the result of the decay of π^0's. Electromagnetic energy (mainly low energy γ's) is predominantly lost in the high Z material. The difference in response to electromagnetic showers and minimum ionizing particles (e/mip ratio) (Ref. 32) is due to the different Z dependence of the cross sections for ionization ($\sim Z^1$), pair production ($\sim Z^2$), bremsstrahlung ($\sim Z^2$), photoelectric effect ($\sim Z^5$) and Compton effect ($\sim Z^1$). In a DU-scintillator sampling structure the e/mip ratio is around 0.6, weakly dependent on the plate thickness (see Fig. 42). Crudely speaking, almost all of the energy of low energy photons is absorbed via the photoelectric effect in the nonactive high Z absorber material.

3. In material with very high Z like DU, hadronic showers produce a significant number of neutrons with energies in the MeV range (Fig. 43). The hydrogen content of an organic scintillator converts

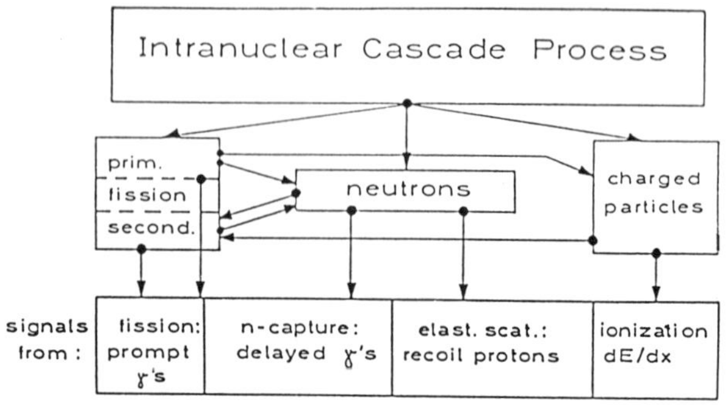

Fig. 41 Interactions contributing to compensation.
Not shown is the electromagnetic component of hadronic showers

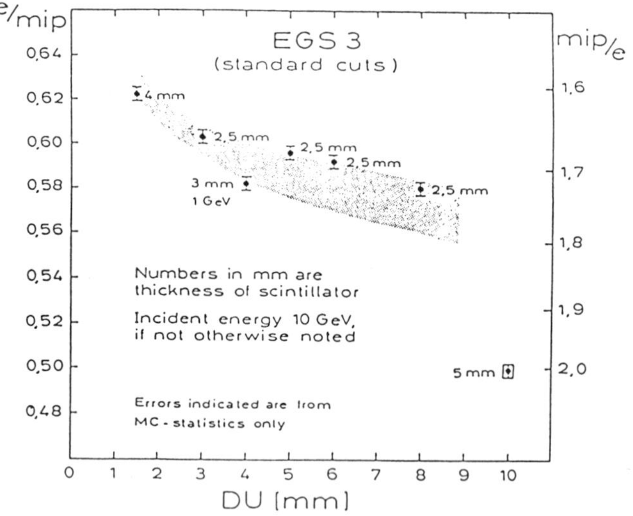

Fig. 42 The e/mip ratio for several DU - scintillator structures,
calculated with EGS (Ref. 7).

most of the kinematic energy of the neutrons into measurable proton-recoil energy. Thus the hadron response is boosted considerably. In contrast, if liquid argon were used as detector material, the energy transfer to the argon nucleus in an n-Ar collision and consequently the boost of the hadron signal would be negligible.

4. The neutrons in the MeV range are moderated in energy mainly by elastic collisions with hydrogen. At very low neutron energies the neutron capture process dominates which yields delayed gamma radiation and hence a delayed boost of the signal.

5. Every spallation and every fission process is accompanied by prompt nuclear gamma deexcitation. Although the γ efficiency of scintillator is very low, the effect still contributes to the compensation. Liquid argon would detect a larger fraction of this component.

We shall express the e/h ratio in terms of the pulse height contributions for the electron and hadron:
Electrons of energy E yield a pulse height $P_e(E)$;
Hadrons of energy E yield:

- π^0's with a total energy of E_1 yielding a pulse height of $P_e(E_1)$,

- charged hadrons with a total energy E_2 yielding via dE/dx a pulse height $P_x(E_2)$,

- neutrons with total energy E_3 yielding via proton recoil a pulse height $P_{rec}(E_3)$,

- energy E_4 lost to binding and neutrinos.

Note, the small contributions from prompt and delayed γ's have been neglected; in the case of scintillator their energy can be included in E_4.

Now make the following assumptions:

1. The pulse height $P_i(E)$ ($i = e,x,...$) is proportional to E, i.e.
   $P_i(E) = p_i E$.

2. The energy carried by neutrons is proportional to the energy lost to the binding energy (plus neutrinos),

$$E_3 = aE_4$$

and
$$E = E_1 + E_2 + E_3 + E_4 \ .$$

This yields
$$\frac{h}{e} = \frac{P_h(E)}{P_e(E)} = \frac{P_e E_1 + P_x E_2 + P_{rec} E_3}{P_e E}$$

$$= 1 - \left[1 - \frac{P_x}{P_e} \right] \frac{E_2}{E} - \left[(1 + \frac{1}{a}) - \frac{P_{rec}}{P_e} \right] \frac{E_3}{E}$$

We shall now make a further and not well founded assumption:

3. The pulse heights produced by electrons and produced via dE/dx by hadrons are equal,

$$P_x = P_e,$$

and obtain
$$\frac{h}{e} = 1 - \left[(1 + \frac{1}{a}) - \frac{P_{rec}}{P_e} \right] \frac{E_3}{E}$$

Compensation can now be achieved for any energy partition E_3/E and therefore separately for each event if

$$\frac{P_{rec}}{P_e} = (1 + \frac{1}{a})$$

Given h/e = 1 for each event we would expect to obtain the best energy resolution.

Figure 44 illustrates how the ratio P_{rec}/P_e might be tuned to achieve compensation. It is the result of a model calculation for a DU-scintillator calorimeter with d_u = 3 mm thick DU plates and variable scintillator thickness d_s (Refs. 7, 34). The air gaps between scintillator and DU plates are assumed to be small (< 0.5 mm). The scintillator is assumed to have a kB factor of $0.85 \ 10^{-2}$ g/MeV cm^2 (as is the case e.g. for SCSN38). The gate length is assumed to be 100 ns. Compensation is predicted for d_u/d_s = 1.4 or a scintillator thickness of 2.1 mm. This is close to the 2.5 mm that has been used by Ref. 30 for their DU-

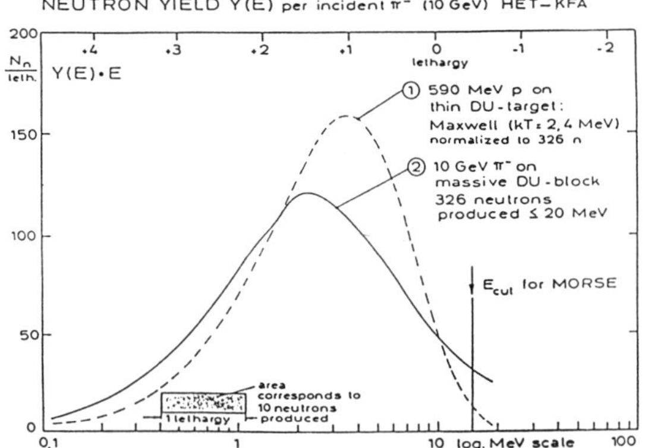

Fig. 43   The neutron yield spectrum for a massive DU block (Ref. 7).

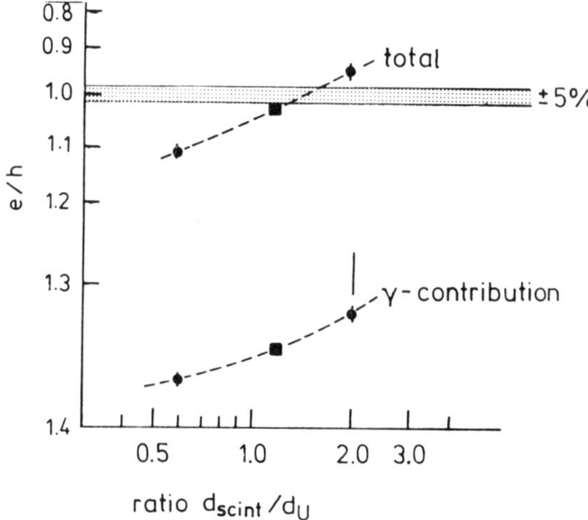

Fig. 44   The e/h ratio as a function of the ratio of the DU to scintillator
thickness (Ref. 7).

scintillator calorimeter. Figure 44 shows that for a fixed DU plate thickness a thicker scintillator will increase the e/h ratio, a thinner one will decrease it.

The energy resolution in a compensated DU-scintillator calorimeter can be written as (Ref. 35):

$$\frac{\sigma(E)}{E} = \frac{22\ \%}{\sqrt{E}} + \frac{0.09\ \Delta E}{\sqrt{E}}$$

where ΔE measures in MeV the energy loss of a mip per absorber plate. The contribution from photon statistics has been neglected.

Figure 45 shows measurements (Ref. 37) from the NA34 experiment at CERN using 3 mm DU plates interleaved with 2.5 mm scintillator plates. The resolution varies between $\sigma/\sqrt{E}$ = 28 % and 36 %, the average being 33 % for pions from 8 to 200 GeV.

7.3   Weighting in a Noncompensating Calorimeter

The energy resolution in a noncompensating calorimeter such as Fe-scintillator is given by

$$\frac{\sigma(E)}{E} \approx \frac{40\ \%}{\sqrt{E}} + \frac{0.09\ \Delta E}{\sqrt{E}} + 5.5\ \%$$

The last term arises from the event-to-event fluctuations in the energy deposited by π^0's and the difference in response to e and h. One can get rid of this term by a crude estimate of the π^0 content of the shower. The idea is the following: The shower of a charged hadron is typically spread over 3 - 5 absorption lengths, λ (e.g. $\lambda(Pb)$ = 17.0 cm) while the energy of a π^0 is absorbed within 5 to 10 radiation lengths, X_0 (e.g. $X_0(Pb)$ = 0.58 cm). The π^0 energy deposits are therefore very localized in a hadronic shower; they form hot spots. By measuring the energy content of hot spots the energy deposited by π^0's can be estimated.

In practice it is sufficient to measure the first and second moment of the energy release along the shower. Figure 46 shows the result from an iron-scintillator calorimeter (Ref. 38) using 2.5 cm thick Fe plates

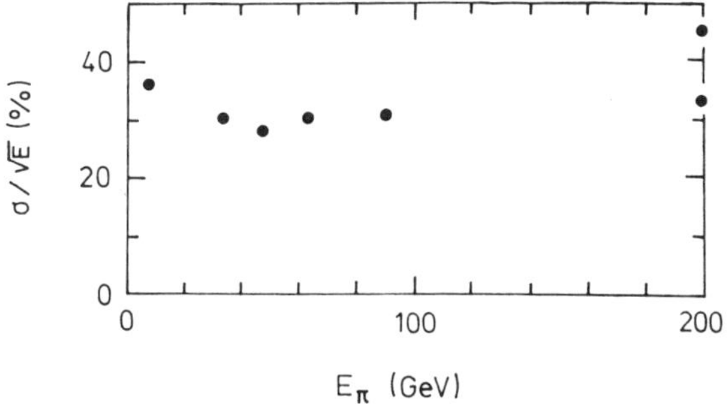

Fig. 45   The energy resolution of a DU-scintillator calorimeter (Ref. 37).

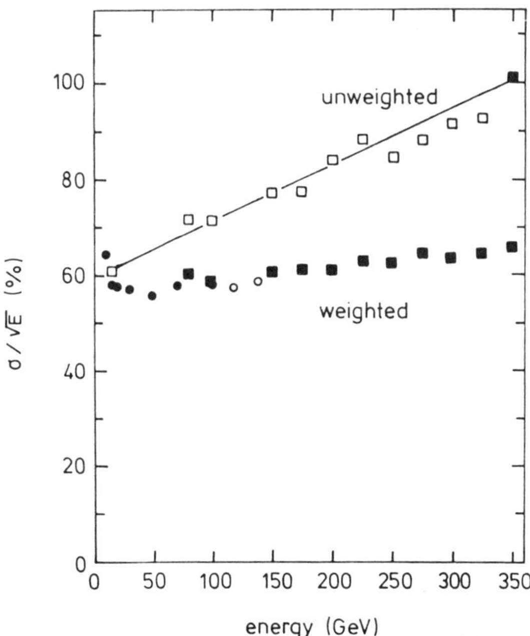

Fig. 46   The energy resolution of a Fe-scintillator calorimeter
before and after weighting (Ref. 38).

(λ(Fe) = 16.8 cm). Every 5 layers were readout by a separate phototube
i yielding the measured energy E^i_{meas}. The corrected energy E^i_{corr} is
found by (Refs. 38, 39)

$$E^i_{meas} = E^i_{corr} (1-\alpha E^i_{corr})$$

$$E_{meas} = \sum_i E^i_{meas} = \sum_i E^i_{corr} - c \frac{\sum_i (E^i_{corr})^2}{\sqrt{\sum_i E^i_{corr}}}$$

The constant c depends on the e/h ratio.
As shown by Fig. 46 the energy resolution in the unweighted case increases from $\sigma(E)/\sqrt{E}$ = 60 % near E = 10 GeV to ~100 % at 200 GeV incident pion energy. After weighting the energy resolution remains 60 %
over the full energy range.

7.4   Depth of the Calorimeter

In order to preserve the optimum energy resolution at all energies,
the calorimeter has to be sufficiently deep. The depth profile of
hadronic showers was measured experimentally in the WA78 experiment at
CERN using a 5.2 λ deep DU-scintillator calorimeter followed by a 7.5 λ
Fe-scintillator calorimeter. Fig. 47 shows the fraction of 30 GeV/c
hadron showers with 95 % energy containment as a function of the calorimeter depth. In order to have 95 % containment for 90 % of 30 GeV/c
hadrons a depth of 6 λ is required.

The maximum jet energy for HERA is shown in Fig. 48 as a function
of the production angle Θ. It reaches 800 GeV near the forward direction,
falling to 300 GeV at 30°, 100 GeV at 60° and less than 50 GeV beyond
90°. Requiring 99 % containment leads to a depth of ~11 λ in the forward
direction, 7 λ at 90° and 6 λ at 180°.

In general, cost considerations prohibit the construction of a
high resolution calorimeter over the full depth. Instead, the calorimeter is divided into a high resolution front part and a backing calorimeter with moderate energy resolution. The backing calorimeter recog-

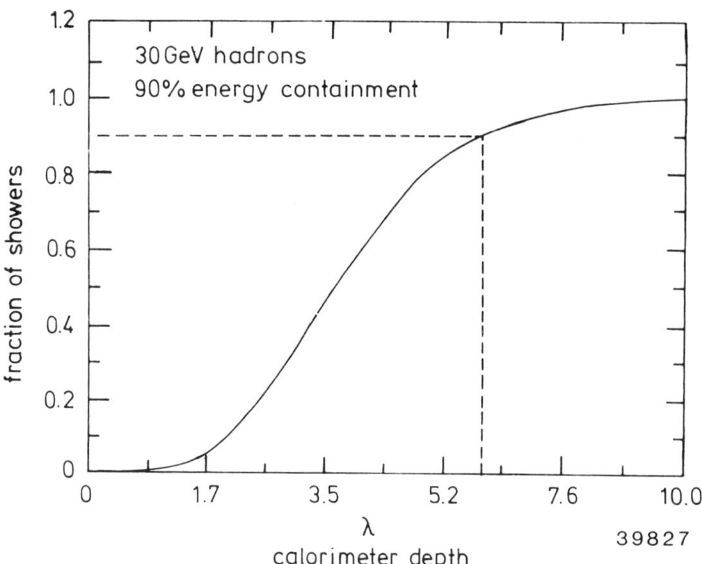

Fig. 47 Fraction of 30 GeV hadronic showers with 95 %
energy containment versus calorimeter depth.

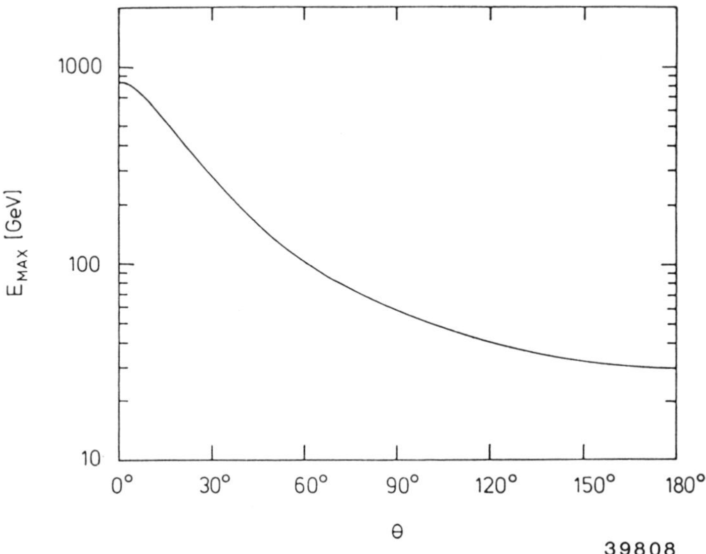

Fig. 48 Maximum jet energy as a function of polar angle.

nizes late showers and can be used to correct for energy leakage or to select a high resolution sample.

8.   PROPOSED HERA EXPERIMENTS

Two collaborations, ZEUS and H1, have proposed to construct detectors for HERA. In 1985 they have submitted letter of intents (Ref. 40, 41) which have been followed in 1986 by technical proposals (Refs. 42, 43). These proposals are presently under review by the DESY Physics Research Committee and approval is expected for summer of this year.

Both detectors require large efforts in terms of personnel and investment. The ZEUS collaboration consists of 268 physicists and estimates the material budget required to build the detector at 135 MDM. The corresponding numbers for H1 are 165 physicists and 92 MDM. Both groups envisage completion of the detector by 1990.

8.1   The H1 Detector

The layout of the detector is shown in Figs. 49, 50. Starting from the interaction point,charged particle tracking in the central and forward region is provided by a central jet-chamber interleaved with two z-chambers, and MWPC's (1) (inner, outer radii: 15, 90 cm) and a forward tracker (2) consisting of a series of radial and planar drift chambers interleaved with three layers of MWPC's and transition radiators. In combination with a magnetic field of 1.2 T the momentum resolution expected for isolated charged particles between 7^0 and 150^0 is $\sigma(P)/P^2 \leq 0.003$

The EM calorimeter utilizes Pb plates with liquid argon as readout in the barrel and forward region (3) and a lead scintillator sandwich in the backward region (5). The liquid argon part of the EM calorimeter is highly segmented. It consists of 3×3 cm^2 towers and 4 longitudinal readouts in the forward direction and of 8×8 cm^2 towers and 3 readouts in the backward barrel region .

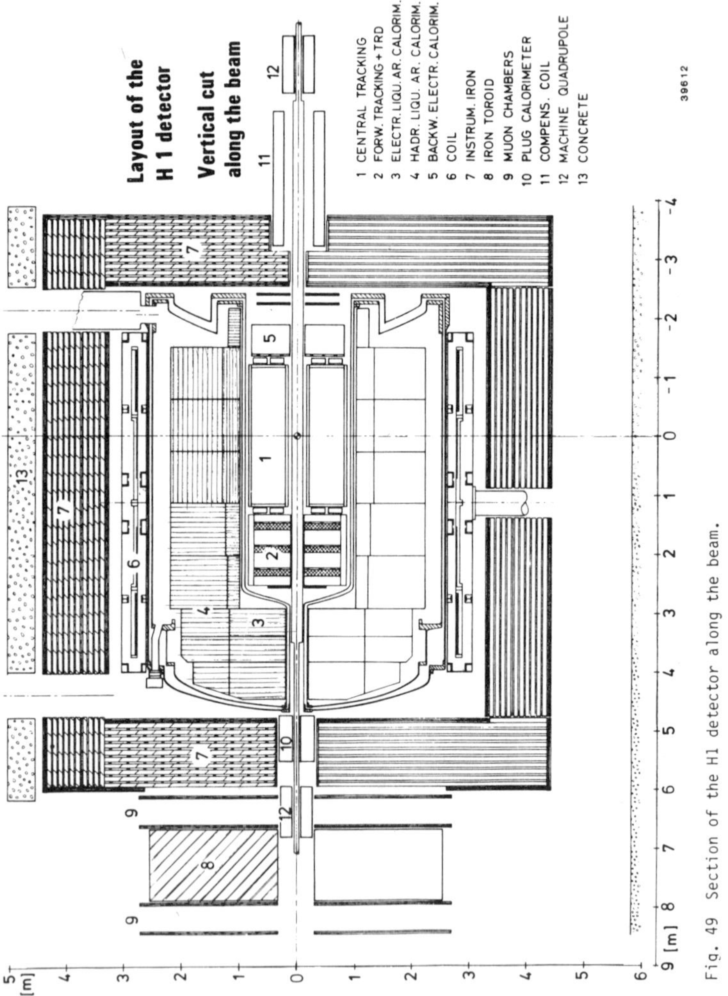

Fig. 49 Section of the H1 detector along the beam.

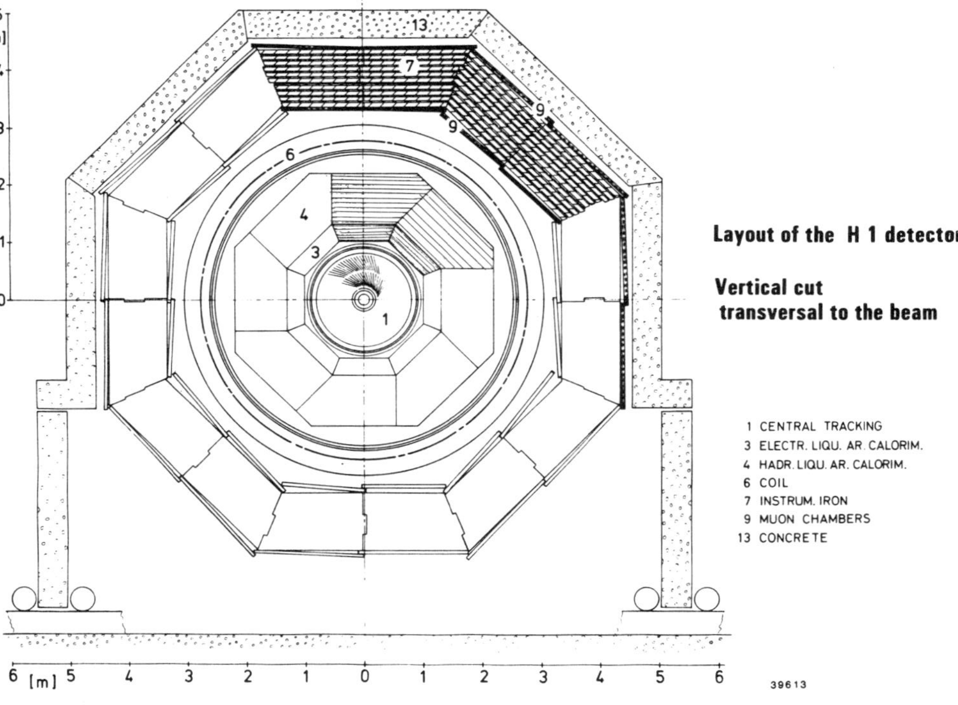

Fig. 50 Section of the H1 detector perpendicular to the beam.

The hadron calorimeter uses stainless steel absorber plates with liquid argon readout. The EM and hadron calorimeters provide an energy resolution of $\sigma(E)/E \leq 10\ \%/\sqrt{E}$ for electrons and, using a weighting procedure $\sigma(E)/E = 55\ \%/\sqrt{E} \oplus 2\ \%$ for hadrons.

A superconducting coil (inner, outer radii 260, 304 cm, length 575 cm) encloses the calorimeter and provides a longitudinal field of 1.2 T.

The solenoid and calorimeters are surrounded by a set of iron plates (7) to contain the return flux. The iron is instrumented with plastic streamer tubes and acts as a backing calorimeter and muon filter.

Muon detection is provided by three layers of muon chambers (9) in the barrel and forward region complemented by a forward muon spectrometer consisting of a magnetized iron toroid (8) and four layers of drift chambers (9).

A plug calorimeter (10) detects hadronic energy at small angles down to 0.7°. It consists of a copper silicon sandwich.

The field produced at the beam by the large coil is compensated by an additional solenoid (11).

8.2   The ZEUS Detector

The layout of the detector is shown in Figs. 51-53. The essential elements are a central track detector (CTD) plus transition radiation (TRD) and planar chambers (FTD, RTD) within a thin magnetic solenoid (SOLENOID), an electromagnetic (EMC) and a hadron calorimeter (HAC) surrounding the coil over the full solid angle, a backing calorimeter (BAC), barrel and rear muon detector (MU), and a forward muon spectrometer (FMU).

The calorimeter consists of depleted uranium plates with plastic scintillator in order to achieve compensation and the best possible energy resolution for hadrons. The scintillator plates form towers which are read out via wave length shifer bars, light guides and photomulti-

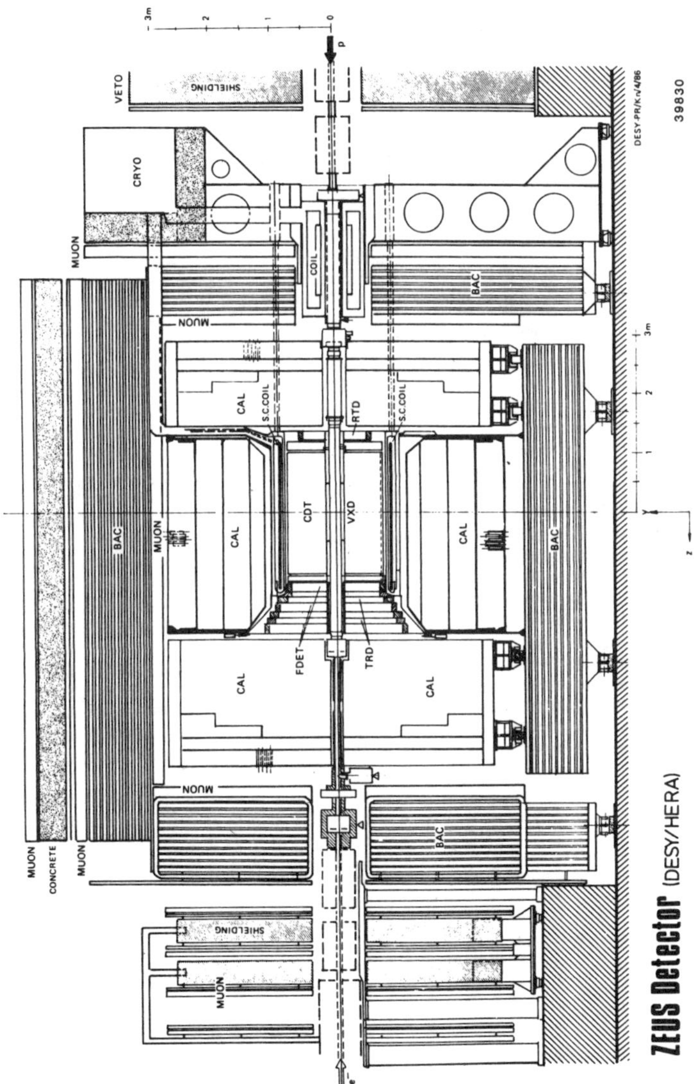

ZEUS Detector (DESY/HERA)

Fig. 51 Section of the ZEUS detector along the beam.

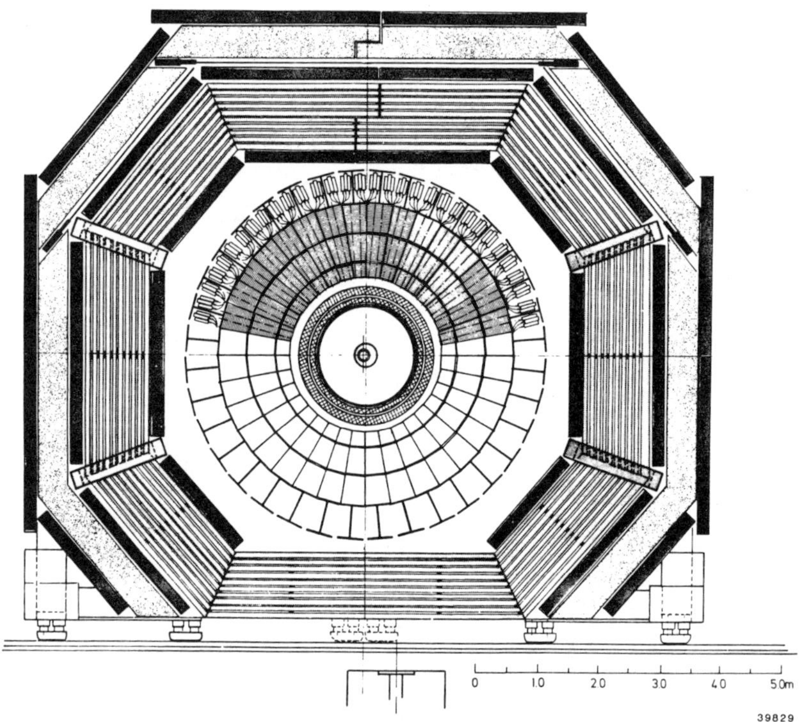

Fig. 52 Section of the ZEUS detector perpendicular to the beam.

pliers. The calorimeter is segmented longitudinally into an electromagnetic and one or two hadronic sections. Typical tower sizes are 5 cm
x 20 cm in the electromagnetic section and 20 cm x 20 cm in the hadronic
section. The calorimeter is divided into a forward, a barrel and a rear
part with 7, 5 and 4 absorption lengths, respectively. The active area
in the forward direction (proton beam direction) starts at about 60 mrad.
The solid angle coverage corresponds to 99.8 % in the forward hemisphere
and 98 % in the backward hemisphere. The expected energy resolutions are
for electrons $\sigma(E)/E = 0.15/\sqrt{E} \oplus 2$ % and for hadrons $\sigma(E)/E = 0.35/\sqrt{E}$
$\oplus 2$ %.

In order to identify electrons within dense jets a silicon pad
detector is under study which could be inserted in the calorimeter at a
depth of 3-5 radiation lengths.

The high resolution calorimeter is surrounded by a backing calorimeter of moderate energy resolution. Its purpose is to measure the
energy of late showering particles. The backing calorimeter uses as
absorber the iron plates which form the magnet yoke. Limited streamer
tube chambers are used for read out. The expected energy resolution for
hadrons is $\sigma(E)/E = 1.0/\sqrt{E}$.

The central track detector consists of a cylindrical jet type drift
chamber with an outer radius of 85 cm and an overall length of 240 cm.
Track position and dE/dx loss are measured in 9 superlayers each with
8 layers of sense wires. Four of the superlayers have stereo wires. A
resolution of 100 μm is expected, leading to a momentum resolution of
$\sigma(p)/p = 0.0018 \cdot p \oplus 0.0027$ (p in GeV/c) for a magnetic field of 1.8 T.
Particle tracking at small forward and backwards angles to the beam is
aided by four planar drift chambers providing a momentum measurement
with $\sigma(p)/p = 0.01 \cdot p$ at a foward angle of 120 mrad.

For the detection of the decays of short lived particles two possible designs for a vertex detector are under study.

Electron identification is performed with dE/dx information from
the tracking detectors and with the calorimeter. In the forward direction

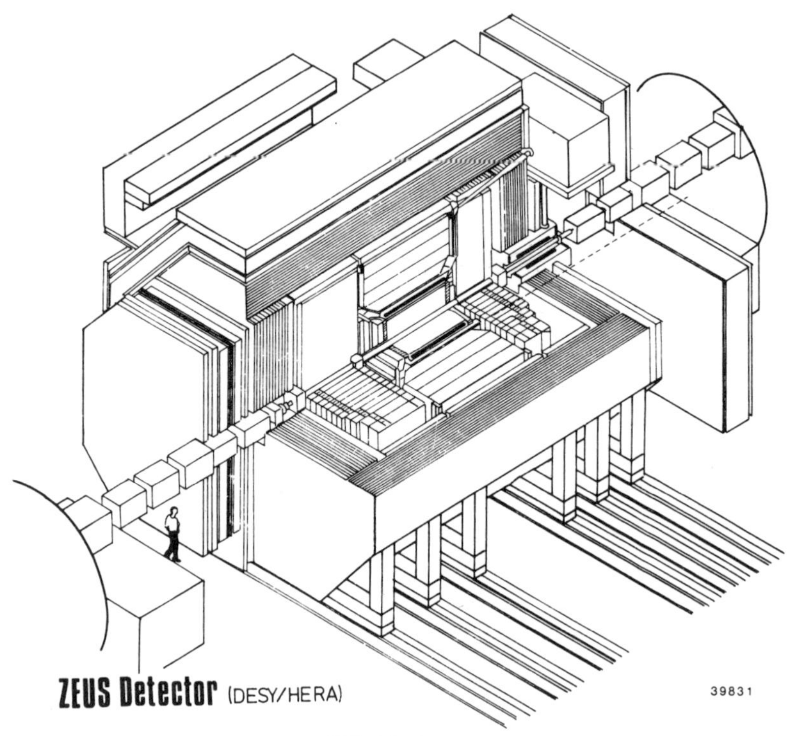

Fig. 53 Isometric view of the ZEUS detector.

a transition radiation detector, consisting of four modules, yields an additional hadron rejection factor of more than 100 for momenta below 30 GeV/c. The combined hadron rejection is well above 10^3.

Muons are detected in the forward direction in a spectrometer using drift and limited streamer tube chambers plus scintillator counters interspersed between the magnetized iron yoke and magnetized iron toroids. The momentum resolution for 100 GeV/c muons is $\sigma(p)/p = 30$ %. In the barrel and rear detectors muons are detected by limited streamer tube chambers before, in between and behind the backing calorimeter and behind the concrete shield. The momentum resolution is 35 % at 20 GeV/c. The pion (kaon) rejection factors are 1000 (100) at 40 GeV/c in the forward direction, and 5000 (50) at 10 GeV/c in the barrel region.

A forward proton spectrometer detects very forward produced protons $x_L > 0.3$. The spectrometer uses proton beam line magnets and five miniature high resolution chambers installed in Roman pots very close to the beam at distances between 20 and 90 m from the interaction point. A momentum resolution of $\sigma(p)/p \sim 1$ % is expected.

Electron and photon detectors are installed some 30 to 100 m downstream in electron beam direction to measure the luminosity and tag small Q^2 processes.

The magnetic solenoid is superconducting and provides a field of 1.8 T. It has an inner radius of 86 cm, a length of 280 cm and is 0.8 radiation lengths thick. A compensating solenoid installed in the rear of the detector compensates the influence of the detector solenoid on the beam.

9. SUMMARY

HERA will give access to a new territory of physics. Structure functions and charged current reactions can be studied for Q^2 values up to 30 000 - 40 000 GeV^2. The Q^2-ν space will be extended by two orders of magnitude in either variable over what presently can be reached. Precise structure function values will provide a stringent test of QCD, will probe for substructure of quarks and leptons down to distances of $3 \cdot 10^{-18}$ cm, and could detect new neutral or charged pieces of the current. In the search for right handed weak currents longitudinally polarized electrons will be extremely useful.

HERA is potentially a rich source of new particles. Basically any state that has electric and/or weak charge can be produced up to masses of 200 - 250 GeV. The sensitivity in mass is shown in Table 7.

Table 7:   Expected sensitivity in mass

new W	800 GeV
new Z^0	800 GeV
right handed W	500 GeV
new quark (t like)	120 GeV
excited q^*, e^*	250 GeV
new quarks, leptons	220 GeV
leptoquarks	180 GeV
supersymmetric quark plus lepton	160 GeV

In many aspects HERA will be complementary to the e^+e^- colliders SLC and LEP which are also under construction. It probes primarily the space like region and gives access to charged currents over a unique kinematic range. For a number of the newly proposed particles the mass range open to HERA experiments exceeds that of LEP II.

The design of an optimal detector for HERA poses a veritable challenge. The large momentum imbalance between incident electron and proton and the nature of space like processes throw most final state particles into a narrow cone around the proton direction. On the other hand, the standard neutral current reaction in HERA, ep → eX, offers an important advantage over $\bar{p}p/pp$ interactions. By tagging the scattered electron the kinematics of the relevant subsystem is determined. It is this handle which provides HERA experiments with a special sensitivity to new exotic particles.

ACKNOWLEDGEMENTS

The organizers of the Winter Institute have provided a very stimulating atmosphere where it has been fun to discuss and learn about new physics ideas. A good fraction of the material presented here has resulted from work of the ZEUS collaboration. I am indebted to P.Schmüser for help and discussions. I am grateful to Mrs. E,Hell for typing the manuscript.

References

1. Wiik, B.H., Electron-Proton Colliding Beams, the Physics Programme and the machine, proc. 10th SLAC Summer Institute, ed.A.Mosher, 1982, p.233.

   Proc. 1984 ICFA Seminar on Future Perspectives in High Energy Physics, May 14-20, 1984, KEK, Japan, eds. S.Ozaki, S.Kurokawa, Y.Unno, p.23.

2. Llewellyn-Smith, C.H. and Wiik, B.H., DESY Report 77/36 (1977).

3. Proc. Study of an ep facility for Europe, ed. U.Amaldi, DESY Report 79/48 (1979);

   HERA, proposal report DESY-HERA 81/10 (1981);

   Résumé Discussion Meeting "Physics with ep Colliders in view of HERA", Wuppertal, report DESY-HERA 81/18 (1981);

   Proc. Workshop "Experimentation at HERA", NIKHEF, Amsterdam, June 9-11, 1983, report DESY-HERA 83/20 (1983);

   Proc. Discussion Meeting on "HERA Experiments", Genova, Oct.1-3, 1984, report DESY-HERA 85/01 (1985);

   Altarelli, G., Mele, B. and Rückl, R., CERN Report TH3932 (1984);

   Bagger, J.A., Peskin, M.E., SLAC-PUB-3447 (1984);

   Cashmore, R.J. et al., "Exotic Phenomena in High Energy ep Collisions", Physics Reports 122, 276 (1985).

   Schmüser, P., "Tief Inelastische Lepton-Nukleon-Streuung", U.Hamburg, 1984, Lecture notes.

4. Bagger, J.A. and Peskin, M., Phys.Rev. D31, 2211 (1985);

   Altarelli, G., Mele, B. and Rückl, R., CERN Report TH3932 (1984);

   Proc. Large Hadron Collider in the LEP tunnel, Vol.2, 549 (1984);

   Rückl, R., Phys.Lett. 129B, 363 (1983); Nucl.Phys. B234, 91 (1984).

5. see Sciulli, F., Proc. 1985 Int.Symp.Lepton Photon Int. High Energies, ed. M.Konuma, K.Takahashi, p.8.

6. see e.g. Lohrmann, E., Mess, K.-H., DESY-HERA 83/08 (1983).

7. The ZEUS detector, Technical Proposal of the ZEUS Collaboration, March 1986.

8. Duke, D. and Owens, J.F., Phys.Rev. D30, 49 (1984)

9. Eichten, E.J., Lane, K.D., Peskin, M.E.,Phys.Rev.Lett. 50,811 (1983);

   Also see Rückl, R., Phys.Lett. 129B, 363 (1983).

10. MAC Collaboration, Fernandez, E. et al.,Phys.Rev.Lett. 50,123 (1983);

    TASSO Collaboration, Althoff, M. et al., Z.Phys. 22, 13 (1984);

Phys.Lett.138B, 441 (1984); MARK J Collaboration, Adeva, B. et al., Phys.Rep. 109, 131 (1984); HRS Collaboration, Bender, D. et al., ANL-HEP-PR-03; 71; PLUTO Collaboration, Berger, Ch. et al., Z.Phys. 27, 341 (1985).

11. Bars, I., Bowick, M.J., Freese, K., Phys.Lett. 138B, 159 (1984).

12. c.f. Pati, J.C., Salam, A., Phys.Rev. D10, 275 (1974); Mohapatra, R.N., Pati, J.C., ibid, 566, 2559 (1975);Mohapatra, R.N., Marshak, R.E., Phys.Rev.Lett. 44, 1316 (1980).

13. Mohapatra, R.N., Senjanovic, G., Phys.Rev.Lett. 44, 912 (1980).

14. Carr, J. et al., Phys.Rev.Lett. 51, 627, ·1222(E) (1983).

15. Senjanovic, G., Proc. SSC Fixed Target Workshop, The Woodlands,Tx, (1984); Stoker, D.P. et al., LBL-18935 (1985) and Phys.Rev.Lett.

16. Beall, G., Bander, M. and Soni, A., Phys.Rev.Lett. 48, 848 (1982); Harari, H. and Leurer, M., Nucl.Phys. B233, 221 (1984); Ecker, G. and Grimus, W., Z.Phys. C30, 293 (1986).

17. Harari, H., private communication.

18. Leveille, J.P., Weiler, T., Nucl.Phys. B147, 147 (1979).

19. Brodsky, S.J., Gunion, J.F., SLAC-PUB-3527 (1984).

20. Altarelli, G., Martinelli, G., Mele, B., Rückl, R., CERN TH 4094 (1985).

21. Ellis, J., Gaillard, M.K., Nanopoulos, D.V., Nucl.Phys. B106, 292 (1976).

22. See e.g. Kühn, J.H., Tholl, H.D., Zerwas, P.M., CERN TH 4131 (1985).

23. Rudaz, S., Vermaseren, J.A.M., CERN TH 2961 (1981).

24. Jones, S.K., Llewellyn-Smith, C.H., Nucl.Phys. B217, 145 (1983).

25. Harrison, D.P., Nucl.Phys. B249, 704 (1985). Marlean, L., SLAC-PUB-3533 (1984).

26. For a recent review see e.g. Wiik, B.H., Progress with HERA, DESY-HERA 85-16 (1985).

27. Dwerstig, B. et al., DESY Report 85-08 (1985) and Proc. 1985 Part. Acc. Conf. Vancouver, IEEE Trans. Nucl. Sci. Vol. NS-32, 3596.

28. See e.g. Buon, J., Steffen, K., DESY report 85-128 (1985).

29. Ford, R.L. and Nelson, W.P., SLAC report 210 (1978).

30. Fabjan, C. and Willis, W., Phys.Rev.Lett. 60B, 105 (1975);
    Fabjan, C. et al., Nucl.Inst.Meth. 141, 61 (1977);
    Akesson, T. et al., Nucl.Inst.Meth. A241, 17 (1985).

31. See e.g. Gabriel, T.A. et al.,IEEE Trans.Nucl.Sci. NS-32, 697 (1985);
    Fesefeldt, H., The simulation of hadronic showers, PITHA 85-02 (1985).

32. Lankford, A.J., Thesis Geneva 1978, CERN Internal Report EP 78-3;
    Kondo, T. et al., 1984 Summer Study on the Design and Utilization of
    the Superconducting Super Collider, Snowmass, Colorado, June 1984;
    Brau, J., Proc. of Workshop on Compensated Calorimetry, Pasadena
    1985;
    Flauger, W., Nucl.Instr.Meth. A241, 72 (1985).

33. See pioneering work Leroy, C., Sirois, Y. and Wigmans, R., CERN
    Report (1986), in preparation.

34. Brückmann, H. et al., private communication.

35. Fabjan, C.W. and Ludlam, T., Ann.Rev.Nuc.Part.Sci. 32,335 (1982).

36. Wigmans, R., Proc. of Workshop on Compensated Calorimetry, Pasadena
    1985.

37. NA-34, private communication.

38. Abramowicz, H. et al., Nucl.Inst.Meth. 180, 429 (1981).

39. WA-78 Collaboration, De Vincenzi, M. et al., CERN Report EP/86-12
    (1986), submitted to Nucl.Inst.Meth.

40. ZEUS: A Detector for HERA, Letter of Intent, June 1985.

41. Letter of Intent for an Experiment at HERA, H1 Collaboration, June
    28, 1985.

42. The ZEUS Detector, Technical Proposal, March 1986.
    Carleton, Manitoba, McGill, Toronto, York, Bonn, DESY, Freiburg,
    Hamburg I, Hamburg II, Siegen, Weizmann Institute, Bologna, Florence,
    Frascati, L'Aquila, Lecce, Milan, Padua, Rome, ENEA-Rome, Torin,
    NIKHEF Amsterdam, Cracow, Warsaw, Madrid, Bristol, Londong (IC),
    London (UC), Oxford, Rutherford, Argonne, Columbia, Illinois, Ohio,
    Pennsylvania State, Virginia, Wisconsin.

43. Technical Proposal for the H1 Detector, March 25, 1986,
    Aachen, Davis, DESY, Dortmund, Ecole Polytechnique, Glasgow, Hamburg,
    Houston, Lancaster, Liverpool, Manchester, Moscow, MPI München,
    Northeastern, Orsay, Paris, Rome, RAL, Saclay, Wuppertal, Zeuthen,
    Zürich.

FROM THE HIGGS TO SUPERSTRING PHENOMENOLOGY

John Ellis

CERN, Geneva, Switzerland

1. SEARCH FOR HIGGS

1.1 Why Higgs?

In the textbooks you usually find quite a fancy, high-brow explanation of the need for a Higgs boson, based on the notion of spontaneous
breakdown of local gauge symmetry. Here, however, we will take a relatively low-brow approach which leads naturally into the motivation for
supersymmetry discussed at the end of this lecture. Our starting-point
is the need to cancel any unruly behaviour at high energies of field
theory tree diagrams[1]. If they do not give cross-sections falling as
$1/E_{cm}^2 \equiv 1/s$ at high energies, as suggested by naïve dimensional analysis,
then higher-order loop diagrams will have uncontrollable non-renormalizable divergences. A first example of this problem was the typical weak
interaction cross-section generated by the old-fashioned four-fermion
interaction of Fig. 1a:

$$\sigma_0 \left(f f \to f f \right) \sim G_F^2 s \tag{1.1}$$

When iterated to form a loop correction diagram as in Fig. 1b, the bad
high-energy behaviour (1.1) led to quadratic divergences in internal
momentum integrals:

$$\sigma_1 \left(f f \to f f \right) \propto \int^\Lambda \frac{d^4 k}{k^2} \propto \Lambda^2 \tag{1.2}$$

The remedy for this disease is of course well known: introduce weak
gauge bosons W to mediate the weak interaction as in Fig. 1c[2]. In this
case the high-energy behaviour (1.1) was modified to

$$\sigma_0 \left(f f \to f f \right) \sim \left(\frac{g^2}{s} \right)^2 s \sim \frac{(g^2)^2}{s} \tag{1.3}$$

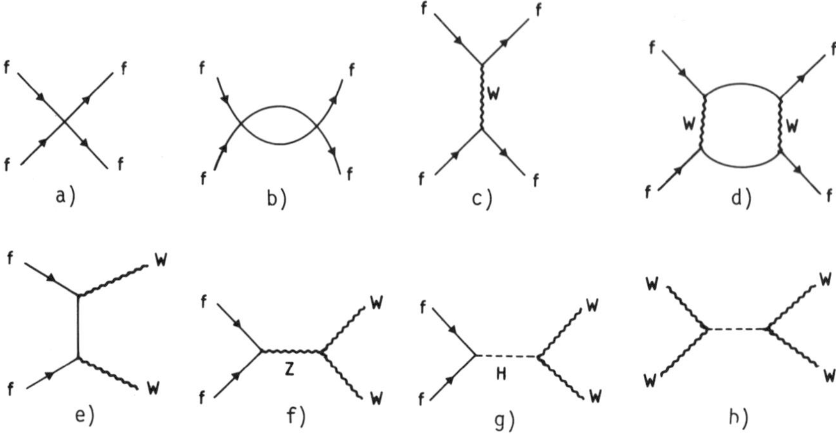

Fig. 1 (a) Four-fermion interaction giving cross-section with bad high energy behaviour, and whose iteration (b) leads to quadratic divergences in loops. This disease is cured (c) by introducing W boson exchange, which (d) gives some finite loops. However, (e) the ff → WW scattering cross-section has bad high-energy behaviour, which (f) is alleviated by the non-Abelian 3-boson coupling, and (g) finally cured by including Higgs boson exchange, which (h) is also needed in WW → WW scattering.

where g is the gauge coupling related to G_F by

$$\frac{G_F}{\sqrt{2}} = \frac{g^2}{8 m_W^2} \qquad (1.4)$$

Then Fig. 1b is replaced by the modified loop diagram of Fig. 1d, which is finite--so all is well. However, the remedy has side-effects: we must now consider the ff → WW scattering process of Fig. 1e, which again has bad high-energy behaviour:

$$\sigma(ff \to WW) \propto \frac{(g^2)^2}{s} \times \left(\frac{s}{m_W^2}\right)^2 \qquad (1.5)$$

The problem is due to the high-energy behaviour $\varepsilon \propto p/m_W$ of the polarization vectors for longitudinally polarized vector bosons[1]. Of course, there is an equally well-known cure for this secondary disease, namely to include the diagram of Fig. 1f with its non-Abelian three-boson coupling[3], which (almost!) restores

$$\sigma\left(f\bar{f}\to WW\right) \quad \propto \quad \frac{\left(g^2\right)^2}{s} \tag{1.6}$$

The good high-energy behaviour (1.6) holds if the fermions are massless; but if they do have masses, there is a residual nasty piece in the high-energy behaviour:

$$\sigma\left(f\bar{f}\to WW\right)_{m_f\neq0} \quad \propto \quad \left(g^2\right)^2 m_f^2 \tag{1.7}$$

The final remedy for this side-side-effect is to cancel off the bad high-energy behaviour (1.7) with the scalar Higgs boson exchange-diagram of Fig. 1g. We can see from Eq. (1.7) that to work this trick we must take

$$g_{Hf\bar{f}} \times g_{HWW} \quad \propto \quad g^2 m_f \tag{1.8}$$

The question then is: How is the product (1.8) shared out between the Hff coupling $g_{Hf\bar{f}}$ and the HWW coupling g_{HWW}? The answer is provided by WW → WW scattering, where we must also exchange the Higgs boson as in Fig. 1h, and cancelling out the bad high-energy behaviour requires

$$g_{HWW} \approx g m_W \quad \Rightarrow \quad g_{Hf\bar{f}} \approx \frac{g m_f}{m_W} \tag{1.9}$$

Thus we have seen that perturbative unitarity and renormalizability not only require the introduction of one or more physical Higgs bosons, but also largely determine their couplings. In the specific case of the minimal Weinberg-Salam model[4] there is a single physical Higgs boson, which is neutral with the couplings

$$g_{Hf\bar{f}} = 2^{1/4}\sqrt{G_F}\, m_f = \frac{g m_f}{2 m_W} \tag{1.10a}$$

$$g_{HWW} = 2^{5/4}\sqrt{G_F}\, m_W^2 = g m_W \tag{1.10b}$$

228

but non-minimal models have at least two more neutrals and two charged Higgs bosons[5], as we will see later.

1.2   Properties of the Higgs Boson

Although the couplings (1.10) of the minimal Weinberg-Salam Higgs boson are completely specified, its mass is not fixed. All one can say is that probably its mass is between

$$1\,TeV \gtrsim m_H \gtrsim 10\,GeV \tag{1.11}$$

The upper limit in (1.11) comes from demanding that the self-coupling λ of the Higgs should not be strong[6]. An extension of the unitarity arguments of subsection 1.1 tells us that the couplings of the Higgs should be those given by a spontaneously broken gauge theory[4] with a Higgs potential of the form

$$V(H) = -\tfrac{1}{2}\mu^2 |H|^2 + \tfrac{1}{4}\lambda \left(|H|^2\right)^2 \tag{1.12}$$

Minimizing the potential (1.12), we find that

$$m_H = \sqrt{2\mu^2} = \frac{2}{g} m_W \sqrt{\lambda} \tag{1.13}$$

Perturbative unitarity is applicable as long as the self-coupling λ is not too strong. Requiring $\lambda \lesssim 1$ in Eq. (1.13) gives an upper bound[6] on m_H of order

$$m_H \lesssim m_W \times O\left(\tfrac{1}{\sqrt{\alpha}}\right) \tag{1.14}$$

which becomes numerically the first bound in (1.11). The second bound[7] comes from considering radiative corrections[8] to the Higgs self-coupling, such as the W boson loop of Fig. 2. It provides a lower bound on the self-coupling:

$$\lambda \gtrsim O(g^4) = O(\alpha^2) \tag{1.15}$$

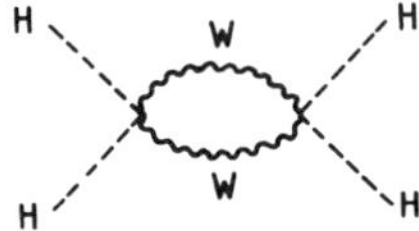

Fig. 2 W boson loop contribution to the Higgs self-coupling, and
hence to the Higgs mass

Putting this into the general expression (1.13) we see that

$$m_H \gtrsim O(\sqrt{\alpha}) \times m_W \qquad (1.16)$$

which becomes numerically the second bound in (1.11). Comparing the two
expressions for m_H in Eq. (1.13) we see that small λ corresponds to small
μ. Indeed, in the limit $\mu \to 0$ the Higgs mass is completely determined[7,8]
by the radiative corrections of Fig. 2:

$$m_H^2 = \frac{3g^2}{32\pi^2} m_W^2 \left[2 + \sec^4\theta_W - O\left(\frac{m_f}{m_W}\right)^4 \right] \qquad (1.17)$$

where the second term comes from fermion loops. We see that Eq. (1.17)
vanishes if $m_f = O(m_W)$, which is a precursor of some of the supersymmetric cancellations we will meet later. However, if $m_f \ll m_W$ as is the
case for all the known fermions except possibly the t-quark, the Higgs
mass is determined by W loops to be about 10 GeV when $\mu \to 0$. This estimate is not greatly changed by higher loops[9].

As is obvious from the couplings (1.10), the Higgs boson prefers to
decay into the heaviest particles which are kinematically available[10], as
seen in Fig. 3. For $H \to f\bar{f}$ one has

$$\Gamma(H \to f\bar{f}) = \frac{N_c G_F m_f^2}{4\sqrt{2}\pi} m_H \times \beta^3 \qquad (1.18)$$

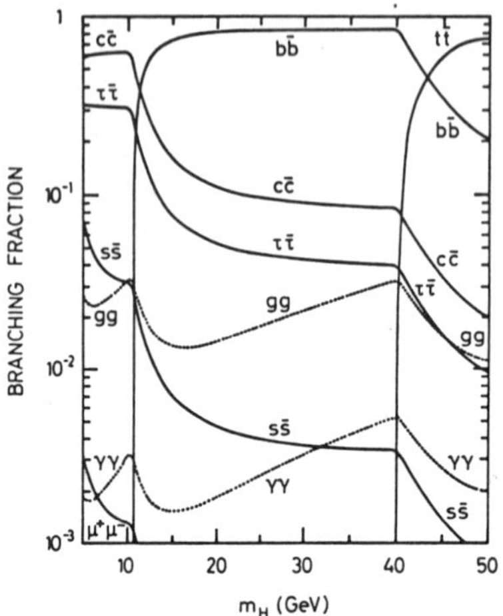

Fig. 3   Decay modes of the neutral Higgs boson[20]

where N_C is the number of fermion colours (1 for leptons, 3 for quarks) and $\beta \equiv (1 - 4m_f^2/m_H^2)^{1/2}$. Numerically, formula (1.18) gives

$$\Gamma\left(H \to f\bar{f}\right) \approx 7.3 \times 10^{-9}\, m_H \times \left(\frac{m_f}{m_\mu}\right)^2 \tag{1.19}$$

As examples, we have

$$\Gamma\left[H(20\,\text{GeV}) \to b\bar{b}\right] \approx 1\,\text{MeV}, \quad \Gamma\left[H(160\,\text{GeV}) \to t\bar{t}\right] \approx 300\,\text{MeV} \tag{1.20}$$

if m_t = 40 GeV. Heavier Higgses have the $H \to VV$ vector-boson decay modes open to them. Using Eq. (1.10b) we find[11]

$$\Gamma(H \to W^+W^-) = \frac{G_F m_W^2 m_H}{8\pi\sqrt{2}} \frac{\sqrt{1-x}}{x}\left(3x^2 - 4x + 4\right) \tag{1.21}$$

where $x \equiv 4m_W^2/m_H^2$, which is about 3 GeV for m_H = 250 GeV. The corresponding $H \to Z^0 Z^0$ decay rate is[11]

$$\Gamma(H \to Z^0 Z^0) = \frac{G_F m_Z^2 m_H}{16\pi\sqrt{2}} \frac{\sqrt{1-x'}}{x'}\left(3x'^2 - 4x' + 4\right) \tag{1.22}$$

where $x' \equiv 4m_Z^2/m_H^2$. Other Higgs decay modes are of higher order and hence generally smaller. Examples are $H \to \gamma\gamma$ [12], which proceeds via fermion and boson loops and is generally insignificant except if $m_H < 2m_\mu$, which seems unlikely in view of the qualitative lower bound $m_H \gtrsim 10$ GeV [see (1.11)], and $H \to gg$ [13] which proceeds via strongly interacting massive particle loops such as quarks. One has

$$\Gamma(H \to gg) = \frac{\sqrt{2}\, G_F\, \alpha_s^2}{8\pi^3}\, \frac{m_H^3}{9}\, |N|^2 \qquad (1.23)$$

where N is the number of quarks heavier than the Higgs. We can see from Fig. 3 that $\Gamma(H \to gg) \lesssim (1 \text{ to } 10)\%\ \Gamma(H \to b\bar{b})$ for $m_H < m_t$ if $m_t = 40$ GeV. Another small decay mode is $H \to Wf\bar{f}'$ [14], which has a branching ratio $\lesssim 3\%$ of $H \to t\bar{t}$ for $m_H \lesssim 160$ GeV if $m_t = 40$ GeV. Above this mass, the dominant decays are $H \to W^+W^-$ and $H \to Z^0Z^0$, as discussed previously.

1.3   Higgs Search Experiments in e^+e^- Annihilation

Here will largely follow the discussion in the CERN Report on Physics at LEP[15].

1.3.1 Toponium $\to H + \gamma$. In leading order one has for any quarkonium $\to H + \gamma$ decay[16]:

$$\frac{\Gamma\left({}^3S_1(q\bar{q}) \to H + \gamma\right)}{\Gamma\left({}^3S_1(q\bar{q}) \to \gamma^* \to \mu^+\mu^-\right)} = \frac{G_F\, m_q^2}{\sqrt{2}\,\pi\alpha}\left(1 - \frac{m_H^2}{m_{{}^3S_1}^2}\right) \qquad (1.24)$$

but this result must be used with care. The denominator has a trivial correction due to the interference of γ^* and Z^{0*} exchanges: for toponium θ,

$$\frac{\Gamma(\theta \to \gamma^*, Z^* \to \mu^+\mu^-)}{\Gamma(\theta \to \gamma^* \to \mu^+\mu^-)} = 1 + \frac{3\left(1 - \frac{2}{3}x_w\right)\left(1 - x_w\right)}{y_w^2}\, \mathrm{Re}\,\chi_Z$$
$$+ \frac{\left(1 - \frac{2}{3}x_w\right)^2\left(1 + (1 - x_w)^2\right)}{\left(\frac{2}{3}\right)^2 y_w^4}\, |\chi_Z|^2 \qquad (1.25)$$

where

$$\chi_z \equiv \frac{m_\theta^2}{(m_\theta^2 - m_z^2) + i\, \Gamma_z m_z}$$

$$\chi_W \equiv 4\sin^2\theta_W \quad , \quad y_W \equiv 4\sin\theta_W \cos\theta_W \tag{1.26}$$

Putting (1.25) into the expression (1.24), one has the branching ratio

$$B(\theta \to H + \gamma) \sim \tfrac{1}{2} B(\theta \to \mu^+\mu^-) \sim 3\% \tag{1.27}$$

for m_θ = 80 GeV, although this type of correction is of course negligible for J/ψ and T decays. Important decay branching modes[17] for 60 GeV < m_θ < 100 GeV are shown in Fig. 4. There is, however, another type of correction to quarkonium $\to$ H + γ decay [Eq. (1.24)] which is not only important for toponium θ, but even more important for the J/ψ and the T, namely QCD radiative corrections[18] as in Fig. 5. They modify the ratio (1.24) by a factor $[1 - (\alpha_s/\pi)\Delta(m_H^2/m_{3S_1}^2)]$, where $\Delta = O(10$ to

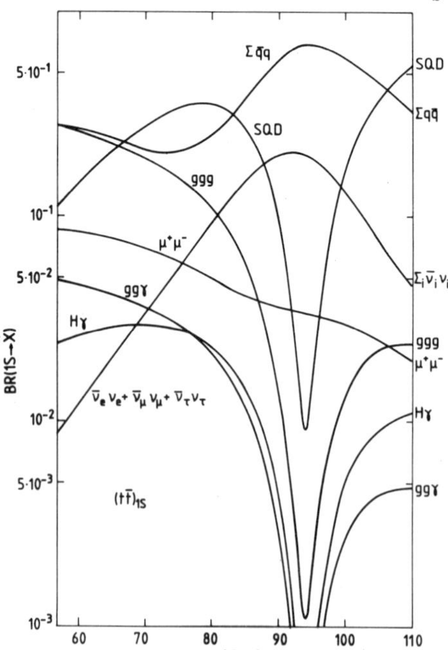

Fig. 4 Important decay modes of toponium[17]

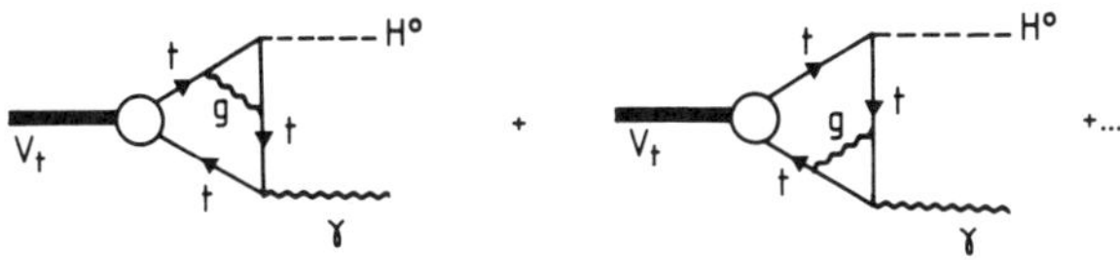

Fig. 5 Diagrams for QCD radiative corrections[18] to J/ψ, Υ or $\theta \rightarrow$ $H^0 + \gamma$ [16] decay

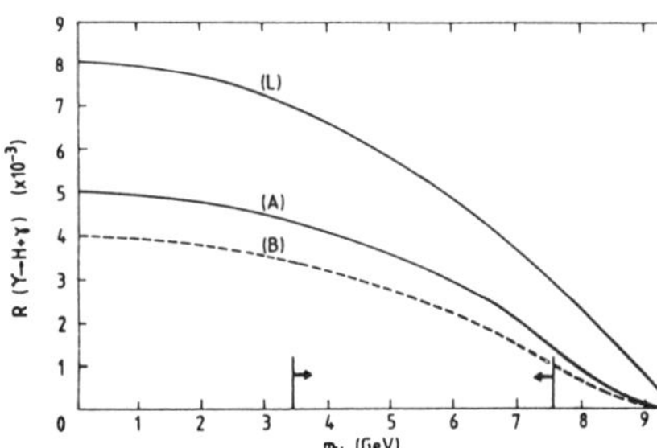

Fig. 6 **Magnitude of QCD radiative corrections to $\Upsilon \rightarrow H^0 + \gamma$ decay[21]:** the curve (L) is leading order, curves (A) and (B) incorporate one-loop corrections using different prescriptions

15) as seen in Fig. 6, which decreases the previous estimate (1.27) of $B(\theta \rightarrow H + \gamma)$ by about 30% [20]. In the case of J/ψ and $\Upsilon \rightarrow H + \gamma$, the radiative corrections suppress the decay rates by factors of 2 or more[21], which means that the CUSB upper limit[22] on $\Upsilon \rightarrow H + \gamma$ cannot be interpreted as excluding any range of Higgs masses. However, the decay of toponium θ to $H + \gamma$ is the best way[17,20] to look for Higgses if $m_H < m_\theta$ and m_θ is O(70 to 90) GeV.

1.3.2 $Z^0 \rightarrow H^0 + \gamma$. This decay occurs via loop diagrams as in Fig. 7. The W boson loop diagrams yield

$$\frac{\Gamma(Z^0 \rightarrow H^0 + \gamma)}{\Gamma(Z^0 \rightarrow \mu^+ \mu^-)} \approx 6.3 \times 10^{-5} \left(1 - \frac{m_H^2}{m_Z^2}\right)^3 \left(1 + 0.14 \frac{m_H^2}{m_Z^2}\right) \tag{1.28}$$

which gives the rates per 10^6 Z^0 (a reasonable target for a full-scale LEP or SLC experiment) shown in Fig. 8 and listed in Table 1. By compari-

Table 1: Numbers of events for $Z^0 \to$ Higgs decays[20]

m_H (GeV)	10	20	30	40	50
$Z^0 \to H + \gamma$ events	1.8	1.7	1.4	1.1	0.7
$Z^0 \to H + (\mu^+\mu^-$ or $e^+e^-)$ events	65	26	11	4	1.5

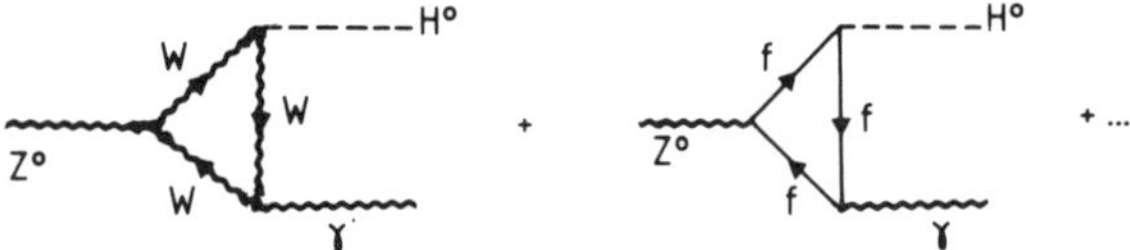

Fig. 7 Loop diagrams for $Z^0 \to H^0 + \gamma$ decay[23]

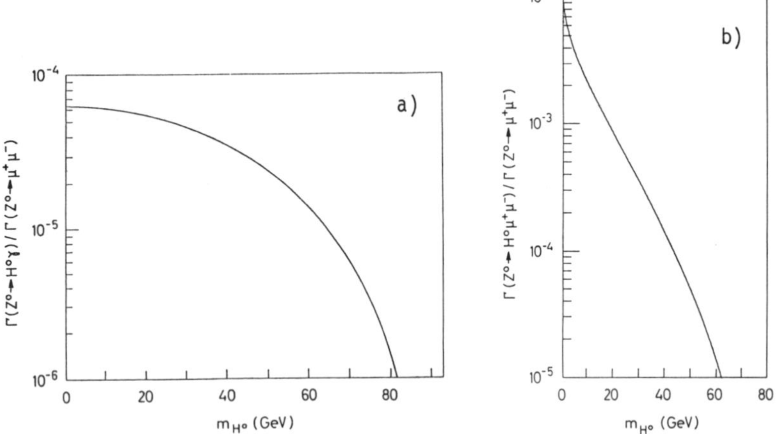

Fig. 8 Rates for a) $Z^0 \to H^0 + \gamma$ [23] and b) $Z^0 \to H^0 + (\mu^+\mu^-$ or $e^+e^-)$ [25]

son with the $Z^0 \to H^0 + (\mu^+\mu^-$ or $e^+e^-)$ decay to be discussed in a moment (see Fig. 8), $Z^0 \to H^0 + \gamma$ decay has a smaller rate and larger backgrounds. Fermion loop diagrams for $Z^0 \to H + \gamma$ tend to interfere destructively with the W-boson loops[23]. A complete calculation of this decay rate in a supersymmetric theory in which many more diagrams contribute would be interesting, but it has yet to be performed[24].

1.3.3 $Z^0 \to H^0 + 1^+1^-$. The dominant diagram for this process is shown in Fig. 9. It gives a rate[25]

$$\frac{1}{\Gamma(Z^0 \to \mu^+\mu^-)} \frac{d\Gamma(Z^0 \to H^0 1^+1^-)}{dx_{H^0}} = \frac{\alpha}{4\pi \sin^2\theta_W \cos^2\theta_W} \times$$

$$\times \frac{\left[1 - x_{H^0} + (x_{H^0}^2/12) + \frac{2}{3}(m_{H^0}^2/m_Z^2)\right]\left[x_{H^0} - 4m_{H^0}^2/m_{Z^0}^2\right]^{1/2}}{(x_{H^0} - m_{H^0}^2/m_{Z^0}^2)^2} \qquad (x_{H^0} \equiv 2E_{H^0}/m_{Z^0})$$

$$(1.29)$$

which is plotted in Fig. 8 and yields the event numbers shown in the bottom row of Table 1. A distinctive feature of this decay is that, because of the virtual Z^0 in the final state, the 1^+1^- pair tends to have a large invariant mass[26], with $m(1^+1^-)$ peaking close to $m_{Z^0} - m_H$ as seen in Fig. 10. It can be seen from Fig. 11 that the rates for $Z^0 \to H + e^+e^-$ are expected[20,27] to be above the backgrounds from $Z^0 \to t\bar{t}$ or $b\bar{b}$ as long as $m_H \lesssim 45$ GeV. A related decay which may also be detectable is $Z^0 \to H + (\nu\bar{\nu})$, which would provide a relatively low-momentum Higgs accompanied by a large amount of missing energy and mass. This decay would have a rate

$$\frac{\Gamma(Z^0 \to H^0 + \nu\bar{\nu})}{\Gamma(Z^0 \to H^0 + (\mu^+\mu^- \text{ or } e^+e^-))} = \frac{6}{1 + (1 - 4\sin^2\theta_W)^2} \approx 6 \qquad (1.30)$$

and could be detected by an experiment with good calorimetry

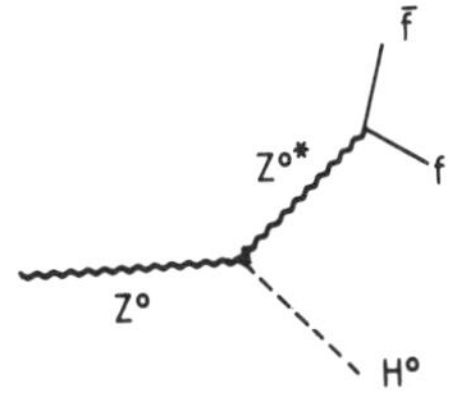

Fig. 9 Dominant diagram for $Z^0 \to H^0 + 1^+1^-$ [25]

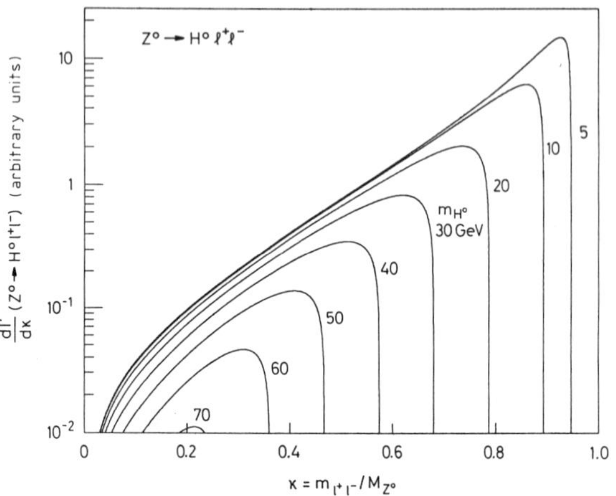

Fig. 10   Shape of $m(l^+l^-)$ spectrum in $Z^0 \to H^0 + l^+l^-$ decay[26]

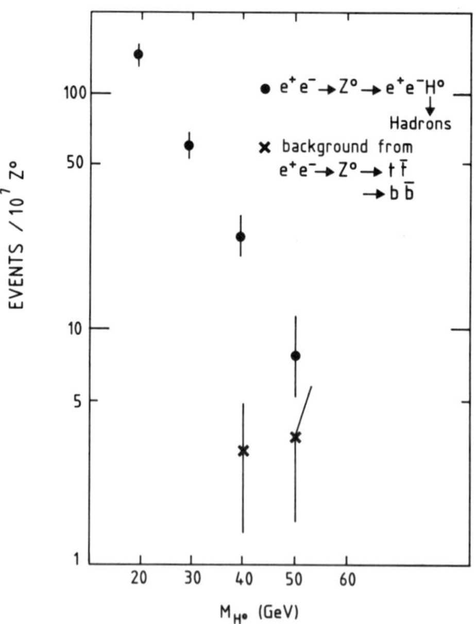

Fig. 11   Signal and background for $Z^0 \to H^0 + e^+e^-$ [20,27]

1.3.4 $e^+e^- \to Z^0 + H^0$. This process is closely related to the previous one, but with the real Z^0 in the inital state replaced by a virtual Z^0 and the virtual Z^0 in the final state replaced by a real Z^0. The cross-section is[11]

$$\sigma(e^+e^- \to Z^0 + H^0) = \frac{\pi\alpha^2}{24}\left(\frac{2k}{\sqrt{s}}\right)\frac{\left(k^2 + 3m_Z^2\right)}{\left(s - m_Z^2\right)^2}\frac{\left(1 - 4\sin^2\theta_W + 8\sin^4\theta_W\right)}{\sin^4\theta_W\cos^4\theta_W} \quad (1.31)$$

where k is the final-state particle momentum. Equation (1.31) gives[20] the numbers of events per day (assuming a luminosity of 10^{31} cm^{-2} s^{-1}) shown in Table 2. This reaction should be observable at LEP II ($\sqrt{s} \to$ 200 GeV). The case where the final Z^0 decays into l^+l^- is almost free of background[28], as seen in Fig. 12a. There is also very little background in the case when $Z^0 \to \nu\bar{\nu}$[28], as seen in Fig. 12b, and the rate is higher by the factor of $\sim$ 6 in Eq. (1.30). There is considerable background from QCD four-jet events in the case when $Z^0 \to q\bar{q}$, although analysis tricks[29] may be able to refine the signal[28] as in Fig. 12c. This problem becomes particularly serious when $m_H \sim m_{W^\pm}$ or m_{Z^0}, and there are large backgrounds from $e^+e^- \to W^+W^-$ and Z^0Z^0.

Table 2: Number of events per day for $e^+e^- \to Z^0 + H^0$
(assuming a luminosity of 10^{31} cm^{-2} s^{-1})

Mass of Higgs (GeV)	10	20	30	40	50	60	70	80	90	100
$\sqrt{s}$ (GeV)										
120	12.33	8.55	0.00	0.00	0.00	0.00	0.00	0.00	0.00	0.00
140	5.17	4.72	3.92	2.51	0.00	0.00	0.00	0.00	0.00	0.00
160	2.70	2.58	2.37	2.08	1.66	1.03	0.00	0.00	0.00	0.00
180	1.60	1.56	1.48	1.37	1.23	1.05	0.83	0.51	0.00	0.00
200	1.04	1.02	0.99	0.94	0.87	0.79	0.70	0.59	0.46	0.28
220	0.72	0.71	0.69	0.67	0.63	0.59	0.54	0.49	0.43	0.35
240	0.53	0.52	0.51	0.49	0.47	0.45	0.42	0.39	0.35	0.31
260	0.40	0.39	0.38	0.38	0.36	0.35	0.33	0.31	0.29	0.26

238

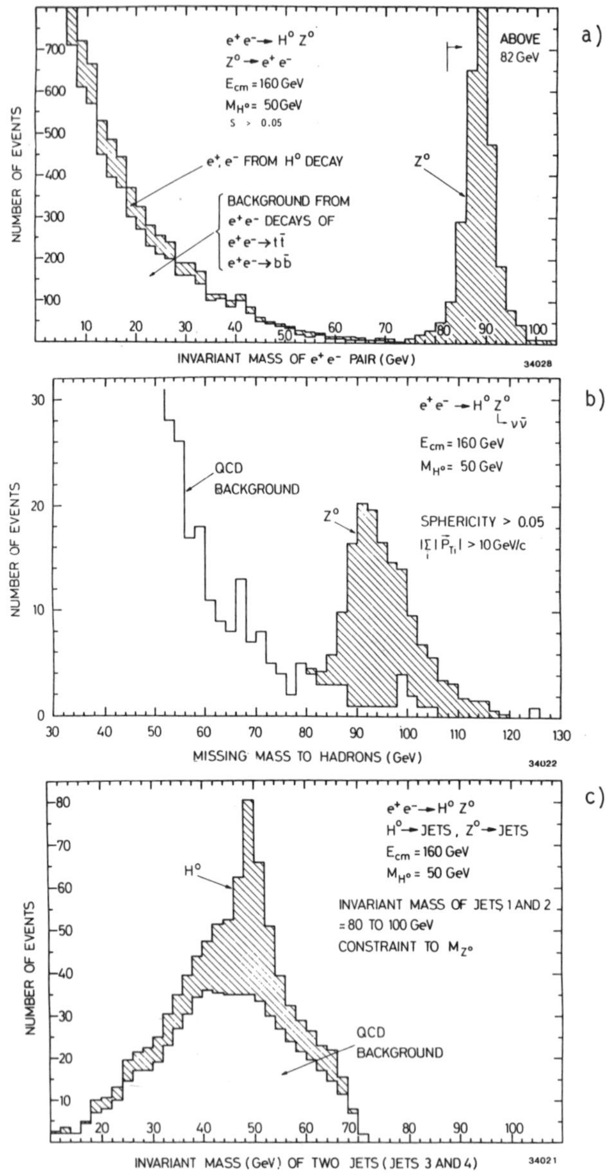

Fig. 12 Signal and background a) for $e^+e^- \to H^0 + (Z^0 \to e^+e^-)$, b) for $e^+e^- \to H^0 + (Z^0 \to \nu\bar{\nu})$, c) for $e^+e^- \to H^0 + (Z^0 \to q\bar{q})$ [20,28]

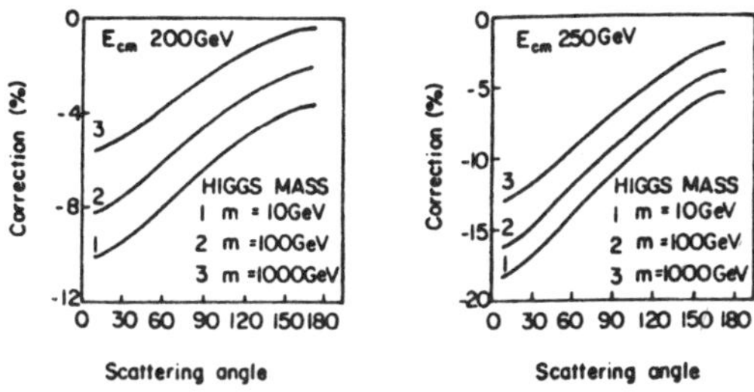

Fig. 13 Radiative corrections to $\sigma(e^+e^- \to W^+W^-)$, showing the effects of varying the Higgs mass[31]

1.3.5 Radiative corrections. There are corrections to the Z^0 mass which are logarithmically sensitive to the Higgs mass[30]:

$$\delta m_Z = 340\,\text{MeV} \times \log_{10}\left(m_H / 100\,\text{GeV}\right) \qquad (1.32)$$

but these are not large, and could be confused with other effects on m_Z^0. There are also radiative corrections[31] to $\sigma(e^+e^- \to W^+W^-)$ which are sensitive to m_H, but the effect is not large, as seen in Fig. 13. A large number of $e^+e^- \to W^+W^-$ events would be required in order to obtain useful indirect information on m_H in this way.

1.3.6 Conclusions

LEP I could be used to search at the Z^0 for Higgses up to 50 GeV, and for Higgses up to $m_H = m_\theta - O(20)$ GeV if the toponium mass m_θ is somewhat below m_Z[17,20]. LEP II at $\sqrt{s}$ = 200 GeV could be used to search for Higgses up to 100 GeV, but searches for heavier Higgses would require a higher-energy e^+e^- collider, the sensitivity being to $m_H \lesssim \sqrt{s} - 100$ GeV.

1.4 Non-Minimal Higgs Bosons

Many theories have more than the minimal set of Higgs fields in the Standard Model. The next-to-minimal possibility is two Higgs doublets[5], as in the smallest supersymmetric extension of the Standard model to be

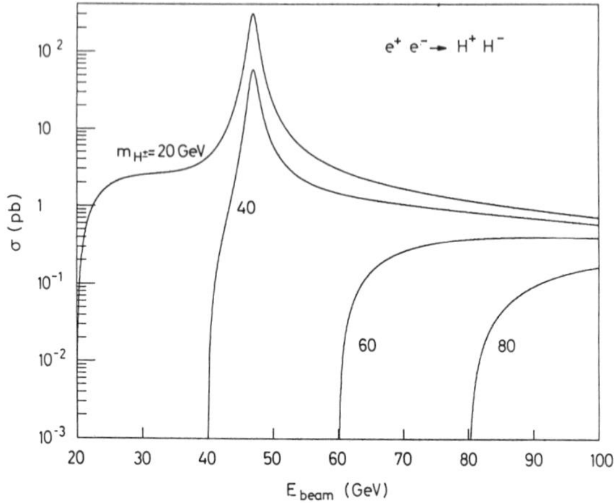

Fig. 14 Cross-section for $e^+ e^- \rightarrow H^+ H^-$ [20]

discussed in the second lecture. In this case, one has two charged Higgs particles $H^\pm$ as well as three neutrals.

The reaction $e^+ e^- \rightarrow H^+ H^-$ has the cross-section [20]

$$\frac{\sigma(e^+ e^- \rightarrow \gamma^*, Z^* \rightarrow H^+ H^-)}{\sigma(e^+ e^- \rightarrow \gamma^* \rightarrow \mu^+ \mu^-)} = \frac{1}{4}\beta^3 \left[1 - 2 c_V c_V' \frac{s(s - m_Z^2)}{(s - m_Z^2)^2 + m_Z^2 \Gamma_Z^2} + \frac{(c_V^2 + c_A^2) s^2 (c_V'^2 + c_A'^2)}{(s - m_Z^2)^2 + m_Z^2 \Gamma_Z^2} \right]$$

where

$$\tag{1.33}$$

$$c_V = \frac{1 - 4\sin^2\Theta_W}{4\sin\Theta_W \cos\Theta_W}, \quad c_A = \frac{-1}{4\sin\Theta_W \cos\Theta_W}, \quad c_V' = \frac{-2 + 4\sin^2\Theta_W}{4\sin\Theta_W \cos\Theta_W}, \quad c_A' = 0$$

where $\beta \equiv (1 - 4m_{H^\pm}^2/s)^{1/2}$. Note that formula (1.33) gives a large rate on the Z^0 peak if $m_{H^\pm} < 1/2\, m_{Z^0}$ as in Fig. 14. Unfortunately, $m_{H^\pm} > m_{W^\pm}$ in the minimal supersymmetric extension of the Standard Model [5,32], although this bound can be avoided in more complicated supersymmetric models [33]. In general, one expects the $H^\pm$ couplings to fermion pairs to be correlated with their masses, analogously to what occurs for neutral Higgs couplings to fermions in the minimal Standard Model (1.10a). Thus, one might expect the $H^\pm$ to prefer to decay into heavy fermion pairs such as $b\bar{t}$, $b\bar{c}$, or $\tau\bar{\nu}$, although this need not be the case if the different neutral Higgs fields have different vacuum expectation values, and

couplings to pairs from different generations may be suppressed by
Cabibbo-Kobayashi-Maskawa angles[32].

The Z^0 can also decay into pairs of unlike neutral Higgses, $Z^0 \rightarrow H^0 H^{0\,\prime}$, although Bose statistics forbids the decay into like pairs $Z^0 \not\rightarrow H^0 H^0$. The rate is[20,21]

$$\frac{\sigma(e^+ e^- \rightarrow Z^* \rightarrow H^0 H^{0\prime})}{\sigma(e^+ e^- \rightarrow \gamma^* \rightarrow \mu^+ \mu^-)} = \frac{1}{4} \times \frac{(C_V^2 + C_A^2)\, s^2\, C_H^2}{(S - m_Z^2)^2 + m_Z^2 \Gamma_Z^2} \times$$

$$\times \left(1 + \left(\frac{m^2}{s}\right)^2 + \left(\frac{m'^2}{s}\right)^2 - 2\left(\frac{m^2}{s}\right) - 2\left(\frac{m'^2}{s}\right) - 2\left(\frac{m^2 m'^2}{s^2}\right)\right)^{3/2} \tag{1.34}$$

if the $Z^0 H^0 H^{0\,\prime}$ coupling is $C_H/2 \cos\theta_w$. This can be large, e.g. in a supersymmetric model with unequal Higgs vacuum expectation values, so that $\Gamma(Z^0 \rightarrow H^0 H^{0\,\prime})/\Gamma(Z^0 \rightarrow H^+ H^-) = 0(1)$. If $m_{H^0} < 10$ GeV and $m_{H^0\,\prime} > 10$ GeV as in some of these models, there is a clean and distinctive event signature[21,34] when $H^0 \rightarrow \tau^+ \tau^-$ and $H^{0\,\prime} \rightarrow b\bar{b}$, as seen in Fig. 15.

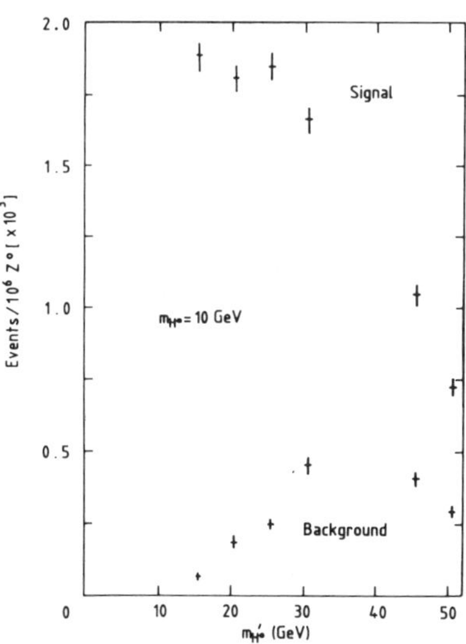

Fig. 15 Signal and background for $Z^0 \rightarrow (H^0 \rightarrow \tau^+ \tau^-) + (H^{0\,\prime} \rightarrow b\bar{b})$ [20,34]

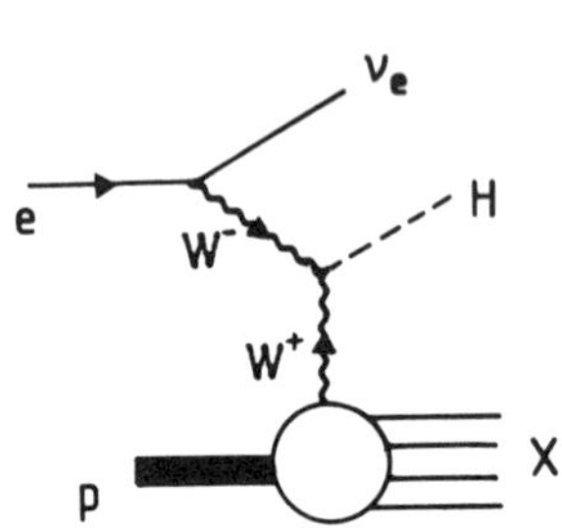

Fig. 16 Dominant diagram
for ep → ν + H^0 + X [35]

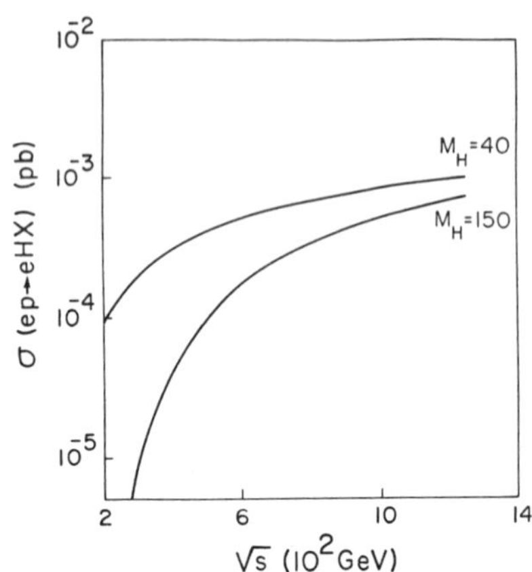

Fig. 17 Rates for
e + p → ν + H^0 + X [35]

1.5 Higgs Production at Other Accelerators

1.5.1 ep collisions. The dominant diagram in this case is two-W collisions as in Fig. 16. Unfortunately, the rates are small[35] as shown
in Fig. 17, and this does not seem a promising way to look for Higgs
bosons.

1.5.2 pp/p$\bar{p}$ collisions. These may well offer the earliest opportunities
to look for massive Higgs bosons. There are several production mechanisms
which have been considered[36,37], as seen in Fig. 18. They include
$gg \to H$ [38], which proceeds via the Hgg coupling (1.23) discussed in
subsection 1.2. The rate it gives depends sensitively on the number N
of heavy flavours, but even in the least favourable case with just a
t-quark of mass 40 GeV, the rate for $gg \to H$ at $\sqrt{s} = 40$ TeV is in principle large enough to be detected for $m_H \lesssim 1$ TeV. However, there is no
clear final-state event signature, and its dominant decays into $t\bar{t}$ (if
$2m_t < m_H < 2m_W$) or W^+W^- (if $m_H > 2m_W$) are likely to be drowned by background sources such as $q\bar{q} + gg \to t\bar{t}$, and $q\bar{q} \to W^+W^-$. The process
$q\bar{q}$ *or* $gg \to t\bar{t} + H$ [39] has an observable rate for $m_H \lesssim 300$ GeV, and the
accompanying $t\bar{t}$ pair might provide a useful event signature. However, it

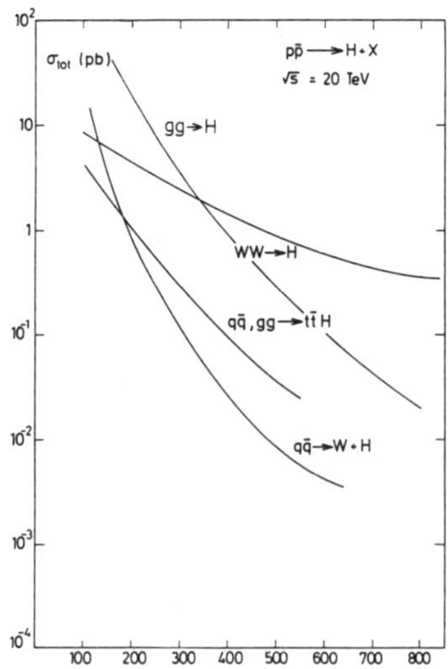

Fig. 18 Cross-sections for different mechanisms producing H^0 in high-energy hadron-hadron collisions[37]

is not to easy to tag $t\bar{t}$ pairs, and there are large backgrounds from $(t\bar{t})$ + $(t\bar{t})$ final-states (if $2m_t < m_H < 2m_W$) and from $t\bar{t}$ + W^+W^- (if $m_H > 2m_W$). The largest rate for large m_H is provided by $W^+W^- \to H$ [40], but this has the same problem as the $gg \to H$ process, namely the absence of a clear event signature. Finally, there is $q\bar{q} \to W^+ + H, Z^0 + H$ [41], which has an observable rate if $m_H \lesssim 300$ GeV. Here one could hope to use the $W^{\pm}$ or Z^0 to tag the event, but one probably cannot tag using the $q\bar{q}$ decays of the vector boson, because of the large QCD backgrounds, and using the leptonic decays involves paying a large price in branching ratio.

It is always hard to look for Higgs bosons in hadron-hadron collisions--even relatively light Higgses would not be easy to find at the CERN or FNAL $p\bar{p}$ Colliders. Nevertheless, the Higgs boson may well be heavier than 100 GeV, in which case one may have to learn to find it at a pp or $p\bar{p}$ collider, despite the problems[42]. Unless, that is, someone can be persuaded to fund a high-energy e^+e^- collider, at which there would be much smaller backgrounds to the detection of a heavy Higgs.

1.6 Rescuing Higgs with Supersymmetry

There are many possible motivations for studying supersymmetric theories[43]: supersymmetry is the only symmetry not yet used, it is beautiful, it can help unify matter with force, etc. However, the most pressing motivation to my mind is the hierarchy problem[44], namely that of reconciling the coexistence of very different small and large mass scales in physics: $m_W/m_P = O(10^{-17})$. This problem appears at two levels. The more profound one is that of creating the hierarchy--maybe one could generate it by quantum corrections,

$$m_W = m_P \exp\left[\frac{-O(1)}{\alpha}\right] \qquad (1.35)$$

much as

$$\Lambda_{QCD} = m_X \exp\left[\frac{-O(1)}{\alpha}\right] \qquad (1.36)$$

in conventional Grand Unified Theories (GUTs). There is also the technical aspect of maintaining the hierarchy despite the destabilizing effects of radiative corrections, which is often called the naturalness problem[45]. One says that a physical parameter is technically natural if radiative corrections to its bare value are no larger than its physical value. An example of a naturally small parameter is a fermion mass, whose radiative corrections take the form

$$\delta m_f = m_f \, O\left(\frac{\alpha}{\pi}\right) \ln\left(\Lambda/m_f\right) \qquad (1.37)$$

where Λ is the effective momentum cut-off in loops. Even if $\Lambda \sim m_P$, $\delta m_f \lesssim m_f$ because the divergence is only logarithmic. This reflects the fact that theory has an additional global symmetry, namely chiral symmetry, when $m_f \to 0$. As discussed in subsection 1.1, in the Standard Model $m_W/m_H = O(1)$, so the hierarchy $m_W/m_P \ll 1$ requires $m_H/m_P \ll 1$ also. But elementary scalar masses are notoriously unstable, and not naturally small, because there is no symmetry to protect them in the minimal Standard Model.

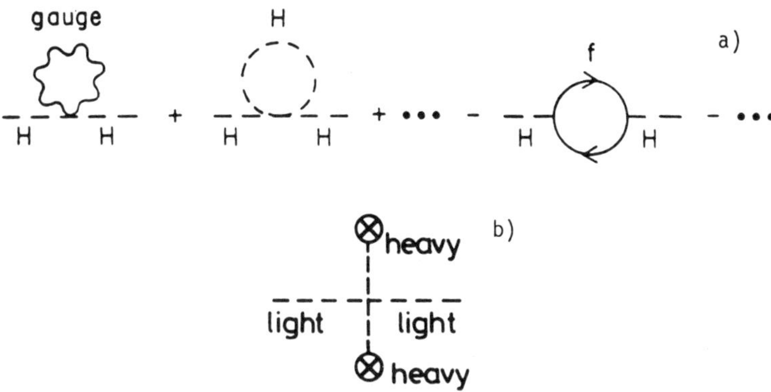

Fig. 19 a) Loop diagrams which may give large contributions to
elementary scalar masses. b) A large contribution to the Higgs mass in
a GUT.

Examples of potentially large corrections to scalar masses are the
loop diagrams of Fig. 19a. These are quadratically divergent, and
provide a shift in the mass squared of the Higgs boson of the form:

$$\delta m_H^2 = g^2 \int^\Lambda \frac{d^4k}{(2\pi)^4 k^2} = O\left(\frac{\alpha}{\pi}\right)\Lambda^2 \qquad (1.38)$$

For $\delta m_H^2 \lesssim m_H^2 \sim (100 \text{ GeV})^2$, one needs a cut-off representing the onset of
new physics:

$$\Lambda \lesssim 1 \text{ TeV} \qquad (1.39)$$

Some physicists do not worry about this problem. They say that the
divergence in Eq. (1.38) is after all renormalizable, and so can be
absorbed into the input parameters. However, I think that the quadratic
divergence in (1.38) is a symptom of a serious disease, which can be
fatal in a more unified model. For example in GUTs there are unavoidable

couplings λ_U between Weinberg-Salam Higgses H and GUT Higgses φ with $\langle 0|\varphi|0\rangle = O(m_X) \gtrsim O(10^{15}$ GeV$)$, as seen in Fig. 19b. These provide[44,46]

$$\delta m_H^2 = \lambda_U \langle 0|\phi|0\rangle^2 = O(m_X^2) \gtrsim O(10^{15} GeV)^2 \qquad (1.40)$$

Even if one imagines finding some unknown symmetry mechanism which would cancel δm_H^2 (1.40) at the tree level, one must also cancel out radiative corrections which regenerate

$$\delta m_H^2 = O\left(\frac{\alpha}{\pi}\right) m_X^2 \gtrsim O(10^{14} GeV)^2 \qquad (1.41)$$

and so on up to $O(\alpha/\pi)^{12}$. As another example of large δm_H^2, it has been argued[47] that quantum gravity effects generate

$$\delta m_H^2 = O(10^{19} GeV)^2 \qquad (1.42)$$

We are in sore need of some symmetry to protect the Higgs boson mass, and avoid the large corrections (1.38), (1.40), (1.41), (1.42).

One proposed strategy[48] has been to dissolve the Higgs boson, so that it is no longer elementary, but is composite on a distance scale

$$R_T \sim \frac{1}{\Lambda_T} \quad : \quad \Lambda_T \sim 1 \, TeV \qquad (1.43)$$

This provides a cut-off $\Lambda \sim \Lambda_T$ in the previously quadratically divergent loop integrals (1.38). The new physics at 1 TeV becomes 'technicolour'-- the dynamics of the strongly interacting constituents of the Higgs bosons. If these are massless, there is a chiral symmetry analogous to that re- flected in (1.37), which guarantees the existence of massless Goldstone bosons which can be eaten by the $W^\pm$ and Z^0, giving them masses. However, to provide quark and lepton masses requires extending technicolour[49], and the proposed models of this type tend to have unacceptably large flavour-changing neutral interactions[50] and/or charged pseudoscalar bosons $P^\pm$ with masses $\lesssim 15$ GeV [51] which have not been seen by experi- ment[52]. Therefore theoretical interest in technicolour models has waned[53].

A second strategy for dealing with the loop corrections to m_H^2 of Fig. 19a is to notice that they can be cancelled. Boson and fermion loops have opposite signs:

$$\delta m_H^2 = O\left(\frac{\alpha}{\pi}\right)\left[\Lambda^2 + O(m_B^2)\right] - O\left(\frac{\alpha}{\pi}\right)\left[\Lambda^2 + O(m_F^2)\right] \qquad (1.44)$$

Therefore, if one has pairs of bosons and fermions with the same quantum numbers and hence the same couplings, and similar masses, one can cancel the loop corrections δm_H^2 [Eq. (1.43)] leaving a residual

$$\delta m_H^2 = O\left(\frac{\alpha}{\pi}\right)\left| m_B^2 - m_F^2 \right| \qquad (1.45)$$

This is 'natural' in the sense that $\delta m_H^2 \lesssim m_H^2$ if the cut-off Λ representing new physics is taken to be the masses of the newly postulated particles:

$$\Lambda^2 \sim \left| m_B^2 - m_F^2 \right| \lesssim \left(1\,\mathrm{TeV}\right)^2 \qquad (1.46)$$

A theory with identical couplings for bosons and fermions, and similar masses, has approximate supersymmetry[43]. Such theories will be the subject of the rest of these lectures.

2. SEARCH FOR SPARTICLES

2.1 Sspectrum of Supersymmetric Particles

Although supersymmetry may appear to be a very cute idea, it is profligate in its requirements for new particles. No known particle can be the supersymmetric partner of any other. Consider, for example, the quark, whose spartner should have spin 0 or 1. No elementary spin-0 particle has even been seen, whilst there is no known spin-1 particle with baryon number B = 1/3 which is a $\underline{3}$ of colour. Similar arguments can be made for all the other known particles, so one must postulate a complete set of new sparticles to accompany them, as shown in Table 3. In fact, the number of elementary particles in the minimal Standard Model must be more than doubled since, as mentioned in Lecture 1, there

248

Table 3: Supersymmetric particles

Particle	Spin	Sparticle	Spin
quark $q_{L,R}$	1/2	squark $\tilde{q}_{L,R}$	0, 0
lepton $l_{L,R}$	1/2	slepton $\tilde{l}_{L,R}$	0, 0
photon γ	1	photino $\tilde{\gamma}$	1/2
gluon g	1	gluino $\tilde{g}$	1/2
W	1	wino $\tilde{W}$	1/2
Z	1	zino $\tilde{Z}$	1/2
Higgs H	0	shiggs $\tilde{H}$	1/2
graviton G	2	gravitino $\tilde{G}$	3/2

must be two Higgs doublets H and $\bar{H}$ in even a minimal supersymmetric
version of the Standard Model. The fact that no charged sparticle has
been seen in e^+e^- collisions at PEP or PETRA[52] means that

$$m_{\tilde{q},\,\tilde{l}^\pm,\,\tilde{W}^\pm,\,\tilde{H}^\pm} \gtrsim 22 \text{ GeV} \tag{2.1}$$

whilst the absence of strongly interacting supersymmetric particles in
fixed-target hadron experiments or quarkonium decays[52] means that

$$m_{\tilde{g}} \gtrsim 3 \text{ GeV} \tag{2.2}$$

and the absence of squarks or gluinos at the CERN $p\bar{p}$ Collider[54] means
that

$$m_{\tilde{q},\,\tilde{g}} \gtrsim (40 \text{ to } 50) \text{ GeV} \tag{2.3}$$

The limits (2.1), (2.2), and (2.3) are discussed in more detail later in
this Lecture, along with ways to improve the limits and/or find sparticles. There is no useful laboratory limit on colourless neutral sparticles such as the photino, for which the best limit,

$$m_{\tilde{\gamma}} \gtrsim \tfrac{1}{2}\ GeV \qquad (2.4)$$

comes from cosmology[55,56], as is also discussed later.

For the sparticle masses to be larger than those of conventional particles, supersymmetry must be broken, presumably spontaneously by analogy with gauge theories. Spontaneous SUSY breakdown would produce non-zero scalar masses m_0 and gaugino masses $m_{1/2}$. The most likely (?) source of these is the super-Higgs mechanism in simple supergravity[57], which works by analogy with the Higgs mechanism in conventional gauge theories. Spontaneous SUSY (gauge) breakdown is accompanied by the appearance of a massless spin-1/2 (spin-0) Goldstone fermion (boson), which is eaten by the gravitino (gauge boson) with states of helicity $\pm 3/2$ (± 1) to form a massive spin-3/2 (spin-1) state with four (three) polarization states. When $m_{3/2} \neq 0$, there are in general also scalar and gaugino masses $m_{0},_{1/2} \propto m_{3/2}$. In minimal supergravity models[58] $m_0 = m_{3/2}$, but both m_0 and $m_{1/2}$ can be less than $m_{3/2}$ in non-minimal supergravity models[59]. The physics responsible for generating these SUSY-breaking masses presumably decouples from the known particles at an energy scale $O(m_p)$. Therefore the physical values of the sparticle masses are subject to renormalization by the known gauge and Yukawa interactions[60], in much the same way as the gauge couplings and fermion masses are renormalized below m_X in conventional GUTs. Typical results in the minimal SUSY extension of the Standard Model include[61]:

$$m_{\tilde{g}} : m_{\tilde{W}} : m_{\tilde{B}} = \alpha_3 : \alpha_2 : \tfrac{5}{3}\alpha_1 \qquad (2.5)$$

implying in particular that

$$m_{\tilde{\gamma}}/m_{\tilde{g}} = \tfrac{8}{3}\,\alpha/\alpha_3 \approx \tfrac{1}{6} \qquad (2.6)$$

One also has[61]

$$m_{\tilde{q}}^2 \approx m_0^2 + 7 m_{1/2}^2 \qquad (2.7a)$$

$$m^2_{\tilde{\ell}_{L,R}} \simeq m^2_0 + \begin{cases} 0 \cdot 5 \\ 0 \cdot 15 \end{cases} m^2_{1/2} \qquad (2.7b)$$

and

$$m_{\tilde{g}} \simeq 3\, m_{1/2} \qquad (2.7c)$$

in this minimal SUSY model.

One comment about the results [Eq. (2.7b)] is perhaps in order. Since quarks and charged leptons are four-component Dirac fermions $f = f_{L,R}$, they have two associated spin-0 fields per flavour: $\hat{f}_{L,R}$. These are often called left- and right-handed squarks and sleptons, respectively. This nomenclature does not refer to their helicities, which are 0 because $\hat{f}_L, \hat{f}_R$ have spin 0, but do describe their gauge interactions which are the same as those of their spartners $f_{L,R}$. In general, both squarks and sleptons could mix in this 'helicity' space as well as in flavour space. Their mixing matrix contains contributions from conventional Yukawa couplings:

$$\lambda\, \hat{f}_L \bar{\hat{f}}_R H \;\Rightarrow\; \lambda^2 \left[\, |\hat{f}_L H|^2 + |\hat{f}_R H^*|^2 \,\right] \qquad (2.8)$$

Using $\langle 0|H|0\rangle \equiv v$ so that $m_f = \lambda v$, we would then find $m_{\tilde{f}} = m_f$ in the absence of SUSY breaking. In the presence of SUSY breaking, there can be additional contributions to the mass matrix from the gauge interactions, called D terms. These are in general different for the f_L and f_R, since they have different gauge interactions, but often vanish in models, so we drop them for now. However, there are always the effects [Eqs. (2.7a) and (2.7b)] of soft SUSY breaking which yield the mass matrix

$$(\hat{f}_L, \hat{f}_R) \begin{pmatrix} m^2_f + m^2_{0_L} & A\, m_f m_0 \\ A\, m_f m_0 & m^2_f + m^2_{0_R} \end{pmatrix} \begin{pmatrix} \hat{f}^*_L \\ \hat{f}^*_R \end{pmatrix} \qquad (2.9)$$

In general, $m_{0L} \neq m_{0R}$ because of their different renormalization by their different gauge interactions. The parameter A in (2.9) is also normally present, taking a model-dependent value of order 1. If the off-diagonal entries in (2.9) are small, as they must be for the spartners of light fermions, $m_f \ll m_0$, the mass eigenstates of (2.9) are almost unmixed $\tilde{f}_L$ and $\tilde{f}_R$ states. However, the off-diagonal entries may be comparable to $|m^2_{0L} - m^2_{0R}|$ in the case of a heavy fermion such as the t-quark. In this case the $\tilde{f}_L - \tilde{f}_R$ mixing could be large, and it could even be[62] that the lighter $\tilde{t}$ mass eigenstate is lighter than the t-quark! This would completely modify[62] the conventional expectations for toponium decays[17].

2.2   Which is the Lightest Sparticle?

In most supersymmetric theories there is one absolutely stable particle which would have been copious in the very early Universe, and present at some level as a cosmological relic today. This arises because most supersymmetric theories contain[63] a multiplicatively conserved quantum number R which takes the values:

$$R = \begin{cases} +1 & \text{for particles} \\ -1 & \text{for sparticles} \end{cases} \tag{2.10}$$

The values of R, and also its conservation in many models, can be deduced from its representation in terms of baryon number B, lepton number L, and spin S:

$$R = (-1)^{2S + 3B + L} \tag{2.11}$$

It is easy to check that R is conserved by the conventional gauge and Yukawa interactions, which generate vertices of the forms

$$\bar{f}fV \to \bar{f}\tilde{f}\tilde{V} , \quad \bar{f}fH \to \bar{f}\tilde{f}\tilde{H} , \quad |\tilde{f}^2|^2 , \text{etc.} \tag{2.12}$$

However, even if the interactions conserve a quantum number such as R parity, the vacuum may not, and this discrete symmetry would then be spontaneously violated. This would occur if some sparticle $\tilde{X}$ had a non-zero vacuum expectation value $\langle 0|\tilde{X}|0\rangle \neq 0$. In the minimal supersymmetric

Standard Model, this could only be the sneutrino $\tilde{\nu}$, and some models exist[64] with $\langle 0|\tilde{\nu}|0\rangle \neq 0$, in which case lepton number is also violated spontaneously. However, such a v.e.v. is very tightly constrained by upper limits on lepton number violation[65,66], particularly for the first two generations. This means that even R-violating theories finish up looking not unlike R-conserving theories[66]. R-parity conservation has three important phenomenological consequences: i) sparticles are always produced in pairs, e.g.

$$e^+e^- \to \tilde{e}^+\tilde{e}^- \quad , \quad p\bar{p} \to \tilde{q}+\tilde{g}+X \qquad (2.13)$$

ii) heavy sparticles decay into lighter sparticles, e.g.

$$\tilde{e} \to e\tilde{\gamma} \;,\; \tilde{q} \to q\tilde{\gamma} \;\text{ or }\; q\tilde{g} \;,\; \tilde{g} \to q\bar{q}\tilde{\gamma} \qquad (2.14)$$

and iii) the lightest sparticle is absolutely stable, because it has no possible decay mode.

One can then argue that this stable relic cannot have either electromagnetic charge or strong interactions[56]. If it did, it would interact with ordinary matter and condense along with it into galaxies, stars, planets, sea-water, etc., forming anomalous superheavy isotopes of conventional elements. Such superheavy isotopes have been looked for unsuccessfully, notably in water[67], and the limit

$$\frac{n(\text{relic})}{n(\text{proton})} < O\left(10^{-20} \text{ to } 10^{-30}\right) \qquad (2.15)$$

established for relics between 3 and 10^3 GeV as seen in Fig. 20. This limit is to be compared with the calculated[68] abundances

$$\frac{n(\text{relic})}{n(\text{proton})} \propto \frac{1}{\sigma(R\bar{R}\to X)} \sim \frac{10^{-10}}{\alpha^2} \sim \begin{cases} 10^{-10} \text{ (strongly interacting)} \\ 10^{-6} \text{ (electroweakly interacting)} \end{cases} \qquad (2.16)$$

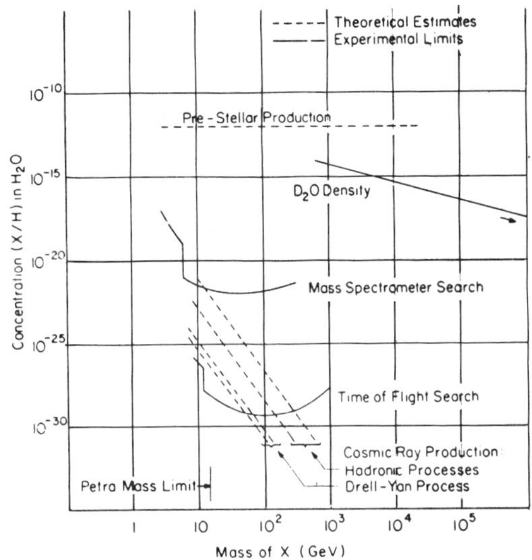

Fig. 20. Limits[67] on stable relics from the Big Bang

This apparent contradiction appears to exclude the existence of a stable strongly or electromagnetically interacting relic weighing between 3 and 10^3 GeV. The situation for relics weighing between 1 and 3 GeV is not so clear; although electromagnetically interacting relics can be excluded on the basis of their high expected density, the relic density is at the borderline of acceptability if they are strongly interacting[69,70]. Although my hunch is that such a relic cannot exist, better calculations and new experiments to exclude this small remaining possibility are desirable. Modulo this small loophole, we conclude that the stable supersymmetric relic must be electromagnetically neutral and only have weak interactions.

The available candidates in Table 3 are therefore the sneutrino of spin 0, some mixture of $\tilde{\gamma}$, $\tilde{H}^0$, and $\tilde{Z}^0$ of spin 1/2, and the gravitino of spin 3/2. In most models the sneutrino mass is of order $m_{\tilde{e}}$, whilst the gravitino may well be even heavier. The best bet is therefore that some $\tilde{\gamma}/\tilde{H}^0$ combination is the lightest supersymmetric particle. These mix in general through the following form of mass matrix in the minimal super-symmetric Standard Model (see ref. 56 and references therein):

$$(\tilde{W}^3, \tilde{B}, \tilde{H}^0, \tilde{\tilde{H}}^0)\begin{pmatrix} M_2 & 0 & -\frac{1}{\sqrt{2}}g_2 v & \frac{1}{\sqrt{2}}g_2 \bar{v} \\ 0 & M_1 & \frac{1}{\sqrt{2}}g_1 v & -\frac{1}{\sqrt{2}}g_1 \bar{v} \\ -\frac{1}{\sqrt{2}}g_2 v & \frac{1}{\sqrt{2}}g_1 v & 0 & \epsilon \\ \frac{1}{\sqrt{2}}g_2 \bar{v} & -\frac{1}{\sqrt{2}}g_1 \bar{v} & \epsilon & 0 \end{pmatrix}\begin{pmatrix} \tilde{W}^3 \\ \tilde{B} \\ \tilde{H}^0 \\ \tilde{\tilde{H}}^0 \end{pmatrix} \qquad (2.17)$$

Here M_2 and M_1 are the SU(2) and U(1) gaugino masses, which have the re-normalized ratio $(5/3)(\alpha_1/\alpha_2)$ [Eq. (2.5)] in unified models with universal gaugino masses at m_p, and ϵ is an $\bar{H}H$ mixing term. The lightest super-symmetric particle χ is in general some complicated mixture:

$$\chi = \alpha \tilde{W}^3 + \beta \tilde{B} + \gamma \tilde{H}^0 + \delta \tilde{\tilde{H}}^0 \qquad (2.18)$$

but it takes simple forms in two interesting limiting cases[56]. When M_2 (and hence also M_1) $\to 0$ for ϵ fixed to be $O(v)$, χ becomes approximately a photino:

$$\tilde{\gamma} = \frac{g_1 \tilde{W}^3 + g_2 \tilde{B}}{\sqrt{g_1^2 + g_2^2}} \; ; \; m_{\tilde{\gamma}} = \frac{8}{3}\frac{g_1^2}{g_1^2 + g_2^2}M_2 = \frac{8}{3}\frac{\alpha_2}{\alpha_3}\sin^2\theta_W \, m_{\tilde{g}} \qquad (2.19)$$

whereas if $\epsilon \to 0$ for M fixed, χ becomes approximately a shiggs:

$$\tilde{H} = \frac{\bar{v}\tilde{H}^0 + v\tilde{\tilde{H}}^0}{\sqrt{v^2 + \bar{v}^2}} \; ; \; m_{\tilde{H}} = \frac{2v\bar{v}}{\sqrt{v^2 + \bar{v}^2}}\epsilon \qquad (2.20)$$

The cosmological density ϱ_χ of such supersymmetric relics can be calculated, and compared with the closure density ϱ_c allowed by observation and required by inflation, which is $\varrho_c = 2 \times 10^{-29}$ g/cm^3 if the present Hubble expansion rate $H_0 = 100$ km s^{-1} Mpc^{-1}. Typical results[56]

for some parameter choices are shown in Fig. 21. If the photino is the
lightest supersymmetric particle, we find[56]

$$m_{\tilde\gamma} \gtrsim \tfrac{1}{2}\,GeV \quad to \quad 5\,GeV \tag{2.21}$$

for squark and slepton masses $m_{\tilde f}$ = 20 to 100 GeV. If the shiggs is the
lightest supersymmetric particle, we find[56]

$$m_{\tilde H} \gtrsim m_b \quad to \quad m_t \tag{2.22}$$

for $m_{\tilde f}$ = 20 to 100 GeV. If the $\tilde H$ is lighter than one of these heavy
quarks, its annihilation in the early Universe is very inefficient, and
its density exceeds ϱ_c. Comparing (2.21) and (2.22) we see that the
photino is likely to be lighter than the shiggs. One possible exception
to this is if $m_{\tilde H} \lesssim$ 100 eV, in which case its cosmological history
resembles that of a massive neutrino, and it does not need to annihilate
to keep its density $< \varrho_c$. However, in this case the $H\bar{H}$ mixing term ε is
so small that there is a light pseudoscalar Higgs boson which is prob-
ably[56] excluded by axion searches.

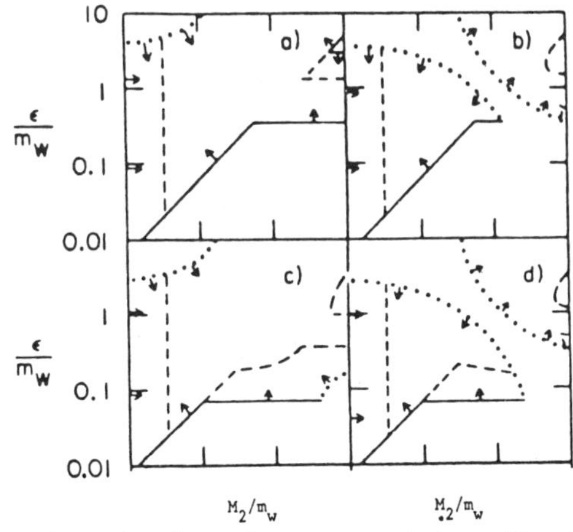

Fig. 21 Cosmological density of supersymmetric relics from the Big
Bang, showing[56] dependence on the gaugino mass M_2, the $H\bar{H}$ mixing term
ε: $\varepsilon > 0$ in (a), (c), $\varepsilon < 0$ in (b), (d), and the ratio of $v \equiv \langle 0|H|0\rangle/\bar{v}$
$\equiv \langle 0|\bar{H}|0\rangle = 1$ in (a), (b), $v/\bar{v} = 2$ in (c), (d).

So far we have not used any prejudice about the relationships (2.5), (2.6), and (2.7) between the photino and other sparticle masses. Using these relations we find the results shown in Fig. 22. The relic density is less than 5×10^{-30} g/cm^3 (the closure density ϱ_c for $H_0 = 50$ km s^{-1} Mpc^{-1}) only if[71]

$$m_{\tilde{q}} \lesssim 0.8\, m_{\tilde{g}} + 40 \text{ GeV} \tag{2.23}$$

Conversely, requiring that the lightest charged sparticle, the $\hat{e}_R$, should have a mass above the experimental limit (2.1) of about 22 GeV [52] tells us that[71]

$$\rho_\chi \gtrsim 0.01\, \rho_c \tag{2.24}$$

Thus there must be a non-negligible supersymmetric relic density in the minimal supersymmetric Standard Model. We will see later that the lower bound (2.24) can even be substantially improved[72] in a low-energy model[33] inspired by the superstring.

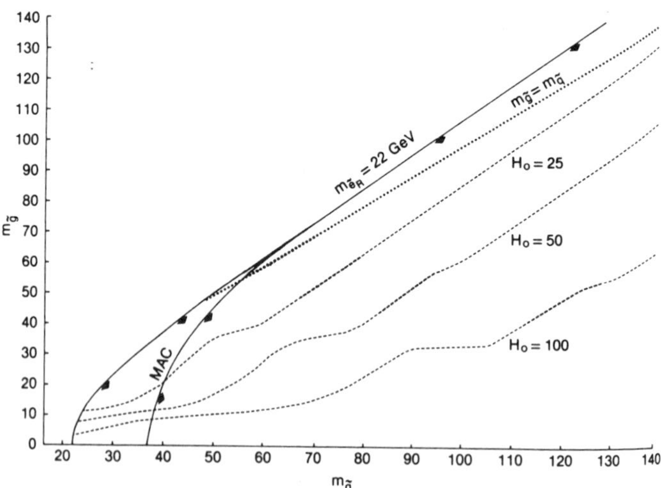

Fig. 22 Cosmological density of supersymmetric relics assuming[71] the minimal supergravity relationships (2.5), (2.6), (2.7) between different sparticle masses

2.3 Sparticle Searches So Far

2.3.1 Light gluinos. Beam-dump experiments have been used to search for the process[73]

$$pN \rightarrow \tilde{g}\tilde{g} + X \, : \, \tilde{g} \rightarrow q\bar{q}\tilde{\gamma} \, \Rightarrow \, \tilde{\gamma} + N \rightarrow \tilde{\gamma} + X \qquad (2.25)$$

The gluino pairs would be produced in the dump, then decay into photinos which traverse the shielding and finally interact in a neutrino detector downstream. The lower bound on $m_{\tilde{g}}$ depends on its lifetime and on $\sigma(\tilde{\gamma}N)$, which in turn depend on $m_{\tilde{q}}$. As seen in Fig. 23a, one finds that[74]

$$m_{\tilde{g}} \gtrsim \begin{cases} 4 \, \text{GeV} \\ 2 \, \text{GeV} \end{cases} \quad \text{for} \quad m_{\tilde{q}} \approx \begin{cases} 50 \, \text{GeV} \\ 250 \, \text{GeV} \end{cases} \qquad (2.26)$$

There have also been searches in hadron-hadron collisions for metastable gluino hadrons with lifetimes longer than about 10^{-10} s, as seen in Fig. 23b, which exclude[75]

$$m_{\tilde{g}} \gtrsim \begin{cases} 2 \, \text{GeV} \\ 4 \, \text{GeV} \end{cases} \quad \text{for} \quad m_{\tilde{q}} \approx \begin{cases} 400 \, \text{GeV} \\ 2000 \, \text{GeV} \end{cases} \qquad (2.27)$$

Finally, the ARGUS Collaboration[76] has looked unsuccessfully for $\Upsilon' \rightarrow {}^3P_1(b\bar{b}) \rightarrow \gamma$ followed by ${}^3P_1 \rightarrow \tilde{g}\tilde{g}g$ [77], when $\tau_{\tilde{g}} \gtrsim 10^{-11}$ s, so that a separated vertex could be detected. As seen in Fig. 23c, this experiment established that

$$m_{\tilde{g}} \gtrsim \begin{cases} 2 \, \text{GeV} \\ 4\frac{1}{2} \, \text{GeV} \end{cases} \quad \text{for} \quad m_{\tilde{q}} \approx \begin{cases} 250 \, \text{GeV} \\ 1600 \, \text{GeV} \end{cases} \qquad (2.28)$$

Thus the gluino mass must certainly exceed 2 GeV, and must exceed 3 GeV for a large range of possible $\tilde{q}$ masses.

258

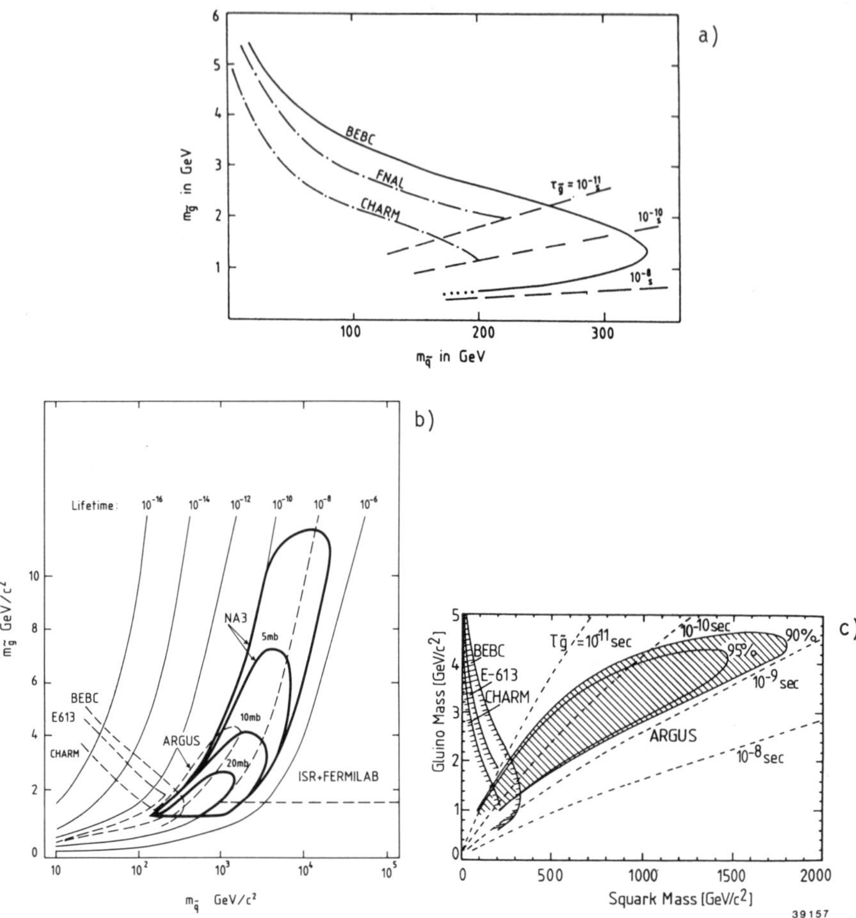

Fig. 23   Compilation of constraints a) on $(m_{\tilde{g}}, m_{\tilde{q}})$ from beam dump experiments[74] and from searches for metastable hadrons, b) in hadron-hadron collisions[75], and c) in $\Upsilon' \to {}^3P_1$ $(b\bar{b}) + \gamma$ decays[76]

2.3.2 Hadronic missing-energy events. The UA1 Collaboration at the CERN $p\bar{p}$ Collider has searched for monojet, dijet, and trijet events accompanied by missing transverse momentum $\not{p}_T$. In their 1983 data they reported[78] only a few events with large $\not{p}_T > 4\sigma$ as seen in Fig. 24, where the measurement error $\sigma \sim 0.7 \sqrt{E_T}$ (GeV). Two monojet events had $\not{p}_T \gtrsim 50$ GeV, and several others had $\not{p}_T \gtrsim 40$ GeV. These few events provided an upper limit on SUSY cross-sections, such as $\sigma(p\bar{p} \to \tilde{q}\tilde{q} + X: \tilde{q} \to q + \tilde{\gamma})$ and $\sigma(pp \to \tilde{g}\tilde{g} + X: \tilde{g} \to q\bar{q}\tilde{\gamma})$. Using QCD calculations of these cross-sections[79] as seen in Fig. 25, which should be reliable to within a factor of 2, one could use these 1983 events to give lower bounds on the $\tilde{q}$ and $\tilde{g}$ masses[80]:

$$m_{\tilde{g}} , m_{\tilde{q}} \gtrsim 40 \, GeV \tag{2.29}$$

Some of us were even more optimistic, and suggested that the few observed events could be due to the pair-production of either $\tilde{q}$ or $\tilde{g}$ with a mass at the lower limit (2.29)[81]. In either case, the minimal supergravity relations (2.5), (2.6), and (2.7) led one to expect $m_{\tilde{\gamma}} \sim$ few GeV, and in the case with a light squark one expected $m_{\tilde{l}} \lesssim 30$ GeV and $m_{\tilde{g}} \sim m_{\tilde{q}}$ as well[82]. In fact, when $m_{\tilde{g}} \sim m_{\tilde{q}}$, the cross-section for $\tilde{q}\tilde{q}$ production is significantly increased above its limit for large $m_{\tilde{g}}$, and the $\tilde{q}\tilde{g}$ and $\tilde{g}\tilde{g}$ production mechanisms also become important, producing a large increase

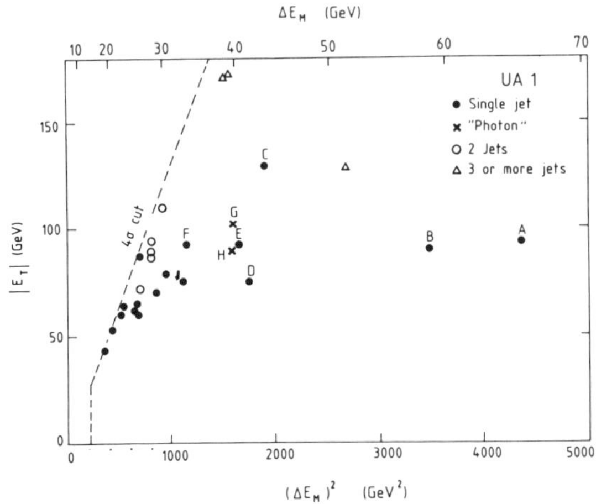

Fig. 24 Scatter plot of missing transverse momentum $\not{p}_T$ versus transverse energy E_T, showing the 1983 data of UA1[78]

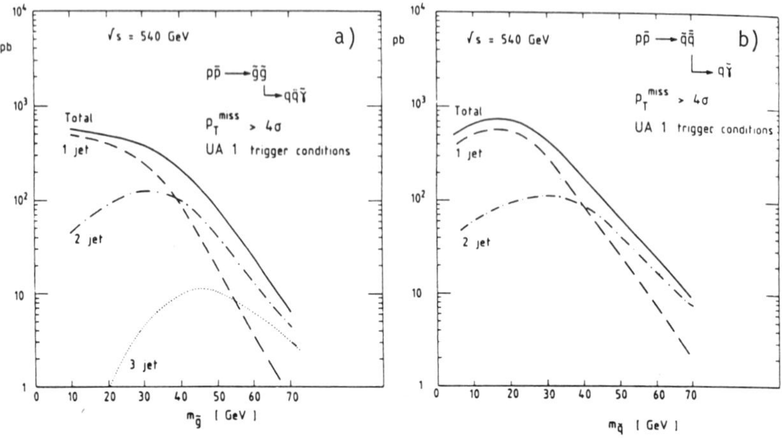

Fig. 25 Cross-sections for $\displaystyle{\not}{p}_T$ events at $\sqrt{s}$ = 540 GeV a) from $\tilde{g}\tilde{g}$ production[80], and b) from $\tilde{q}\bar{\tilde{q}}$ production

in the total SUSY $\displaystyle{\not}{p}_T$ signature[83]. To be compatible with the experimental upper limit on the $\displaystyle{\not}{p}_T$ cross-section, one would actually need[83]

$$m_{\tilde{g}} = m_{\tilde{q}} \gtrsim 60 \ \mathrm{GeV} \tag{2.30}$$

if the $\tilde{g}$ and $\tilde{q}$ masses were approximately equal, and one would have expected as many (or more) dijet + $\displaystyle{\not}{p}_T$ events as monojet + $\displaystyle{\not}{p}_T$ events, as seen in Fig. 26.

The 1984 data of UA1 revealed[54] no new monojet events with $\displaystyle{\not}{p}_T \gtrsim$ 50 GeV, and about the same number of events with $\displaystyle{\not}{p}_T \sim$ 40 to 50 GeV as had been seen in the 1983 data, as shown in Fig. 27. This corresponded to a reduction in the apparent effective rate by a factor of $\sim$ 3, since the 1984 data had a higher integrated luminosity and a higher centre-of-mass energy than the 1983 data. This meant that the gross number of events was now compatible with the expected Standard Model backgrounds due to $W \to \tau\nu$, $(W \to e\nu$ or $\mu\nu)$ + jets, $(Z^0 \to \bar{\nu}\nu)$ + jets, $t\bar{t}$, $b\bar{b}$, and $c\bar{c}$

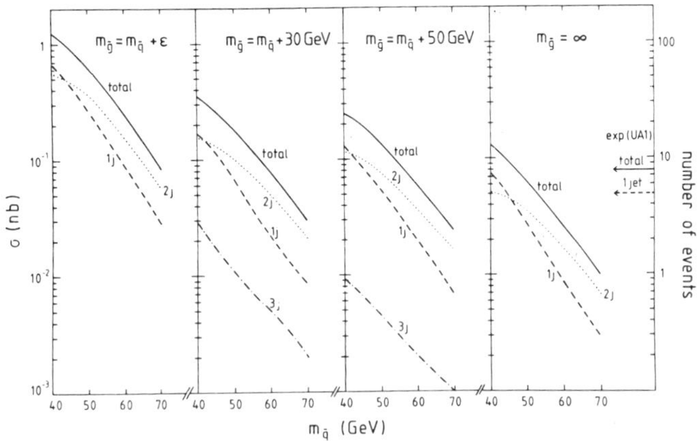

Fig. 26 Cross-sections for $\not{p}_T$ events when $m_{\tilde{g}} \approx m_{\tilde{q}}$ [83]

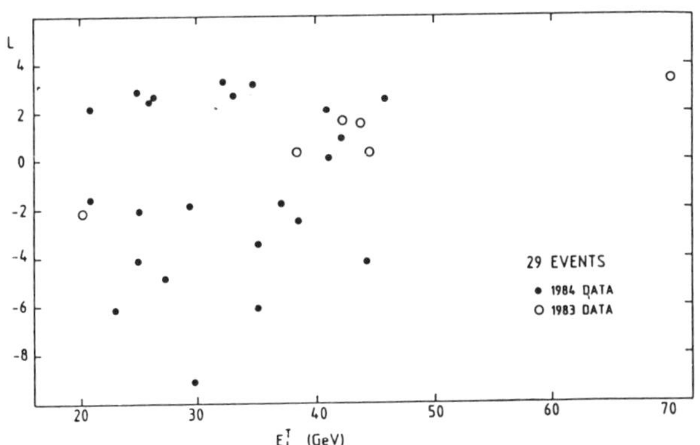

Fig. 27 Scatter plot of unbalanced jet energy versus jet thinness, showing the 1984 data of UA1[54]. Events close to the top of the plot resemble taus.

production followed by semileptonic decays, etc.[84]. This was already true for events with $\not{p}_T \lesssim 40$ GeV in the 1983 data, but now became true for the events with $\not{p}_T \gtrsim 40$ GeV also, as seen in Table 4. One could still look for more detailed differences between the monojets produced

Table 4: Backgrounds to monojets
(per 100 nb^{-1} at $\sqrt{s}$ = 540 GeV)

	Ref. 84	Ref. 54
jet + ($Z^0 \to \nu\bar{\nu}$)	0.36	
W → e lost	0.12	] ~ 1
W → μ lost	< 0.01	
W → τ lost	0.25	] ~ 0.6
W → τ seen	0.50	

by the various sources[54], and look to see if some subsample of the $\not{p}_T$ events could be due to new physics such as supersymmetry. For example, one could look at the 'fatness' of the jets as measured by

$$F \equiv \frac{\text{energy in cone of opening angle } 0.4}{\text{energy in cone of opening angle } 1.0} \tag{2.31}$$

and combine it with other measures of a jet's thinness. Monte Carlo cal-culations[54] suggest that τ jets are thinner than conventional QCD jets, with jets from $\hat{q}$ decay intermediate in thickness. However, no clear evidence of a subset of monojet events with a source beyond the Standard Model has yet emerged. One can use the improved 1984 bounds on $\not{p}_T$ cross-sections to quote[85] improved lower bounds on sparticle masses: based on Fig. 28 one concludes,

$$m_{\tilde{g}} \gtrsim (45 \text{ to } 50)\,\text{GeV} \qquad \text{if } m_{\tilde{q}} \text{ large}$$

$$m_{\tilde{q}} \gtrsim (50 \text{ to } 60)\,\text{GeV} \qquad \text{if } m_{\tilde{g}} \text{ large}$$

$$m_{\tilde{g},\tilde{q}} \gtrsim (60 \text{ to } 70)\,\text{GeV} \qquad \text{if } m_{\tilde{g}} = m_{\tilde{q}} \tag{2.32}$$

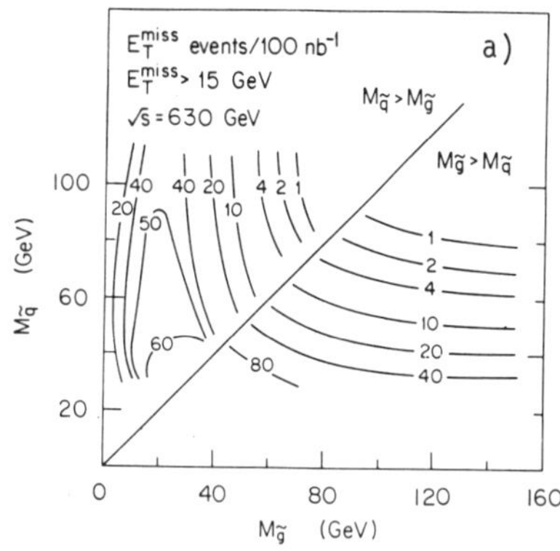

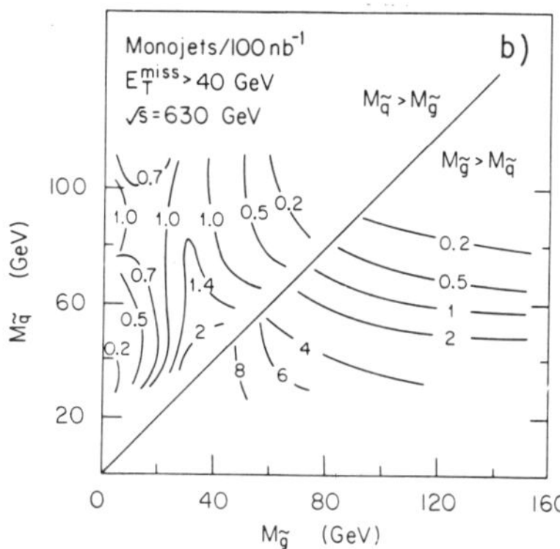

Fig. 28   Cross-sections for $\not{p}_T$ events at $\sqrt{s}$ = 630 GeV for general values of $m_{\tilde{g}}$ and $m_{\tilde{q}}$ [85]

2.3.3 Leptonic missing-energy events. These could be produced by $W^{\pm}$ and/or Z^0 decays into sleptons or winos and/or zinos[86]. The latter are discussed at this meeting by Howie Baer[87], so here we concentrate on $W^{\pm} \to \tilde{l}^{\pm}\tilde{\nu}$ and on $Z^0 \to \tilde{l}^+\tilde{l}^-$. The decay $W^{\pm} \to \tilde{l}^{\pm}\tilde{\nu}$ [88] followed by $\tilde{l}^{\pm} \to l^{\pm} + \tilde{\gamma}$ can give an excess of $l^{\pm} + \not{p}_T$ events with $|\not{p}_T| \lesssim$ 30 GeV and $\cos \theta_{e^-p}^{lab} < 0$ preferred. For $m_{\tilde{l}} \lesssim$ 30 GeV one finds[89]

$$\sigma(W \to \tilde{e} \to e)/\sigma(W \to e) \approx 0.1$$

$$\sigma(W \to \tilde{\mu} \to \mu)/\sigma(W \to \mu) \approx 0.15 \tag{2.33}$$

after including UA1 cuts and resolutions. Note that $W^{\pm} \to e_L^{\pm}$ or $\mu_L^{\pm}$ only, not $e_R^{\pm}$ or $\mu_R^{\pm}$. The UA1 Collaboration has looked for such an excess of $e + \not{p}_T$ events in addition to its established $W \to e\nu$ signal. It has been able to bound[54]

$$m_{\tilde{e}_L} \begin{cases} \gtrsim 26\,GeV & \text{if } m_{\tilde{\gamma}} = m_{\tilde{e}_L} \\ \gtrsim 33\,GeV & \text{if } m_{\tilde{\gamma}} = 0 \end{cases} \tag{2.34}$$

The decay $z^0 \to \hat{l}^+\hat{l}^-$ [90] followed by $\hat{l}^\pm \to \tilde{l}^\pm + \tilde{\gamma}$ gives $l^+l^- + \not{p}_T$ events with

$$|\not{p}_T| \sim |p_T(l^+l^-)| \sim 20\,\text{GeV} \ , \quad m(l^+l^-) \sim 40\,\text{GeV} \qquad (2.35)$$

at a rate [89]

$$\sigma(z^0 \to \tilde{l}^+\tilde{l}^- \to l^+l^- + \not{p}_T)/\sigma(z^0 \to l^+l^-) \approx 0.1 \qquad (2.36)$$

for $m_{\tilde{l}^\pm} \lesssim 30$ GeV, after including the effects of UAn cuts and resolutions The large $\not{p}_T$ signature (2.36) means that such events would be readily distinguishable from the Drell-Yan background. In contrast to $w^\pm \to \tilde{e}^\pm$ decay, in this case one is sensitive to $\tilde{l}^\pm_R$ as well as $\tilde{l}^\pm_L$. Although UA1 and UA2 have not quoted upper limits on such $l^+l^- + \not{p}_T$ cross-sections, they can probably be used to establish bounds at least as sensitive as (2.34).

2.3.4 $e^+e^- \to \gamma + nothing$. In this reaction, one uses a bremsstrahlung γ to tag events in which missing neutral particles escape from the apparatus undetected. The Standard Model process of this type is $e^+e^- \to \gamma + \nu\bar{\nu}$ [91], which has the low-energy cross-section

$$\frac{d^2\sigma}{dx_\gamma d\cos\theta_\gamma} = \frac{s}{4x_\gamma \sin^2\theta_\gamma}\left[(1-x_\gamma)\left(1-\frac{x_\gamma}{2}\right)^2 + \frac{x_\gamma^2}{4}(1-x_\gamma)\cos^2\theta_\gamma\right]$$

$$\times \left[\frac{G_F^2\alpha}{12\pi^2}(4+N_\nu)\right] \qquad (2.37)$$

where $x_\gamma = 2E_\gamma/\sqrt{s}$, θ_γ is the angle between the outgoing γ and the beam direction, and N_ν is the number of neutrino flavours. The term proportional to N_ν in Eq. (2.37) is due to z^0 exchange, and extrapolates to give a large cross-section when $\sqrt{s} > m_{z^0}$ as discussed in the next section. The other piece in Eq. (2.37) is due to the $w^\pm$ exchange graph for producing $\nu_e\bar{\nu}_e$ pairs. The ASP experimental upper limit on $\sigma(e^+e^- \to \gamma +$ nothing) at $\sqrt{s} = 29$ GeV has been used [92] to set the limit

$$N_\nu < 14 \qquad (90\% \text{ conf.}) \qquad (2.38a)$$

More recently (R. Prepost, private communication) it has been shown that the data of ASP, MAC, and CELLO can be combined to yield

$$N_\gamma < 8 \qquad (90\% \text{ conf.}) \tag{2.38b}$$

In the case of $e^+e^- \to \gamma + (\tilde{\gamma}\tilde{\gamma})$ [93], one replaces the second row in equation (2.37) by the factor

$$[\quad] \to \frac{2}{3} \alpha^3 \left[\frac{1}{m_{\tilde{e}_L}^4} + \frac{1}{m_{\tilde{e}_R}^4} \right] \times (\text{phase space}) \tag{2.39}$$

Using the appropriate elaboration of this formula for finite, non-zero $m_{\tilde{e}}$ and $m_{\tilde{\gamma}}$, the ASP Collaboration[92] has established

$$m_{\tilde{e}_L = \tilde{e}_R} \gtrsim \begin{cases} 51\,\text{GeV} & \text{if} \quad m_{\tilde{\gamma}} = 0 \\ 45\,\text{GeV} & \text{if} \quad m_{\tilde{\gamma}} = 5\,\text{GeV} \end{cases} \tag{2.40}$$

which is by far the most stringent lower limit on $m_{\tilde{e}}$ for small $m_{\tilde{\gamma}}$, as seen in Fig. 29. (Combining with CELLO and MAC data yields $M_{\tilde{e}L=R} \gtrsim 58$ GeV if $m_{\tilde{\gamma}} = 0$.) One can also consider $e^+e^- \to \gamma + (\tilde{\nu}\tilde{\nu})$ [94], but in this case the cross-section at the energies at present accessible is suppressed by

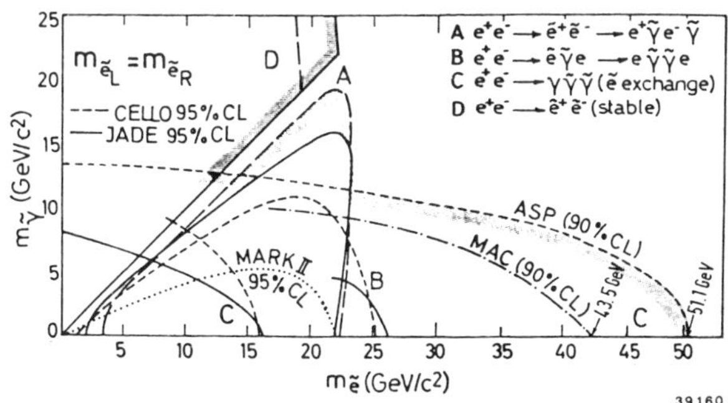

Fig. 29 Bound on $(m_{\tilde{e}}, m_{\tilde{\gamma}})$ from the ASP experiment[92], compared with other e^+e^- experiments

phase space, since typically one expects $m_{\tilde{\nu}} \sim m_{\tilde{e}} \gtrsim 20$ GeV. Another reaction which can give γ + nothing events at higher energies is $e^+e^- \rightarrow Z^0 \rightarrow {}'\tilde{H}' + {}'\tilde{\gamma}'$ [94], followed by $'\tilde{H}' \rightarrow + \gamma \, '\tilde{\gamma}'$ decay, which can occur via the $\tilde{H}/\tilde{\gamma}$ mixing discussed in subsection 2.2. In many models, $m_{\tilde{H}}$ and $m_{\tilde{\gamma}}$ are both $\lesssim 30$ GeV, which can lead to an observable rate for $e^+e^- \rightarrow \gamma +$ nothing at the Z^0 peak.

2.3.5 Compilation. Shown in Fig. 30 is a compilation[95] of the available limits on the SUSY-breaking parameters m_0 and $m_{1/2}$ of the minimal supergravity models discussed in subsection 2.1. It is noteworthy that e^+e^- experiments, $\bar{p}p$ collider experiments, and cosmology all provide the most sensitive bounds in some region of the $(m_0, m_{1/2})$ plane. One should emphasize, however, that the e^+e^- limits are in general the cleanest and the most model-independent.

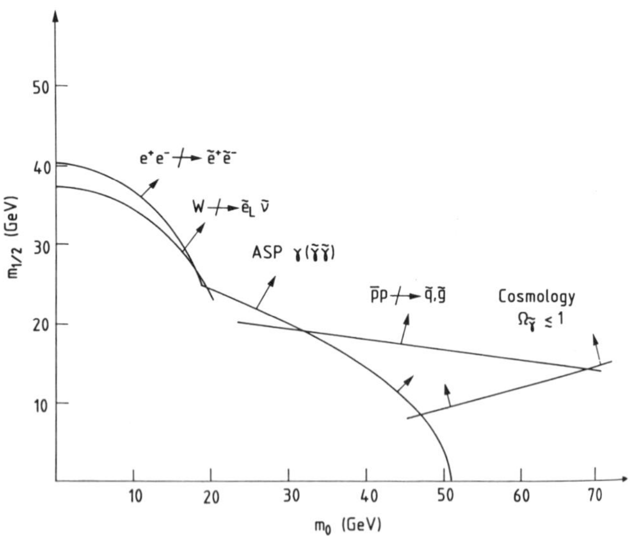

Fig. 30 Compilation[95] of available constraints on the SUSY-breaking parameters $(m_0, m_{1/2})$ in minimal supergravity models

2.4   Future e^+e^- Sparticle Searches

The general strategy to look for sparticles outlined in Ref. 96 is
i) to look for all two-particle production processes $e^+e^- \to \tilde{A} + \tilde{B}$, which
have the highest cross-sections and enable one to search for all sparti-
cles with masses $\tilde{m} \lesssim E_{beam}$, and then ii) to look for some processes with
≥ 3 particles $e^+e^- \to \tilde{A} + \tilde{B} + C$, etc., which have smaller cross-sections
but may in some cases allow one to search for sparticles with masses $\tilde{m} >$
E_{beam}. We start with weakly interacting sparticles such as sleptons and
winos, whose decays have the general signature: $e^+e^- \to$ acoplanar dileptons + large $\not{p}_T$. The principal backgrounds are the higher-order QED
processes $e^+e^- \to 1^+1^-\gamma$ and $1^+1^-1'^+1'^-$, where some final-state particles
are unobserved. In what follows, we assume[96] a detector which can see
electrons in the polar angle range $5° < \theta_e < 175°$ and muons in $10° < \theta_\mu <$
$170°$, with a resolution $\Delta E/E = 1\%$ for e, μ, and γ. We see in Fig. 31
that the background due to higher-order QED processes is negligible for
$\not{p}_T \gtrsim 5$ GeV, compared with the signal from $e^+e^- \to \tilde{e}^+\tilde{e}^-$. Weakly interacting sparticles will be very easy to see at SLC or LEP, if they have

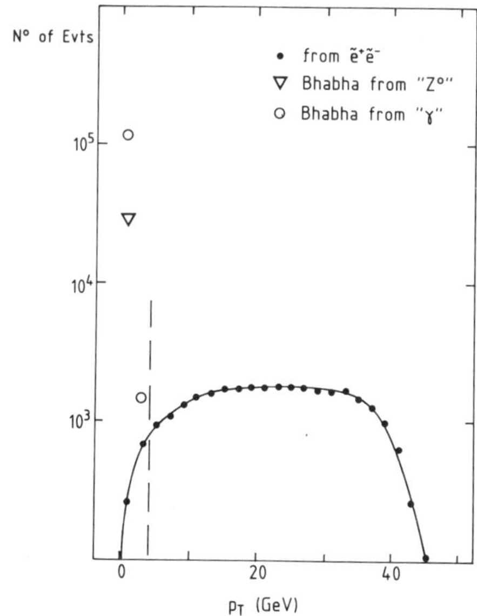

Fig. 31   QED background to $e^+e^- \to (\tilde{e}^+ \to e^+\tilde{\gamma})(\tilde{e}^- \to e^-\tilde{\gamma})$ [96]

masses within the kinematical ranges of these accelerators. Producing enough of these weak sparticles to observe them at hadron-hadron colliders is very difficult unless $\tilde{m} < m_Z/2$ as discussed in section 2.3.3.

In the case of $e^+ e^- \to \tilde{e}^+ \tilde{e}^-$, there are direct-channel γ and Z^0 diagrams as well as crossed-channel $\tilde{\gamma}$, $\tilde{Z}^0$, and $\tilde{H}$ diagrams. The cross-section due to γ, Z^0, and massless $\tilde{\gamma}$ exchanges is

$$\frac{d\sigma}{d\Omega} = \frac{\alpha^2}{8s}\,\beta^3 \sin^2\Theta\left[1 + \left(1 - \frac{4}{1 - 2\beta\cos\Theta + \beta^2}\right)^2\right]$$
$$+ \text{ corresponding } Z^* \text{ terms} \tag{2.41}$$

The cross-section for $m_{\tilde{\gamma}} \neq 0$ is given in Ref. 96: it is quite insensitive to $m_{\tilde{\gamma}} \lesssim 5$ GeV. Figure 32a shows the observable $e^+ e^- + \not{p}_T$ cross-section assuming the detector characteristics described above. We conclude that $m_{\tilde{e}} \lesssim 80$ GeV should be observable at LEP II. In the case of $e^+ e^- \to \tilde{\mu}^+ \tilde{\mu}^-$, there is no $\tilde{\gamma}$ exchange diagram, and the cross-section is somewhat smaller:

$$\frac{\sigma(e^+ e^- \to \gamma^*, Z^* \to \tilde{\mu}^+ \tilde{\mu}^-_{L,R})}{\sigma(e^+ e^- \to \gamma^* \to \mu^+ \mu^-)} = \frac{1}{4}\beta^3\left[1 + \frac{2C_V C_{V_{L,R}} s(s - m_Z^2)}{(s - m_Z^2)^2 + m_Z^2 \Gamma_Z^2} + \frac{(C_V^2 + C_A^2)s^2 C_{V_{L,R}}^2}{(s - m_Z^2)^2 + m_Z^2 \Gamma_Z^2}\right]$$

where $\tag{2.42}$

$$C_{V_L} = \frac{-2 + 4\sin^2\Theta_W}{4\sin\Theta_W \cos\Theta_W} \quad , \quad C_{V_R} = \frac{4\sin^2\Theta_W}{4\sin\Theta_W \cos\Theta_W}$$

but the process should nevertheless be observable at LEP II if $m_{\tilde{\mu}} \lesssim 80$ GeV, as seen in Fig. 32b. In the case of $e^+ e^- \to \tilde{W}^+ \tilde{W}^-$ there are direct-channel γ and Z^0 diagrams, and crossed-channel $\tilde{\nu}$ diagrams. A complete formula for the cross-section can be found in Refs. 96 and 97. Figure 33 shows that it is large for $m_{\tilde{W}^\pm} < \sqrt{s}/2$. The dominant decays of the $\tilde{W}^\pm$ would be into $l^\pm \nu \tilde{\gamma}$ if $m_{\tilde{l}} < m_{\tilde{W}} < m_{\tilde{q}}$, giving $l^\pm l'^\mp + \not{p}_T$ as a final-state event signature. For other mass ratios, $\tilde{W} \to q\bar{q}'\tilde{\gamma}$ could dominate over $\tilde{W} \to l\nu\tilde{\gamma}$, and the final states would be combinations of 4-jet $+ \, p_T$, 2-jet $+ \, l^\pm + \not{p}_T$ and $l^\pm l'^\mp + \not{p}_T$. Studies of $W^\pm$ decay at the CERN $p\bar{p}$ Collider may well be able to rule out $m_{\widetilde{W}} \lesssim 40$ GeV [86], as discussed at this meeting by Howie Baer[87]. It is apparent from Fig. 33 that $m_{\widetilde{W}^\pm} \lesssim 80$ GeV should be detectable at LEP II. The process $e^+e^- \to \widetilde{Z}\widetilde{\gamma}$, which can proceed

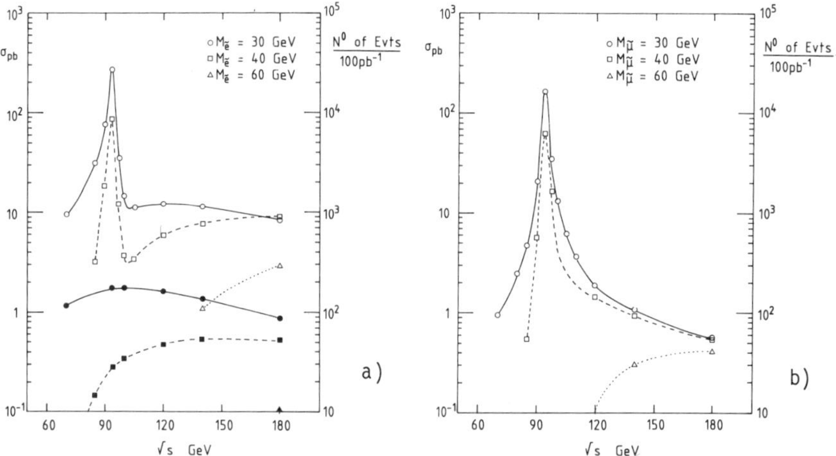

Fig. 32 a) Observable $\sigma(e^+e^- + \not{p}_T)$ from $e^+e^- \to \widetilde{e}^+\widetilde{e}^-$, and b) observable $\sigma(\mu^+\mu^- + \not{p}_T)$ from $e^+e^- \to \widetilde{\mu}^+\widetilde{\mu}^-$ [96]

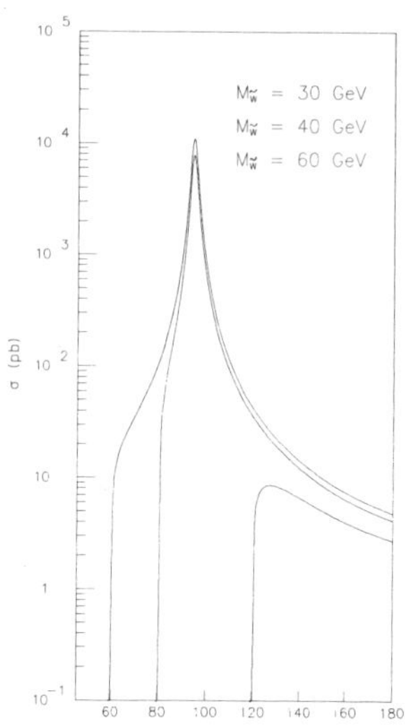

Fig. 33 The cross-section for $e^+e^- \to \widetilde{W}^+\widetilde{W}^-$ [96]

via $\tilde{e}$ exchange, is detectable[96] at SLC and LEP for a wide range of $\tilde{Z}$ and $\tilde{e}$ masses, as seen in Fig. 34. Because one expects $m_{\tilde{\gamma}} \ll m_{\tilde{Z}}$, this process can be used to look for $m_{\tilde{Z}} > \sqrt{s}/2$. The decay $\tilde{Z} \to l\tilde{l} \to l^+l^-\tilde{\gamma}$ would be dominant for $m_{\tilde{l}} < m_{\tilde{Z}} < m_{\tilde{q}}$, and provides an $l^+l^- + \not{p}_T$ signature with a different angular distribution from those of the processes $e^+e^- \to \tilde{l}^+\tilde{l}^-$. In particular, if $m_{\tilde{Z}} \ll \sqrt{s}$ the l^+l^- pair will emerge in one hemisphere.

We now turn to some three-body final states which might enable one to probe sparticle masses beyond $\sqrt{s}/2$. The reaction $e^+e^- \to e^-\tilde{W}^+\tilde{\nu}$ has 12 contributing diagrams, whose interferences yield 78 terms in the cross-section. The results of a complete calculation can be found in Refs. 96 and 97. It is possible to remove essentially all the background using the detector cuts mentioned previously. The process can be used to probe $m_{\tilde{W}} > \sqrt{s}/2$ if $m_{\tilde{\nu}} \approx 0$, as seen in Fig. 35a, but one cannot reach significantly beyond $m_{\tilde{W}} \sim \sqrt{s}/2$ if $m_{\tilde{\nu}} \sim m_{\tilde{e}}$ as expected in many models (see

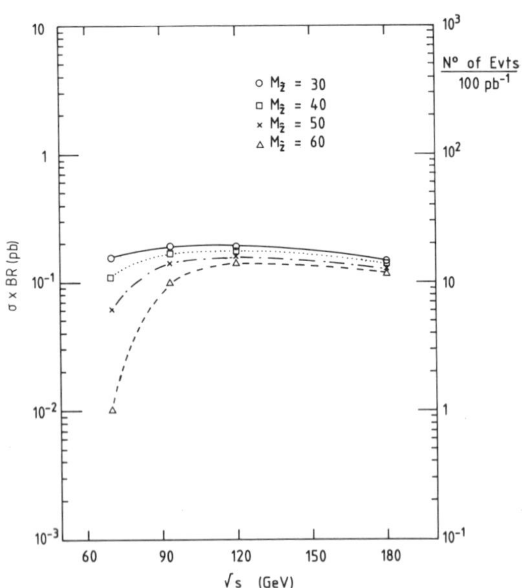

Fig. 34 The cross-section for $e^+e^- \to \tilde{Z}^0\tilde{\gamma}$ [96]

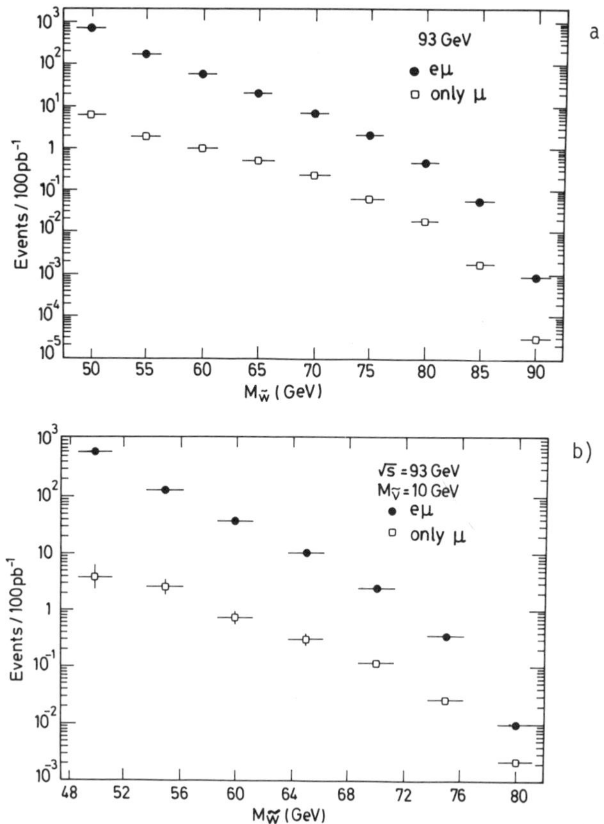

Fig. 35 The cross-section for $e^+e^- \to e^{\pm}\tilde{W}^{\mp}\tilde{\nu}\bar{\nu}$ a) for $m_{\tilde{\nu}} = 0$, and b) for $m_{\tilde{\nu}} = 10$ GeV [96] to give $e\mu$ + nothing and μ + nothing events

Fig. 35b). The decay $Z^0 \to e^{\pm}\tilde{e}^{\mp}\tilde{\nu}\bar{\nu}$ [98] has a branching ratio in excess of 10^{-6} for $m_{\tilde{e}} \lesssim 0.6 m_{Z^0} \simeq 57$ GeV, so one can use this reaction to probe some-what beyond $m_{\tilde{e}} = \sqrt{s}/2$. The processes $e^+e^- \to \gamma + (\tilde{\gamma}\tilde{\gamma})$ and $e^+e^- \to \gamma + (\tilde{\nu}\bar{\tilde{\nu}})$ are a SUSY background to neutrino counting. As seen in Fig. 36, the Standard Model process $e^+e^- \to \gamma + (Z^0 \to \nu\bar{\nu})$ is dominant at SLC and LEP energies, and $[e^+e^- \to \gamma + (\tilde{\gamma}\tilde{\gamma})]$ is too small to be interesting at SLC or LEP [93,96]. However, if one can measure $B(Z^0 \to$ missing neutrals) with a

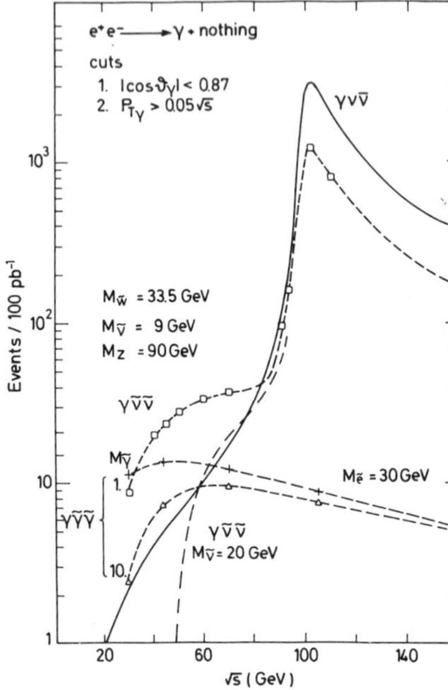

Fig. 36 Cross-sections for $e^+ e^- \rightarrow \gamma$ + nothing as functions of $\sqrt{s}$ [96]

precision of better than half a conventional neutrino flavour, one begins to be sensitive to $Z^0 \rightarrow \tilde{\nu}\bar{\tilde{\nu}}$ decays as well:

$$\frac{\Gamma(Z^0 \rightarrow \tilde{\nu}\bar{\tilde{\nu}})}{\Gamma(Z^0 \rightarrow \nu\bar{\nu})} = \frac{1}{2}\beta^3 \; ; \quad \beta \equiv \sqrt{1 - \frac{4m_{\tilde{\nu}}^2}{m_Z^2}} \tag{2.43}$$

Finally, as mentioned previously, $Z^0 \rightarrow {'\tilde{\gamma}'} + ({'H'} \rightarrow \gamma'\gamma')$ is another possible source of γ + nothing events at the Z^0 peak[94].

We now turn to the basic mechanism for the production of strongly interacting sparticles, namely $e^+ e^- \rightarrow \tilde{q}\bar{\tilde{q}}$. This proceeds via direct-channel γ and Z^0 exchange and has the cross-section

$$\frac{\sigma(e^+ e^- \rightarrow \gamma^*, Z^* \rightarrow \tilde{q}\bar{\tilde{q}})}{\sigma(e^+ e^- \rightarrow \gamma^* \rightarrow \mu^+ \mu^-)} = \frac{1}{4}\beta^3 \left[Q^2 - 2Q C_V C_V' \frac{(s - m_Z^2)}{(s - m_Z^2)^2 + m_Z^2 \Gamma_Z^2} \right.$$

$$\left. + \frac{(C_V^2 + C_A^2) s^2 C_V'^2}{(s - m_Z^2)^2 + m_Z^2 \Gamma_Z^2} \right] \tag{2.44}$$

where the couplings

$$C_V' = \frac{-2+\frac{8}{3}\sin^2\theta_W}{4\sin\theta_W\cos\theta_W} \underset{(\tilde{u}_L)}{} , \quad \frac{\frac{8}{3}\sin^2\theta_W}{4\sin\theta_W\cos\theta_W} \underset{(\tilde{u}_R)}{} , \quad \frac{2-\frac{4}{3}\sin^2\theta_W}{4\sin\theta_W\cos\theta_W} \underset{(\tilde{d}_L)}{} , \quad \frac{-\frac{4}{3}\sin^2\theta_W}{4\sin\theta_W\cos\theta_W} \underset{(\tilde{d}_R)}{} \quad (2.45)$$

As can be seen from Fig. 37, the cross-section (2.44) is large enough to be observable at LEP II if $m_{\tilde{q}} \lesssim 80$ GeV. The final-state event signature is missing energy from $\tilde{q} \to q\tilde{\gamma}$ or $\tilde{q} \to q\tilde{g}$ followed by $\tilde{g} \to q\bar{q}\tilde{\gamma}$. Monte Carlo studies[96] indicate that it should be possible to distinguish $e^+e^- \to \tilde{q}\bar{\tilde{q}}$ from $e^+e^- \to q\bar{q}$ backgrounds at the Z^0 peak. Toponium decays into sparticles are also possible if $m_{\tilde{t}} < m_t$ [62]: they are discussed in Ref. 17. For other searches for strongly interacting sparticles in e^+e^- collisions, see Refs. 96 and 99.

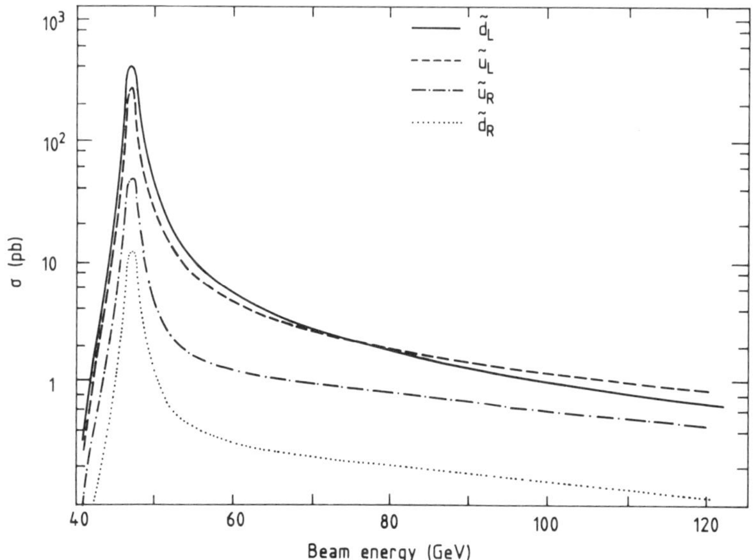

Fig. 37 Cross-sections for $e^+e^- \to \tilde{q}\bar{\tilde{q}}$ [96]

274

2.5 Sparticle Searches at Other Colliders

2.5.1 Hadron-hadron collisions. One will be able to search, at future
hadron-hadron colliders, for strongly interacting sparticles such as
squarks and gluinos, by using missing-energy signatures analogous to
those studied already at the CERN $p\bar{p}$ Collider[36,37]. Decays such as $\tilde{q} \to$
$q\tilde{\gamma}$ or $\tilde{g} \to q\bar{q}\tilde{\gamma}$ give missing-energy signatures with $\not{p}_T \propto m_{\tilde{q}}$ or $m_{\tilde{g}}$. On the
other hand, measurement errors in calorimeters increase only as $\sqrt{E_T} \propto$
$\sqrt{m}$. Moreover, many Standard Model sources of $\not{p}_T$, such as $W \to \tau\nu_\tau$, $Z^0 \to$
$\nu\bar{\nu}$, or heavy-flavour decays, only have limited kinematics which give
$|\not{p}_T| \lesssim 0(50)$ GeV. For these two reasons, searching for the larger $\not{p}_T$
signature at future $p\bar{p}$ and pp colliders will be easier than it has been
at the CERN $p\bar{p}$ Collider. We have studied[37] hadron-hadron colliders with
$\sqrt{s} = 10$ to 40 TeV and luminosities of 10^{32} to 10^{33} cm^{-2} s^{-1}. Typical
cross-sections for $p\bar{p} \to \tilde{g}\tilde{g} + X$ are shown in Fig. 38a, and the spectra of

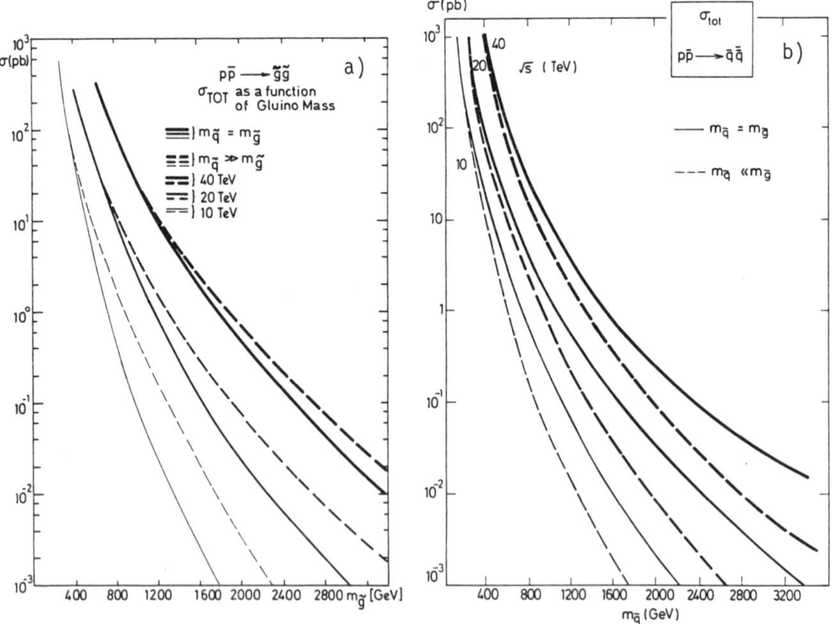

Fig. 38 High-energy cross-sections a) for $p\bar{p} \to \tilde{g}\tilde{g} + X$, b) for $p\bar{p} \to$
$\tilde{q}\tilde{\bar{q}} + X$ [37]

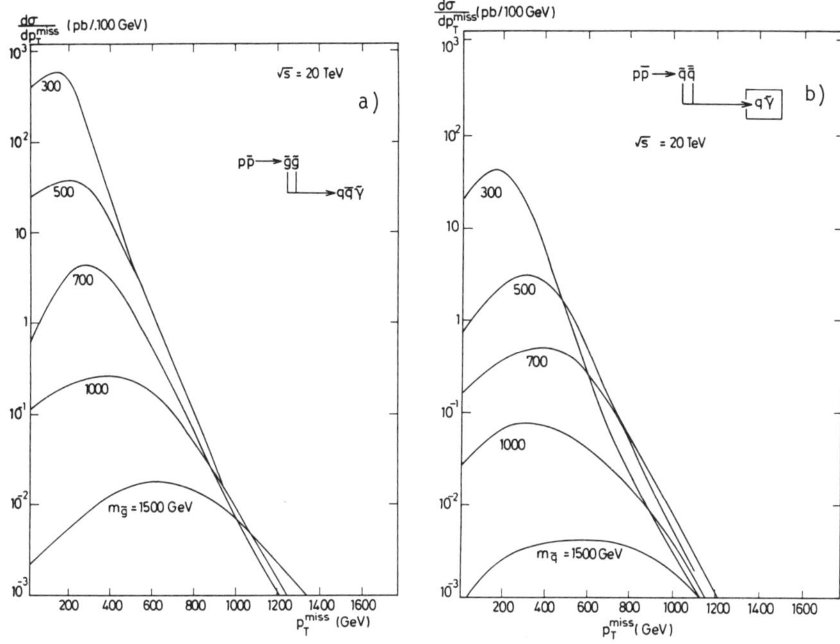

Fig. 39 Spectra of $\not{p}_T$ a) from $\bar{p}p \rightarrow \tilde{g}\tilde{g} + X$, b) from $\bar{p}p \rightarrow \tilde{q}\bar{\tilde{q}} + X$ [37]

$\not{p}_T$ for different $m_{\tilde{g}}$ at $\sqrt{s}$ = 20 TeV are shown in Fig. 39a [37]. We conclude that the colliders considered can reach $m_{\tilde{g}} \lesssim$ (1 to 3) TeV: similar
conclusions apply to $\tilde{q}$ searches, as seen in Figs. 38b and 39b [37]. If no
strongly interacting sparticle showed up in these mass ranges, the idea
of using SUSY to stabilize the weak gauge hierarchy would in my opinion
be dead.

2.5.2 *ep colliders*. Here I will be brief, since Günther Wolf[100] has
discussed in some detail at this meeting the capabilities of HERA for
sparticle searches. Among the possible reactions which can be studied are
$ep \rightarrow \tilde{e}\tilde{q}X$, $\tilde{v}\tilde{q}'X$, $\tilde{q}\bar{\tilde{q}}X$, and $\tilde{q}\tilde{g}X$ [101]. HERA, with its centre-of-mass energy
$\sqrt{s}$ = 314 GeV, can search for the first two of these processes for $m_{\tilde{1}}$ +
$m_{\tilde{q}} \lesssim$ 180 GeV [100]. The final-state event signatures are very clean, e.g.
$e + p \rightarrow e + \text{jet} + \not{p}_T$, and can easily be distinguished from Standard Model

backgrounds[101]. Since HERA is expected to come into operation before
LEP II, it will have first shot at a large range of possible sparticle
masses.

3.   SIGNATURES OF THE SUPERSTRING

3.1   Basic Ideas

     Why are theorists so excited about the superstring? Because it is
an apparently finite[102] theory in ten dimensions which does everything
except make coffee (and perhaps that too). i) It contains gravity. The
mass spectrum of the superstring is of the form

$$m_n^2 = m_0^2 + n M^2 \tag{3.1}$$

where the excitation energy $M = O(m_p)$. The zero modes which have $m_0 = 0$
constitute a ten-dimensional supergravity theory which includes a graviton of spin 2, a gravitino of spin 3/2, gauge bosons of spin 1, matter (?)
particles of spin 1/2, and elementary (Higgs, etc.?) scalars of spin 0.
If the superstring is indeed finite, it would be the first acceptable
quantum theory of gravity. ii) It has a grand unification group. Green
and Schwarz[102] showed that the one-loop superstring diagrams were finite
and anomaly-free if and only if the internal gauge group was O(32) or
$E_8 \times E_8'$. In the most promising of early scenarios for compactification[103] using Calabi-Yau manifolds the choice of O(32) cannot give
a four-dimensional theory which violates parity, whereas $E_8 \times E_8'$ can.
It gives[103] a four-dimensional gauge group

$$E_8 \to E_6 \supset SO(10) \supset SU(5) \tag{3.2}$$

with chiral fermion representations. The groups E_6, SO(10), and SU(5)
have long been the favoured candidates for GUTs. iii) It may solve the
Hierarchy Problem, because the superstring contains conventional
supersymmetry and supergravity. iv) It may solve the Family Problem,

since topology controls[103] the number of generations N_g: in Calabi-Yau compactification

$$N_g = \tfrac{1}{2} |\chi| \tag{3.3}$$

where χ is the Euler number of the manifold of compactification from ten to four dimensions. Also, the magnitudes of Yukawa couplings[104], which are related to the overlaps of the wave functions of zero modes in the six-dimensional compactification manifold, can be related to topological quantities. v) Nothing else has worked. This is perhaps the main reason why theorists have descended on the superstring in such numbers.

How does the superstring make contact with reality? Here we define 'reality' to be physics in four dimensions at energies $\lesssim$ 1 TeV. A general strategy based on the (less un)realistic $E_8 \times E_8'$ superstring is illustrated in Fig. 40. *As a first step*, one has to throw away the massive higher modes in the spectrum (3.1) by taking $M^2 \to \infty$ and doing a formal expansion in powers of $1/M^2$. This has traditionally been called taking the zero-slope limit, since in the old days when string theories were interpreted as models for hadrons, the slopes of the Regge trajectories were

$$\alpha' = 1/M^2 \tag{3.4}$$

After taking this limit, one arrives at a simple supergravity theory in ten dimensions, which contains in addition to the basic Lagrangian[105] a sequence of terms of higher order in $1/M^2$, some of which are necessary for the consistency of the theory at the quantum level[102]. *Next*, one reduces the supergravity theory to four dimensions by compactifying the six surplus dimensions. One wishes to do this without breaking supersymmetry, so as to be able to use it to stabilize the weak gauge hierarchy in four dimensions. The first favoured ansatz[103] for doing this is to compactify on a Calabi-Yau manifold: although other solutions preserving supersymmetry exist, useful explicit examples have not yet been exhibited. Compactification on a Calabi-Yau manifold requires[103]

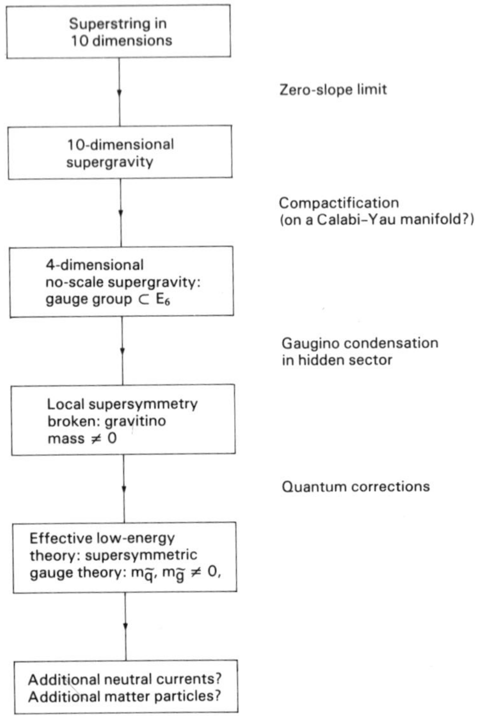

Fig. 40 Sketch of the strategy for making contact between the
superstring and reality

the breaking of one of the E_8 gauge group factors to SU(3) $\times$ E_6, with
the observable gauge group being some subgroup of the E_6. The effective
four-dimensional theory, which may also be obtainable directly from the
superstring without stepping through an intermediate ten-dimensional
supergravity stage[107], is a simple supergravity theory with a rather
special structure. The effective scalar potential in four dimensions is
completely flat, until one starts to evaluate loop corrections and/or the
effects[108] of topologically non-trivial string configurations[109] analogous to the instantons of QCD. The flat potential is only possible
because the effective four-dimensional supergravity theory has a highly
specific non-compact group invariance[59,110,111]. Such supergravity

theories are called no-scale models. Once supersymmetry is broken, loop corrections may destroy the flatness of the potential and generate a dynamically preferred scale for the weak interactions, thereby creating[112] the weak gauge hierarchy:

$$ m_W/m_P = \exp\left[-\frac{O(1)}{\alpha} \right] \tag{3.5} $$

The preferred ansatz to *break supersymmetry* is dynamical condensation of the gauginos in the second 'shadow' E_8 gauge sector[113], which generates a non-zero gravitino mass $m_{3/2}$. Loop corrections are then expected to feed this primordial supersymmetry-breaking down to the observable sector sparticles. It may well be that the window through which the supersymmetry-breaking enters the observable sector is via a primordial gaugino mass $m_{1/2}$ [114], although the physical sparticle masses, $m_{\tilde{q}}$ and $m_{\tilde{g}}$, $m_{\tilde{l}}$, $m_{\tilde{W}}$, and m_H are probably all comparable and $\lesssim 10^3$ GeV. Thus *one finishes up* with a low-energy effective gauge theory with softly broken supersymmetry. Calabi-Yau compactification leads one to expect at least one neutral gauge boson beyond those present in the $SU(3)_C \times SU(2)_L \times U(1)_Y$ Standard Model[115], and also additional matter particles and sparticles beyond those contained in the minimal supersymmetric extension of the Standard Model.

Let us now look in more detail at these possible additional features of low-energy physics. It was originally mentioned in the previous paragraph that Calabi-Yau compactification necessarily breaks down the observable sector gauge group E_8 at least as far as E_6. There is a mechanism[116] for breaking the gauge group down further, which exploits the non-trivial topology of most Calabi-Yau manifolds and is reminiscent of the Aharonov-Bohm effect. The idea is that if one has a hole in one's manifold, one may pick up a non-trivial non-Abelian phase factor

$$ U_c \equiv \langle 0 | \exp i \oint_c A^\mu dx_\mu | 0 \rangle \neq 0 \tag{3.6} $$

as one follows a contour round the hole, much as in electromagnetism one picks up a non-trivial Abelian phase as one follows a contour round a solenoidal magnet. In the non-Abelian case, the residual low-energy gauge group G is that which commutes with all the phase factors U_c [Eq. (3.6)]:

$$[G, U_c] = 0 \qquad (3.7)$$

There are many possibilities for G, depending on the topology of the Calabi-Yau manifold and on the amount of non-Abelian 'twisting' [see Eq. (3.6)] around each hole[115]. However, all the possible residual gauge groups which contain the $SU(3)_C \times SU(2)_L \times U(1)_Y$ Standard Model have rank 5 or 6, and therefore have one or two additional neutral gauge bosons, and possibly other charged vector bosons as well. To avoid catastrophically rapid proton decay, the low-energy gauge group cannot contain any factor larger than SU(4), and then the available possibilities are[115]

$$G = SU(4) \times SU(2)^2 \times U(1), \; SU(4) \times SU(2) \times U(1)^2, \; SU(4) \times SU(2) \times U(1),$$
$$SU(3) \times SU(2)^2 \times U(1)^2, \; SU(3) \times SU(2) \times U(1)^3, \qquad (3.8)$$
$$\text{and} \quad SU(3) \times SU(2) \times U(1)^2$$

The last of these is the uniquely minimal choice with rank 5, just one extra neutral gauge boson Z_E, and no additional charged vector bosons.

Although E_6 can be broken in this way, each generation of particles must[103,115] contain 27 particles with the quantum numbers of a $\underline{27}$ representation of E_6. It is convenient to decompose this via its well-known SO(10) and SU(5) subgroups, although one does not expect these to be good gauge symmetries in four dimensions:

$$\underline{27}\Big|_{E_6} = \left[\underline{16} + \underline{10} + \underline{1}\right]\Big|_{SO(10)} \qquad (3.9a)$$
$$= \left[(\underline{10} + \underline{\bar{5}} + \underline{1}) + (\underline{5} + \underline{\bar{5}}) + \underline{1}\right]\Big|_{SU(5)} \qquad (3.9b)$$

In this way we can understand more easily the particle content of the theory. We see in Eq. (3.9b) the conventional SU(5) generation of $\underline{10}$ +

$\underline{5}$, as well as an SU(5) singlet which is seen in Eq. (3.9a) to be linked with these fields in a $\underline{16}$ of SO(10). It therefore has the right properties to be a right-handed neutrino ν^c. The additional $\underline{5} + \underline{\bar{5}}$ of SU(5) in Eq. (3.9b) contain SU(2) doublets with the quantum numbers of conventional supersymmetric Higgses H and $\bar{H}$. These are partnered by the left- and right-handed components of an additional colour triplet 'quark' of charge -1/3: D and D^c. Finally, Eq. (3.9b) contains a singlet of SU(5) which is also seen as a singlet of SO(10) in Eq. (3.9a), which we denote by N. Thus each conventional generation must be accompanied by the following new matter particles:

$$\nu^c_{(1,1)} \quad ; \quad H_{(1,2)} + \bar{H}_{(1,2)} \quad ; \quad D_{(3,1)} + D^c_{(\bar{3},1)} \quad ; \quad N_{(1,1)} \qquad (3.10)$$

where we have exhibited the SU(3) × SU(2) content of each new particle. In addition to these there could be other light particles which are conjugate subsets of the fields in $\underline{27} + \underline{\overline{27}}$ combinations of irreducible E_6 representations. There is no need to have such fields, and they seem to play no useful phenomenological role, so we will not consider them further.

The matter fields have trilinear Yukawa couplings such as those responsible for quark masses:

$$\lambda Q Q^c H \quad \Rightarrow \quad m_Q = \lambda \langle 0|H|0 \rangle \qquad (3.11)$$

which are obtained[104] from gauge couplings in D = 10 dimensions:

$$g_{10} \int d^{10}z \, \bar{\psi} \gamma^\mu \psi A_\mu \rightarrow \left(\int d^4x \, \bar{\psi}_0 \psi_0 \phi_0 \right) \times \left(g_{10} \int d^6y \, \bar{\chi}_0 \chi_0 A_0 \right) \qquad (3.12)$$

In Eq. (3.12) we have projected out the zero modes of the original ten-dimensional fields, $\psi(z) \rightarrow \psi_0(x)\chi_0(y)$, $A(z) \rightarrow \varphi_0(z)A_0(y)$, where x and y are coordinates in conventional four-dimensional space and in the com-

282

pactification manifold, respectively. We read off from formula (3.12) a
representation

$$\lambda = g_{10} \int d^6 y \, \bar{\chi}_0 \chi_0 A_0 \qquad (3.13)$$

of the Yukawa coupling as a ten-dimensional gauge coupling multiplied by
a threefold overlap integral. Thus Yukawa couplings are allowed only[104]
if they could have been present among the original E_6-invariant coup-
lings $\underline{27} \times \underline{27} \times \underline{27}$ (or $\overline{\underline{27}} \times \overline{\underline{27}} \times \overline{\underline{27}}$: but no trilinear mixtures of $\underline{27}$ and
$\overline{\underline{27}}$). However, if one has the Aharonov-Bohm type of E_6 breaking [Eqs.
(3.6) and (3.7)], these different Yukawa couplings need not be inter-
related by E_6 Clebsch-Gordan coefficients[104]. A complete catalogue of
the couplings allowed by E_6 is

$$\underbrace{Q d^c \bar{H}}_{} + \underbrace{Q u^c H}_{m_q} + \underbrace{L l^c \bar{H}}_{m_l} + \underbrace{L \nu^c H}_{m_\nu ?}$$

$$+ \underbrace{\bar{H} H N}_{\text{Higgs mixing}} + \underbrace{D D^c N}_{m_D} \qquad (3.14)$$

$$+ \underbrace{(D Q Q + D^c u^c d^c)}_{B(D) = -2/3} \quad \text{OR} \quad \underbrace{(D^c L Q + D l^c u^c + D \nu^c d^c)}_{B(D) = +1/3 \,, \quad L(D) = +1}$$

where their possible phenomenological roles are also indicated. The
first two terms give quark masses, and the third gives charged lepton
masses. The fourth term would give neutrinos a Dirac mass if it were
present. This would not be an embarrassment if there were some mechanism
to give the ν^c a large Majorana mass. Unfortunately, the Yukawa couplings do not contain any possible source for such a Majorana mass, nor
has any other source of a large Majorana mass for the ν^c yet been con-
vincingly demonstrated to exist, so we are forced to assume that the
fourth term in formula (3.14) is either very small or strictly zero, for
some reason. The first term in the second row of formula (3.14) is
necessary in dynamical models for weak gauge-symmetry breaking[33]. If
there were no $H\bar{H}$ coupling, having $\langle 0|H|0 \rangle \equiv v \neq 0$ because $m_H^2 < 0$ [in many

models this is driven negative by a large Yukawa coupling to the t-quark[60,117]] would not by itself lead to $\langle 0|\bar{H}|0\rangle \equiv \bar{v} \neq 0$, and hence the leptons and the charge -1/3 quarks would be massless. Even if one could somehow achieve both v and $\bar{v} \neq 0$, the spectrum would contain a traditional axion of the type thought to be excluded by accelerator experiments. So we conclude that there must be an $H\bar{H}$ coupling, and the only one offered in formula (3.14) is the $H\bar{H}N$ term. The second term in the second row of formula (3.14) is the only possible source of masses for the D 'quarks', which would be proportional to $\langle 0|N|0\rangle \equiv x$. If either or both of the first two terms in the last line of (3.14) are present, they will conserve baryon number B if we assign B = -2/3 (+2/3) for the D (D^C) 'quarks'. If any one or more of the last three terms in (3.14) are present, they will conserve B and lepton number L if we assign B = +1/3 (-1/3) and L = +1 (-1) to the D (D^C) 'quarks'. Clearly the different assignments proposed in the last two sentences are mutually incompatible. This means that B and L would be violated by D and D^C exchanges and we would have catastrophically rapid proton decay, unless we allow couplings in *either* the first *or* the second set of parentheses to be non-zero, but *not* both. Thus, to make sense of formula (3.14) we must assume arbitrarily that some couplings vanish if we are to avoid problems with neutrino masses and proton decay. Two mechanisms are available which may generate the required zeros. The more familiar one is to postulate a simple set of discrete symmetries (e.g. $Z_2 \otimes Z_2'$: SU(3) $\underline{3}$, $\underline{\bar{3}} \rightarrow$ (-1) $\times$ SU(3) $\underline{3}$, $\underline{\bar{3}}$, $v^C \rightarrow$ (-1) $\times v^C$ to forbid the B(D) = -2/3 couplings, or (L, e^C, v^C) $\rightarrow$ (-1) $\times$ (L, e^C, v^C), $v^C \rightarrow$ (-1) $\times v^C$ to forbid the B(D) = +1/3 couplings). More innovatively, the overlap integral (3.13) may vanish for topological reasons: it is proportional to the number of intersections of three complex hypersurfaces, and vanishes if they miss each other[104]. Thus the required zeros may be disquieting, but cannot be regarded as impossible *a priori*.

Although most of the issues and methodology discussed in the next three sections arise and are applicable for models with more than one additional neutral gauge boson beyond the Standard Model Z^{0}, we will work explicitly with the minimal rank-5 model at the end of the catalogue (3.8). This choice is motivated by the following phenomenological argument[33]. If one starts from a group of rank 6, one must reduce the rank by 2 to get down to the Standard Model group $SU(3)_C \times SU(2)_L \times U(1)_Y$. This can only be done by having vacuum expectation values for $SU(3)_C \times SU(2)_L \times U(1)_Y$ singlet fields in the $\underline{27}$ representation (3.9), and the only possible choices are the $\tilde{v}^C$ and N fields. Having $\langle 0|N|0 \rangle \equiv x \neq 0$ is not only desirable for $H\bar{H}$ mixing, but also can be generated by radiative corrections to the effective potential via the $H\bar{H}N$ or DD^CN couplings, much as $\langle 0|H|0 \rangle \equiv v \neq 0$ has been generated by the $Q_3 t^C H$ coupling in the minimal supersymmetric Standard Model[60,117]. On the other hand, $\langle 0|\tilde{v}^C|0 \rangle \equiv y \neq 0$ could only be generated by a large LHv^C or Dd^Cv^C coupling. The former would appear to give an unacceptably large neutrino mass, but even if this difficulty could be avoided it would give an unacceptable contribution to the lepton/shiggs mass matrix, and in general the vacuum would have $\langle 0|\tilde{v}|0 \rangle \neq 0$ implying spontaneous violation of lepton number. If the Dd^Cv^C coupling were large enough to drive $\langle 0|\tilde{v}^C|0 \rangle \neq 0$, one would have unacceptable flavour-changing interactions[118] as well as mixing for charge -1/3 quarks[119]. We conclude that we do not know how to get $\langle 0|\tilde{v}^C|0 \rangle \neq 0$, but can get $\langle 0|N|0 \rangle \neq 0$, which would reduce the rank of the low-energy gauge group by one only. Therefore we focus[33] on the unique minimal gauge group $SU(3)_C \times SU(2)_L \times U(1)_Y \times U(1)_E$. Notice that even in this case $x \equiv \langle 0|N|0 \rangle$ cannot be much larger than $v \equiv \langle 0|H|0 \rangle$ and $\bar{v} \equiv \langle 0|\bar{H}|0 \rangle$, because the $H\bar{H}$ mixing term $H\bar{H} \langle 0|N|0 \rangle$ cannot be much larger than m_W. Therefore the extra U(1) is fated to be relevant to contemporary phenomenology, as described in the next section.

There is one proviso to the above discussion. There may be manifolds satisfying our other consistency requirements which enable compactification to break E_8 directly down to SO(10) or SU(5), i.e. rank 1 or 2 smaller than the E_6 case discussed above[120]. The non-Abelian Aharanov-Bohm effect can also be invoked as above, so the four-dimensional gauge group might have rank 4 without needing to invoke conventional vacuum expectation values for scalar fields[121]. In these cases, matter fields would have the quantum numbers of $\underline{16}$, $\underline{10}$, and singlet representations of SO(10), or of $\underline{10}$, $\underline{\bar{5}}$, $\underline{5}$, and singlet representations of SU(5).

However, it is not clear that the dynamical ideas applied successfully to models based on E_6 could also be applied to $SO(10)$ or $SU(5)$ models. Indeed, it is not clear that any plausible $SO(10)$ or $SU(5)$ models even exist. Therefore we will concentrate on the E_6 models in general, and on the minimal $SU(3)_C \times SU(2)_L \times U(1)_Y \times U(1)_E$ model in particular.

3.2 Additional Neutral Currents

We have seen that the low-energy gauge group after compactification on a Calabi-Yau manifold is at least as large as $SU(3)_C \times SU(2)_L \times U(1)_Y \times U(1)_E$, where the extra hypercharges Y_E of the known particles are completely fixed as follows[112]:

$$
\begin{array}{c|c|c|c|c|c}
(\nu,e)_L & e^c_L & (u,d)_L & u^c_L & d^c_L & \\
\hline
-\tfrac{1}{6} & +\tfrac{1}{3} & +\tfrac{1}{3} & +\tfrac{1}{3} & -\tfrac{1}{6} & \times\sqrt{3/5}
\end{array}
\tag{3.15}
$$

The normalization factor of $\sqrt{3/5}$ is fixed by the requirement that the $U(1)_E$ gauge coupling α_E be equal to α_2 and α_3 at the grand unification scale, which is close to the compactification scale. The renormalization group then tells us the magnitude of the $U(1)_E$ gauge coupling at presently accessible energies: $\alpha_E = 0.016$. Thus the couplings of the extra neutral boson Z_E are completely fixed, but not its mass -- although it cannot be much heavier than the $W^{\pm}$ and Z^0, for the reasons discussed in the previous subsection. In fact, the Z_E mixes with the conventional Standard Model Z^0 via the mass-squared matrix[33,122]:

$$
M^2_{Z^0}
\begin{pmatrix}
1 & \sin^2\theta_w \dfrac{4v^2-\bar{v}^2}{3(v^2+\bar{v}^2)} \\[3mm]
\sin^2\theta_w \dfrac{4v^2-\bar{v}^2}{3(v^2+\bar{v}^2)} & \sin^2\theta_w \dfrac{16v^2+\bar{v}^2+25x^2}{9(v^2+\bar{v}^2)}
\end{pmatrix}
\tag{3.16}
$$

which has the following approximate mass-squared eigenvalues in the limit $x \gg v, \bar{v}$:

$$m_Z^2 = m_{Z^0}^2 \left(1 - \frac{\sin^2\theta_w \left(\frac{4v^2 - \bar{v}^2}{v^2 + \bar{v}^2} \right)^2}{\sin^2\theta_w \left(\frac{25x^2 + 16v^2 + \bar{v}^2}{v^2 + \bar{v}^2} \right) - 9} + \cdots \right)$$

$$m_{Z'}^2 = m_{Z^0}^2 \, \frac{\sin^2\theta_w}{9} \left(\frac{25x^2 + 16v^2 + \bar{v}^2}{v^2 + \bar{v}^2} \right) + \cdots \tag{3.17}$$

We now turn to phenomenological constraints on the parameters of the model, which are x/v and $\bar{v}/v$.

3.2.1 $\bar{p}p$ *Collider.* The production cross-section for the Z_E relative to the Z^0 is[122,123]

$$\frac{\sigma(Z_E)}{\sigma(Z^0)} \simeq 0.22 \, f\left(m_{Z^0}^2 / m_{Z_E}^2 \right) \tag{3.18}$$

where the kinematic factor f is 1 for equal Z^0 and Z_E masses, and decreases for increasing m_{Z_E}/m_{Z^0}. The branching ratio for the Z_E varies between

$$B(Z_E \to e^+ e^-) \approx \begin{cases} 3\% & \text{if decays into SU(5) fermions only} \\ 1\% & \text{if decays into 3 generations} \\ & \text{of } E_6 \text{ fermions + bosons} \end{cases} \tag{3.19}$$

The fact that no more than one Z_E event has been seen by UA1 or UA2 tells us that

$$m_{Z_E} \gtrsim \begin{cases} 140 \text{ GeV} & \text{if only SU(5) fermion decays} \\ 110 \text{ GeV} & \text{if all } E_6 \text{ decays} \end{cases} \tag{3.20}$$

Clearly the latter bound is the more conservative one -- perhaps too conservative.

3.2.2 Δm_Z^2. So far we have ignored the (Z^0, Z_E) mixing (3.16), which means that the observed Z mass should be somewhat lower than that expected in the Standard Model. We analyse this possibility by using[33] the $W^{\pm}$ mass to fix $\sin^2 \theta_W$:

$$\sin \theta_W \equiv \frac{38.65 \text{ GeV}}{m_W} \tag{3.21}$$

in preference to low-energy neutral-current experiments[122,125] which could be polluted by Z_E exchange. We then compare $\sin^2 \theta_W$ defined by Eq. (3.21) with

$$\sin^2 \overline{\theta}_W \equiv 1 - m_W^2 / m_Z^2 \tag{3.22}$$

Since m_Z is shifted downwards in general by its mixing with the Z_E, we expect an effective $\sin^2 \overline{\theta}_W < \sin^2 \theta_W$. UA1 and UA2 tell us that measurements of m_W yield[124]

$$\sin^2 \theta_W = \begin{cases} 0.214 \pm 0.005 \pm 0.014 & : \text{UA1} \\ 0.227 \pm 0.005 \pm 0.008 & : \text{UA2} \end{cases} \tag{3.23}$$

whereas their measurements of m_W/m_Z yield[124]

$$\sin^2 \overline{\theta}_W = \begin{cases} 0.194 \pm 0.031 & : \text{UA1} \\ 0.229 \pm 0.030 & : \text{UA2} \end{cases} \tag{3.24}$$

We infer[33] that

$$\Delta\left(\sin^2 \theta_W\right) \equiv \sin^2 \theta_W - \sin^2 \overline{\theta}_W = 0.012 \pm 0.023 \tag{3.25}$$

Figure 41a shows the bounds on $(x/v, \bar{v}/v)$ corresponding to different upper

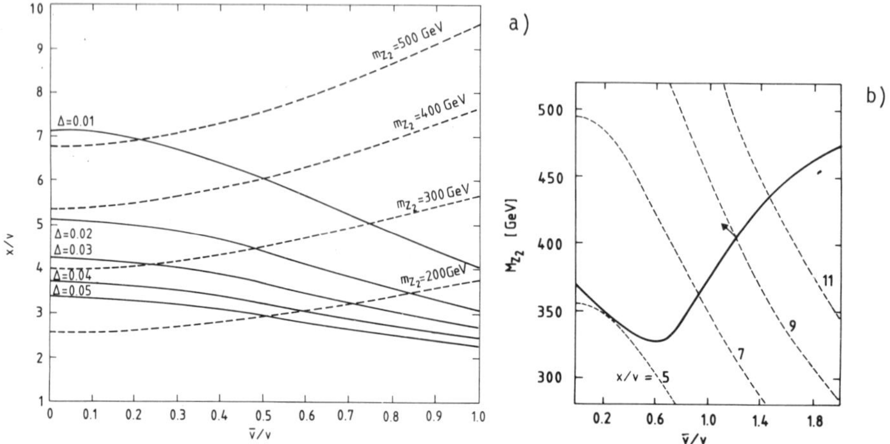

Fig. 41 Bounds on the ratios of vacuum expectation values x/v and $\bar{v}/v$,
a) from the upper bound on $\Delta(\sin^2 \theta_w)$ [124], b) from primordial nucleosynthesis[126] assuming 5.5 equivalent left-handed neutrinos are permitted

bounds on $\Delta(\sin^2 \theta_w)$. For example, if one takes $\Delta(\sin^2 \theta_w) < 0.035$ [i.e.
the 1σ limit from Eq. (3.25)], one finds $x/v \gtrsim 3.5$ for the value of
$\bar{v}/v = 1/2$ which is typical of our dynamical calculations, corresponding[33] to

$$m_{z'} \gtrsim 230 \text{ GeV} \tag{3.26}$$

This bound is significantly more stringent than the $p\bar{p}$ Collider bound
(3.20).

3.2.3 $\nu N \rightarrow \nu X$. An analysis[122] of these data, neglecting (Z^0, Z_E)
mixing, yielded

$$\left(g_E^2/m_{Z_E}^2\right) / \left(g_2^2/m_W^2\right) = -0.12 \pm 0.24 \tag{3.27}$$

corresponding to

$$m_{Z_E} \gtrsim 150 \text{ GeV} \tag{3.28}$$

which is also not as stringent as Eq. (3.26) (see also Ref. 125).

3.2.4 e^+e^- annihilation. The forward-backward asymmetry for $e^+e^- \to \mu^+\mu^-$ and $\tau^+\tau^-$ is not significantly different from the Standard Model prediction. This implies[122] that

$$\left(g_E^2/m_{Z_E}^2\right) \bigg/ \left(g_2^2/m_W^2\right) = 0.57 \pm 0.89 \tag{3.29}$$

which does not give an interesting constraint on m_{Z_E}.

3.2.5 $ve \to ve$. The constraint from this process is at present no stronger than those listed above. Indeed, one finds $\Delta\sigma/\sigma \sim 2\%$ for $x/v \sim 5$ and $\bar{v}/v \sim 1/2$, which is unobservably small[126].

3.2.6 Primordial nucleosynthesis. In the previous subsection we saw that each E_6 generation contains a right-handed neutrino v^c, that there was no source of a large Majorana mass for it, and hence that the Dirac neutrino mass given by the fourth term in formula (3.14) should be either zero or at least very small. In this case both the v and the v^c are light degrees of freedom which could show up in neutrino-counting experiments. The most precise of these is the measurement by UA1 and UA2 of the cross-section times branching ratio for the Z at the $p\bar{p}$ Collider[124], which yields $N_v \leq 5.4 \pm 1.0$. However, this result does not apply to the v^c, which has no $SU(2)_L \times U(1)_Y$ interactions, and hence cannot show up in Z decay except via Z^0/Z_E mixing. The v^c would make a small contribution to $\sigma(e^+e^- \to \gamma + \text{nothing})$ through Z_E exchange, but since the bound[92] from this reaction is $N_v < 14$ there is no important constraint on the Z_E mass. The most stringent bound quoted for N_v is that from primordial nucleosynthesis[126]. It is usually stated[127] that astrophysical limits on the primordial abundance of the light elements constrain $N_\nu < 4$. We now ask ourselves two questions: Do we believe this upper limit? and if so, how much heavier than the Z^0 must the Z_E be in order to suppress the density of the ν^c sufficiently far below that of the conventional ν that the three flavours of ν^c together have a density less than one conventional neutrino flavour?

There are three crucial ingredients in the usual cosmological bound on N_ν: i) The abundance of ^{4}He, which was taken[127] to be less than 0.254 by mass; ii) the abundance of $D + {}^3$He, which was taken[127] to be less than 10^{-4} by mass; and iii) the neutron half-life, which was taken[127] to be more than 10.4 min. One can question each of these input bounds. i) Although the statistical error in measuring the ^{4}He abundance can be very low, there are substantial systematic errors which cloud the interpretation of the observations. Even in the case of I Zwicky 18, which is one of the most primordial objects known and also one of the cleanest to observe, there is a systematic scale error of $(1.1)^{0 \pm 1}$ on the measured abundance of 0.212 ± 0.013 [128]. We prefer to take the more conservative upper bound of 0.26 on the ^{4}He abundance. ii) Observations of some HII regions suggested[129] that the abundance of $D + {}^3$He might be anticorrelated with the abundance of heavier elements, indicating that $D + {}^3$He were destroyed in the first generation of stars, rather than created as had previously been assumed. We therefore took[126] 5×10^{-4} as a conservative upper bound on $D + {}^3$He. It now seems[130] that some of these observations may have been in error, in which case the apparent conflict with the conventional wisdom on early stellar processing would be reduced though not removed, and a conservative upper bound on $D + {}^3$He would perhaps[130] be 2×10^{-4}. iii) There is at least one experiment on the neutron halflife which gives $\tau_{1/2} \sim 10.2$ min [131], and this value cannot yet be excluded by an on-going experiment[132]. Therefore we take 10.2 min as a lower bound on $\tau_{1/2}$. Combining ^{4}He < 0.26, $D + {}^3$He $< 5 \times 10^{-4}$, and $\tau_{1/2} > 10.2$ min, we arrived[126] at the conservative bound

$$N_\nu < 5.5$$

(3.30)

for the maximum number of equivalent left-handed neutrinos, and we could not convince ourselves that $N_v = 6$ was impossible, although we agree that $N_v = 3$ or 4 [127] is the value favoured by the astrophysical data.

How does a bound on N_v translate into a bound on the parameters of the (Z^0, Z_E) mass matrix (3.16)? The density of the v^c will be suppressed relative to that of the conventional v if they decouple at a high enough temperature[133]: $T_d \gtrsim 1/2\, m_\mu$ to get $N_v < 5.5$, $T_d \gtrsim 300$ MeV to get $N_v < 4$. Since they couple to other particles through the Z_E, the cross-sections for annihilations to $v^c \bar{v}^c$ will be suppressed and T_d higher if $m_{Z_E} > m_{Z^0}$. For example, $T_d \gtrsim 1/2\, m_\mu$ if[126]

$$\langle \sigma(e^+ e^- \to v^c \bar{v}^c)\, v \rangle = \alpha T^2 : \quad \alpha < 7.0 \times 10^{-15}\, \mathrm{GeV}^{-4} \tag{3.31}$$

to be compared with

$$\langle \sigma(e^+ e^- \to v \bar{v})\, v \rangle \approx 0.27\, G_F^2 T^2 \tag{3.32}$$

The upper bound (3.31) yields the allowed domain of parameters shown in Fig. 41b, including the lower bound $x/v \gtrsim 5$ for plausible values of $\bar{v}/v \lesssim 1/2$, resulting[126] in

$$m_{Z'} \gtrsim 330\ \mathrm{GeV} \tag{3.33}$$

If one demands $T_d \gtrsim 300$ MeV to get $N_v < 4$, one finds

$$m_{Z'} \gtrsim 780\ \mathrm{GeV} \tag{3.34}$$

Although this type of cosmological bound on $m_{Z'}$ is apparently the most stringent, particle physicists could be forgiven for finding it less certain than, say, the Δm_Z^2 bound (3.26). Moreover, quite apart from the cosmological and astrophysical uncertainties, perhaps one can create a model with large Majorana masses for the v^c.

3.3 Future Neutral-Current Phenomenology

Here we discuss some of the effects[134] of the extended gauge interactions on the phenomenology of neutral currents at forthcoming accelerators, namely the SLC, LEP, and HERA. Since the weak interaction parameter which will be measured with the highest precision in the near future is m_Z, we compare different theories with the same value of m_Z. Since any model with extended neutral currents pushes down the Z mass by comparison with the Standard Model for the same value of $\sin^2 \theta_W$ [cf. Eq. (3.17)], in order to compare two models with the same value of m_Z we must compensate by using a different input value of $\sin^2 \theta_W$ in the extended neutral-current model[134].

Figure 42 shows[134] the difference between $\sin^2 \theta_W^{SM}$ obtained from the Z mass in the Standard Model:

$$m_Z = \frac{38.65}{\sin \theta_W^{SM} \cos \theta_W^{SM}} \tag{3.35}$$

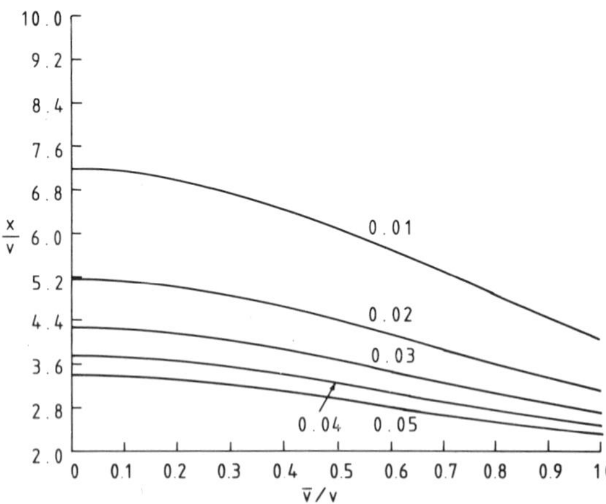

Fig. 42 The change in $\sin^2 \theta_W$ which must be made in order that the lighter neutral boson Z in the extended superstring model (3.16) has the same mass as the Z^0 in the Standard Model[134]

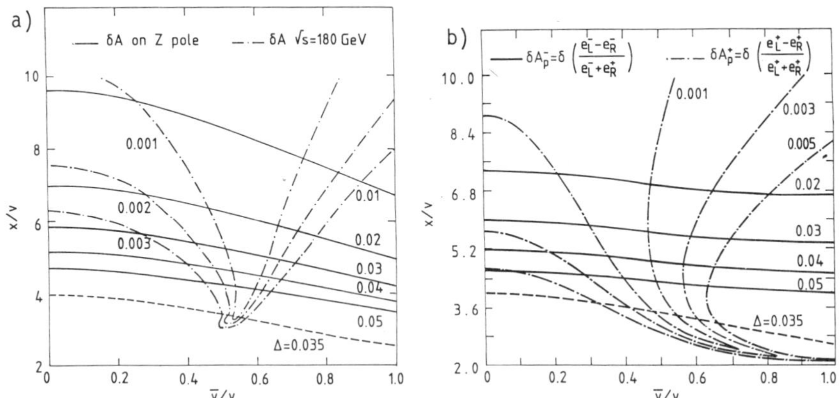

Fig. 43   a) The changes in the forward-backward asymmetry for $e^+e^- \to \mu^+\mu^-$ or $\tau^+\tau^-$ at the Z^0 peak and at $\sqrt{s}$ = 180 GeV, and b) the changes in the parity-violating asymmetries $A_p^\pm$ in ep scattering[134]

and the value of $\sin^2\theta_W$ which must be put into the extended model mass-squared matrix (3.16) to obtain the same value of m_Z, as a function of x/v and $\bar{v}/v$.

It is then clear that other electroweak measurements will discriminate between the extended model and the Standard Model, thanks to the presence of the second Z' and also the difference in $\sin^2\theta_W$. Some such effects are shown in the following figures. Figure 43a shows[134] the difference in the forward-backward asymmetry

$$A \equiv \frac{\int_0^1 d\cos\theta\, \frac{d\sigma}{d\cos\theta} - \int_{-1}^0 d\cos\theta\, \frac{d\sigma}{d\cos\theta}}{\int_0^1 d\cos\theta\, \frac{d\sigma}{d\cos\theta} + \int_{-1}^0 d\cos\theta\, \frac{d\sigma}{d\cos\theta}} \tag{3.36}$$

between the extended neutral-current model and the Standard Model, for $e^+e^- \to \mu^+\mu^-$ or $\tau^+\tau^-$ at the Z peak and at $\sqrt{s}$ = 180 GeV. We see that the effect is much bigger on the Z peak than at $\sqrt{s}$ = 180 GeV, essentially

because a line of accidental zeros trails across the interesting domain
of parameter space at $\sqrt{s}$ = 180 GeV. Since event rates will be much
larger at the Z peak, and since data there will be available sooner,
this will be the best place to look for a discrepancy with the Standard
Model. With a sample of 10^6 Z decays it should be possible to measure
the asymmetry A [Eq. (3.36)] to better than 1%, enabling one to constrain the parameters of the extended neutral-current model much more
tightly than with existing data. Figure 43b shows[134] the changes in
the following asymmetries in ep scattering:

$$A_P^\pm = \frac{d\sigma/dy\,dx\,(e_L^\pm p) - d\sigma/dy\,dx\,(e_R^\pm p)}{d\sigma/dy\,dx\,(e_L^\pm p) + d\sigma/dy\,dx\,(e_R^\pm p)} \tag{3.37}$$

at $\sqrt{s}$ = 314 GeV, x = 0.25, and y = 0.5. Whilst it would be difficult to
measure the asymmetry $A_P^\pm$ [Eq. (3.37)] with such high precision at this
particular value of x and y, we have not tried to estimate the power of
a global fit to the asymmetry data over the full kinematic range of x
and y, which would undoubtedly be more sensitive to the extended neutral-
current model parameters.

3.4   Superstring Dark Matter

The fact that most supersymmetric theories contain a stable relic
particle which could constitute the dark matter in the Universe[55,56,71]
was already discussed in subsection 2.2. Here we discuss briefly the
specific variant[72] of this general feature which appears in the minimally extended neutral-current model inspired by the superstring. As
discussed in subsections 3.1 and 3.2, this model contains a single additional neutral gauge boson Z_E' accompanied by its gaugino partner $\tilde{B}_{E'}$
and additional matter bosons N, accompanied by their fermionic partners
$\tilde{N}$, which can mix with the conventional sparticles ($\tilde{W}^3$, $\tilde{B}$, $\tilde{H}^0$, $\tilde{\bar{H}}^0$) in an
extension of the matrix (2.17). since there are three generations of ($\tilde{H}^0$,
$\tilde{\bar{H}}^0$, $\tilde{N}$) in this type of model, the full neutralino mixing matrix is 12 × 12
instead of 4 × 4 [Eq. (2.17)]! However, we can work in a field basis

where only one each of the H^0, $\bar{H}^0$, and N scalar fields acquires a vacuum expectation value[72]. It is then the fermionic partners of these scalars that mix with the gauginos $\tilde{W}^3$, $\tilde{B}$, and $\tilde{Z}_E$ in a 6 × 6 submatrix, whilst the other two generations of $\tilde{H}^0$, $\bar{\tilde{H}}^0$, and $\tilde{N}$ fields mix among themselves in a disconnected 6 × 6 submatrix. The first of these submatrices is

$$\left(\tilde{W}^3,\tilde{B},\tilde{H}^0,\bar{\tilde{H}}^0,\tilde{B},\tilde{N}\right)\begin{pmatrix} M_2 & 0 & -\frac{1}{\sqrt{2}}g_2 v & \frac{1}{\sqrt{2}}g_2\bar{v} & 0 & 0 \\ & M_1 & \frac{1}{\sqrt{2}}g_1 v & -\frac{1}{2}g_1\bar{v} & 0 & 0 \\ & & 0 & \lambda x & -\frac{2}{3}\frac{g_1 v}{\sqrt{2}} & \lambda\bar{v} \\ & & & 0 & -\frac{1}{6}\frac{g_1\bar{v}}{\sqrt{2}} & \lambda v \\ & \text{Symmetric} & & & M_1 & \frac{5 g_1 x}{6\sqrt{2}} \\ & & & & & 0 \end{pmatrix}\begin{pmatrix} \tilde{W}^3 \\ \tilde{B} \\ \tilde{H}^0 \\ \bar{\tilde{H}}^0 \\ \tilde{B} \\ \tilde{N} \end{pmatrix} \quad (3.38)$$

where $M_1/M_2 = (5/3)(\alpha_1/\alpha_2)$ as in the original matrix (2.17). Using an empirical correlation:

$$\frac{x}{v} = \frac{(m_{1/2} + 75\,\text{GeV})}{88\,\text{GeV}} \; : \; M_2 = 0.33\, m_{1/2} \quad (3.39)$$

between M_2 and x which we have found[33] in dynamical calculations, and taking $\lambda = 0.10$ to 0.25 as also favoured by these calculations, we find[72] that the lightest Majorana mass eigenstate χ of the 6 × 6 matrix (3.38) has the field content

$$\chi = \alpha\tilde{W}^3 + \beta\tilde{B} + \gamma\left(\tilde{S} \equiv \frac{\bar{v}\tilde{H}^0 + v\bar{\tilde{H}}^0}{\sqrt{v^2+\bar{v}^2}}\right) + \delta\left(\tilde{A} \equiv \frac{v\tilde{H}^0 - \bar{v}\bar{\tilde{H}}^0}{\sqrt{v^2+\bar{v}^2}}\right)$$
$$+ \epsilon\tilde{B} + \phi\tilde{N} \quad (3.40)$$

shown in Fig. 44. The χ state is mainly a $\tilde{W}^3/\tilde{B}$ combination for small values of $m_{1/2}$, but acquires larger $\tilde{H}$ components when $m_{1/2}$ is increased. It never contains large $\tilde{B}_E$ or $\tilde{N}$ components. We have also studied[72,135] the other 6 × 6 submatrix of $\tilde{H}^0$, $\bar{\tilde{H}}^0$, and $\tilde{N}$ fields, and verified that all its mass eigenvalues are heavier than the χ for large ranges of coupling constants and vacuum expectation values. Therefore we focus on the state χ of Fig. 44 as our best supersymmetric dark-matter candidate.

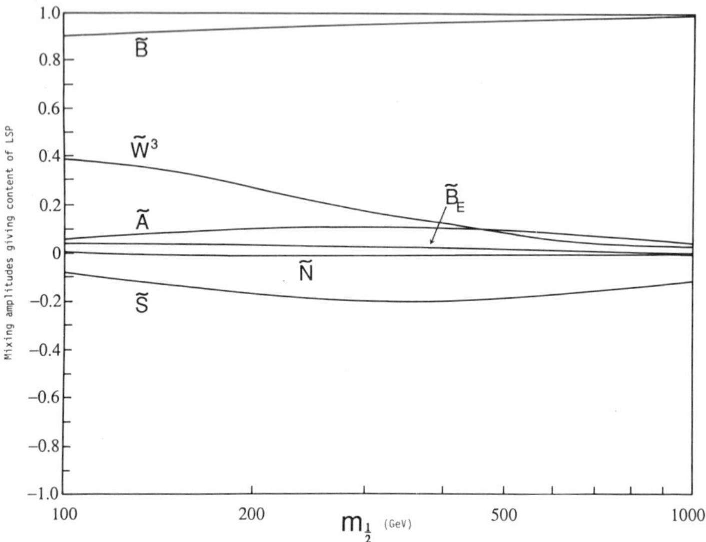

Fig. 44 Mixing amplitudes showing the composition of the lightest super-symmetric particle χ [72]

When we compute its cosmological relic density ϱ_χ, we find[72] that ϱ_χ is within a factor of 2 of the closure density (assuming a present Hubble constant $H_0 = 50$ km s^{-1} Mpc^{-1}) for a wide range of the parameters $(\lambda, m_{1/2})$, as shown in Fig. 45. The relic density would be two or more orders of magnitude too large if the lightest supersymmetric particle were predominantly made out of $\tilde{H}^0$, $\tilde{\bar{H}}^0$, and $\tilde{N}$. We have also computed[72] the rates of capture and annihilation of χ sparticles in the Sun, and the resultant flux of solar neutrinos that they produce at the Earth's surface. As shown in Fig. 46, this flux is within range of the upper limits acceptable to proton decay experiments. In particular, if $m_\chi > m_t$ the neutrino flux jumps and may become too large. If this is indeed the case, one has a remarkable correlation between the t-quark mass and the Hubble expansion rate[72]. To get the critical density of χ sparticles (as suggested by inflation) for $H_0 \gtrsim 50$ km s^{-1} Mpc^{-1} (as suggested by observation), we need $m_\chi \gtrsim 45$ GeV. If proton stability experiments provide the solar neutrino constraint, this means that $m_t \gtrsim m_\chi \gtrsim 45$ GeV--

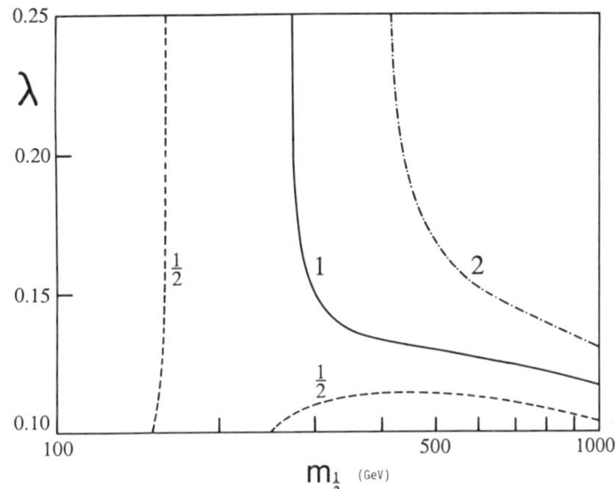

Fig. 45 The cosmological relic density ϱ_χ compared with the closure density for $H_0 = 50$ km/(s·Mpc) [72]: λ is a Yukawa coupling parameter.

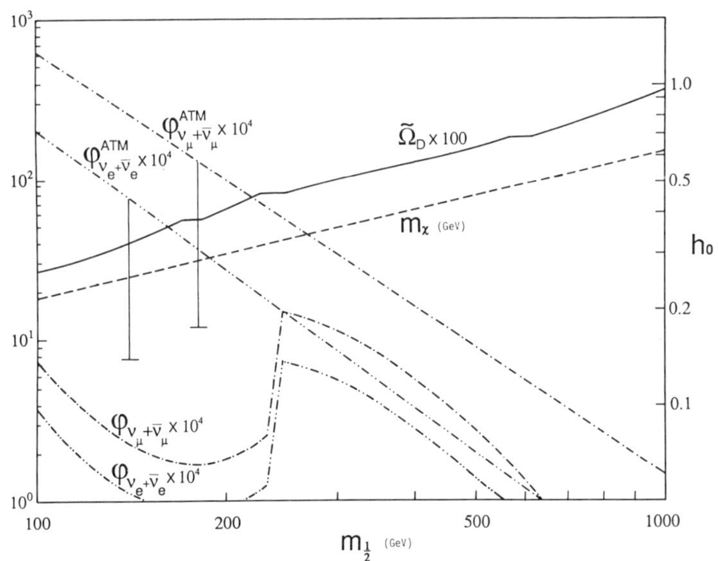

Fig. 46 Compilation of values of m_χ, ϱ_χ and neutrino fluxes as functions of $m_{1/2}$ [72]

a cosmological lower bound on the t-quark mass! Conversely, if it were
confirmed that $m_t \approx 40$ GeV, one would predict that $H_0 \lesssim 45$ km s^{-1} Mpc^{-1}--
a particle physics upper bound on the Hubble constant! Une affaire à
suivre! Note, finally, that the superstring relic χ could be expected[72]
to have $O(10^{-2})$ elastic collisions per kilogram per day in a block of
aluminium, making experimental detection of dark matter in the laboratory
feasible if not trivial.

3.5   Observable Superstring Matter

In the type of low-energy theory inspired by the superstring that
we have been discussing, the conventional matter sparticles are subject
to certain constraints, and one expects the additional superstring
matter particles and sparticles shown in Eq. (3.10). As in the minimal
supersymmetric extension of the Standard Model [Eqs. (2.5)-(2.7)], the
conventional sparticle masses are renormalized, but with some differences.
One is that there is additional renormalization due to the extra $U(1)_E$;
another is that all the gauge β-functions are modified by the presence
of additional matter fields; and a third is that there are reasons to
expect the dominant source of soft supersymmetry-breaking in the observable sector to be $m_{1/2}$, not m_0. We find[122] in the minimally extended
rank-5 model of subsection 3.1 that

$$m_{\tilde{q}} : m_{\tilde{\ell}_L} : m_{\tilde{\ell}_R} : m_{\tilde{g}} : m_{\tilde{\gamma}} \simeq 1.95 : 0.7 : 0.4 : 1 : \frac{1}{7} \qquad (3.41)$$

In addition, there are D-term corrections to the masses which can be
important, particularly for the right-handed selectron[33]. We can impose
on the superstring sspectrum the phenomenological constraints on sparticle
masses discussed in Section 2, obtaining[33] the results shown in Fig. 47
and Table 5. In so doing, some correlations between parameters which
follow from our dynamical model for the generation of the weak interaction scale have been used, so the results should be taken as indicative
rather than absolute. The UA1 constraint (2.34) on the $\tilde{e}_L$ and $\tilde{\nu}$ masses[54],
and the ASP constraint (2.40) on the $\tilde{e}$ and $\tilde{\gamma}$ masses[92] are not useful.
Nor is the UA1 constraint (2.32) on $m_{\tilde{g}}$, nor a fortiori the constraint on

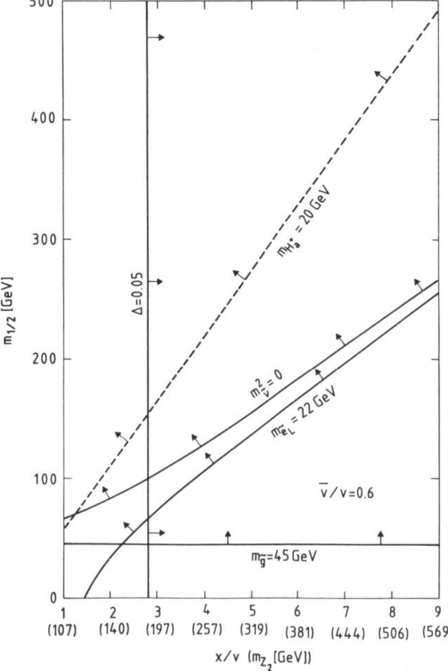

Fig. 47 Phenomenological constraints on particle masses in a minimal superstring model[33]

$m_{\tilde{q}}$ [54]. The most significant constraints are the banal e^+e^- bound[52] $m_{\tilde{e}_L}$ > 22 GeV (2.1) and the trivial requirement that $m_{\tilde{\nu}}^2$ > 0 so as to avoid spontaneous lepton number violation, as seen in Fig. 47. Also shown is a somewhat more tentative constraint following from $m_{H^\pm}$ > 20 GeV, which could perhaps be evaded by juggling with the Yukawa couplings in the model. Finally, the constraint[33] that $\Delta(\sin^2 \theta_W)$ < 0.05 [cf. Eq. (3.25)] is also applied. In this way we obtain the 'absolute' lower bounds in Table 5. Also shown are very model-dependent 'indicative' upper bounds on the Z' and sparticle masses which follow from requiring that the outputs of our dynamical calculations should not be excessively sensitive to the precise values of the input couplings[33]. We see that although the sparticles and the Z' are expected to be beyond the reach of the present CERN $p\bar{p}$ Collider, some of them may well be within the reach of the FNAL Tevatron Collider and LEP, and certainly within the reach of the SSC and/or LHC.

Table 5: Limits on the masses of unseen particles in the minimal low-energy superstring models considered in the text.

Particle	Lower limit	Upper limit
t	–	70
Z_2	185	320
$\tilde{g}$	100	250
$\tilde{\gamma}$	15	35
$\tilde{e}_L$	55	150
$\tilde{e}_R$	90	170
$\tilde{\nu}$	0	150
$\tilde{q}_{1,2}$	180	500

The first column gives lower limits, using experimental data, the assumption that supersymmetry breaking in the observable sector is triggered by a universal primordial gaugino mass and the result $0.2 < \tilde{v}/v < 0.6$ of our dynamical calculations. The second column gives more tentative and model-dependent upper limits, obtained by imposing on our hybrid no-scale models the 'naturalness' constraint[33]. All the masses are expressed in GeV units.

How easy would it be to see these novel matter particles and sparticles? As was already remarked in subsection 3.2, the ν^c does not couple to the Z^0, and will be very difficult to see before the Z' is located. The N and $\tilde{N}$ mix with the neutral Higgs scalars and neutral shiggses and gauginos, respectively. Since the N and $\tilde{N}$ do not couple directly to conventional matter, it will not be easy to deduce their admixtures in any

elementary scalar or neutralino found in one of the searches discussed
in Lectures 1 or 2, respectively. The best bets for easy discovery[136]
are the new 'quarks' D, D^C and their scalar partners $\hat{D}$, $\hat{D}^C$. These
couple to gluons, the photon, and the Z^0, and there is one per generation.
The Dirac D-fermion masses are given by unknown Yukawa couplings to the
$\acute{N}$ scalar field, and could be quite low. Analysing the $\hat{D}$ mass-squared
matrix analogous to that (2.9) for the conventional squarks, we find[136]
that either of the hierarchies

$$m_{\tilde{D}_1} < m_D < m_{\tilde{D}_2}$$

or

$$m_D < m_{\tilde{D}_1} < m_{\tilde{D}_2}$$

$$(3.42)$$

is possible for two bosonic $\hat{D}$ eigenstates and one Dirac D fermion expected
in each generation, so that some scalar $\hat{D}$ could be lighter than the corresponding D fermion. We have

$$\sigma\left(\bar{p}p \to D\bar{D} + X\right) = N_D \, \sigma\left(\bar{p}p \to \bar{t}t + X\right) \qquad (3.43)$$

for similar t and D-fermion masses, and

$$\sigma\left(\bar{p}p \to \tilde{D}\bar{\tilde{D}} + X\right) = \frac{N_{\tilde{D}}}{N_{\tilde{q}}} \, \sigma\left(\bar{p}p \to \hat{q}\bar{\hat{q}} + X\right) \qquad (3.44)$$

for similar $\hat{D}$ and $\hat{q}$ scalar masses. If the D 'quarks' have the couplings
shown in the first set of parentheses in Eq. (3.14), then single $\hat{D}$ production is also possible[136] in $p\bar{p}$ collisions:

$$\sigma\left(p\bar{p} \to \tilde{D} + X\right) = \frac{4\pi}{9\, m_{\tilde{D}}^3} \times$$

$$\int dx_1 \, dx_2 \, x_1 x_2 \, \delta(x_1 x_2 - \tau) \sum_{a,b} \Gamma(\tilde{D} \to ab)\left(q_{a/p}(x_1) q_{b/\bar{p}}(x_2) + a \leftrightarrow b\right) \qquad (3.45)$$

$$: \quad \tau \equiv m_{\tilde{D}}^2 / s$$

whereas if the D 'quarks' have the couplings shown in the second set of parentheses in Eq. (3.14), then single $\overset{\smile}{D}$ production is possible[136] in ep collisions:

$$\sigma\left(ep \to \overset{\smile}{D} + X\right) = \frac{4\pi}{9m_{\overset{\smile}{D}}^3} \times$$

$$\int dx \; \delta(x - \tau) \sum_a \Gamma(\tilde{D} \to ea) \, q_{a/p}(x) \quad : \tau \equiv \frac{m_{\tilde{D}}^2}{s} \qquad (3.46)$$

In the former case, $\tilde{D}$ decays to $\bar{q} + \bar{q}$ and $\overset{\smile}{D}$ decays to $\bar{q} + \bar{q} + \chi$, whilst in the latter case $\tilde{D} \to q + l$ and $\overset{\smile}{D} \to q + l + \chi$. Thus possible signatures for the new matter particles and sparticles are

$$\text{bump in jet} + l^{\pm} \quad : \quad \tilde{D} \to q + l^{\pm}$$

$$\text{bump in jet} + \text{jet} \quad : \quad \tilde{D} \to \bar{q} + \bar{q}$$

$$\text{missing energy} \quad : \quad \begin{cases} \tilde{D} \to q + \nu \; (cf \; \tilde{q} \to q + \overset{\sim}{\gamma}) \\ D \to \bar{q} + \bar{q} + \chi \; (cf \; \tilde{g} \to q + \bar{q} + \overset{\sim}{\gamma}) \end{cases}$$

$$l^{\pm} + \text{jet} + \text{missing energy} : \quad D \to q + l^{\pm} + \chi \qquad (3.47)$$
$$(cf \; t \to b + l^{\pm} + \nu)$$

These signatures, together with the existence of a second neutral gauge boson Z', an apparent shift in the mass of the first Z, and/or anomalous deviations from the neutral-current asymmetries predicted by the Standard Model, seem to be the best possible low-energy signatures for the superstring--at least until next week.

REFERENCES

1) Bell, J.S., Nucl. Phys. $\underline{B60}$, 427 (1973).

   Llewellyn Smith, C.H., Phys. Lett. $\underline{46B}$, 233 (1973).

   Cornwall, J.M., Levin, D.N., and Tiktopoulos, G., Phys. Rev. Lett.
   $\underline{30}$, 1268 (1973) and Phys. Rev. $\underline{D10}$, 1145 (1974).

2) Schwinger, J., Ann. Phys. $\underline{2}$, 407 (1957).

   Glashow, S.L., Nucl. Phys. $\underline{B22}$, 579 (1961).

3) Yang, C.N., and Mills, R., Phys. Rev. $\underline{96}$, 191 (1954).

4) Weinberg, S., Phys. Rev. Lett. $\underline{19}$, 1264 (1967).

   Salam, A., Proc. 8th Nobel Symposium, Aspenäsgården, 1968
   (ed. N. Svartholm) (Almqvist and Wiksell, Stockholm, 1968), p. 367.

5) Flores, R.A., and Sher, M., Ann. Phys. (NY) $\underline{148}$, 95 (1983).

6) Veltman, M., Acta Phys. Pol. $\underline{B8}$, 475 (1978).

   Vayonakis, C., Univ. Athens preprint (1978).

   Lee, B.W., Quigg, C., and Thacker, H.B., Phys. Rev. $\underline{D16}$, 1519 (1977).

7) Weinberg, S., Phys. Rev. Lett. $\underline{36}$, 294 (1976).

   Linde, D.A., Zh. Eksp. Teor. Fiz. Pis'ma $\underline{23}$, 73 (1976).

8) Coleman, S., and Weinberg, E., Phys. Rev. $\underline{D7}$, 1888 (1973).

9) Mahanthappa, K.T., and Sher, M., Phys. Rev. $\underline{D22}$, 1711 (1980).

10) See, for example,

   Ellis, J., Gaillard, M.K., and Nanopoulos, D.V., Nucl. Phys. $\underline{B106}$,
   292 (1976).

   Gaillard, M.K., Comments Nucl. Part. Phys. $\underline{8}$, 31 (1978).

   Ansel'm, A.A., Ural'tsev, N.G., and Khoze, V.A., Sov. Phys. Usp. $\underline{28}$,
   113 (1985).

11) Lee, B.W., et al., Ref. 6.

12) Ellis, J., et al., Ref. 10.

13) Wilczek, F., Phys. Rev. Lett. $\underline{39}$, 1304 (1977).

   Georgi, H., Glashow, S.L., Machacek, M., and Nanopoulos, D.V., Phys.
   Rev. Lett. $\underline{40}$, 692 (1978).

14) Keung, W., and Marciano, W., Phys. Rev. $\underline{D30}$, 248 (1984).

15) Ellis, J., and Peccei, R. (eds.), Physics at LEP, CERN 86-02 (1986).

16) Wilczek, F., Ref. 13.

17) Buchmüller, W., et al., Ref. 15, Vol. 1, p. 203.

18) Vysotsky, M.I., Phys. Lett. $\underline{97B}$, 159 (1980).

19) Nason, P., Columbia University preprint CU-TP-346 (1986), has recently checked the result of Ref. 18 for $^3S_1 \to$ scalar $+ \gamma$, and also calculated the QCD radiative corrections for $^3S_1 \to$ pseudoscalar $+ \gamma$, which are comparable although somewhat smaller.

20) Baer, H., et al., Ref. 15, Vol. 1, p. 297, Section 2.

21) Ellis, J., Enqvist, K., Nanopoulos, D.V., and Ritz, S., Phys. Lett. $\underline{158B}$, 417 (1985).

22) CUSB Collaboration, Franzini, P., et al., Paper 312 submitted to the Int. Symp. on Lepton and Photon Interactions, Kyoto, 1985.

23) Cahn, R.N., Chanowitz, M.S., and Fleishon, N., Phys. Lett. $\underline{82B}$, 113 (1979).

24) Bergström, L., Phys. Lett. $\underline{167B}$, 332 (1986).

25) Bjorken, J.D., Proc. SLAC Summer Institute on Particle Physics, Stanford, 1976, ed. Zipf, M.C. (SLAC-198, Stanford, 1977), p. 1.

26) Finjord, J., Phys. Scripta $\underline{21}$, 143 (1980).

27) Mikenberg, G., Yagil, A., and Yekutieli, G., Weizmann Institute preprint WIS-86/3/Feb-PH (1986).

28) Barklow, T., Rudolph, G., Wu, S.L., and Zobernig, G., ALEPH Note 158 (1986).

29) Wu, S.L., Z. Phys. $\underline{C9}$, 329 (1981).

30) Altarelli, G., et al., Ref. 15, Vol. 1, p. 1.

31) Lemoine, M., and Veltman, M., Nucl. Phys. $\underline{B164}$, 445 (1980).

32) Gunion, J., and Haber, H.E., SLAC preprint PUB-3404 (1984) and U.C. Davis preprint UCD-86-12 (1986).

33) Ellis, J., Enqvist, K., Nanopoulos, D.V., and Zwirner, F., CERN preprints TH-4323 (1985) and TH-4350 (1986).

34) Yagil, A., and Mikenberg, G., Weizmann Institute preprint WIS-86/4/Feb-PH (1986).

35) R. Bates and J.N. Ng, TRIUMF preprint TRI-PP-57 (1985).

36) Eichten, E., Hinchliffe, I., Lane, K., and Quigg, C., Rev. Mod. Phys. $\underline{56}$, 579 (1984).

37) Ellis, J., Gelmini, G., and Kowalski, H., Proc. ECFA-CERN Workshop on a Large Hadron Collider in the LEP Tunnel, Lausanne and Geneva, 1984 (Report ECFA 84/85/CERN 84-10, Geneva, 1984), p. 393.

38) Georgi, H., et al., Ref. 13.

39) Kunszt, Z., Univ. Berne preprint BUTP 84/10-BERN (1984).
Gunion, J.F., Kalyniak, P., Soldate, M., and Galison, P., Phys. Rev. Lett. $\underline{54}$, 1226 (1985).

40) Cahn, R.N. and Dawson, S., Phys. Lett. $\underline{136B}$, 196 (1984).

41) Glashow, S.L., Nanopoulos, D.V., and Yildiz, A., Phys. Rev. $\underline{D18}$, 1724 (1978).
Golden, M., and Sharpe, S., Berkeley preprint LBL 19431 (1985).

42) Donaldson, R., and Morfin, J.G. (eds.), Design and utilization of the SSC, Snowmass, 1984 (APS, New York, 1984), p. 87.

43) Gol'fand, Yu., and Likhtman, E.P., Sov. Phys.-JETP Lett. $\underline{13}$, 323 (1971).
Volkov, D.V., and Akulov, V.P., Phys. Lett. $\underline{46B}$, 109 (1973).
Wess, J., and Zumino, B., Nucl. Phys. $\underline{B70}$, 39 (1974).

44) Gildener, E., and Weinberg, S., Phys. Rev. $\underline{D13}$, 333 (1976).
Gildener, E., Phys. Rev. $\underline{D14}$, 1667 (1976).

45) 't Hooft, G., Recent developments in field theories, eds. 't Hooft, G., et al. (Plenum Press, Inc., New York, 1980).

46) Buras, A.J., Ellis, J., Gaillard, M.K., and Nanopoulos, D.V., Nucl. Phys. $\underline{B135}$, 66 (1978).

47) Hawking, S., Page, D.N., and Pope, C.N., Nucl. Phys. $\underline{B170}$, 283 (1980).

48) For reviews, see
Farhi, E., and Susskind, L., Phys. Rep. $\underline{74C}$, 277 (1981).
Ellis, J., Proc. SLAC Summer Institute on Particle Physics, Stanford, 1981, ed. Mosher, A. (SLAC-245, Stanford, 1982), p. 621.

49) Dimopoulos, S., and Susskind, L., Nucl. Phys. $\underline{B155}$, 237 (1979).
Eichten, E., and Lane, K.D., Phys. Lett. $\underline{90B}$, 125 (1980).

50) Dimopoulos, S., and Ellis, J., Nucl. Phys. $\underline{B182}$, 505 (1981).

51) Binétruy, P., Chadha, S., and Sikivie, P., Nucl. Phys. $\underline{B207}$, 65 (1982), and references therein.

52) Komamiya, S., Proc. Int. Symp. on Lepton and Photon Interactions at High Energies, Kyoto, 1985, eds. Konuma, M., and Takahashi, K. (Kyoto University, Kyoto, 1986), p. 612.

53) Peskin, M., *loc. cit.*, p. 714.

54) Rubbia, C., *loc. cit.*, p. 242.

55) Goldberg, H., Phys. Rev. Lett. 50, 1419 (1983).

56) Ellis, J., Hagelin, J.G., Nanopoulos, D.V., Olive, K., and Srednicki, M., Nucl. Phys. B238, 453 (1984).

57) Polonyi, J., Budapest preprint KFKI-1977-93 (1977).
Cremmer, E., et al., Phys. Lett. 79B, 28 (1978) and Nucl. Phys. B147, 105 (1979).

58) Ellis, J., and Nanopoulos, D.V., Phys. Lett. 116B, 133 (1982).

59) Ellis, J., Kounnas, C., and Nanopoulos, D.V., Nucl. Phys. B247, 373 (1984).

60) Ellis, J., Nanopoulos, D.V., and Tamvakis, K., Phys. Lett. 121B, 123 (1983).

61) Kounnas, C., Lahanas, A.B., Nanopoulos, D.V., and Quirós, M., Phys. Lett. 132B, 95 (1983) and Nucl. Phys. B236, 438 (1984).

62) Ellis, J., and Rudaz, S., Phys. Lett. 128B, 248 (1983).

63) Fayet, P., Unification of the fundamental particle interactions, eds. Ferrara, S., Ellis, J., and Van Nieuwenhuizen, P. (Plenum Press, Inc., New York, 1980), p. 587, and references therein.

64) See, for example,
Ross, G.G., and Valle, J.W.F., Phys. Lett. 151B, 375 (1985).

65) Hall, L., and Suzuki, M., Nucl. Phys. B231, 419 (1984).
Lee, I.H., Phys. Lett. 138B, 121 (1984).

66) Ellis, J., Gelmini, G., Jarlskog, C., Ross, G.G., and Valle, J.W.F., Phys. Lett. 150B, 142 (1985).

67) Smith, P.F., et al., Nucl. Phys. B206, 333 (1982), and references therein.

68) Wolfram, S., Phys. Lett. 82B, 65 (1979).
Dover, C.B., Gaisser, T.K., and Steigman, G., Phys. Rev. Lett. 42, 1117 (1979).

69) Barbieri, R., and Maiani, L., Nucl. Phys. B243, 429 (1984).

70) Okun', L.B., and Voloshin, M.B., Yad. Fiz. 43, 779 (1986), and references therein.

71) Ellis, J., Hagelin, J.S., and Nanopoulos, D.V., Phys. Lett. 158B, 26 (1985).
Nandi, S., Phys. Rev. Lett. 54, 2493 (1985).

72) Campbell, B.A., et al., Phys. Lett. 173B, 270 (1986).

73) Farrar, G., and Fayet, P., Phys. Lett. <u>76B</u>, 575 (1978).

74) Bergsma, F., et al., Phys. Lett. <u>121B</u>, 429 (1983).

Ball, R.C., et al., Phys. Rev. Lett. <u>53</u>, 1314 (1984).

Cooper-Sarkar, A.M., et al., Phys. Lett. <u>160B</u>, 212 (1985).

75) Badier, J., et al., preprint CERN-EP/86-22 (1986).

76) Albrecht, H., et al., Phys. Lett. <u>167B</u>, 360 (1986).

77) Campbell, B.A., Ellis, J., and Rudaz, S., Nucl. Phys. <u>B198</u>, 1 (1982).

78) Arnison, G., et al., Phys. Lett. <u>139B</u>, 115 (1984).

79) Kane, G.L., and Léveillé, J.P., Phys. Lett. <u>112B</u>, 227 (1982);

Harrison, P., Llewellyn Smith, C.H., Nucl. Phys. <u>B213</u>, 223 (1982);

   E <u>B223</u>, 542 (1983).

80) Ellis, J., and Kowalski, H., Phys. Lett. <u>142B</u>, 441 (1984).

Reya, E., and Roy, D.P., Phys. Lett. <u>141B</u>, 442 (1984).

81) Ellis, J., and Kowalski, H., Nucl. Phys. <u>B246</u>, 189 (1984) and <u>B259</u>,

   109 (1985).

Reya, E., and Roy, D.P., Phys. Rev. Lett. <u>52</u>, 881 (1984) and Phys.

   Rev. <u>D32</u>, 645 (1985).

82) Ellis, J., and Sher, M., Phys. Lett. <u>148B</u>, 309 (1984).

Hall, L., and Polchinski, J., Phys. Lett. <u>152B</u>, 355 (1985).

Enqvist, K., Lahanas, A.B., and Nanopoulos, D.V., Phys. Lett. <u>155B</u>,

   83 (1985).

83) Delduc, F., Navelet, H., Peschanski, R., and Savoy, C.A., Phys. Lett.

   <u>155B</u>, 173 (1985).

Glück, M., Reya, E., and Roy, D.P., Phys. Lett. <u>155B</u>, 284 (1985).

Barnett, R.M., Haber, H.E., and Kane, G.L., Phys. Rev. Lett. <u>54</u>,

   1983 (1985).

84) Ellis, S.D., Kleiss, R., and Stirling, W.J., Phys. Lett. <u>158B</u>, 341

   (1985).

85) Barnett, R.M., Haber, H.E., and Kane, G.L., Nucl. Phys. <u>B267</u>, 625

   (1986).

Reya, E., and Roy, D.P., Phys. Lett. <u>166B</u>, 223 (1986).

86) Baer, H., and Tata, X., Phys. Lett. <u>155B</u>, 278 (1985).

87) Baer, H., Hagiwara, K., and Tata, X., Argonne preprint ANL-HEP-

   PR-86-07 (1986).

Baer, H., These proceedings.

88) Barbieri, R., Cabibbo, N., Maiani, L., and Petrarca, S., Phys. Lett. 127B, 458 (1983).

Barnett, R., Haber, H.E., and Lackner, K., Phys. Rev. Lett. 51, 176 (1983) and Phys. Rev. D29, 1381 (1984).

89) Baer, H., Ellis, J., Nanopoulos, D.V., and Tata, X., Phys. Lett. 153B, 265 (1985).

90) Cabibbo, N., Maiani, L., and Petrarca, S., Phys. Lett. 132B, 195 (1983).

91) Dolgov, A.D., Okun', L.B., and Zakharov, V.I., Nucl. Phys. B41, 197 (1972).

Ma, E., and Okada, J., Phys. Rev. Lett. 41, 287 (1978) and Phys. Rev. D18, 4219 (1978).

Gaemers, K.J.F., Gastmans, R., and Renard, F.M., Phys. Rev. D19, 1605 (1979).

92) Bartha, G., et al., Phys. Rev. Lett. 56, 685 (1986).

93) Fayet, P., Phys. Lett. 117B, 460 (1982).

Ellis, J., and Hagelin, J.S., Phys. Lett. 122B, 303 (1983).

Kobayashi, T., and Kuroda, M., Phys. Lett. 139B, 208 (1984).

Grassie, K., and Pandita, P.N., Phys. Rev. D30, 22 (1984).

Ware, J.D., and Machacek, M.E., Phys. Lett. 147B, 415 (1984).

94) Komatsu, H., and Kubo, J., Nucl. Phys. B263, 205 (1986).

95) Ellis, J., same Proc. as Ref. 52, p. 850.

96) Baer, H., et al., in Ref. 15, Vol. 1, p. 297, Section 3.

97) Dionisi, C., et al., Phys. Rep. in preparation (1986).

98) Johnson, J.E., and Rudaz, S., Phys. Rev. D30, 1590 (1984).

99) Campbell, B.A., Scott, J.A., and Sundaresan, M.K., Phys. Lett. 126B, 376 (1983).

Kane, G.L., and Rolnick, W.B., Nucl. Phys. B217, 117 (1983).

100) Wolf, G., These proceedings.

101) Cashmore, R., et al., Phys. Rep. 122C, 275 (1985).

102) Green, M.B., and Schwarz, J.H., Phys. Lett. 149B, 117 (1984).

103) Candelas, P., Horowitz, G.T., Strominger, A., and Witten, E., Nucl. Phys. B258, 46 (1985).

104) Strominger, A., and Witten, E., Commun. Math. Phys. 101, 341 (1985).

105) Chamseddine, A.H., Nucl. Phys. B185, 403 (1981).

Bergshoeff, F., de Roo, M., de Wit, B., and Van Nieuwenhuizen, P., Nucl. Phys. B195, 97 (1982).

Chapline, G.F., and Manton, N., Phys. Lett. 120B, 105 (1983).

106) Calabi, E., in Algebraic Geometry and Topology, a Symposium in Honour of S. Lefschetz (University Press, Princeton, 1957), p. 78.

Yau, S.-T., Proc. Nat. Acad. Sci. 74, 1798 (1977).

107) Ellis, J., Gómez, C., and Nanopoulos, D.V., Phys. Lett. 171B, 203 (1986).

108) Ellis, J., Gómez, C., Nanopoulos, D.V., and Quirós, M., Phys. Lett. 173B, 59 (1986).

109) Wen, X.-G., and Witten, E., Phys. Lett. 166B, 397 (1986).

110) Cremmer, E., Ferrara, S., Kounnas, C., and Nanopoulos, D.V., Phys. Lett. 153B, 61 (1983).

111) Ellis, J., Kounnas, C., and Nanopoulos, D.V., Nucl. Phys. B241, 406 (1984).

112) Ellis, J., Lahanas, A.B., Nanopoulos, D.V., and Tamvakis, K., Phys. Lett. 134B, 429 (1984).

113) Derendinger, J.-P., Ibáñez, L., and Nilles, H.P., Phys. Lett. 155B, 65 (1985).

Dine, M., Rohm, R., Seiberg, N., and Witten, E., Phys. Lett. 156B, 55 (1985).

114) Cohen, E., Ellis, J., Enqvist, K., and Nanopoulos, D.V., Phys. Lett. 161B, 85 (1985).

115) Witten, E., Nucl. Phys. B258, 75 (1985).

Dine, M., Kaplunovsky, V., Mangano, M., Nappi, C., and Seiberg, S., Nucl. Phys. B259, 519 (1985).

Breit, J.D., Ovrut, B.A., and Segré, G., Phys. Lett. 158B, 33 (1985).

116) Hosotani, Y., Phys. Lett. 129B, 193 (1983).

117) Ibáñez, L., and Ross, G.G., Phys. Lett. 110B, 215 (1982).

Ellis, J., Ibáñez, L., and Ross, G.G., Phys. Lett. 113B, 283 (1982) and Nucl. Phys. B221, 29 (1983).

Inoue, K., Kakuto, A., Komatsu, H., and Takeshita, S., Prog. Theor. Phys. 66, 927 (1982) and 71, 413 (1984).

Ibáñez, L., and López, C., Phys. Lett. 126B, 54 (1983).

Alvarez-Gaumé, L., Polchinski, J., and Wise, M.B., Nucl. Phys. B221, 495 (1983).

310

118) Campbell, B.A., Ellis, J., Enqvist, K., Gaillard, M.K., and
     Nanopoulos, D.V., CERN preprint TH-4473 (1986).

119) Rosner, J., Enrico Fermi Institute preprint EFI 85-34 (1985).
     Robinett, R.W., Univ. of Massachusetts preprint UMHEP-239 (1985).

120) Witten, E., Nucl. Phys. $\underline{B268}$, 79 (1986).

121) Green, B.R., Kirklin, K.H., and Miron, P.J., Oxford University
     Department of Theoretical Physics preprint (1985).

122) Cohen, E., Ellis, J., Enqvist, K., and Nanopoulos, D.V., Phys. Lett.
     $\underline{165B}$, 76 (1985).

123) Barger, V., Deshpande, N.G., and Whisnant, K., Phys. Rev. Lett. $\underline{56}$,
     30 (1986).

124) DiLella, L., same Proc. as Ref. 52, p. 280.

125) Durkin, L.S., and Langacker, P., Phys. Lett. $\underline{166B}$, 436 (1986).

126) Ellis, J., Enqvist, K., Nanopoulos, D.V., and Sarkar, S., Phys. Lett.
     $\underline{167B}$, 457 (1986).

127) Yang, J., Turner, M.S., Steigman, G., Schramm, D.N., and Olive, K.A.,
     Ap. J. $\underline{281}$, 493 (1984).

128) Davidson, K., and Kinman, T.D., Ap. J. Suppl. $\underline{58}$, 321 (1985).

129) Rood, R.T., Bania, T.M., and Wilson, T.L., Ap. J. $\underline{280}$, 629 (1984).

130) Steigman, G., Olive, K.A., Schramm, D.N., and Turner, M.S.,
     Threatened preprint (1986).

131) Bondarenko, L.N., et al., Sov. Phys.-JETP Lett. $\underline{28}$, 303 (1978).

132) Green, K., private communication (1985).

133) Steigman, G., Olive, K.A., and Schramm, D.N., Phys. Rev. Lett. $\underline{43}$,
     239 (1979).
     Olive, K.A., Schramm, D.N., and Steigman, G., Nucl. Phys. $\underline{B180}$, 497
     (1981).

134) Angelopoulos, V., Ellis, J., Nanopoulos, D.V., and Tracas, N., CERN
     preprint TH-4408 1986).

135) Ellis, J., Nanopoulos, D.V., Petcov, S.T., and Zwirner, F., Gauginos
     and Higgs particles in superstring models, CERN preprint TH-4454
     (1986).

136) Angelopoulos, V., Ellis, J., Kowalski, H., Nanopoulos, D.V.,
     Tracas, N., and Zwirner, F., CERN preprint in preparation (1986).

A SUPERSTRING PRIMER

Michael Dine

Department of Physics, The City College
of City University of New York
New York, New York 10031

and

The Institute for Advanced Study
Princeton, New Jersey 08540

ABSTRACT

These lectures present a brief introduction to
the theory and phenomenology of superstrings.

1. INTRODUCTION

Since the summer of 1984, the particle physics community has been
in a state of great ferment. You need only be a reader of the <u>New York
Times</u>, <u>Time</u>, <u>Newsweek</u>, or the <u>Economist</u>, or to examine a recent issue
of the SLAC preprint list, to see that most particle theorists are
working on string theory. One can try and explain this away by saying
that this is a period of relatively few exciting experimental results,
or by deriding theorists as followers of fashion, but, as I hope to
make clear, there are some very good reasons for the excitement. Superstring theories[1,2] are, in fact, the first plausible candidates that
we have ever had for truly unified theories of all interactions. By
this I mean that, while these theories possess no free parameters,[3] it
is plausible that they make predictions--perhaps correct predictions--
for all the phenomena we observe. As I will at least briefly explain
in these lectures, string theories address every question we have ever

thought to ask in particle physics: the origin of the spectrum of quarks and leptons, of the low energy gauge interactions, of the quark and lepton mass matrix, of weak symmetry breaking and the large hierarchy between the W mass and the Planck scale; the smallness of the cosmological constant and the problems of quantum gravity. Not that string theory has yet provided a complete answer to any of these questions, and, of course, it may turn out to provide the wrong answers to some of them. But it is amazing that we have in our hands a theory which truly has pretensions to be, in a phrase that I believe is due to John Ellis, a theory of everything.

Of course, any theory with enough structure to address all of these questions is necessarily rather complicated. To cover it adequately would require a full year course (at least!), not three lectures. This is due, I should stress, not merely to the technical complications of the subject, but also to the present rudimentary state of our understanding. Unlike the case of, say, Yang-Mills theories, we cannot state in a simple way what are the basic principles of string theory. We possess instead a sort of large cookbook, filled with many different recipes for the same items, with relations between them poorly understood. I say all of this, not to discourage you with the prospect of the field's complexity, but to provide you some comfort as you try to learn the subject. You are not alone in finding it difficult, and those questions which you cannot answer after some effort are probably suitable research problems. I should also say that, while highly complex, the subject has many beautiful aspects, and is filled with bizarre and presently poorly understood connections.

What I will attempt to do in these lectures is provide an overview of string theory, with particular emphasis on how these theories might make contact with nature. The remainder of this section will provide a brief introduction, just stating some of the basic facts of string theory: the various types of theories, their critical dimensions, and some features of their spectra. In order to guide our thinking about phenomenology, the second section will contain an introduction to supersymmetry in four[4] and ten dimensions.[5]

The third section will be an introduction to real string theory. We will consider an unadorned, bosonic string propagating in space-time, by analogy to the corresponding problem of a particle. We will derive its spectrum and discover the critical dimension. We will explain how interactions are introduced, and scattering amplitudes computed. At least very briefly, we will discuss what it might mean to have a second-quantized theory of strings.[6,7] Then we will describe how fermions are included in string theory, and the construction of the various superstring theories, including the heterotic string theory.[2] We will say a few words about anomalies[8] and anomaly cancellations[9] (the discovery of which ignited the present flood of interest), and the question of finiteness. Finally, we will explain what are the coupling constants of string theory, and why there are not any.[3,10]

In the fourth section, we return to "phenomenology". Of course, superstring theories are theories in ten dimensions, so we must consider how six dimensions might curl up, or "compactify", while four remain flat. To understand this we will briefly review the original (and simplest) Kaluza-Klein theory, in which space-time is five dimensional, with the fifth dimension a tiny circle. I will then mention what has been the biggest obstacle to implementing the Kaluza-Klein program: the problem of obtaining chiral fermions,[11] and, more generally, the lack of a complete theory of gravity.

Then I will go on to discuss the problem of compactification in string theory. At the classical (i.e. weak coupling) level, there is a well-defined procedure for investigating the possible compactifications of the theory.[12] In fact, at the classical level string theory has an infinite set of ground states, including an infinity of states with six compact dimensions. Among these, we will see, are some which closely resemble our world.[12,13] They have gauge group SU(3) x SU(2) x U(1), three or four standard generations of light quarks and leptons, and light Higgs doublets.[13,14] It is possible, in principle, to compute the Yukawa couplings of these particles.[15] These vacua also have unbroken supersymmetry.

The discovery of these compactifications leaves many open questions. First, one would like to find a vacuum or set of vacua in which the various couplings of the light fields agree with those which are observed. At a gross level, these couplings should not allow too rapid proton decay, or large neutrino masses. At a finer level, one would like to obtain the correct values for the quark and lepton masses. This requires, however, an understanding of the mechanism of supersymmetry breaking.

In fact, the problem of how supersymmetry gets broken is perhaps the most serious one facing string theory. It is intimately connected with the question of how the true vacuum state is selected. While our present understanding of string dynamics is quite incomplete, it is possible to make some statements about this problem on quite general grounds.[17] The results, as we will see, show that, if the string theory does describe nature, we will almost certainly have to face up to solving a strong coupling problem.

1.1 Lightning Introduction to String Theory

String theories were originally proposed as theories of strong interactions. They possessed certain remarkable features which reproduced some aspects of the hadronic spectrum and hadronic scattering amplitudes. Strings are precisely that, one-dimensional objects, propagating through space-time. They come in certain basic types: open strings, with ends (Fig. 1a), and closed strings (Fig. 1b). They can possess an intrinsic orientation or not, and they can possess only bosonic states (the bosonic string) or fermionic states as well (the fermionic string or superstring). The motion of a free, classical, relativistic string can be described in terms of normal modes. Upon quantization, each normal mode is identified with a particle type of some mass. There are, in fact, an enormous number of states; the density of states grows as e^m for large mass, m. Strings interact by splitting and joining (Fig. 2). It turns out to be easiest to calculate S matrix elements, and these splitting and joining interactions translate into an intricate set of rules for the calculation of S

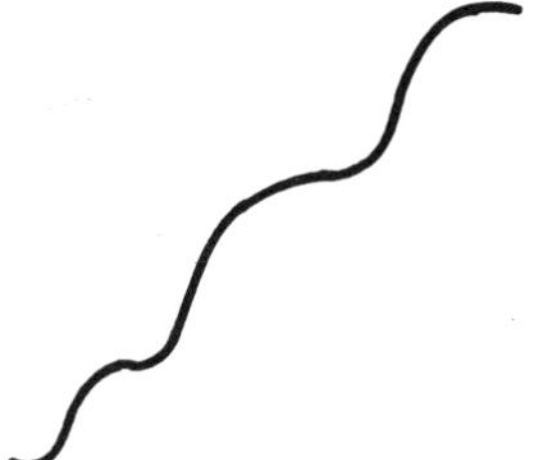

Figure 1a. Open string

Figure 1b. Closed string

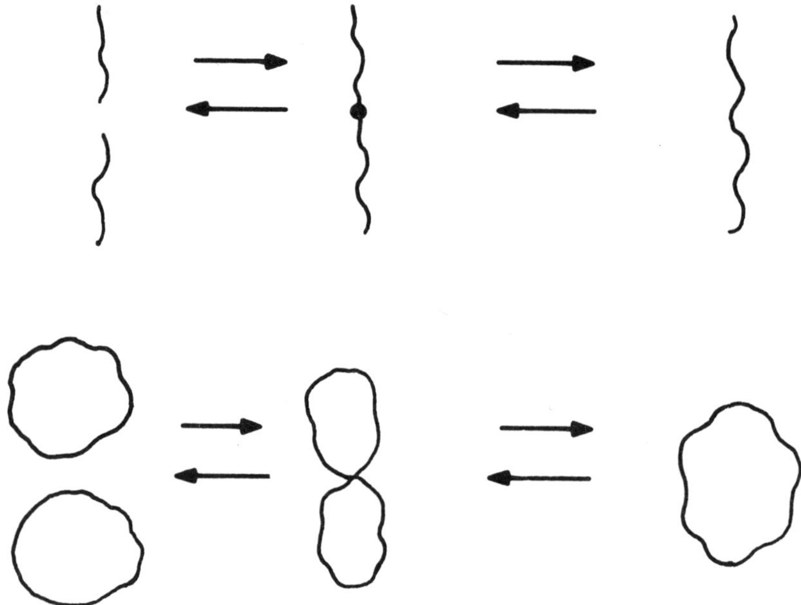

Figure 2. String interactions

matrix elements. The facts that strings can only interact in such limited numbers of ways, and that strings are extended objects, raise the possibility that these theories may be finite. The question of finiteness, however, is not yet completely settled.

String theories cannot be theories of hadrons, for several reasons. First, the various string theories only make sense for certain values of the dimension of space-time. In other dimensions, they are not unitary or Lorentz invariant. For the bosonic theories, this "critical dimension" turns out to be 26, while for fermionic theories the critical dimension is 10. Second, string theories have massless particles unlike anything which appears in the hadronic spectrum. They contain massless spin-2 particles; Lorentz invariance guarantees that these couple precisely like gravitons in Einstein's theory. They also contain massless spin-one particles. Lorentz invariance, again, guarantees that these particles couple precisely as massless Yang-Mills fields. Of course, there is nothing like this in the hadronic spectrum. Finally, the bosonic string contains another catastrophe: a tachyon.

QCD eventually provided us with a theory of hadrons. However, the fact that string theories automatically possessed gauge and general coordinate invariance was quite intriguing. Moreover, Gliozzi, Scherk, and Olive discovered that the fermionic string could be made supersymmetric[17] (more on supersymmetry below), and that in this case the theory contained no tachyon. Schwarz and Green developed a formulation in which the supersymmetry was manifest,[1] and which was more practical for involved calculations. However, the prospects for using superstring theories as theories of nature seemed dim. The problem was not merely their dimensionality. More serious was that, on genneral grounds, these theories were expected to possess anomalies which would spoil their consistency.[8] The recent explosion of interest in string theory was launched by the discovery by Green and Schwarz that string theory could be consistent with two gauge groups, $O(32)$ and $E_8 x E_8$ (and only these groups.[18] At the time of their discovery, no string theory with gauge group $E_8 x E_8$ was known; such a theory, the "heterotic" string theory, was quickly discovered.[2] In fact, it is the $E_8 \times E_8$ theory which seems most promising for phenomenology.

2. SUPERSYMMETRY IN 0+1, 3+1, AND 9+1 DIMENSIONS

One of the very exciting features of the superstring is that it is supersymmetric. It is impossible, in so short a space, to give a detailed introduction to supersymmetry, but it is possible to give some examples which illustrate the main concepts.

A supersymmetry is a symmetry which, like any symmetry, commutes with the Hamiltonian. What is striking is that the corresponding charge (or charges) obey anticommutation relations with one another. We can construct examples in ordinary quantum mechanics of two component spinors,[19]

$$\Psi(x) = \begin{bmatrix} \psi_1(x) \\ \psi_2(x) \end{bmatrix} \tag{1}$$

These illustrate almost all of the important concepts relevant in four or ten dimensions.

Consider the "supercharges"

$$Q_1 = \frac{1}{2}(\sigma_1 p + \sigma_2 \omega x)$$

$$Q_2 = \frac{1}{2}(\sigma_2 p - \sigma_1 \omega x) \tag{2}$$

where σ_i are the Pauli matrices and $p = - i\hbar \, d/dx$. Then

$$\{Q_i, Q_j\} = \delta_{ij} H = \delta_{ij} \, i\hbar \, \frac{d}{dt} \tag{3}$$

and

$$[Q, H] = 0 \tag{4}$$

where

$$H = \frac{1}{2}(p^2 + \omega^2 x^2 + \hbar\sigma_3\omega). \tag{5}$$

The Q's are like fermions, in that they obey anticommutation relations. More precisely, we can define a "fermion number operator" to
be simply σ_3. σ_3 commutes with the Hamiltonian.   It anticommutes
with the Q's. If we call $\begin{pmatrix} \psi_1 \\ 0 \end{pmatrix}$ a boson and $\begin{pmatrix} 0 \\ \psi_2 \end{pmatrix}$ a fermion, then

$$Q_i |B\rangle = \sqrt{E} |F\rangle \quad \text{and} \quad Q_i |F\rangle = \sqrt{E} |B\rangle \tag{6}$$

i.e. the Q's take bosons to fermions. Also, consider the ground state
(or "vacuum") energy. It is a sum of two pieces: $1/2\hbar\omega$, from the
zero point motion of the oscillator, and $-1/2\hbar\omega$ from the $\hbar\sigma_3\omega$ term.
This is the simplest example of the famous boson-fermion cancellations
for which supersymmetric theories are famous.

It is possible to extend all of this to four dimensions.[4] Here,
there are four supersymmetry charges, Q_α, transforming as a spinor
under the Lorentz group. Again, they obey anticommutation rules among
themselves, while commuting with the Hamiltonian and the momenta:

$$\{\overline{Q}_\alpha, Q_\beta\} = 2p^\mu (\gamma_\mu)_{\alpha\beta} \tag{7}$$

$$[Q_\alpha, p^\mu] = 0 \tag{8}$$

where γ_μ are the Dirac γ-matrices. (My precise conventions for the
γ-matrices will not be important here). There is a straightforward
recipe for constructing field theories with this symmetry. First note
that, just as in our quantum mechanics example, the action of the
supercharges on a bosonic state is a fermionic one, and vice versa:

$$Q_\alpha |B\rangle = \sqrt{E} |F\rangle \qquad Q_\alpha |F\rangle = \sqrt{E} |B\rangle. \tag{9}$$

Similarly, the action of Q_α on a fermionic field is to produce a
bosonic field, and vice versa. Typical commutation relations are

$$[Q_\alpha, \phi] = \psi_\alpha \qquad \{Q_\alpha, \psi_\beta\} = \varepsilon_{\alpha\beta} m\phi^* \tag{10}$$

where ϕ is a scalar field, ψ_α is a two-component fermion, and

$$\varepsilon_{\alpha\beta} = \begin{bmatrix} 0 & 1 \\ -1 & 0 \end{bmatrix}. \tag{11}$$

As for an ordinary bosonic symmetry, like isospin, if supersymmetry is unbroken, particles and fields will fall into supersymmetry multiplets. The interesting types of multiplets are chiral supermultiplets and gauge supermultiplets. A chiral supermultiplet contains a complex scalar, ϕ and a chiral fermion, ψ. Gauge supermultiplets contain gauge bosons and chiral fermions in the adjoint representation of the gauge group, λ^a. Given the gauge symmetry and particle content of a theory, there is a straightforward recipe for obtaining the corresponding Lagrangian. For all the various fields, one has the usual gauge invariant kinetic terms:

$$\mathscr{L}_{kE} = \sum_a \left(- \frac{1}{4} F^{a\,2}_{\mu\nu} + i\, \bar{\lambda}^a \, \not{D} \lambda^a \right) + \Sigma\left(|D_\mu\phi|^2 + i\, \bar{\psi} \not{D} \psi\right) \tag{12}$$

where the sums run over all of the various gauge groups and chiral supermultiplets, and the covariant derivatives are those appropriate to the representations of the ϕ's and ψ's. There are also Yukawa couplings of λ's, ψ's, and ϕ's with strength determined by the gauge couplings, and also certain quartic scalar couplings:

$$\mathscr{L}_Y = \Sigma\left(\sqrt{2}\, g^a\, \phi^*\, T^a\, \psi\, \lambda^a + \frac{1}{2}\, g^{a\,2} (\phi^*\, T^a\, \phi)^2 \right). \tag{13}$$

Again, the sums are over gauge group generators and the various chiral multiplets.

The remaining interactions are specified by a function of the chiral scalars, $W(\phi)$, called the superpotential. There is a potential for the scalars

$$V = \sum_i \left| \frac{\partial W}{\partial \phi_i} \right|^2, \tag{14}$$

mass terms for fermions

$$\mathcal{L}_m = \frac{\partial^2 W}{\partial\phi^i \partial\phi^j} \, \psi^i \psi^j \tag{15}$$

and Yukawa couplings

$$\mathcal{L}_{mY} = \frac{\partial^3 W}{\partial\phi^i \partial\phi^j \partial\phi^k} \, \psi^i \psi^j \phi^k . \tag{16}$$

In these theories, the Hamiltonian is given by

$$H = \Sigma |Q_\alpha|^2 . \tag{17}$$

This follows from the fundamental commutation relations, Eq. (7). As a result, the ground state energy is always positive. Spontaneous supersymmetry breaking means

$$Q_\alpha |0\rangle \neq 0 \tag{18}$$

and thus

$$\langle 0|H|0\rangle \neq 0 . \tag{19}$$

If the symmetry is spontaneously broken, particles need not lie in supermultiplets (fermions need not be degenerate with bosons). This is just what happens in the case of chiral symmetry, where particles are not "parity doubled" but, instead, there are (pseudo) Goldstone bosons, the pions. Similarly, in the standard electroweak theory, the SU(2) symmetry is not manifest. In the case of spontaneous breaking of supersymmetry, there is a Goldstone fermion.

Why should we be interested in supersymmetry at all? It is a rather pretty mathematical structure, but there is no evidence, at present, that it has anything to do with nature. Perhaps the principle reason to suspect that the laws of nature might be supersymmetric, and that relics of this supersymmetry might show up around the W mass, has to do with the "hierarchy problem", or the problem of the "naturalness" of fundamental scalars.[20] This problem is basically a modern formulation of the self-energy problem of the classical electron theory. In the classical electron theory, the electron mass diverges linearly with the electron size. In modern field theory parlance, we would say that the mass diverges linearly with the cutoff, and we would think of this cutoff as representing some new scale in physics, such as a scale connected with some new interactions, or where our ideas about space-time break down. Weisskopf discovered long ago that in quantum electrodynamics the divergence of the electron mass is only logarithmic. Thus the sensitivity of low energy parameters to this new physics is very mild. This occurs because QED acquires a new symmetry as $m_e \to 0$, a chiral symmetry under which

$$\psi \to e^{i\alpha\gamma_5} \psi. \tag{20}$$

However, in field theories with scalars, the scalar mass-squared (the natural quantity in a relativistic theory) depends quadratically on the cutoff, i.e.

$$M_s^2 = M_s^{0\,2} + \Lambda^2(a\alpha + b\alpha^2 + \ldots). \tag{21}$$

Here M_s^0 is some "bare" parameter, and Λ is the cutoff. If Λ is a number like M_p, then any scalar field should have a mass of order M_p. In the Weinberg-Salam model, fundamental scalars are responsible for the breaking of the electroweak symmetry. Their mass is necessarily not much different than M_w. The question is: how can they be so light? Alternatively, it is clear that the cutoff, Λ, in Eq. (21) cannot be much different than M_w. What new physics does this represent?

There are presently two candidates for this new physics that we can suggest. The first is technicolor,[21] in which the Higgs fields are bound states of some new strong interaction; Λ is then the binding scale. The second is supersymmetry.[14,22] Supersymmetry "solves" the problem since, if the laws of nature were exactly supersymmetric, the scalars of the Weinberg-Salam model would be exactly degenerate with certain fermions; since fermions can be light "naturally", it is natural for the bosons to be light as well. If the laws of nature are supersymmetric, supersymmetry must be spontaneously broken. Λ is then, essentially, the scale of this breaking, the typical splitting of the particles we know from their superpartners. By our arguments above, Λ must not be too different from the W mass. It is, in fact, possible to construct more or less realistic models in which these ideas are realized.[23,24] One major task for the next generations of accelerators will be to determine whether any of the new particles suggested by supersymmetry in fact exist.[24,25]

Another aspect of the hierarchy problem is the question: why do large hierarchies of mass scales exist at all? Again, supersymmetry offers some hope of providing an answer.[19,22] Recall the criterion (in global supersymmetry) for spontaneous supersymmetry breaking: the vacuum energy must be non-zero. It turns out that if a theory, at the classical level, does not break supersymmetry, then this is also true to all orders of perturbation theory. This is a consequence of what are called non-renormalization theorems.[26] (Recall the vanishing of the vacuum energy in our quantum mechanics problem; this generalizes to perturbation theory in less trivial models.) However, non-perturbative effects sometimes do lead to breaking of supersymmetry.[27,28] If M represents some typical mass scale in the theory, and λ is a typical coupling, one may generate in this way a scale, e.g., of order $\exp(-c/\lambda^2)$ M, for some constant c. If λ is small, this can generate an enormous hierarchy. Building realistic theories of this type, in practice, is possible but not so easy.[28]

The extension of all these ideas to the case where gravity is present is known.[23] In such "supergravity" theories, supersymmetry becomes a local symmetry. The graviton lies in a multiplet with a spin-3/2 field called the gravitino, which interacts with gravitational strength. All of the ideas discussed above can be extended to

324

this case. In fact, in most popular supersymmetric models, (super)
gravitational couplings play a crucial role.

Supersymmetric theories can be constructed in dimensions other
than four. Of particular interest to us, of course, are theories in
ten dimensions.[5] Here, I will be content to list the particle content of gauge and gravitational theories, and some of the low dimension terms in the Lagrangian. These particles and this Lagrangian
corresponds, roughly, to the low energy limit of superstrings.

The particle content can almost be determined by simple counting
arguments; we want the number of bosonic and fermionic degrees of
freedom to be equal. To count bosonic degrees of freedom for massless fields such as the metric tensor, there is a simple recipe:
count all the fields with different indices as independent, but work
in d-2 dimensions. Thus for the metric, $g_{\mu\nu}$, we have $\frac{8 \times 9}{2} - 1$ degrees
of freedom (d.o.f.) (the -1 arises because the graviton is traceless.
Note that the same counting, in four dimensions, yields two degrees of
freedom.) There is also an antisymmetric tensor field, $B_{\mu\nu}$, with
$\frac{8 \times 7}{2} = 28$ d.o.f., and a scalar field, the "dilaton", ϕ, with one degree of freedom (for a total of 64). For the fermions, there is a
Majorana-Weyl (i.e. real of self-conjugate and of definite handedness)
spin-1/2 field. Spinors in 0(8) have eight components, so this gives
8. Then there is a spin-3/2 field, (also Majorana-Weyl) ψ_{μ}^{α}, with
$8 \times 8 = 64$ d.o.f. This field, though, is subject to a constraint,
$\gamma^{\mu} \psi_{\mu} = 0$, which subtracts eight degrees of freedom. Thus we have a
total of 64 fermionic degrees of freedom, as well.

When gauge fields are present, there is a similar counting. We
then have gauge bosons, A_{μ}^{a}, in the adjoint representation of the
group. For each gauge generator, we have 8 bosonic degrees of freedom
here. We also have a set of Majorana-Weyl gluinos, each in the adjoint representation of the group. Again, these give eight degrees of
freedom.

For the supersymmetric theory, the lowest dimension terms in the
Lagrangian are known. For the bosonic piece,

$$\mathcal{L}_B = -\frac{1}{2\kappa^2} R - \frac{1}{4}\frac{\phi^{-1/2}}{g_{10}^2} F^2_{\mu\nu} - \frac{3}{4}\frac{\kappa^2}{g_{10}^4}\phi^{-1} H^2_{\mu\nu\rho} - \frac{1}{4\kappa^2}\left[\frac{\partial\phi}{\phi}\right]^2. \qquad (22)$$

Here ϕ is the dilaton field, R is the curvature scalar, $F_{\mu\nu}$ the Yang-Mills field strength, and $H_{\mu\nu\rho}$ the corresponding field strength for the field $B_{\mu\nu}$ ($B_{\mu\nu}$ is also associated with a gauge symmetry, $B_{\mu\nu} \rightarrow B_{\mu\nu} + \partial_\mu \varepsilon_\nu - \partial_\nu \varepsilon_\mu$). H is given by, for a U(1) theory,

$$H_{\mu\nu\rho} = \partial_{[\mu} B_{\nu\rho]} + A_{[\mu} F_{\nu\rho]} \qquad (23)$$

and a slightly more complicated expression for a non-Abelian theory. Note that in our Lagrangian, ϕ does not have the usual kinetic term. This is easily remedied. If we call $\phi = e^D$, then the field D has a standard kinetic term, $(\partial_\mu D)^2$. In L, g is the gauge coupling (with dimensions m^{-3}) while κ is the gravitational coupling, with dimension m^{-4}.

The fermionic terms in the Lagrangian are known, as well as the supersymmetry transformation laws. As we have remarked above, the massless states of the superstring are precisely those appearing in this Lagrangian. Also, since the superstring is supersymmetric, at low energies it is described by precisely this Lagrangian. This is a powerful result, which will allow us to make many phenomenological statements without too deep an understanding of the full string theory.

3. AN INTRODUCTION TO REAL STRING THEORY

A string, as it moves through space-time, sweeps out a world sheet, just as the trajectory of a particle defines a world line.

The motion of a particle is described, in classical special relativity, by specifying the four coordinates as a function of a parameter, τ. This parameter is usually taken to be the proper time, but this is not essential. Physics should not--and does not--depend on how this parameter is chosen. Correspondingly, the world sheet swept out by the string in the course of its motion is described by specifying the coordinates X^μ of the string as a function of two parameters, σ and τ. Roughly, τ can be thought of as time, while σ labels the points of the string at any fixed time. Any choice of the parameterization (σ, τ), is rather arbitrary. Physics should not depend on this choice. This "reparameterization invariance" plays a crucial role in our present understanding of string theory. Due to lack of time, we will, in this talk, choose a particular parameterization in which the physics is reasonably transparent. The reader should be aware, however, that the full structure of string theory only begins to emerge if one exhibits this large symmetry. In choosing a parameterization, we will essentially be making a gauge choice. It will be as if I taught you electrodynamics exclusively in axial gauge. On the other hand, I should stress that our understanding of string theory is still rather limited, and is best developed in this gauge, called "light cone gauge". In fact, at present we know how to write down the heterotic string theory only in this gauge.

Before choosing the gauge, we can at least write down the action in a reparameterization invariant way, taking our cue from general relativity. Defining a two-dimensional metric, $g_{\alpha\beta}$ (σ,τ), and recalling that our coordinates, x^μ, are functions of σ and τ, we take for the action

$$ S = \frac{1}{4\pi\alpha'} \int d^2\xi\sqrt{-g}\ g^{\alpha\beta}\ \partial_\alpha X^\mu \partial_\beta X_\mu \tag{24} $$

where $\xi^\alpha = (\tau,\sigma)$, and α' is the inverse of the string tension, with dimensions of mass-squared. From the point of view of the two dimensional coordinates, σ and τ, this action is generally coordinate invariant, i.e. reparameterization invariant. The action has another symmetry, which plays a crucial role in the theory. This is called the Weyl symmetry, $g_{\alpha\beta} \to e^{\lambda(\xi)}\ g_{\alpha\beta}$, $X^\mu \to X^\mu$. (To see that this is a symmetry recall that $\sqrt{-g} = \sqrt{-\det g_{\alpha\beta}}$, and $g^{\alpha\beta} = g_{\alpha\beta}^{-1}$). This rescaling is essentially a rescaling of length and mass scales in two dimensions. No time derivatives of $g_{\alpha\beta}$ appear in the action, so the two

dimensional metric is not a dynamical degree of freedom; it can be
eliminated in terms of the other fields, the X^μ's. Varying the action
with respect to $g_{\alpha\beta}$ yields the equation (and using

$$\frac{\delta}{\delta g_{\alpha\beta}} \det g = g^{\alpha\beta}(\det g)), \tag{25}$$

$$\partial_\alpha X^\mu \partial_\beta X_\mu - \frac{1}{2} g_{\alpha\beta} g^{\gamma\delta} \partial_\gamma X^\mu \partial_\delta X_\mu = 0. \tag{26}$$

Viewing the theory as a two-dimensional field theory on the σ, τ space,
with fields X^μ, this equation says that the energy-momentum tensor of
the theory vanishes (recall that, in general relativity, the variation
of the matter action with respect to the metric is the energy-momentum
tensor).

Up to a Weyl transformation, this equation has the solution

$$g_{\alpha\beta} = \partial_\alpha X^\mu \partial_\beta X_\mu. \tag{27}$$

This means that the two-dimensional metric is the metric induced on
the two-dimensional surface, since

$$(ds)^2 = \eta_{\mu\nu} dX^\mu dX^\nu = (\eta_{\mu\nu} \partial_\alpha X^\mu \partial_\beta X^\nu) d\xi^\alpha d\xi^\beta. \tag{28}$$

Plugging back in the action yields

$$S = \frac{1}{2\pi\alpha'} \int d^2\xi \sqrt{-g}. \tag{29}$$

The equations for X^μ following from this action are

$$\partial_\tau p^\tau_\mu + \partial_\sigma p^\sigma_\mu = 0 \tag{30}$$

where

$$p^\alpha_\mu = \frac{\delta \mathcal{L}}{\delta(\partial_\alpha X^\mu)} = \frac{1}{2\pi\alpha'} \sqrt{-g}\, g^{\alpha\beta} \partial_\beta X_\mu. \tag{31}$$

328

These equations look rather complicated, but they can be drastically
simplified by a suitable choice of gauge, i.e. by a clever choice of
the parameters, σ and τ. To do this we first choose a set of coordinates in our d-dimensional space-time called light cone coordinates.
These are defined by

$$X^{\pm} = \frac{1}{\sqrt{2}} \, (X^0 \pm X^1)$$

$$X^i = X^i \qquad i = 2, \ldots, d-1. \tag{32}$$

In these coordinates, the dot product of two vectors is

$$x \cdot y = x^+ y^- + x^- y^+ - \vec{x} \cdot \vec{y} \tag{33}$$

where $\vec{x}$ refers to the d-2 transverse coordinates.

Before going on, we must make a choice. We need to decide
whether we are studying open or closed strings. For our discussion
here, we will focus on closed strings. This means, thinking of σ as
space and τ as time, that the X's will be periodic functions of σ.
We will take σ to run from 0 to π.

To go to light cone gauge, we first set

$$X^+ = 2\alpha' p^+ \tau; \tag{34}$$

This fixes our τ parameterization. The quantity

$$P^+ = \int d\sigma \, P^{\tau+} \tag{35}$$

is conserved. To see this, just integrate Eq. (30) over σ, and recall
the periodicity in σ. By a choice of σ parameterization, we can take
$P^{+\tau}$ constant. Then $P^{+\sigma}$ is a constant, which we can set to zero by a
choice of the σ origin as a function of τ. From Eq. (31), we see that
since $P^{\sigma+} = 0$, $g^{\sigma\tau} = 0$. Since $P^{\tau+}$ is constant, we also have that
$\sqrt{-g} \, g^{\tau\tau}$ is constant, or

$$|g_{\tau\tau}| = |g_{\sigma\sigma}|. \tag{36}$$

By a Weyl transformation, we can set these quantities equal to one, i.e.

$$g_{\alpha\beta} = \begin{bmatrix} 1 & 0 \\ 0 & -1 \end{bmatrix}. \tag{37}$$

Now

$$p^{\mu}_{\tau} = -\frac{1}{2\pi\alpha'}\, \partial_{\tau}X^{\mu}; \qquad p^{\mu}_{\sigma} = \frac{1}{2\pi\alpha'}\, \partial_{\sigma}X^{\mu} \tag{38}$$

and

$$p^{+}_{\tau} = \frac{p^{+}}{\pi}; \qquad p^{+} = p^{+}. \tag{39}$$

Note that the equations of motion for the transverse coordinates are just

$$\partial^2_{\tau} X^i - \partial^2_{\sigma} X^i = 0, \tag{40}$$

i.e. the equations of motion for d-2 massless fields. We can easily solve these equations, subject to the periodicity requirement. We have (from now on we choose $2\alpha' = 1$)

$$X^i(\sigma,\tau) = \hat{X}^i + \hat{P}^i\, \tau + \frac{i}{2} \sum_{n \neq 0} \left[\frac{1}{n}\, \alpha^i_n\, e^{in(\tau-\sigma)} + \frac{1}{n}\, \tilde{\alpha}^i_n\, e^{in(\tau+\sigma)} \right]. \tag{41}$$

Hermiticity requires

$$\alpha^{i+}_n = \alpha^i_{-n}. \tag{42}$$

The two terms in the sum correspond to waves moving "to the right" ("right movers") and "to the left" ("left movers") along the string. We can quantize this system following the standard canonical quantization procedure. The canonical momenta are just the P^i_{τ}'s. One finds, for the α's, the commutation relations

330

$$[\alpha_n^i, \alpha_m^i] = n\, \delta_{n,-m} \delta^{ij} \tag{43}$$

while
$$[\hat{X}, \hat{P}] = i. \tag{44}$$

These normal modes are identified with d-dimensional particles. To make this identification, recall that in light cone gauge, $g_{\tau\tau} + g_{\sigma\sigma} = 0$, or

$$\partial_\tau X^\mu \partial_\tau X_\mu + \partial_\sigma X^\mu \partial_\sigma X_\mu = 0. \tag{45}$$

Integrating this equation over σ, substituting our explicit expression for the X^i's, recalling that $X^+ = P^+ \tau$, and calling

$$P^- = \int d\sigma\, \partial_\tau X^- \tag{46}$$

we have,

$$0 = P^+ P^- - 1/2\, \vec{P}^2 - :\hat{N}: + c = 0 \tag{47}$$

Here, $\hat{N}$ is like a number operator

$$:\hat{N}: \ = \ :\hat{N}_L: + :\hat{N}_R: \ = \ \sum_n (\alpha_n^{i+} \alpha_n^i + \tilde{\alpha}_n^{i+} \tilde{\alpha}_n^i). \tag{48}$$

Eq. (47) may be written as

$$m^2 = 2(\hat{N}-c). \tag{49}$$

Note that $\hat{N}$ has been normal ordered (annihilation operators all stand to the right). This is the origin of the constant c. In fact, c turns out to equal $\frac{d-2}{12}$.

Let us now consider the states of this theory. The lowest state is the vacuum, $|0,p\rangle$, with

$$m^2 = 2p^+ p^- - p^{i\,2} = -1/\alpha' . \tag{50}$$

It is thus a tachyon. The first excited states are the states

$$\varepsilon_{ij}\, \tilde{\alpha}^i_{-1}\, \alpha^j_{-1}\,|0\rangle \tag{51}$$

(There is actually a further constraint on the system which requires $\langle N_L \rangle = \langle N_R \rangle$. This is a consequence of the fact that the _states_ must be invariant under shifts of σ by a constant). These states have $\alpha' m^2 = 2 - \frac{d-2}{12}$. Clearly there are $(d-2)^2$ such states. This is just the number of states, in a Lorentz invariant theory, to form a massless graviton, antisymmetric tensor, and dilaton. So we need, for Lorentz invariance, $m^2 = 0$, or $d = 26$. This requirement can be shown more rigorously by examining in detail the structure of the Lorentz algebra.

In this "critical dimension", there is clearly an infinite tower of massive states. One can show, in fact, that the density of states as a function of mass, m, grows as e^m. Of course, the tachyon is an unpleasant surprise, but it will disappear in the superstring. Finding a massless spin-2 particle, however, is quite exciting. General theorems show that in a Lorentz invariant theory, this particle must couple as a graviton. Similarly, in theories with open strings, one finds Yang-Mills gauge particles.

Space does not permit us any detailed discussion of interactions. The basic interaction, for closed strings is a splitting or joining interaction. (Fig. 2) Essentially, there is nothing else one can write. In terms of the two-dimensional fields we have been discussing, this interaction translates into the following rule: in constructing the S-matrix element for the scattering of N particles with momenta $k_1, \ldots k_n$, one must evaluate the vacuum expectation value of N "vertex operators", $V_i(k_i, \xi_i)$

$$\int_\pi d^2\xi^i \langle 0|V_1(k_1\xi_1)\ldots V_n(k_n, \xi_n)|0\rangle. \tag{52}$$

Here the V_i's are functions of the x^i's. For example, for the tachyon,

$$V(\sigma) = g : e^{ik \cdot X} :$$ (53)

Here g is the coupling constant of string theory.

Computation of the scattering amplitudes is reasonably straightforward, but, for reasons of time, I must simply refer the reader to the literature. The amplitudes have fascinating analyticity properties and the remarkable property of duality (which ignited the original interest in the days when string theories were supposed to be theories of hadrons). This means that, say, a four-point amplitude can be thought of as arising as a sum of an infinity of massive particle exchanges in the s channel, or in the t channel, but that one does not sum both s and t channel processes. (Fig. 3)

The vertex operators for all the states are known. For the graviton, it is

$$V_G = \int d^2\xi \; \varepsilon_{\mu\nu} \frac{\partial X^\mu}{\partial \xi^\alpha} \frac{\partial X^\nu}{\partial \xi^\alpha} e^{ik \cdot X},$$ (54)

where $\varepsilon_{\mu\nu}$ is the gravitational polarization tensor.

Most of the subtlety and beauty of string theory, especially the intricate workings of conformal invariance, only becomes apparent through the study of interactions. One important point: string theory, as it stands now, only allows efficient calculation of S-matrix elements. Off-shell amplitudes, so useful in conventional field theory, can only be calculated with great difficulty. One would like to have something like a "field theory of strings", a theory with field operators which create and destroy strings.[6,7] The fields of this theory would, in particular, create all the particle states described previously. This theory would hopefully make clear the origin of the general covariance and gauge invariance of string theories, allow a systematic derivation of the computational rules, and perhaps permit exploration of non-perturbative phenomena. The search for this field theory, or some alternative formulation, is presently a subject of intense research.

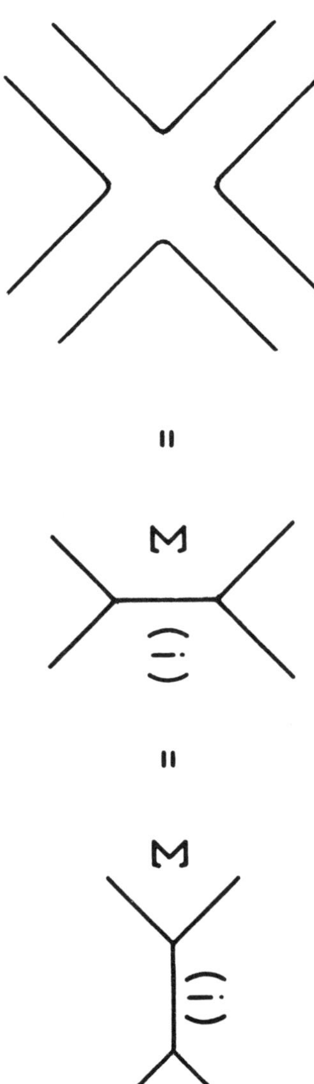

Figure. 3 Duality scattering amplitude

    - a sum of S-channel or t-channel exchanges

334

So far, we have described a beautiful mathematical structure, but one hopelessly removed from nature. The theory has given us, for free, general coordinate and gauge invariance, but only in 26 dimensions and with a tachyon. It also possesses no fermions. The superstring gets rid of the tachyon, introduces fermions--and in fact supersymmetry--and reduces the dimensionality to 10, while keeping all of the other remarkable features we have described. We have space, here, for only the most cursory introduction.

Recall that the bosonic string in light cone gauge had the structure of a two-dimensional field theory with d-2 bosonic fields. In the superstring of Schwarz and Green, we essentially add d-2 more fields, which are fermions both from the point of view of two and d (hereafter 10) dimensions, $S^a_\alpha(\xi)$. Here a is an eight-dimensional spinor index; α a two dimensional one. The action for these fields is just that of a free two-dimensional field theory:

$$\mathcal{L}_F = \int d^2\xi \; \overline{S}^i \; i\partial_\alpha \gamma^\alpha \; S^i. \tag{55}$$

It is a straightforward exercise in 10-dimensional (really d-2 = 8 dimensional) Fierz identities to show that this action is supersymmetric; for this I must send you to the literature. Here let us just note that the counting is right: we have eight bosonic and eight fermionic degrees of freedom. The fermions are again to be expanded in normal modes and quantized, with periodic boundary conditions. There is one subtlety here. The fermions have a constant mode, with coefficients $S^a_{o\alpha}$. These coefficients satisfy, upon quantization

$$\{S_{o\alpha}, S_{o\beta}\} = \delta^{ij} \delta_{\alpha\beta} \tag{56}$$

(twiddled and untwiddled variables, again, refer to left and right movers). This cannot be satisfied by the usual Fock-space type construction, with the S_o's, say, annihilating states. Instead, one must take the ground state to be 128-fold degenerate, and the S_o's to be matrices, similar to γ-matrices, acting in the space of these states.

These states will be identified with the massless particles of the superstring, the graviton and the gravitino. The full Hilbert space of the string is then constructed as a direct product of the states of the bosonic and fermionic fields. Now, however, when one studies the mass operator for this theory, one discovers that there is no tachyon in this theory, and that the ground states are, in fact, massless.

All of this is for closed strings. One can also consider open strings. In this case, just as for bosonic strings, one finds gauge bosons, now with fermionic partners. Such a theory can incorporate all $O(N)$ and $S_p(N)$ gauge groups. Closed string theories make sense by themselves. Open string theories must be combined with closed string theories to obtain unitary models. In both cases, it is known how to construct the relevant vertex operators, and many computations have been done.

At one loop, in most of these theories, a problem arises. Almost all of these theories suffer from anomalies in both gauge currents and energy and momentum conservation. Such anomalies spoil the invariances of the theory, and lead ultimately to a breakdown of such crucial features as Lorentz invariance and unitarity. This result was anticipated from features of the low energy field theory.[8] The surprise[9] was that, in fact, one string $(O(32))$ theory did not possess anomalies. When Green and Schwarz discovered this fact, they re-thought the field theory arguments, and noticed that this phenomenon could occur, not only in theories with gauge group $O(32)$, but also in theories with gauge group $E_8 \times E_8$. At that time, however, no such theory existed.

Stimulated by this observation, Gross, Harvey, Martinec and Rohm[2] constructed two new string theories, one with gauge group $O(32)$ and one with group $E_8 \times E_8$. These theories were referred to as "heterotic", since they combine certain features of the bosonic and fermionic strings. We cannot detail here their intricate construction, and so must content ourselves with listing a few of their basic features. They are, again, Lorentz invariant and unitary, they have no tachyons, and their low energy limits are described by the field

theory (with appropriate gauge groups) that I have described earlier.
These theories are theories of closed strings only. As such, they
possess only one type of interaction: the joining of two closed
strings into one (a breaking of one closed string into two).

This brings us to the final point of this section. So far, I
have not said much about the coupling constant which appears in the
vertex opeators, which is the coupling constant of string theory.
This coupling, for example, is directly related to the effective gauge
couplings of the low energy field theory. Just as in field theory, if
this coupling constant is in some appropriate sense small, the perturbation expansion is reliable; if not, the expansion is not. Unlike
the case of field theory, it is believed that there are no infinities
in these theories; all calculations should give finite results. This
is, essentially, because of the uniqueness of the interaction; the
infinities of ordinary field theories (especially non-renormalizable
ones) arise because of the many possible interactions.

However, in string theory, not only are there no infinities;
there are not really any parameters.[10] To see this, we can reexamine
our ten-dimensional Lagrangian, focusing particularly on the dilaton
field. This field, in flat ten-dimensional space at least, has no
potential. Thus its expectation value is undetermined, at least in
the classical theory. This statement is not modified even if we
include higher order terms in the effective Lagrangian arising from
exchange of massive states of the string. The Lagrangian we have
written can be rewritten, by rescalings of the fields, as:

$$\mathcal{L}^{(10)} = \phi^{-2}[-\frac{1}{2} R - \frac{1}{4} F^2 - \frac{3}{4} H^2 + 2(\frac{\partial\phi}{\phi})^2 + ...]. \qquad (57)$$

Here, we have set the string tension to one; by our rescalings, all
the dimensionless ratios have been absorbed in ϕ (which, as defined,
is dimensionless). ϕ^{-2} now sits out in front of the whole Lagrangian.
Thus changing the expectation value of ϕ changes the effective coupling of string theory.

The coupling constant of this theory is in fact the expectation value of a dynamical field. It is thus a _dynamical_ question what the coupling of string theory in fact is, and whether a weak coupling calculation is valid. We will in fact argue at the end of the next section that, on quite general grounds, if superstring theory does describe our world, it must be strongly coupled.[17] This is worrisome, since our present understanding of string theory is based exclusively on perturbation theory.

4. SUPERSTRING PHENOMENOLOGY

Despite our somewhat pessimistic remarks at the end of the last section, it is in fact possible to go quite a long way towards the development of a plausible superstring phenomenology. Not only can one suggest how we might get down from ten to four dimensions, but it is even possible to see how the parameters of the low energy theory might be determined, and how string theory might solve certain problems of four-dimensional theories, which have so far escaped solution in conventional field theories. In this lecture, I will focus exclusively on the $E_8 \times E_8$ heterotic string theory, since this theory seems the most promising phenomenologically.

Of course, the first experimental result we must come to terms with is the fact that our world is four, not ten, dimensional. This, as we will see, is not a trivial problem, but it is by no means insurmountable, either. What must happen, for the theory to be viable, is that six dimensions must curl up or "compactify" to some very small size (say M_p^{-1}) while four remain flat. Such a possibility was suggested long ago by Kaluza and Klein, in an attempt to understand the origin of electromagnetism and electromagnetic gauge invariance. In order to understand in concrete terms what compactification means, let us consider their example. We suppose that space-time is five dimensional. Four dimensions, $X_0, \ldots, X_3$ form the usual Minkowski space-time. The fifth dimension, y, is compact,

$$0 \leqslant y \leqslant 2\pi R, \tag{58}$$

where R is the radius of the fifth dimension. We will take y to have the topology of a circle. This means that all fields must be periodic in y. We will also take the background metric to be flat, i.e.

$$g_{\mu\nu} = \text{diag } (-1,1,1,1,1). \tag{59}$$

plus small fluctuations.

Before discussing the modes of the metric tensor, let us first consider a scalar field, ϕ, such as will appear in string theory. In this background, the ϕ equation of motion is (ignoring interactions)

$$\Box_4 \phi + \frac{\partial^2}{\partial y^2} \phi = 0. \tag{60}$$

Here $\Box_4$ is the usual four-dimensional Laplacian. This equation is easily solved, subject to the periodicity condition. We can take

$$\phi(x,y) = \sum_n e^{iny/R} \psi_n(x) \tag{61}$$

where $\psi_n(x)$ is a function of $X_0,\ldots,X_3$ obeying

$$\Box \psi_n(x) = \left(\frac{n^2}{R^2}\right) \psi_n(x) \tag{62}$$

i.e. it is a massive, four-dimensional field with mass n/R. In other words we have obtained an infinite tower of massive four-dimensional fields, with Compton wavelength less than or of order the size of the compact dimension. Note that, for n=0, there is a massless scalar field. From the point of view of the Laplace operator on the internal space, $\frac{\partial^2}{\partial y^2}$, this is a zero mode. Such massless particles are the only fields which can play any role at low energies in our four-dimensional theory; thus the search for zero modes in Kaluza-Klein theories is of extreme importance. Similar expansions can be performed for the other fields which may appear in our theory. Of particular interest is the metric, $g_{\mu\nu}$. It is natural to break up $g_{\mu\nu}$ into various pieces, according to their transformation properties under the four-dimensional

Lorentz group. g_{55} is like a scalar, $g_{5\alpha}$, $\alpha = 0,\ldots, 3$ is like a vector, and $g_{\alpha,\beta}$, $\alpha,\beta = 0,\ldots, 3$ is like the usual metric tensor. Thus we label:

$$\rho(x,y) = g_{55}(x,y)$$

$$A_\alpha(x,y) = g_{5\alpha}(x,y)$$

$$g_{\alpha\beta}(x,y) = g_{\alpha\beta}(x,y). \tag{63}$$

Just as for our scalar field example, each of these fields has an expansion in terms of eigenfunctions of an appropriate Laplace operator on the internal space, corresponding to four-dimensional fields of higher and higher mass. Of particular interest are the zero modes, since they correspond to massless fields. In the present example, all of the fields have zero modes. The zero modes of $g_{\alpha\beta}$ correspond to the four-dimensional metric tensor (graviton). Those of A_α correspond to a four-dimensional photon, and the massless mode of g_{55} is a "dilaton", corresponding to increasing or decreasing the size of the internal dimension, L, everywhere in space-time. To see this recall that

$$L = \int_0^R dy\sqrt{g_{55}} . \tag{64}$$

The masslessness of this mode arises because there is nothing in the classical theory which determines the length of the extra dimensions. Recall that Einstein's Lagrangian involves the curvature, which vanishes here. In general this feature will not be preserved by quantum corrections, or by the higher dimension operators which we must include in this non-renormalizable theory.

The masslessness of the photon, here, can also be understood in simple terms. There is no preferred point in our fifth dimension. Thus the metric is unchanged if we make a translation,

$$y \rightarrow y + \epsilon \qquad (65)$$

If ϵ is a function of x, the only component of the metric which changes under this general coordinate transformation is

$$A_\alpha = g_{5\alpha} \rightarrow A_\alpha + \partial_\alpha \epsilon(x). \qquad (66)$$

This is a gauge transformation from the point of view of four-dimensions. It is clearly a symmetry of the theory.

Efforts have been made to construct Kaluza-Klein theories in which all the gauge interactions we know arise in this way. These have been thwarted by two problems. First, one has not had any sensible theory of gravity which might explain the origin of compactification (or perhaps, more seriously, the flatness of four dimensions). Second, it turns out that, whatever one assumes about the dynamics, one always obtains a left-right symmetric spectrum of chiral fermions.[11] To understand the problem, remember that massless four-dimensional fields arise as zero modes of appropriate operators on the internal space. In the cases of interest, these spaces may be quite non-trivial, with non-vanishing curvature. For fermions, one needs the spectrum of zero modes of a certain Dirac operator. This is a well studied problem in mathematics, about which many general results, which go under the heading of "index theorems", are known. These may be used to show that one never obtains fermions in a left-right asymmetric representation of the gauge group.[11]

There is an exception to this theorem.[11] If the theory contains, in addition to gravity, fundamental gauge fields, it is possible to obtain chiral fermion representations. For the Kaluza-Klein purist, this is anathema, but for the string enthusiast, it is quite promising. After all, in string theory, we just want to get to four dimensions. We already have fundamental gravitons, gauge bosons, and fermions. Also, since we believe we have a finite, parameter-free theory, it should be possible to address the dynamical question of compactification in a sensible way.

As we discussed above, at the classical level, the coupling constant of string theory is undetermined, and we are free to work in a regime of weak coupling. In such a regime, we expect that the effective Lagrangian of Eq. (22) should provide a good description of the dynamics of the light fields. For compactifications with $R \gg M_p^{-1}$, it should be enough to keep only the massless modes of the ten-dimensional theory to consider the problem of compactification. In weak coupling, the system is essentially classical; in the classical approximation, one searches for ground states of the theory by looking for solutions of the classical equations of motion.

Thus we are led to seek solutions of the classical equations of motion with six dimensions compact,[12] where for the equations of motion we take those which follow from the variation of the Lagrangian of Eq. (22). It is easy to see that the equations will be satisfied if $A_\mu = 0$, $B_{\mu\nu} = 0$, and the Ricci tensor, $R_{\mu\nu} = 0$, since then Einstein's equation,

$$R_{\mu\nu} - \frac{1}{2} g_{\mu\nu} R = 8\pi T_{\mu\nu} \qquad (67)$$

is satisfied. The condition $R_{\mu\nu} = 0$ can be trivially satisfied if we take the compact dimensions to be flat, i.e. tori, as in our Kaluza-Klein example. However, in this case, we obtain too much supersymmetry to obtain a realistic low energy theory (there are four supersymmetries, rather than one). However, non-trivial "Ricci-flat" manifolds exist; these are the so-called Calabi-Yau manifolds. It is not hard to show that they yield effective four-dimensional theories with only one supersymmetry.

To understand this, consider ten-dimensional spinors. Our compactification is to $M^4 \times K$, where K is a six-dimensional Ricci-flat manifold. Locally, space-time is like $M^4 \times$ (6 dimensional flat space), and it is natural to decompose spinors (and other quantities) according to their behavior under the four-dimensional Lorentz group, $L^{(4)}$, $\times$ O(6). It is a simple fact of group theory that O(6) is isomorphic to SU(4), and the spinor representation of O(6) corresponds to the fundamental 4 of SU(4). So it is equivalent to give the decomposition under $L^{(4)}$ x SU(4). Our 16-component Majorana-Weyl spinors then decompose as

$$(16) = (1/2, 0) \text{ x } 4 + (0, 1/2) \text{ x } \bar{4} . \tag{68}$$

((1/2, 0) and (0, 1/2) are just left and right-handed two component spinors).

Now, suppose there exists on the six-dimensional manifold a 4-component spinor, ε, which is unaffected by parallel transport about K, i.e. which is covariantly constant. On SU(4) spinors, the covariant derivative is[29]

$$D_\mu \varepsilon^i = \partial_\mu \varepsilon^i + \omega^i_{\mu j} \varepsilon^j \tag{69}$$

where $\omega^i_{\mu j}$ is called the spin connection. It can be constructed (under the circumstances which are relevant to us) from the metric tensor. It takes its values, in this case, in the Lie algebra of SU(4), and it behaves essentially like an SU(4) gauge field. It is in fact the gauge field of the Lorentz group. Now choose a gauge in which ε everywhere has the form

$$\varepsilon^i = \begin{bmatrix} 0 \\ 0 \\ 0 \\ \varepsilon \end{bmatrix} \tag{70}$$

where ε is a constant. Then the statement that ε is covariantly constant means that ω takes values only in SU(3), i.e.

$$\omega = \begin{bmatrix} \omega & \vdots & 0 \\ \cdots\cdots & \vdots & 0 \\ 0 & 0 & 0 \end{bmatrix} . \tag{71}$$

A manifold with such a spin connection is said to be a "manifold of SU(3) holonomy" or a Calabi-Yau manifold. It is not hard to show that

for such a spin-connection the Ricci tensor vanishes (see, e.g.,
Ref. 29 for the relevant formulae).

The supersymmetry of the heterotic string is an N=1 supersymmetry
in 10 dimensions; the supersymmetry charges are ten-dimensional
Majorana-Weyl spinors and as such they have a decomposition under
SU(4) x L just as in Eq. (68), i.e.

$$Q^a = Q^i_\alpha \tag{72}$$

where i is an SU(4) index and α a four-dimensional spinor index. Only
the components which are singlets under the SU(3) subgroup of Eq. (71)
survive to four dimensions, i.e. the four dimensional theory has N=1
supersymmetry.

We now have much of what we want: a set of ground states with
four flat and six compact dimensions, and N=1 supersymmetry in the low
energy theory as a bonus (recall that such a symmetry might hold the
key to the solution of the hierarchy problem). However, we also need
chiral fermions, and, to obtain those, we need a non-vanishing
background gauge field. It turns out, that, in fact, we have not yet
satisfied all of the equations of motion. In the dilaton equation,
there is a term proportional to [12,30]

$$(R^2 - F^2) \; . \tag{73}$$

Thus we in fact require that there be a non-vanishing gauge field.
The simplest way to satisfy this equation appears to be the following.
E_8 has a subgroup SU(3) x E_6. Choose one of the E_8's, and take the
gauge field in this SU(3) subgroup, $A^i_{\mu j}$ to equal the spin connection
(recall that the spin connection is an SU(3) gauge field):

$$A^i_{\mu\, j} = \omega^i_{\mu j} \; . \tag{74}$$

This breaks E_8 down to E_6. Of course, we need to break E_6 further
down to something like the standard model. We will come to that
shortly. The other E_8, at this stage, is untouched.

First, what about massless fermions? Recall that originally, our only massless fermions were gluinos in the adjoint representation of the two E_8's. Under $SU(3) \times E_6$, the adjoint representation (248) decomposes as

$$248 = (8,1) + (1, 78) + (3,27) + (\bar{3}, \overline{27}). \qquad (75)$$

Here 8 is the adjoint of $SU(3)$, 78 is the adjoint of E_6. Now the index theorems I referred to earlier tell us that in fact, the number of 27's on the manifold minus the number of $\overline{27}$'s on the manifold is given purely by a topological invariant of the manifold, essentially half the Euler number. This means that we have chiral fermions. The 27, in fact, contains all the fields to make a single generation of the standard model, (15), plus some extras (12).

Now, as I remarked earlier, we still need to break E_6. Since we now have solutions of the equations of motion, we would seem to have little room for further symmetry breaking. However, many of the Calabi-Yau vacua are not simply connected. In such a case, one can have non-vanishing vector potentials, $A^i_{\mu j}$, while the field strengths, $F_{\mu\nu}$ are zero.[31] Since the equations of motion, and also the index theorems, involve only the field strengths, all of our previous discussion is unaffected. However, $\langle A_\mu \rangle$ breaks E_6 further. For reasons of time, we will not be able to explain this beautiful mechanism in detail. Let me just illustrate the idea with a simple example. Suppose our space is two-dimensional, with a circular hole as in Fig. 4. Take the gauge group to be $SU(3)$, and take A_μ to be

$$A_\mu = i\, g^{-1}\, \partial_\mu g \qquad (76)$$

where, in cylindrical coordinates, centered in the middle of the hole

$$g = e^{\,i\, \frac{\lambda_3}{2}\, \phi}. \qquad (77)$$

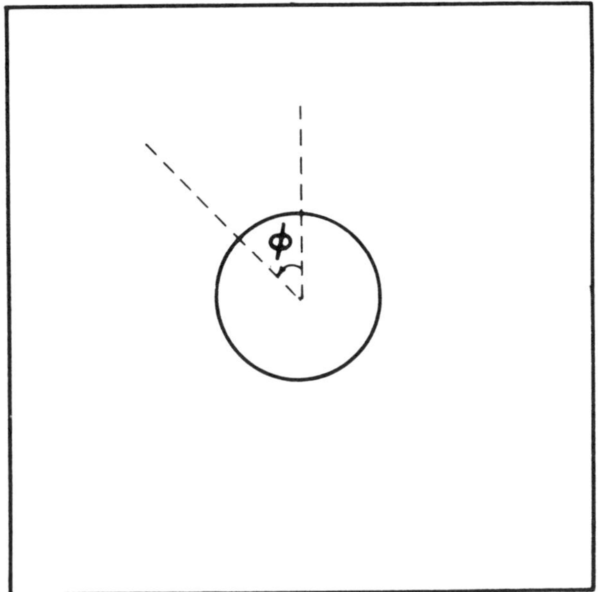

Figure 4. Two-dimensional space-time with a hole

Explicitly,

$$A_\phi = \frac{\lambda_3}{2} = \frac{1}{2}\begin{bmatrix} 1 & & \\ & -1 & \\ & & 0 \end{bmatrix}. \qquad (78)$$

This looks like a pure gauge field, but we cannot simply gauge it away. The problem is that g is not single valued even though A_μ is,

$$g(2\pi) = \begin{bmatrix} -1 & & \\ & -1 & \\ & & 0 \end{bmatrix} \neq g(0). \qquad (79)$$

The gauge transformation, acting on single valued fields, such as fermions in the fundamental representation, would produce non-single valued fields. In fact, consider a closed path C around the hole. Then the Wilson line,

$$U = P\, e^{\,i\oint dx_\mu A^\mu} = \begin{bmatrix} -1 & & \\ & -1 & \\ & & 0 \end{bmatrix}. \qquad (80)$$

where P denotes path ordering, is gauge invariant and non-trivial for this configuration. This shows that A cannot be gauged away. In fact, A acts precisely like a Higgs field.

Returning to the six-dimensional problem, for a given manifold, it turns out that there are only a discrete set of choices for the expectation values of A (or equivalently of the matrices U).[12] It is possible in this way to obtain groups which resemble the standard model, for example $SU(3)_C$ x $SU(2)_L$ x $U(1)_Y$ x (additional U(1)'s), and a small number (e.g. 3 or 4) of generations.[13] It is also possible, in principle, to compute all of the relevant low energy parameters for any given manifold. For example, one can give an explicit formula for the Yukawa couplings,[15] which can be evaluated if the metric is known.

So far, what we have found amounts to a huge set of classical solutions of string theory, or, from a quantum mechanical viewpoint, a huge set of potential ground states. These states carry a variety of labels. There are discrete numbers which label the topological character of the manifold. There are, invariably, continuous parameters which label the size and shape of the manifold. There are discrete numbers which label the Wilson lines. At the classical level, all of these are perfectly good vacua for string theory. One can, in fact, argue more directly in string theory that all of these vacua are solutions of the full (classical) string equations of motion,[12] without going through the intermediate field theory steps. In addition, new types of solutions have recently been discovered.[32] And, of course, there are vast numbers of additional ground states: flat ten dimensions, compactifications to other numbers of dimensions, compactifications on tori, and presumably more. We seem to be suffering from an embarrassment of riches.

There are two approaches we might try to adopt at this stage. First, we could try and examine all of these vacua, evaluate their couplings, spectra, and so forth, to determine if any of them resemble the world we see. Second, we could attempt to discover what features of string dynamics determine the true ground state. In fact, both approaches have been, and are being, tried, and I will comment briefly on each of them.

The search for phenomenologically viable models focusses on certain problems. First, one seeks models with a reasonably small number of generations. Otherwise, the gauge couplings (and especially the QCD coupling) are not asymptotically free and become large well below the Planck mass.[33] Second, one must find models in which proton decay is adequately suppressed.[13] The problem in these theories is that the extra colored particles in each generation may be light and can mediate rapid proton decay. Remarkably, unlike ordinary grand unified theories, it is easy in string theory to obtain light Higgs doublets without triplet partners,[13,14] which could also mediate proton decay.[12] But a complete solution of the problem has not yet been proposed. The new class of manifolds which have recently been discovered, in fact, have the structure of SU(5) or O(10) gauge theories, and may ameliorate some of these difficulties.[32] But this remains to be seen.

At a slightly finer level, are a series of other problems.

(1) In many of these vacua, one obtains right-handed neutrinos, which may lead to unacceptably large Dirac masses for neutrinos.[32,33]

(2) Standard renormalization group calculations can rule out many prospective ground states, but some pass the tests of: reasonable Weinberg angle and perturbative unification.[33]

(3) In many ground states, the θ-parameter of QCD is too large,[35] leading to too large a value of the neutron electric dipole moment. (The whole subject of the strong CP problem in string theory is a fascinating one which we do not have space to touch on here.)

Of course, at a more ambitious level, one would like to check that the entire spectrum of quarks and leptons turns out correctly. This question is, to some extent tied up with the question of supersymmetry breaking, which will be discussed below.

Of course, one would like to understand a variety of dynamical questions in these theories. Is this enormous vacuum degeneracy lifted in some way, and how? Is supersymmetry spontaneously broken, and how? And perhaps most importantly, if supersymmetry is broken, does the cosmological constant vanish?

These questions are all intimately related. For example, we have seen that there are sets of vacua described by continuous parameters, corresponding to the size and shape of the manifold. Variations in the size and shape correspond to variations in the metric, a dynamical field. Since it costs no energy to change these parameters, with each such parameter is associated a massless scalar field in four dimensions. When supersymmetry is broken, we expect that these fields will acquire mass (more generally potentials). We would like to know where the minima of these potentials lie, to see if at least some of the degeneracy is lifted in a suitable way. Similarly, the dilaton field, which controls the size of the string coupling, is also classically undetermined, but should be fixed once supersymmetry is broken.

One of the most impressive experimental facts about our universe, is that it is basically flat. This does not appear surprising, until one recalls that gravity couples to energy--all forms of energy. In a field theory, the ground state energy is certainly just that--energy, and gravity will couple to it. In conventional field theories, this energy is in fact infinite. For a free theory, for example

$$E_0 = \Sigma \; \frac{1}{2} \; (-1)^F \int^\Lambda \frac{d^3k}{(2\pi)^3} \; \sqrt{k^2 + m^2} \tag{81}$$

when the sum is over particle types and helicity states. The factor $(-1)^F$, where F is 0 for bosons and 1 for fermions, arises because for bosons we get a contribution $1/2 \; \hbar\omega$ for the zero point oscillations of each mode, while for fermions we get $- \; 1/2 \; \hbar\omega$ from filling a state in the Dirac sea. Λ is some cutoff. If Λ is, say, 100 GeV, then

$$E_0 \sim (100 \text{ GeV})^4 \ . \tag{82}$$

The corresponding experimental limit is

$$E_0 \lesssim 10^{-47} \text{ GeV}^4 \ . \tag{83}$$

In a theory with unbroken supersymmetry, E_0 is in fact zero; the fermionic and bosonic contributions cancel in Eq. (81). But, if supersymmetry is broken at same scale, Λ_s E_0 tends to be of order Λ_s^4. Of course, one can add a constant to E_0 to give $E_0 = 0$, but this constant must somehow be adjusted to extraordinary precision (at least one part in 10^{55}). No plausible solution has ever been suggested to this problem.

Superstring theory has no adjustable parameters, so it must somehow produce zero cosmological constant, even when supersymmetry is broken, if it is to describe nature.

In fact, it turns out that supersymmetry is broken for most of the vacua we have been discussing.[16] The source of the breaking is the dynamics of the "other E_8". The results are rather disappointing. A potential is in fact generated for the various light fields, but it has no minimum, at least in the regime of weak coupling, where the analysis is understood. Instead, the potential is such that the fields role to zero coupling and flat ten dimensions (i.e. infinitely large "compact" spaces).

After a little thought, this result is no surprise.[17] The coupling constants of this theory are dynamical fields. At zero coupling, the theory is free and has vanishing E_0. Thus, as the coupling tends to zero, the energy must tend to zero; correspondingly, the potentials for these fields must vanish in this regime. This argument, with slight refinement, applies both to the dilaton and to the field which describes the size of the internal space.

It is possible that at strong coupling there does exist a ground state which resembles our world. If so, it will be necessary to discover some general arguments why it is there, and what principle forces the cosmological constant to vanish.

REFERENCES

1. Reviews of superstrings:
    Schwarz, J.H., Phys. Rep. 89, 223 (1982).
    Green, M.B., Surveys in H.E. Physics 3, 127 (1982).
    Reviews of earlier work:
    Jacob, M., editor, "Dual Theory", Phys. Reports (reprint) Vol. 1
    (North-Holland, 1974).
    Scherk, J., Rev. Mod. Phys. 47, 123 (1975).

2. Gross, D.J., Harvey, J.A., Martinec, E., and Rohm, R., Phys. Rev.
    Lett. 54, 502 (1985); Nucl. Phys. B256, 253 (1985); and Princeton
    preprint (1985).

3. Witten, E., Phys. Lett. 149B, 351 (1984).

4. Texts for supersymmetry in four dimensions:
    Wess, J. and Bagger, J., "Supersymmetry and Supergravity",
    (Princeton University Press, Princeton, 1983).
    Gates, Jr., S.J., Grisaru, M.T., Rocek, M., and Seigel, W.,
    "Superspace", (Benjamin, Reading, 1983).

5. Chamseddine, A.H., Nucl. Phys. B185, 403 (1981).
    Bergshoesff, E., de Roo, M., de Wit, B. and van Nieuwenhuizen, P.,
    Nucl. Phys. B195, 97 (1982).
    Chapline, G.F. and Manton, N.S., Phys. Lett. 120B, 105 (1983).

6. Kaku, M. and Kikkawa, K., Phys. Rev. D10, 1110 (1974).

7. Banks, T. and Peskin, M., SLAC Preprint (1985).
    Kaku, M., City College preprint (1985).
    Kaku, M. and Lykken, J., City College preprint (1985).
    Siegel, W. and Zwiebach, B., U.C. Berkeley preprint (1985).
    Witten, E., Princeton preprint (1985).

8. Alvarez-Gaume, L and Witten, E., Nucl. Phys. B234, 269 (1983).

9. Green, M.B. and Schwarz, J.H., Phys. Lett. $\underline{149B}$, 117 (1984).

10. Dine, M. and Seiberg, N., Phys. Rev. Lett. $\underline{55}$, 366 (1985).

11. Witten, E., Nucl. Phys., $\underline{B186}$, 412 (1981).

12. Candelas, P., Horowitz, G.T., Strominger, A., and Witten, E., Nucl. Phys. $\underline{B258}$, 46 (1985).

13. Witten, E., Nucl. Phys. $\underline{B258}$, 75 (1985).

14. Breit, J., Ovrut, B., and Segre, G., University of Pennsylvania preprint (1985).
    Sen, A., Fermilab preprint (1985).

15. Witten, E. and Strominger, A., IAS preprint (1985).

16. Dine, M., Rhome, R., Seiberg, N., and Witten, E., Phys. Lett. $\underline{156B}$, 55 (1985).

17. Dine, M. and Seiberg, N., IAS preprint (1985); Weizmann Institute preprint (1985).

18. Gliozzi, F., Scherk, J., and Olive, D., Nucl. Phys. $\underline{B122}$, 253 (1977).

19. Witten, E., Nucl. Phys. $\underline{B188}$, 513 (1981).

20. Wilson, K., Phys. Rev. $\underline{D3}$, 1818 (1971).

21. Technicolor reviews:
    Sikivie, P., "Theory of Fundamental Interactions", Costa, G. and Gato, R.R., editors (North-Holland, Amsterdam, 1982).
    Farhi, E. and Susskind, L., Phys. Rep. $\underline{74C}$, 277 (1981).

22. Dimopoulos, S. and Raby, S., Nucl. Phys. $\underline{B192}$, 353 (1981).
    Dine, M., Fischler, W., and Srednicki, M., Nucl. Phys. $\underline{B189}$, 575 (1981).

23. Nilles, H.P., Phys. Rep. $\underline{110}$, 124 (1984).

24. Haber, H.E., Kane, G.L., Phys. Rep. 117, 75 (1985).

25. Eichten, E., Hinchcliffe, I., Lane, K., and Quigg, C., Rev. of Modern Phys., 56, 579 (1984).

26. Grisaru, M.J., Siegel, W., and Rocek, M., Nucl. Phys. B159, 429 (1979).

27. Affleck, I., Dine, M., and Seiberg, N., Nucl. Phys. B241, 493 (1984); Phys. Rev. Lett. 52, 1677 (1984).

28. Affleck, I., Dine, M. and Seiberg, N., Nucl. Phys. B256, 557 (1985).

29. Weinberg, S., "Gravitation and Cosmology" (Wiley, New York, 1972).
Misner, C.W., Thorne, K.S., and Wheeler, J.A., "Gravitation" (Freeman, San Francisco, 1973).

30. Gross, D.J. and Witten, E., to appear.

31. Hosotani, Y., University of Pensylvania preprint (1983).

32. Witten, E., Princeton preprint (1985).

33. Dine, M., Kaplunovsky, V., Nappi, C., Mangarno, M., and Seiberg, N., Nucl. Phys. B259, 519 (1985).

34. Mohapatra, R., University of Maryland preprint (1985).

35. Dine, M. and Seiberg, N., City College preprint (1986).

SEARCHES FOR LEPTON FLAVOR VIOLATION

Douglas Bryman
TRIUMF, 4004 Wesbrook Mall, Vancouver, B.C., Canada, V6T 2A3

I. INTRODUCTION

In accordance with the standard electroweak model, all observed reactions involving leptons are consistent with the existence of separately conserved additive flavor numbers for each of the three known lepton pairs (e,ν_e), (μ,ν_μ) and (τ,ν_τ).* However, the present theoretical framework is based on only one generation and there is virtually no understanding of the observed multiplicity of lepton flavors or the relationship between the lepton pairs. Searches for processes which exhibit lepton flavor violation (LFV) such as

$$\mu \rightarrow e\gamma$$
$$\mu \rightarrow eee$$
$$\mu^- + nucleus \rightarrow e^- + nucleus$$
$$K_L^0 \rightarrow \mu e$$
$$K^+ \rightarrow \pi\mu e$$
$$\pi^0 \rightarrow \mu e$$
$$\tau \rightarrow e\gamma \text{ and}$$
$$\tau \rightarrow \mu\gamma,$$

may provide new information or new constraints on the development of alternate approaches to the standard model. Recently, unprecedented levels of sensitivity have been achieved in experiments looking for evidence of LFV.

Direct searches for LFV are related to other quests for hypothetical new processes. These processes include neutrinoless double beta decay which involves absolute lepton number (L) violation ($\Delta L=2$), proton decay which in grand unified theories conserves B-L with $\Delta B=\Delta L=1$, where B is baryon number, neutrino oscillations which could occur if neutrinos are massive and if weak eigenstates differ from mass eigenstates (small masses and large mixings) and direct heavy neutrino

---

*The existence of ν_τ has not been directly confirmed but there is strong circumstantial evidence.

production (large masses and small mixings) in, for example, π_{e2} decay.

Although flavor changing reactions occur in the quark sector through Cabibbo-Kobayashi-Maskawa mixing and although the existence of at least three generations allows the accomodation of CP violation, all mixing in the standard model is absent in the lepton sector if neutrino masses are zero. Absolute lepton flavor conservation presents a conceptual problem in attempting to define what a generation is if both quarks and leptons are included as members of the same generation. For example, this could imply different inherent mixing of quarks and leptons. Even if the neutrinos had masses comparable to their present experimental upper limits, all LFV reactions (except, perhaps, some neutrino oscillations) would have unobservably small rates.

Instead of being due to an unbroken underlying symmetry as would be the case for a strict conservation law, the suppression of LFV processes could be the result of mechanisms which arise in various extensions of the standard model. Table 1 lists some theoretical approaches presently under discussion along with the associated particles which would generate LFV interactions. Fig. 1 shows some

Table 1

Some Theoretical Extensions to the Standard Model

Model	Flavor Violator-
Additional generations	Heavy Neutral Leptons
L-R symmetric models	Majorana particles
Extended Higgs sector	H scalar
Horizontal gauge models	Gauge bosons, leptoquarks
Extended technicolor	Gauge bosons (exotic bound states), leptoquarks
Supersymmetry	Scalar partners of fermions
Substructure models	Leptoquarks, gauge bosons
Family symmetry	familon

related diagrams. In minimal models where additional massive neutral
leptons arise[1]), such as left-right symmetric theories, LFV reactions
could occur through a leptonic G.I.M. mechanism.[2]) If a local gauge
theory were responsible for generation splitting, then horizontal gauge
bosons,[3]) additional Higgs scalars[4,5]) or Goldstone bosons might
mediate LFV interactions. In extended technicolor[6]) (ETC) models in
which fundamental Higgs scalars are dispensed with, LFV processes are
difficult to suppress. In supersymmetric[7]) theories, the bosonic
partners to fermions could generate LFV decays. Finally, in substructure models[8]) in which constituent rearrangement leads to bosons with
leptoquark quantum numbers, LFV observables also occur. As will be
indicated below, present results imply that mass scales up to hundreds
of TeV could be involved if LFV processes were to exist. It is also
important to note that the rates for LFV processes generally are dependent on a factor m_H^{-4}, where m_H is the mass of the exchanged particle.

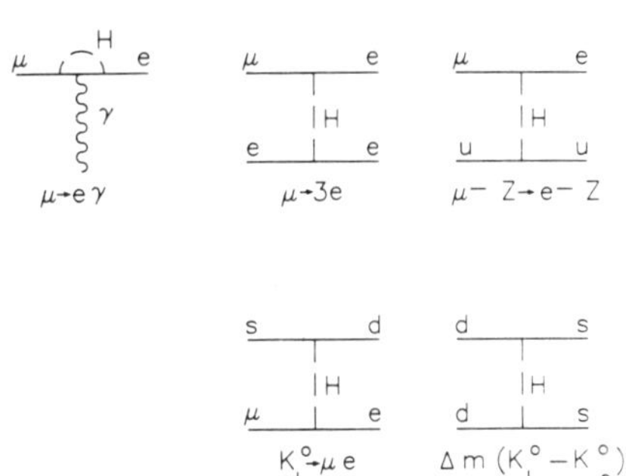

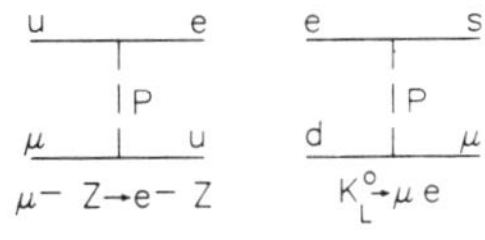

Fig. 1

Diagrams for some LFV processes a) Higgs exchange, b)
leptoquark exchange (from Ref. 5)

Although at some level LFV is a common feature in approaches which go beyond the standard model, absolute predictions are few. Nevertheless, particular theories tend to favor individual or classes of LFV reactions, making searches for a wide variety of processes desirable. When examining the usefulness of the various processes in searching for new effects, the experimental situation and the prospects for improved sensitivity are of primary importance. In the following, the present status of a selection of important LFV processes mainly involving muon number volation will be briefly discussed, with an attempt to indicate the special attractiveness of each.

II. LEPTON FLAVOR VIOLATING PROCESSES

a) $\mu \to e\gamma$

Historically $\mu \to e\gamma$ was the first LFV process to be searched for and it remains one of the most interesting because of its simplicity. Experiments at the meson factories have continued to reduce the limits for $\mu \to e\gamma$ leading to the best published value[9])*of

$$R_\gamma = \frac{\Gamma(\mu^+ \to e^+\gamma)}{\Gamma(\mu^+ \to e^+\nu\bar{\nu})} < 1.7 \times 10^{-10}. \tag{1}$$

In a recent effort,[10]) the group at LAMPF has improved on their earlier work using the crystal box shown in Fig. 2. Produced by $\pi^+ \to \mu^+\nu_\mu$ decay from the edge layer of the production target a 26 MeV/c 'surface' muon beam of average intensity 3×10^5 s^{-1} was stoppped in a target surrounded by a cylindrical drift chamber which in turn was surrounded by the crystal box which consisted of ~12 L_0 of NaI segmented into 336 crystals. Positron tracks were measured in the drift chamber and the photon conversions were detected in the NaI crystals. An energy resolution of 8% (FWHM) at 55 MeV and a photon conversion position resolution of $\sigma \sim 0.7$ cm, crucial to traceback of the event origin, were obtained. The main background came from random coincidences of positrons from the upper edge of the normal $\mu^+ \to e^+\nu\bar{\nu}$ decay with photons from the radiative decays $\mu^+ \to e^+\nu\bar{\nu}\gamma$ of other muons. Potentially, there is also a resolution limited intrinsic

---

*All limits quoted will be at the 90% confidence level.

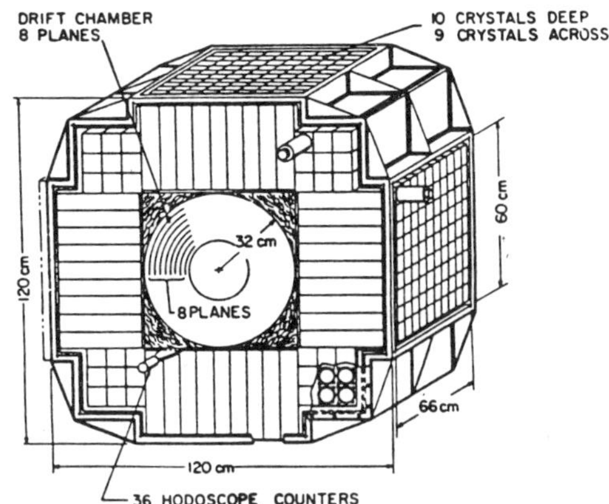

Fig. 2
Crystal box detector used in the search for $\mu \to e\gamma$ (Ref. 10)

background from the extreme phase space limits of the radiative decay. Based on a maximum likelihood analysis, a new upper limit of $R_\gamma < 4.9 \times 10^{-11}$ has been reported.[10] This experiment will result in an improved search for the process $\mu \to e\gamma\gamma$, as well as in a limit for the decay $\mu^+ \to e^+e^+e^-$.

A major new experiment MEGA[11] has been proposed at LAMPF to continue the search for $\mu \to e\gamma$. It aims to reach the level $R \leqslant 10^{-13}$.* It would consist of a large magnetic spectrometer housed in a 1.8 m diameter $\times$ 2 m long superconducting solenoid (the former 'LASS' magnet at SLAC) operated at 1.5 T. The decay positrons from muons incident at an average rate $>10^7$ s^{-1} ($>10^8$ s^{-1} instantaneous rate due to the 7% duty factor) would be confined by the field and detected in the inner region by cylindrical wire chambers. The photons would be detected in cylindrical sets of pair spectrometers surrounding the target and inner chambers, allowing accurate vertex reconstruction and the good energy resolution required for a high level of background rejection. The aim is to achieve a background free experiment not limited by the background statistics of previous efforts.

b) $\mu \to 3e$

---

*A letter of intent to study $\mu \to e\gamma$ using a similar technique has also been submitted to SIN by H.K. Walter et al.

$\mu \to 3e$ is generally a radiative correction to $\mu \to e\gamma$, although in some models the rate for $\mu \to 3e$ is larger. In 1984 the SINDRUM group at SIN reported the branching ratio upper limit[12]

$$\frac{\Gamma(\mu \to 3e)}{\Gamma(\mu \to e\nu\bar{\nu})} < 2.4 \times 10^{-12}. \tag{2}$$

The experiment shown in Fig. 3 used a surface μ^+ beam of momentum 28 MeV/c stopped in a thin hollow double-cone target at a rate of 10^7 s^{-1}. A solenoidal field of 3.3 kG contained the electron tracking apparatus, which consisted of a 64 element scintillation counter telescope and 5 cylindrical multiwire proportional chambers. The geometrical acceptance for $\mu \to 3e$ was 24%. Also, the branching ratio for the allowed process $\mu^+ \to 3e\nu_e\bar{\nu}_\mu$ was measured to be[12]

$$\frac{\Gamma(\mu \to 3e2\nu)}{\Gamma(\mu \to e\nu\bar{\nu})} = (3.2 \pm 0.5) \cdot 10^{-5}, \tag{3}$$

in good agreement with theoretical expectations.

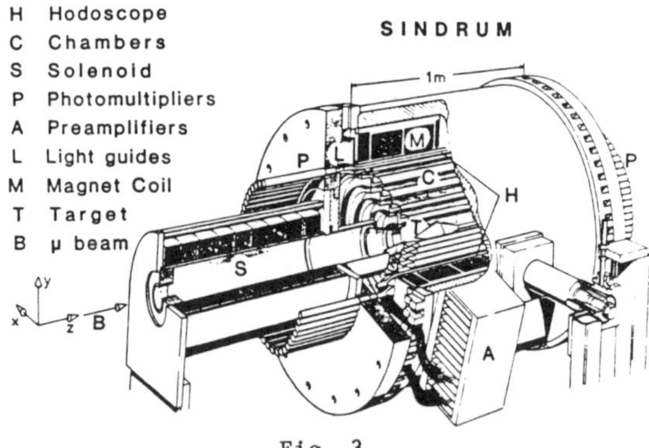

Fig. 3

SINDRUM detector (Ref. 12)

The SINDRUM group is continuing to run with an improved system and higher rates. A final sensitivety $<10^{-12}$ is expected in the future.

c) μe conversion

Muon-electron (μe) conversion in the field of a nucleus

$$\mu^- + \text{nucleus} \to e^- + \text{nucleus}$$

is an attractive experimental reaction with which to search for LFV.
Coherent action of the nuclear quarks leads to a unique signature for
μe conversion, in which the nucleus is left in its ground state and an
electron is emitted at a unique energy $E_e \sim m_\mu c^2 - b \approx 104$ MeV/c, where b is
the binding energy of the muonic atom. Incoherent μe conversion
leading to excited nuclear states is suppressed by an order of magnitude by Pauli blocking. The potential backgrounds are the small upper
tail in the electron energy spectrum from bound μ decay $\mu_B^- \to e^- \nu_e \bar\nu_\mu$,
asymmetric conversion following radiative muon capture, pion contamin-
ations in the muon beam, and cosmic rays. Previous limits on the
branching ratio for μe conversion were reported for Cu[13]) and
S[14]) targets as $R_{\mu e}(Cu) < 1.6 \times 10^{-8}$ and $R_{\mu e}(S) < 7 \times 10^{-11}$.

An experiment was performed recently using the TRIUMF time projection chamber (TPC) magnetic spectrometer system shown in Fig. 4.[15,16,17])

Fig. 4

TRIUMF TPC used to search for
μ⁻Ti → e⁻+Ti. The numbered
elements are: 1) the magnet
iron, 2) the coil, 3a) and
3b) outer trigger trigger
scintillators, 4) outer
trigger proportional
counters, 5) end cap support
frame, 6) inner electric
field cage wires, 7) central
high voltage plane, 8) outer
electric field cage wires,9)
inner trigger scintillators,
10) inner trigger cylindrical
proportional wire chamber,
and 11) end cap proportional
wire modules for track
detection.

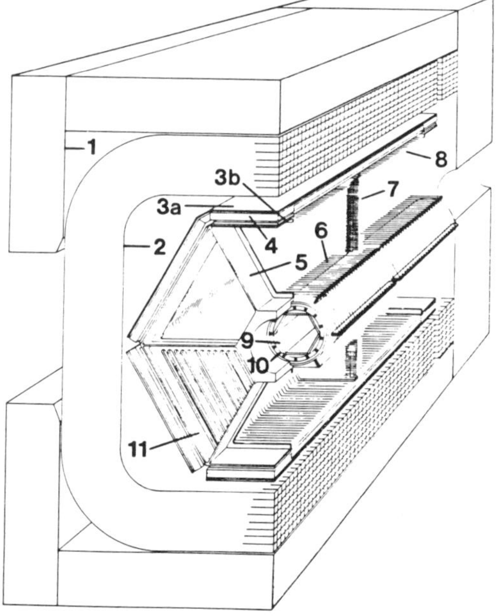

A 73 MeV/c μ^- beam was stopped at a rate 10^6 s^{-1} in a natural Ti target. Pions and electrons were reduced to the level $\pi/\mu \sim 10^{-4}$ and $e/\mu \sim 10^{-2}$ with an rf particle separator.[18]

Events firing both inner and outer trigger detectors were accepted during a 900 ns wide time gate following a muon stop in the target. The magnetic field of 9 kG prevented most $\mu \rightarrow e\nu\bar{\nu}$ decay electrons from reaching the outer trigger counters. Backgrounds due to cosmic rays and beam pions were identified by external counters and beam scintillators, respectively.

The decay $\pi^+ \rightarrow e^+\nu$ ($E_e \sim 70$ MeV/c) was used to monitor the efficiency and resolution of the system. The acceptance for $\pi \rightarrow e\nu$ decay was approximately 11%. The expected resolution for coherent μe conversion was determined by Monte Carlo simulation to be 2.3 MeV/c (rms).

No events were found in the momentum region 97 to 106 MeV/c, which was expected to contain 86% of the μe conversion peak centered at 101 MeV/c (after energy loss). A limit on R(Ti) was determined from Poisson statistics as

$$R(Ti) \equiv \frac{\Gamma(\mu^- + Ti \rightarrow e^- + Ti)}{\Gamma(\mu^- + Ti \rightarrow capture)} < \frac{2.3}{\Sigma N_i \Omega_i} ,$$

giving the <u>preliminary</u> result

$$R(Ti) < 6 \times 10^{-12}, \tag{4}$$

where N_i is the number of μ captures in run period i and Ω_i is the efficiency. The average efficiency determined with preliminary cuts was $\bar{\Omega} \sim 5\%$ and $\Sigma_i N_i \cong 9 \times 10^{12}$ μ captures.

What are the prospects for further improvement? The TRIUMF experiment was limited primarily by the flux of the incident μ^- beam. The potential backgrounds arising from bound μ decay, from radiative processes such as radiative muon and pion capture, and from cosmic rays can be suppressed further by modestly improving the momentum resolution and by reducing the beam pion contamination to an effective level of order 10^{-10}. Such a proposal has recently been discussed by a group[19] at SIN for the improved SIN machine, which could provide a low energy

μ^- beam of $>10^7$ s^{-1}. If proceeded with, the new experiment could lead to a sensitivity $<10^{-13}$ later in this decade. Further improvement might come along with another order of magnitude in μ^- beam intensity at the proposed 30 GeV KAON Factory.[20]

d) $\mu \to eX^0$

Flavor violation could occur in reactions involving the emission of real particles such as axions, Higgs particles or familons as discussed by Wilczek.[21] Such processes include $K^+ \to \pi^+X^0$ and $\mu^+ \to e^+X^0$, where X^0 denotes a neutral, weakly interacting scalar or pseudoscalar boson. Experiments at KEK[22] have set limits

$$B(K^+ \to \pi^+X_0^0) < 3.8 \times 10^{-8} \tag{5}$$

for a zero mass X_0^0 and $B(K^+ \to \pi^+X_h^0) < 10^{-5}$ to 10^{-6} for X_h^0 in the mass range 15 to 280 MeV[23], exclusive of the region around $m_{X^0} = m_{\pi^0}$.

The decay $\mu^+ \to eX^0$ involves a purely leptonic current and provides complementary information to that obtained in K decay. A by-product of the Berkeley-Northwestern-TRIUMF experiment which searched for right handed currents in $\mu^+ \to e^+\nu_e\bar{\nu}_\mu$ decay[24] allowed a limit to be obtained for $\mu \to eX_0^0$

$$\frac{\Gamma(\mu \to eX_0^0)}{\Gamma(\mu \to e\nu\bar{\nu})} < 2.6 \times 10^{-6},$$

based on the non-observation of a peak at the kinematic maximum in the polarized muon decay spectrum shown in Fig. 5. This limit applies to an interaction which would produce an isotropic angular distribution for $\mu \to eX_0^0$ with respect to the muon polarization direction.

Another experiment at TRIUMF[25] has been re-examined for evidence of the decay $\mu \to eX_h^0$ for $30 < M_{X_h^0} < 90$ MeV. The $\mu^+ \to e^+\nu\bar{\nu}$ positron spectrum as shown in Fig. 6(a), was used as normalization in the measurement of the $\pi \to e\nu/\pi \to \mu\nu$ branching ratio. This spectrum has been searched looking for evidence of peaks due to $\mu \to eX_h^0$, with the resulting upper limits shown in Fig. 6(b). Other muon decay experiments are presently being investigated for similar effects.

Heavy scalars, such as axions or Higgs particles, might be expected to decay to observable particles, such as $X^0 \to e^+e^-$ or $X^0 \to \gamma\gamma$. Prompt decays would not invalidate the experimental limits indicated

Fig. 5

Measured positron spectra from muon decay a) Spin-processing mode and b) Spin-holding mode (from Ref. 24)

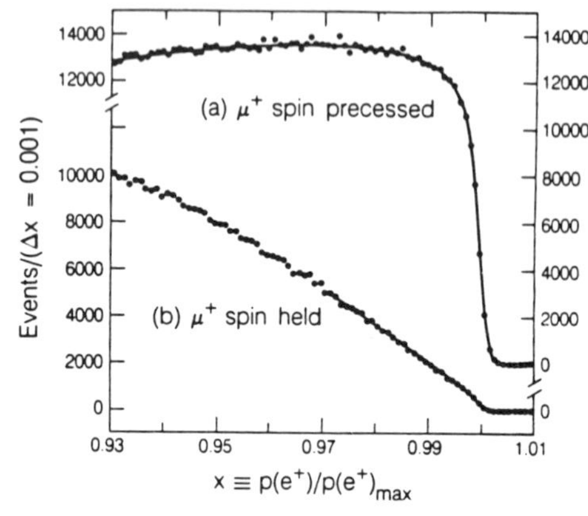

Fig. 6a and 6b

a) $\mu \to e\nu\bar{\nu}$ spectrum from Ref. 25, b) Limits on $\mu \to eX_h^0$ peak area (left scale) and branching ratio (right scale) vs $M_{X_h^0}$.

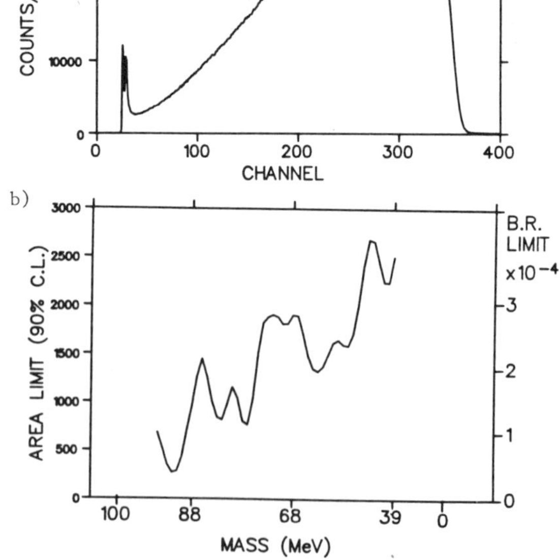

above, with the exception of Eq. (5), since only in this study were events with accompanying particles eliminated.

e) $K_L^O \to \mu e$

The best published limit for $K_L^O \to \mu e$ ($R_{K\mu e} = \Gamma(K_L^O \to \mu e)/\Gamma(K \to all) < 1.6 \times 10^{-9}$) comes from Clark et al.[26], who failed to observe $K_L^O \to \mu\mu$ which occurs at the 10^{-8} level. The previous limit was $R_{K\mu e} < 6 \times 10^{-6}$.[27] Here, $R_{K\mu e} < 10^{-8}$ will be used, under the assumption that this level could likely have been accessible in Ref. 26. Thus, $K_L^O \to \mu e$ is a process for which a considerable gain in sensitivity may be possible using modern techniques.

Three groups are taking advantage of this situation by proposing searches for $K_L^O \to \mu e$ at branching ratio levels 10^{-10} to 10^{-12}. $K_L^O \to \mu^+\mu$ and $K_L^O \to e^+e^-$ will be studied also. Set-ups for two experiments at the AGS of Brookhaven National Laboratory E780[28] and E791[29] are shown in Fig. 7 and Fig. 8. E780, aimed at the sensitiviry $<10^{-10}$, is

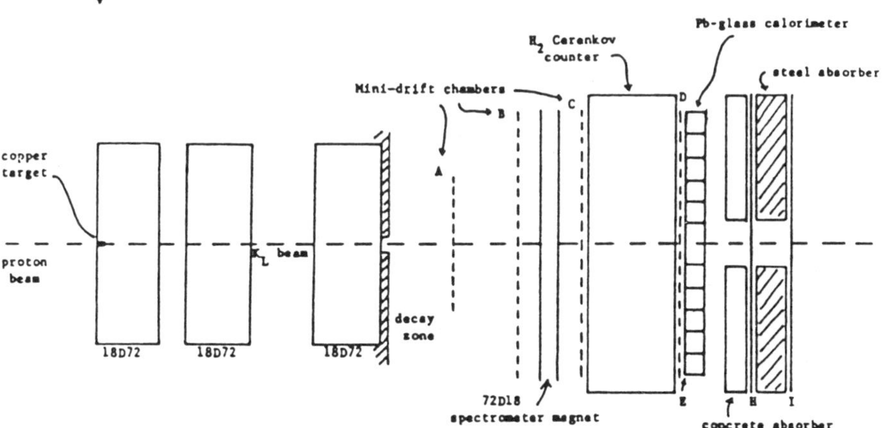

Fig. 7.

Set-up of search for $K_L^O \to \mu e$ for E780 at BNL

Table 2 $K_L \to \mu e$ Experiments (from Ref. 30)

	BNL-E780	BNL-E791	[KEK]
Proton beam momentum	30 GeV/c	28.5 GeV/c	13 GeV/c
Proton beam intensity	5×10^{11} ppp	1×10^{13} ppp	1×10^{12} ppp
K_L production angle	0°	0°	0°
Beam solid angle	15 μstr	100 μstr	314 μstr
K_L momentum range	4 - 12 GeV/c	4 - 20 GeV/c	2 - 8 GeV/c
Beam line length	8 m	10 m	10 m
Decay volume	Helium 2 m	vacuum 8 m	vacuum 10 m
Spectrometer	single-arm single-stage	single-arm double-stage	two-arm double-stage
P_t kick	200 MeV/c	±300 MeV/c	119 MeV/c×2
Chambers	16 layers	20 layers	26 layers
Position resolution	144 μm	100 μm	200 μm
Mass resolution	1.6 MeV	1.2 - 1.6 MeV	1.5 MeV
K_L production yield	1.6×10^7/pulse	4.7×10^8/pulse	3.0×10^7/pulse
Decay rate	0.01	0.03	0.08
K_L decay yield	1.6×10^5/pulse	1.4×10^7/pulse	2.3×10^6/pulse
Geometrical acceptance		0.06	0.03
Event selection eff.	0.06	0.85	0.8
Reconstruction eff.		1.0	0.7
Machine time	700 hours	2000 hours	2400 hours
Total K_L decay acc.	10^{10}	2×10^{12}	10^{11}
Sensitivity	10^{-10}	5×10^{-13}	10^{-11}

approaching the data taking stage and E791, an even more ambitious
the data taking stage and E791, an even more ambitious endeavour with a
goal 10^{-12}, will begin in late 1986. Fig. 9 is a diagram of the experiment proposed at KEK[30]). It will reach for the 10^{-11} level.

The ultimate background for $K_L^0 \to \mu e$ is expected to arise from

$K_L^0 \rightarrow \pi^\pm e^\mp \nu$ followed by $\pi^+ \rightarrow \mu^+ \nu$, in which the two neutrinos have minimal lab energy. The three approaches are similar in that the K_L^0 decay zone is followed by ultra-thin tracking chambers, analyzing magnets (one for E780 and two for E791 and KEK), Cerenkov counters for particle identification, Pb-glass electron detectors and redundant massive muon energy detectors. In addition, E791 and KEK have a

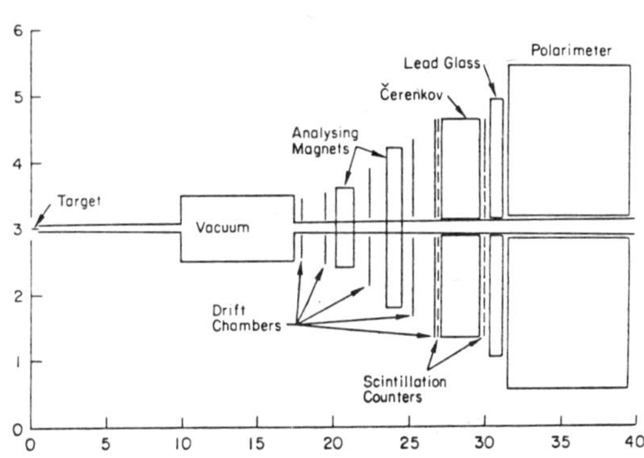

Fig. 8

Set-up for $K_L^0 \rightarrow \mu e$ from E791 proposal for BNL

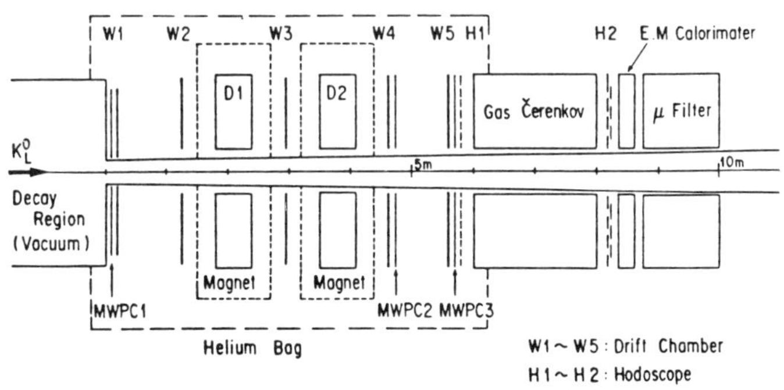

Fig. 9

Proposed apparatus for the $K_L^0 \rightarrow \mu e$ experiment at KEK (Ref. 30) central vacuum chamber and a segmented muon range-stack which is designed to obtain a muon momentum measurement limited only by range straggling. The KEK experiment involves two separate spectrometer arms. E780 also will study polarization effects in $K_L^0 \rightarrow \mu^+\mu^-$. Table 2 from Ref. 30 compares some parameters of the three experiments.

$K_L^0 \rightarrow \mu e$ is an especially attractive process with which to search for LFV, since it involves both quarks and leptons, has a large phase space available, and occurs at a favorable rate in many models compared to some other LFV processes, such as $K^+ \rightarrow \pi^+\mu^+e^-$.* The hadronic current for $K_L^0 \rightarrow \mu e$ must be either axial-vector or pseudoscalar unless the process is mediated by leptoquarks. $K_L^0 \rightarrow \mu e$ could also occur by means of constituent rearrangement in some substructure models.

f) $K^+ \rightarrow \pi^+\mu^+e^-$

Even if $K_L^0 \rightarrow \mu e$ is absent, $K^+ \rightarrow \pi^+\mu^{\pm}e^{\mp}$ could occur since it can be generated by a vector current. It could also be the result of lepto-quark exchange or constituent rearrangement. The best published results are[31]

$$\frac{\Gamma(K^+ \rightarrow \pi^+\mu^+e^-)}{\Gamma(K^+ \rightarrow \text{all})} < 5 \times 10^{-9} \text{ and } \frac{\Gamma(K^+ \rightarrow \pi^{\pm}\mu^{\mp}e^+)}{\Gamma(K^+ \rightarrow \text{all})} < 7 \times 10^{-9}.$$

In E777[32] at BNL, it is proposed to search for $K^+ \rightarrow \pi^+\mu^+e^-$ at a level 10^{-11} using the apparatus shown in Fig. 10. The experiment which is in progress uses a 6 GeV/c unseparated K^+ beam at a rate of approximately $1.5 \cdot 10^7$ K^+/s^{-1}. Since both initial and final states are all charged, strict momentum and energy conservation constraints can be imposed. A potentially limiting background comes from $K^+ \rightarrow \pi^+\pi^+\pi^-$ followed by $\pi^+ \rightarrow \mu^+\nu_\mu$ decay and misidentification of the π^- as an e^-. Thus, considerable attention has been paid to the e^- detection arm involving gas Cerenkov counters and a segmented Pb-glass shower array. The aim is to reject negative pions at the level of 10^{-7}. The proposed acceptance for $K^+ \rightarrow \pi^+\mu^+e^-$ is 5%.

---

*$K^+ \rightarrow \pi^+\mu e$ has a reduced phase space, due to the additional pion, and the K^+ has a lifetime four times shorter than the K_L^0.

Table 3

Limits on Rare τ Decay

Decay Mode	Upper Limit on Branching Ratio (90% C.L.)	Reference
$\tau^- \to \mu\gamma$	5.5×10^{-4}	35
$e\gamma$	6.4×10^{-4}	35
$\mu\mu\mu$	4.9×10^{-4}	35
$e\mu\mu$	3.3×10^{-4}	35
μee	4.4×10^{-4}	35
eee	4.0×10^{-4}	35
$\mu^-\pi^0$	8.2×10^{-4}	35
$e^-\pi^0$	21.0×10^{-4}	35
μ^-K^0	10.0×10^{-4}	35
e^-K^0	13.0×10^{-4}	35
$\mu^-\rho$	4.4×10^{-4}	35
$e^-\rho$	3.7×10^{-4}	35
μX^0_0	0.02	36
$e X^0_0$	0.007	36

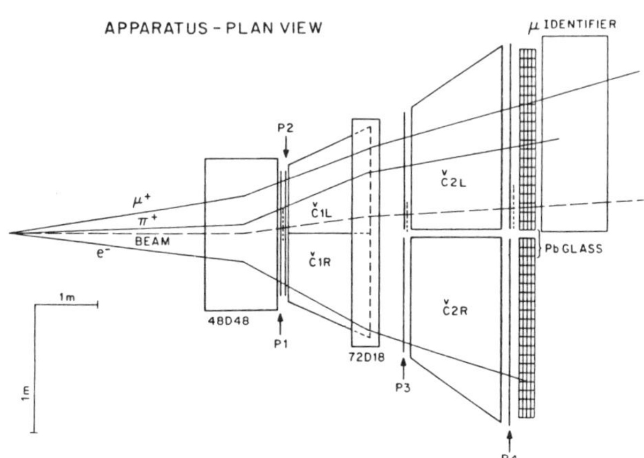

Fig. 10 Set-up for $K^+ \to \pi^+\mu^+e^-$ search in E777 at BNL

This experiment will also study some other interesting processes, including $K^+ \rightarrow \pi^+ e^+ e^-$, $\pi^0 \rightarrow e^+ e^-$, and $\pi^0 \rightarrow \mu e$. $\pi^0 \rightarrow \mu e$ is a LFV process for which the present branching ratio limit[33] is $B(\pi^0 \rightarrow \mu e) < 7 \times 10^{-8}$. It can be generated by the same interactions as incoherent $\mu^- \rightarrow e^-$ conversion, which is expected to provide more stringent limits on any new interactions[33,34].

g)   Rare τ Decays

The limits on rare τ decays, listed in Table 3, come from the SLAC MARK II[35] group at the SPEAR $e^+ e^-$ collider working in the 3.8–6.9 GeV range. Tagged τ's were produced in $e^+ e^- \rightarrow \tau^+ \tau^-$ reactions. The limits on decays to known particles, reported in 1980, are in the 10^{-3} to 10^{-4} range. Some improvement may be expected from experiments at the upgraded PEP ring at SLAC. A particularly interesting decay is is $\tau^- \rightarrow K^0 \mu^-$, which has a net $\Delta G = 0$ with generation number change of -1 for the leptons and $+1$ for the quarks.

Recently, the Mark III group has reported[36] the first limits on decays $\tau \rightarrow \mu X^0$ and $\tau \rightarrow e X^0$ for $M_{X^0} < 0.1$ GeV. The limits are shown in Table 3. These results allow limits $M > 3 \times 10^6$ GeV to be established for the symmetry breaking scale for a spontaneously broken global family symmetry[37] for leptons based on $U(1)_{e-\mu+\tau}$ or $U(1)_{e+\mu-\tau}$. Analysis of existing data might also yield interesting results on $\tau \rightarrow e(\mu) X^0_h$ for $0.1 < M_{X^0_h} < 1.8$ GeV.

h) $M^0 \rightarrow \mu e$

Decays $M^0 \rightarrow \mu e$, where $M^0 = D^0$ or B^0, might prove to be significant because of the speculation that there could be larger couplings of exotic LFV mediating particles to heavy quarks. M. Halling of the CLEO group at Cornell has reported[38]

$$\frac{\Gamma(B^0 \rightarrow \mu e)}{\Gamma(B^0 \rightarrow \text{all})} < 2 \times 10^{-4}.$$

It was futher suggested by J. Ellis that $D^0 \rightarrow \mu e$ (for which there are no known limits) might also be worth investigating. Another process

Table 4

Experimental Searches for Lepton Flavor Violation

Process	Present Upper Limit on Branching Ratio (90% C.L.)	Reference	Proposed Sensitivity (Date)
$\mu \to e\gamma$	4.9×10^{-11}	10	10^{-13} (1989)
$\mu \to 3e$	2.4×10^{-12}	12	10^{-12} (1987)
$\mu \to e\gamma\gamma$	3.8×10^{-10}	10	
$\mu \to eX_0^0$	2.6×10^{-6}	24	
$\mu^- Z \to e^- Z$	6.0×10^{-12}	15	10^{-13} (1990)
$K_L^0 \to \mu e$	10^{-8}	(see text)	10^{-10} (1987)
			10^{-11}
			10^{-12} (1988)
$K^+ \to \pi^+ \mu^+ e^-$	5.0×10^{-9}	31	10^{-11} (1987)
$\pi^0 \to \mu e$	7.0×10^{-8}	33	10^{-9} (1987)
$\tau \to \mu\gamma$	5.5×10^{-4}	35	
$\tau \to \mu X_0^0$	0.02	36	
$B^0 \to \mu e$	2.0×10^{-4}	38	

that has been discussed[39]) is $Z^0 \to \tau\mu$, although the stringent limits from $\mu \to 3e$ would probably lead to unobservable levels, even at the Z^0 factories.

III. DISCUSSION

Table 4 summarizes the status of various searches for lepton flavor violation and indicates the presently proposed sensitivity for new experiments and their estimated completion dates.

Calculations of the rates for LFV processes have been done in a wide variety of models. Firm predictions of absolute rates are rare, due to unknown couplings and masses of the hypothetical exchange particles which transmit the LFV interactions. However, comparison of the rates of LFV reactions will be important clues to their origin if such interactions are eventually observed. At this time, while only upper limits are known, one can at best examine the constraints on

masses and couplings from the most sensitive processes in each type of
model. LFV reactions could occur at observable rates if additional
neutral massive leptons exist. Altarelli et al.[40]) put forward a model
representative of those involving the exchange of a hypothetical
massive neutral lepton ν_1 with mixings U_{ei} and $U_{\mu i}$ to electrons and
muons, respectively. This model involved only left-handed interactions. All the relevant diagrams involving neutral and charged
currents (box diagrams) were evaluated for the processes $\mu \to e\gamma$, $\mu \to 3e$
and $\mu^- Z \to e^- Z$. The results of the calculations along with the experimental limits lead to constraints on the products of the mixing
coefficients $|U_{ei}U_{\mu i}|$ as a function of m_{ν_1}, the mass of ν_1, as shown in
Fig. 11[41]). The coherence effect in μe conversion leads to its high

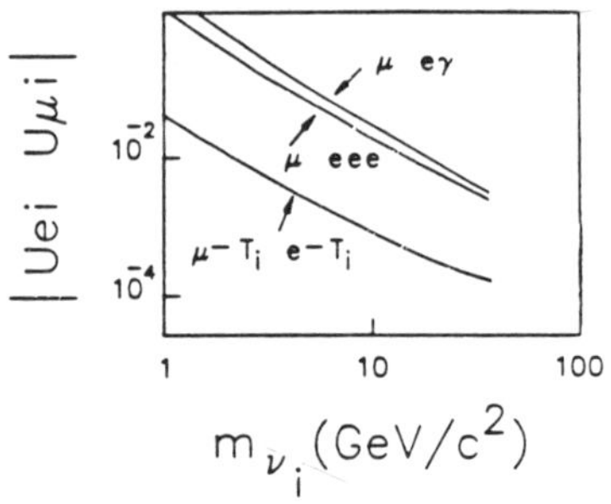

Fig. 11

Limits on e,μ mixing $|U_{ei}U_{\mu i}|$ with a heavy neutral lepton ν_1
vs m_{ν_1}

sensitivity. It should be noted that rare K decays are not expected to
provide important constraints on the mixing coefficients U_{ei}, because
of G.I.M. suppression.

Shanker[5]) and subsequently, Ellis[42]) examined models which
involved LFV interactions of additional Higgs scalars (beyond those
required in the standard model) and of leptoquarks. The first column
in Table 5 gives the mass limits on new Higgs particles derived from
the present branching ratio limits of various processes (see Table 4).
The extreme model dependence of such limits is illustrated by the fact
that Shanker and Ellis chose mass scales that differed by 10^3 depending

Table 5   Lower Bounds on Masses from Various Processes based on (Refs. 5 and 42)

	Higgs Scalars (GeV (Ref.42)/ TeV (Ref.5))	Pseudoscalar Leptoquarks (TeV)	Vector Leptoquarks (TeV)
$\mu \to e\gamma$	0.3	---	---
$\mu \to 3e$	2.0	---	---
$\mu^-Ti \to e^-Ti$	23.0	22.0	120.0
$K_L^0 \to \mu e$	5.0	3.5	62.0
$K_L^0 \to \mu^+\mu^-$	5.0	4.0	62.0
$K_L^0 \to e^+e^-$	2.0	2.0	30.0
$K^+ \to \pi\mu e$	1.0	0.3	4.0
$\Delta m(K_L - K_s)$	150.0	---	---

on assumptions made about Higgs-lepton couplings. Nevertheless, it is clear that the potential contribution to the $K_L^0 - K_S^0$ mass difference of a flavor changing Higgs exchange is highly significant and provides the best limits. It is possible that this latter contribution is suppressed somehow and that the other processes listed will proceed at a higher rate. In that case, μe conversion would give the best constraint at present, whereas if the proposed 10^{-12} level is reached for $K_L^0 \to \mu e$, another factor of 2 could be achieved. An earlier model by Bjorken and Weinberg[4]) considered two Higgs doublets and very weak Higgs-lepton coupling leading to the dominance of higher order effects.

Shanker and Ellis also considered what limits could be obtained on the masses of pseudoscalar and vector leptoquarks. Leptoquarks, which can initiate lepton to quark (or vice versa) transitions, arise in technicolor[6]) and other theories. μe conversion gives the best present limits $M_{PLQ} > 22$ TeV and $M_{VLQ} > 120$ TeV for pseudoscalar and vector leptoquarks, respectively. Rare kaon decays represent the most promising future improvement.

In another approach, Hall and Randall[44]) considered a model incor-
porating spontaneous CP violation which is transmitted by scalar leptoquarks (SL). Here again, the results on μe conversion provide the best
present constraints $M_{SL} \gtrsim 50$ TeV.

Next, limits on the masses of horizontal gauge bosons are
considered in light of a general discussion presented by Cahn and
Harari.[3]) They entertained the hypothesis that the generation
pattern could be the result of some underlying symmetry and they arbi-
trarily assigned generation numbers $G = $ 1, 2, and 3 to (e, ν_e, u, d),
(μ, ν_μ, c, s) and (τ, ν_τ, t, b) groups respectively. Then, processes
like $K_L^0 \to \mu e$, $K^+ \to \pi\mu e$ and $\tau \to K\mu$ have no net generation number change
(i.e. $\Delta G = 0$). $\mu \to e\gamma$, $\mu \to 3e$ and $\mu^- Z \to e^- Z$ are $\Delta G = 1$ processes, etc.
Thus, for example, if $\Delta G = 0$ processes were favored, a prime process
would be $K_L^0 \to \mu e$. With the assumption $g_h \sim g_\omega$, where g_h and g_ω are the
couplings of the horizontal gauge bosons and the standard model weak
coupling, and using the updated experimental results (Table 4), Table 6
gives the constraints on horizontal gauge boson masses. β_L, β_{LQ} and β_Q
are lepton (L = e or μ) and quark (Q = u or d) mixing angles and
Δ is a mass parameter related to gauge boson masses (see Ref. 3) which
vanishes if all horizontal boson masses are degenerate. The processes
$K_L^0 \to \mu e$ and $K^+ \to \pi\mu e$ give reasonably unambiguous and comparable limits.
Unless β_{LU} and β_{LD} are extremely small or conspire to be equal, μe
conversion is extremely sensitive to the presence of horizontal gauge
bosons. The contribution to the $K_L^0 - K_S^0$ mass difference dominates the
constraints on Δ, although the usual caveat regarding mixing angles
obviously applies here.

The absence of flavor changing radiative processes like $\mu \to e\gamma$ and
$\mu \to e\gamma\gamma$ has also been used to obtain limits on the scale of possible
lepton substructure. For instance, using the approach of Tomozawa[44]),
the combined limits on $\mu \to e\gamma$ and $\mu \to e\gamma\gamma$ lead to a lower limit on the
substructure scale of $\Lambda > 5\times10^5$ TeV, assuming the matrix element is
proportional to $\left(\frac{m_\mu}{\Lambda}\right)\alpha$. However, as discussed by Lyons,[8]) such
estimates are very model dependent and rely on the assumptions that LFV

Table 6 Limits on Horizontal Boson Masses Based on the Work of Cahn and Harari[3]) Assuming $g_H \sim g_\omega$

Process	m_H(TeV)
$K_L^0 \to \mu e$	18
$K^+ \to \pi\mu e$	18
$\mu^- Ti \to e^- Ti$	$200\left|\sin\beta_{LU} - \sin\beta_{LD}\right|^{1/2}$
	Λ (TeV)
$\mu \to e\gamma$	$15\left|\sin\beta_L \cos\beta_L\right|^{1/2}$
$\mu \to 3e$	$70\left|\sin\beta_L \cos\beta_L\right|^{1/2}$
$\Delta m(K_L - K_S)$	$400\left|\sin\beta_D\right|$

processes are not forbidden or suppressed by a G.I.M. type mechanism, small mixing angles, etc. For example, if the matrix element for the the transition were proportional to $\left(\frac{m_\mu}{\Lambda}\right)^2 \alpha$ or $\left(\frac{m_e m_\mu}{\Lambda^2}\right)\alpha$ rather than $\left(\frac{m_\mu}{\Lambda}\right)\alpha$ used in Ref. 44, then the scale limit would become $\Lambda > 500$ TeV or $\Lambda > 30$ TeV. Similar limits on Λ could be obtained from other LFV processes, such as $\mu^- Z \to e^- Z$ and $K_L^0 \to \mu e$ with additional assumptions covering the relationships of quark and lepton subconstituents.

IV. CONCLUSION

As illustrated in Table 4, the search for lepton flavor violation has reached considerable sensitivity, but with only null results so far. The experiments are sensitive to new particles in the 1 to 100 TeV range arising in a variety of theories, although the constraints on the masses of such particles improve only as the inverse fourth power of the branching ratios. Presently, neutrinoless μe conversion in the field of a nucleus provides the most serious constraints for many models. New experiments on rare kaon decays, μe conversion and $\mu \to e\gamma$ will result in improved sensitivity in the next few years. Ignoring theoretical prejudice, it is important to study many different processes in the hope of uncovering some new effects.

374

REFERENCES

[1] Riazuddin, in Weak Interactions as Probes of Unification, G.B. Collins et al. ed., American Inst. of Physics Conference Proceedings 72, (AIP, New York, 1981) P.21

[2] Marciano, W. and Sanda, A.J., Phys. Rev. Lett. 38, 1512 (1977)

[3] Cahn, R.N. and Harari, H., Nucl. Phys. B176, 135 (1980)

[4] Bjorken, J.D. and Weinberg, S., Phys. Rev. Lett. 38, 622 (1977)

[5] Shanker, O., Nucl. Phys. B206, 253 (1982)

[6] Nanopoulos, S., Ellis, J., Nucl. Phys. B182, 505 (1981)

[7] Ellis, J. and Nanopoulos, D., Phys. Lett. 110B, 44 (1982)

[8] Lyons, L., Prog. Part. Nucl. Phys. 10, ed. D. Wilkinson, Oxford Pergamon Press (1983), P.227

[9] Kinnison, W.W. et al., Phys. Rev. D25, 2846 (1982)

[10] Bolton, R.D. et al., LAMPF preprint LAUR-86-446 (1986) For $\mu \to e\gamma\gamma$, Grosnick, D. et al., Proc. of Santa Fe meeting, ed. T., Goldman, M.M. Nieto (1985).

[11] Cooper, M.D. et al., LAMPF proposal 969 (1986)

[12] Walter, H.K. et al., Nucl. Phys. A434, 409c (1985)

[13] Bryman, D.A. et al., Phys. Rev. Lett. 28, 1469 (1972)

[14] Badertscher, A. et al., Nucl. Phys. A377, 406 (1982)

[15] Bryman, D.A. et al., Phys. Rev. Lett. 55, 465 (1985)

[16] Hargrove, C.K. et al., Nucl. Instr. Meth., 219, 461 (1984)

[17] Bryman, D.A. et al., Nucl. Instr. Meth., 234, 42 (1985)

[18] Blackmore, E.W. et al., Nucl. Inst. Meth., 234, 235 (1985)

[19] Badertscher, A. et al., SIN letter of intent R85-07.0

[20] TRIUMF KAON Factory Proposal (1986)

[21] Wilczek, F., Phys. Rev. Lett. 49, 1549 (1982)

[22] Asano, Y. et al., Phys. Lett. 107B, 159 (1981)

[23] Yamazaki, T. et al., Proc. XI Intl. Conf. Neutrino Physics and Astrophysics, Nordkirchen, June 11-16, 1984 - UTMSL-94.

[24] Carr, J. et al., Phys. Rev. Lett. 51, 627 (1983); Jodidio, A.E., LBL thesis (1986)

[25] Bryman, D.A., et al., Phys. Rev. D33, 1211 (1986) and Clifford, E.T.H., et al., to be published.

[26] Clark, A.R. et al., Phys. Rev. Lett. 26, 1667 (1971)

[27] Fitch, V. et al., Phys. Rev. 164, 1711 (1967)

[28] AGS experiment 780, Schmidt, M.P. and Morse, W.M. et al.

[29] AGS experiment 791, Wojcicki, S.G. et al.

[30] Inagaki, T. et al., KEK proposal 85-1 (1985)

[31] Diamant-Berger, A.M. et al., Phys. Lett. 62B, 485 (1976)

[32] AGS experiment 777, Zeller, M. et al.

[33] Bryman, D.A., Phys. Rev. D26, 2538 (1982)

[34] Herczeg, P. and Hoffman, C.M., Phys. Rev. D29, 1954 (1984)

[35] Hayes, K.G. et al., Phys. Rev. D25, 2869 (1982)

[36] Baltrusaitis, R.M. et al., Phys. Rev. Lett. 55, 1842 (1985)

[37] Grinstein, B., Preskill, J. and Wise, M.B., CALT-68-1266 (1985)

[38] Halling, M., Proc. DPF Meeting, Santa Fe, New Mexico Nov. 12, 1984

[39] Kuo, T.K. and Nakagawa, N., Phys. Rev. D32, 306 (1985)

[40] Altarelli, G. et al., Nucl. Phys. B125, 285 (1977)

[41] Numao, T., private communication.

42) Ellis, J., in Proc. IMESS Conference, AIP Conf. Proc. $\underline{102}$, 191 (1983)

43) Hall, L.J. and Randall, L.J., Harvard University preprint HUTP-86/A002.

44) Tomozawa, Y., Phys. Rev. $\underline{D25}$, 1488 (1982)

PHYSICS RESULTS FROM UNDERGROUND EXPERIMENTS

John M. LoSecco

Department of Physics
University of Notre Dame
Notre Dame, Indiana 46556
U.S.A.

Abstract

Physics studied using massive underground detectors is reviewed. Bound proton and neutron decay limits are quoted and compared. The range of interest is now $\tau \sim 10^{32}$ years. Neutron-antineutron oscillations and free proton decay results are examined. Magnetic monopole signals as observable in underground detectors are considered. The best limits do not come from massive detectors. A number of results are already well below the Parker bound of $10^{-15}/cm^2/sec/str$. Energetic muons arising from atmospheric sources or extraterrestrial sources are examined. We comment on the current status of energetic muons from Cygnus X3. Energetic muons from atmospheric neutrino interactions are considered. Limits on any excess attributable to celestial point sources are quoted. No sources are currently seen in the data. Intermediate energy neutrino interactions are examined. The observations are consistent with an atmospheric source. Limits are quoted on celestial, stellar and terrestrial sources. Less than 5.5% of the total flux can be attributed to any localized source. Expected geomagnetic modulations are indicated in the neutrino data. The consistency and uniformity of the data have been used to place limits on neutrino oscillations with baselines as long as 10^7 meters.

The future of such experiments and their capabilities is mentioned.

1 INTRODUCTION

Underground physics experiments have emerged in the last 5 years as unique tools to study a wide variety of problems in physics and astrophysics. The scale and location of these experiments was initially motivated by a search for proton decay.

From prior experiments the proton was known to live at least 10^{29} years. A number of theories suggest a value for the lifetime in the range of 10^{30} years. To study such lifetimes in a reasonable period requires detectors in the range of 1000 metric tons. A metric ton contains 6.02×10^{32} nucleons and so a lifetime of 10^{31} years would yield about 60 proton and bound neutron decays per year. Optimistically one could hope to do an experiment with less matter if the lifetime were shorter but errors in the lifetime calculation indicated that values as high as 10^{32} years would still be possible.

In conducting such a sensitive experiment one must take great care in shielding the detector from spurious signals. Such background could be mistaken for proton decay and at the very least makes the task of finding candidates time consuming because of all the background that must be rejected. In general an experiment's sensitivity is determined by its mass and by its ability to reject background. Smaller detectors with good background rejection may be competitive with larger coarser detectors. This statement is decay mode dependent since some modes such as $P \to e^+ \pi^0$ produce spectacular, unique signatures while modes with missing neutrals such as $P \to \nu K^+$ have much greater ambiguity.

To eliminate the hadronic and electromagnetic component of cosmic rays proton decay detectors are located deep underground. The surviving components of cosmic rays are very energetic muons that penetrate from the surface and neutrinos formed by cosmic ray interactions in the atmosphere that propagate unattenuated through the earth. The neutrinos form the ultimate background to proton decay.

One expects about 120 neutrino interactions in a 1000 ton detector in one year of operation. To a large extent these events are distinguishable from proton decay. Neutrinos have a momentum equal to their energy. Any interaction will conserve the momentum so any neutrino events with enough energy to be confused with proton decay ($E =938$ MeV) will have a momentum much higher than could be attributed to a nucleon at rest or with fermi motion ($P \approx 250$ MeV/c) bound in a nucleus. In practice ambiguities can arise. Protons can decay into modes with missing neutrals and neutrinos can interact and produce unobserved particles which carry off the momentum. Both of these effects imply that even in the best detector the momentum can only be approximately measured.

One of the major tasks of proton decay experimenters is to make neutrino background estimates that are consistent with known physics and the behavior of their detector.

As we shall see in section 4 below the environment required for proton decay experiments is ideal for a large number of other physics objectives.

2 CLASSES OF EXPERIMENTS

In general there have been two approaches to the above problem. One can build a very large detector with modest spatial and momentum resolution or one can construct a fine grain detector where the ultimate size is limited by costs. Both classes of detectors exist.

The most massive detectors, in excess of 1000 tons, are water Cherenkov detectors. Charged particles with $E > 1.52mc^2$ will emit Cherenkov light in water. The light is emitted in a cone along the track direction. This light can be collected on arrays of photomultipliers and the event reconstructed from the structure of the light output. The technique cannot see neutral particles or particles below the Cherenkov threshold. But photons will readily convert and produce a shower that represents the photon energy very well.

Water Cherenkov detectors have sizes of 10 to 20 meters on a side. The attenuation of light in water is not a problem on this scale. The spatial resolution is about 30 centimeters and it depends on the topology of the event. Angular resolution for long tracks traversing the detector is from 2° to 3.5° on the average.

Fine grain detectors employ flash chamber or ionization technology to measure tracks. The bulk of material comes from iron plates placed between these counters. The spatial resolution is from millimeters to 4 cm depending

Table 1: Current Proton Decay Experiments

Detector	Mass (fid.)	Location	Depth (mwe)	Exposure Kt-yr.	# Evts.
Kolar Gold Fields	60 ton Fe	3°N	7000	.22	19
Nusex (Mt. Blanc)	100 ton Fe	50°N	5000	.25	32
IMB	3300 ton H_2O	52°N	1570	3.77	401
Komiokande	880 ton H_2O	26°N	2400	.83	103
Frejus	912 ton Fe	50°N	4800	.19	13

on the counter size. To some extent the resolution is limited by the thickness of the steel plates and the effect of multiple scattering on particles of modest momentum. The angular resolutions of these devices is $< 1°$ for long tracks. In general the masses range from 100 to 912 tons.

3 CURRENT EXPERIMENTS

Table 1 summarizes some information about current underground experiments. The largest detector operating is the 3300 metric ton water IMB detector. The largest of the iron calorimeter type detectors is the 912 ton Frejus detector which has been in full operation for less than a year. Both the smaller KGF and Nusex detectors are of the iron calorimeter types and have been operating for years. These detectors gave a quick early look at the lifetime of greatest interest $\sim 10^{30}$ years but do not have the mass to be competitive with latter detectors for lifetimes on the order of $10^{31} - 10^{32}$ years. The 880 ton water Kamioka detector while smaller than the two largest detectors has better light collection than IMB and a much longer exposure at full mass than Frejus.

The depths of the detectors is listed to give a crude comparison of the energy of muons penetrating from the surface. For IMB this is $E_\mu > 400$ GeV. For Nusex this is $E_\mu > 2$ TeV.

The Geomagnetic latitude is listed since there is a small dependence of the low energy neutrino flux on the local magnetic field. The flux is lower near the equator than near the poles since low energy primaries are deflected away from the earth near the equator.

The exposures listed in the table do not reflect the total exposure of the detector. The exposure listed is the largest that has been published by the research group. This is the exposure I will try to summarize in later sections.

4 PHYSICS CAPABILITIES

These detectors are all well designed to observe a number of proton decay modes. Modes that result in only observable particles such as $P \rightarrow e^+\pi^0$ can be strongly constrained by the observations. On the other hand modes such as $P \rightarrow \nu K^+$ are a challenge to the experimenter since one must first prove that the single observed track is a K^+ and then by assuming energy and momentum conservation that the unseen recoiling track was a ν with the correct energy. One is limited by fermi motion and nuclear rescattering effects to how well one can determine the initial state of the decaying proton.

A number of physics considerations limit the sensitivity of even the well constrained modes. Many modes result in the production of energetic hadrons in a nucleus. The probability of the hadron interacting before it leaves the parent nucleus is quite high. The problem is accentuated since the momenta involved place the hadrons in the resonance region and they originate in a nucleus, a region of high nucleon density. As many as 40% in oxygen and 60% in iron may interact on leaving.

In general such distortions can not be compensated and are reflected in an efficiency for finding proton decay considerably below the instrumental efficiency. To a great extent these effects offset any major gains to be gained by making major new detector designs. A better detector will not be able to see a better signal but it may have better background rejection capabilities.

The best proton decay detector has the conflicting requirements of high mass and fine granularity while at the same time having low density to reduce particle interactions. In general all realized detectors are a compromise of these requirements.

The $\Delta B = 2$ process of neutron-antineutron oscillations yields a large number of final states following the annihilation of the antineutron. The general signature is a highly isotropic multitrack event. Since the **neutrino** background is falling with energy it is often possible to find a clean selection criterion for the $n \to \bar{n}$ signal.

Intense interest in magnetic monopoles has led to a number of searches in underground detectors. In general one looks for a massive (10^{15} GeV) slowly moving object. People have looked for slowly moving objects in scintillation counter arrays and for unusual ionization in gas counters. Proton decay catalysis in the presence of a magnetic monopole provides a direct link of proton decay experiments to magnetic monopole fluxes. Magnetic monopoles in grand unified theories are believed to accelerate proton decay. Cross sections on the order of 100's of millibarns are possible. The signature for a monopole would be multiple interactions in a short interval of time.

If monopoles do cause proton decay a number of astrophysical limits become relevant. Results on excess neutrinos from the sun or xray emission from neutron stars give stronger limits than terrestrial experiments.

The neutrinos that make up the major background to proton decay can also yield new information. In general the results from underground experiments are complementary to accelerator or reactor neutrino experiments. The atmospheric neutrinos observed underground are primarily in the 500 MeV to 2 GeV energy range. They originate from all directions and are a mixture of $\nu_\mu, \nu_e, \bar{\nu}_\mu$ and $\bar{\nu}_e$. Since the source is distributed at large distances and over a range of from 8 km to 13,000 km experiments that depend on long lived neutrinos can be performed. Neutrino oscillations can be studied with unprecedent sensitivity down to $10^{-5} eV^2$. In addition, since the source is a mixture of neutrino types one can potentially study many components of the 3×3 mixing matrix.

The matter enhanced oscillations believed to account for the solar neutrino problem[24] may also be studied. A strong effect is expected in the 150-550 MeV region. Oscillation of low energy ν_μ into ν_e would greatly increase the event rate because of the lower threshold for ν_e.

While we expect neutrinos to originate from cosmic ray interactions there are many other potential sources. Many stars that are sources of energetic photons are expected to be sources of energetic neutrinos. The photons and neutrinos are assumed to come from pion decays. One can use the underground data to search for a signal in the face of the atmospheric background. Sensitivity is improved by making directional cuts on the data and using the off source data to estimate the background.

The energy range over which neutrinos can be studied is limited to $<$ 2 GeV. Above this two factors reduce the rate. Energetic events exit the detector and so only a lower bound on the energy can be measured. In addition the flux falls to levels more than a decade below that at lower energies. Both of these problems can be circumvented by studying upward going muons. These originate from muon neutrino interactions in the rock surrounding the detector. The huge mass of rock gives a much larger target for the smaller flux and the energetic muons produced can travel through the rock and penetrate the detector. Information about the neutrino energy is lost but the neutrino direction is better known at these energies than at lower values due to smaller scattering angles. In general only upward muons are chosen to be certain of having muons from neutrino interactions and not just penetrating muons from the surface. Some of the detectors have a directional ambiguity and so they cannot tell upward moving from downward moving tracks. They keep events near the horizonal as neutrino interactions.

Surface muons themselves provide a window on multi TeV cosmic ray primaries. The earth filters out all but the most energetic muons and these can be studied for very energetic point sources. Several energetic sources are known to be pulsed or periodic and this information can be used to enhance a possible signal and to reduce background. A number of groups have reported signals using this technique.

5 PHYSICS RESULTS

All experiments in recent times[1),2),3),4),5)] that have searched for proton decay have reported signals that are compatible with such decays. The rate of occurance of such decays is about the same[7)] for all experiments. This rate corresponds to about $\tau = 10^{32}$ year for the proton lifetime. Table 2 summarizes the data. But this result is highly ambiguous. At this level neutrino backgrounds are significant. Quantitative estimates range from the total signal to 1/3 of the total signal. Many of the candidate events

Table 2: Summary of Candidates

Experiment	Exposure	Number of Candidates	τ if eff. $= 100\%$
KGf	.22 Kt - yr.	4	$3.0 \pm 1.5 \times 10^{32}$
Nusex	.25 Kt - yr.	3	$2.0 \pm 1.2 \times 10^{32}$
IMB	3.77 Kt - yr.	21	$.93 \pm .20 \times 10^{32}$
Kamiokande	.83 Kt - yr.	4	$.93 \pm .20 \times 10^{32}$
Frejus	.19 Kt - yr.	0	$> .18 \times 10^{32}$
Total	5.26 Kt - yr.	32	$1.0 \pm .18 \times 10^{32}$

have more than one interpretation and there is no consistent picture of the dominant decay mode or modes.

In general results have been reported as lifetime limits (90% c.l.) on a mode by mode basis. The major reason for this is to clarify the role of background in the problem. There are very few candidates for the decay mode $e^+\pi^0$ and none in the large detectors. Two factors are significant in interpreting the data. All experiments have an efficiency for finding a specific decay if it has occured. A number of factors contribute to the efficiency. Actual physics effects such as charge exchange and scattering can distort an event so that it no longer looks like a candidate. Beyond that fermi motion and unstable particle width smear the allowed range of momenta for the final state particles.

The other major factor is the background to a specific decay mode as seen in a given detector. While real neutrino interactions can never have the same kinematics as proton decay (since $E = P$) most detectors lose information and so some fraction of the highly inelastic neutrino events will look like proton decays. Since most detectors have poor neutral particle detection a substantial part of the background rate will be common to all detectors. But specific details of energy and track resolution can vary the ambiguous region from detector to detector.

Results are usually quoted as 90% confidence limits on the lifetime assuming all candidates to be real events, that is without background subtraction. Figure 1 shows a comparison of several experiments. The IMB

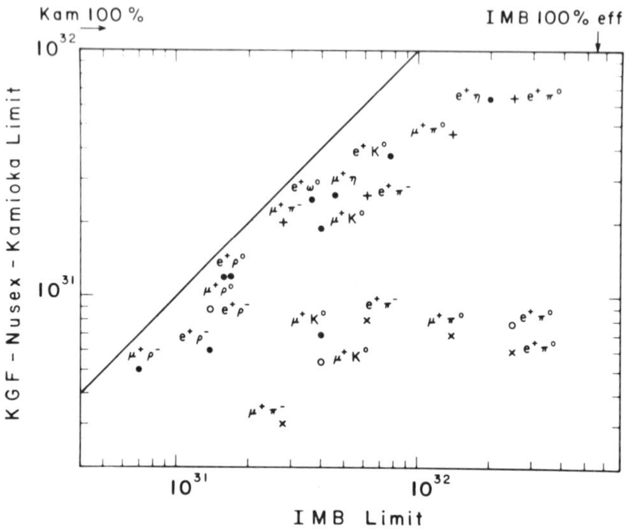

Figure 1: Comparison of Proton Lifetime limits for specific modes from 4 experiments

limit[1] is plotted on the horizontal scale. The limit from Kamioka[2], (+,.) Nusex[4], (x) or Kolar Gold Fields[3] (o) is plotted on the vertical scale. If a clear consistent signal were present the points would lie on the diagonal line. This would also occur if the experiments were background limited. For the modes shown it is clear that a longer exposure has produced a larger lifetime limit so the experiments as a whole are not background limited.

The lifetime limit for $e^+\pi^0$ decay is 2.5×10^{32} years from IMB[1]. Combining experiments will not significantly increase the limit. IMB has a limit of 4×10^{31} years for the μ^+K^0 decay. Most experiments have reported candidates for this mode. It is likely that the background is greater for this mode.

20% of all protons in water detectors are in the form of hydrogen. These protons give a sample of unbound nucleons free of nuclear effects and fermi motion. IMB has published limits[1] on free proton decay, assuming all

Table 3: Proton Decay Limits (10^{31} yrs.) (90%) c.l.

Mode	KGF	Nusex	Kamioka	IMB	IMB Free
$e^+\pi^0$	.77	.6	6.4	25	8.2
μ^+K^0	.54	.7	1.9	4.0	.67
νK^+	.98	.8	-	1.0	.18
$e^+\pi^-$	-	.8	2.6	3.1	-
e^+K^0	-	-	3.8	7.7	2.4
$\mu^+\pi^0$	-	.7	4.6	10.	5.9

bound decays are surpressed. Due to increased efficiency for detection the limits are about 1/3 (not 1/5) of those for all protons. These results are significant from a theoretical point of view since they are insensitive to decay supression mechanisms in the nucleus. Lifetime limits for a number of modes are listed in Table 3.

IMB has extended their analysis to full reconstruction of low multiplicity proton decay candidates[25]. The neutron-antineutron oscillation time can be extracted from these experiments. A calculation[17] is needed to convert results in the nucleus to a free oscillation time. Kamioka[18] quotes a limit of 1.2×10^8 seconds with no candidates. The IMB limit is comparable.

The lowest limit for terrestrial magnetic monopole searches comes from an examination of old mica deposits[9]. A monopole can form a bound state with an aluminum nucleus and as the state traverses a medium it should be heavily ionizing. By examining samples with very long exposure times sensitivity to fluxes as low as $2 \times 10^{-19}/cm^2/sec/str$ at $\beta = 10^{-3}$ can be achieved. Objections are raised since a monopole may already be bound to another nucleus and so be unable to penetrate, or to pick up another nucleus. Such arguments can reduce the sensitivity by two orders of magnitude.

The lowest terrestrial limit for monopole catalysis comes from IMB[10]. A flux limit of $4 \times 10^{-14}/cm^2/sec/str$ is found for an assumed cross section of 100 millibarns.

Kamioka has found limits based on a lack of 35 MeV neutrinos from the sun. Such neutrinos would be produced in a decay chain following monopole catalysis. A limit of $10^{-21}/cm^2/sec/str$ at $\beta = 10^{-3}$ is found.

If the sun can capture magnetic monopoles so can other celestial objects. A limit of $10^{-23}/cm^2/sec/str$ at $\beta = 10^{-3}$ is achieved by considering the effect of monopole catalysis on the xray emission from a neutron star[11].

There has been no evidence for a magnetic monopole flux from any underground experiment.

Most proton decay detectors have reported a neutrino flux as measured in their detectors[4),5),8),12)]. In general the agreement with expected fluxes is good. Both the Kamioka detector[18] and the Nusex detector[4] can distinguish ν_e from ν_μ by shower development. They quote a ν_e/ν_μ flux ratio of 0.36 ± 0.08 and 0.28 ± 0.11 respectively. These are lower than the expected value[6] of 0.64. The IMB group has studied the fraction of their contained events resulting in a muon decay[8]. The 26% observed can be converted to a ν_e/ν_μ ratio with a number of assumptions about muon capture in water. If 40% of the ν_μ interactions do *not* result in a muon decay signal the observed value corresponds to $\nu_e/\nu_\mu \approx 1.3$.

The problem of the ν_e/ν_μ ratio is still under active study. There is no directional dependence of the muon rate.

IMB[8] has used their sample of 401 contained interactions to study atmospheric neutrinos[6),8),16)]. Neutrino oscillations with a base line of 10^7 meters and a mean neutrino energy of 920 MeV can be studied. These parameters give sensitivities on the order of $10^{-5}eV^2$ for maximal mixing (if matter effects[19] are neglected). By varying the zenith angle neutrino path lengths can run from 10^7 meters to 2.6×10^6 meters. The neutrino energies run from 400 MeV to 2 GeV for ν_μ and from 200 MeV to 2 GeV for ν_e. Hence E/L spans a range of from $4 \times 10^{-5} MeV/m$ to $7.7 \times 10^{-4} MeV/m$. Sensitivity depends on the statistics available. Since the spectrum is falling with energy one is less capable of seeing effects of small mixing angles at the larger value of E/L, and hence Δm^2. Figure 2 summarizes their results for oscillation of ν_e and ν_μ into noninteracting states. The region at maximum mixing $2.2 \times 10^{-5}eV^2 < \Delta m^2 < 7.6 \times 10^{-4}eV^2$ is excluded by this analysis. Results have been obtained for mixing angles $\eta > 22.5°$.

These results may be extended to just ν_μ oscillations. If ν_μ oscillates to ν_τ or a sterile neutrino state than matter effects will be irrelevant. If ν_μ oscillates into ν_e the data rate will not change significantly (unless the ν_μ were below the detection threshold and above detection threshold for ν_e)

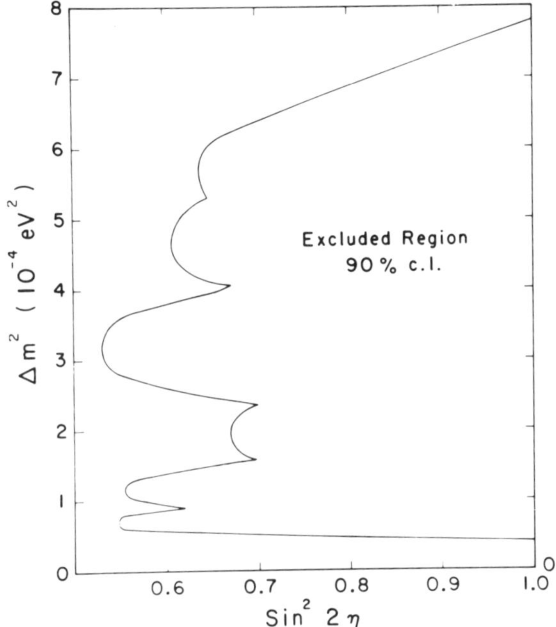

Figure 2: Neutrino oscillation limits from IMB for oscillation into sterile neutrinos

but the fraction of events with a muon decay will decrease. No decrease is observed either in total event rate or in muon decay fraction as the zenith angle varies. Restricting the analysis to events with a muon decay signature gives the limits in Figure 3. These are not quite as low as those for all neutrinos due to the slightly higher energy threshold for ν_μ. The range $4.2 \times 10^{-5} eV^2 < \Delta m^2 < 5.4 \times 10^{-4} eV^2$ is excluded at maximum mixing. Since only 26% of the events can be used in this analysis the statistical errors are larger than for the sample as a whole.

These results have ignored matter effects on oscillations. In spite of this they may be interpretted as the degree of distortion in the neutrino flux from different directions and as such used to set limits[26] on possible matter effects.

Searches[20] for point sources of neutrinos with energies below 2 GeV yields no excess in regions of 60°, consistent with the angular resolution at these energies. No more than 5.5% of the total observed rate can be

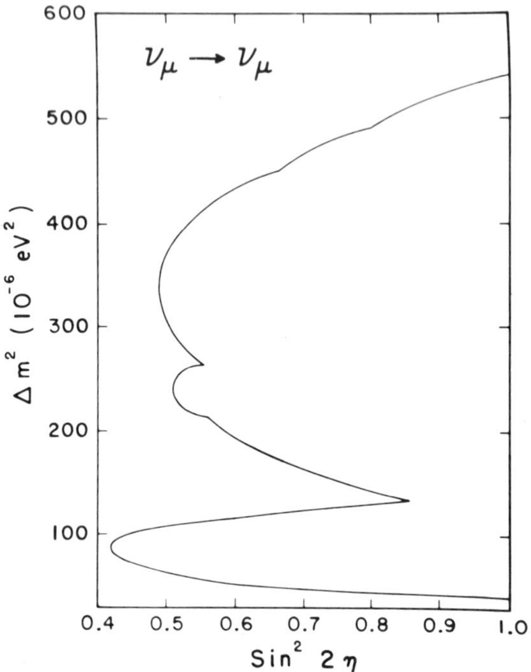

Figure 3: Neutrino oscillation limits from IMB for any oscillations of ν_μ

attributed to either stellar sources or sources on earth. Less than 3.5% of the total rate can be attributed to a source in the sum.

At high energies results have been reported by 3 groups[18),20)]. Caution is needed in comparing results on neutrino interactions in the rock. Groups are located at different lattitudes and have different selection criteria in their local coordinate system. Celestial objects will pass in and out of the local selection region giving rise to reduced sensitivity and perhaps spurious periodic effects. For example[21)] if one requires that the source be below the horizon then some sources are never visible and some are visible 100% of the time. Comparison should be done with regions at the same declination to avoid any normalization difficulties. Table 4 lists limits for some sources. Upward muons have also been used to search for neutrino oscillations[22)]. This involves comparing the observed distribution with an expected one. Since the neutrino energies are higher the limits are not as low in Δm^2 as for the contained events.

Table 4: Upward Muon Flux Limits (90% c.l.) ($\nu/cm^2/sec.$)

Source	IMB Limit	Kamiokande Limit
LMC X-4	3.5×10^{-14}	1.1×10^{-13}
Her X-1	6.9×10^{-14}	6.5×10^{-13}
Cyg X-3	2.1×10^{-13}	4.6×10^{-13}
Crab	8.8×10^{-14}	2.8×10^{-13}
SS 433	5.7×10^{-14}	2.3×10^{-13}
Galactic Center	3.8×10^{-14}	1.6×10^{-13}
Vela X-1	3.5×10^{-14}	1.3×10^{-13}

Two groups[13],[14] have reported evidence for underground muons associated with the celestial source Cygnus X3. Both groups have selected muons in the direction of Cygnus X3 and analyzed their time structure using the characteristic 4.8 hour period observed in other high energy experiments. A signal emerges at a phase of about .7-.8.

The Soudan group [13] reports a flux of $7 \times 10^{-11} \mu/cm^2/sec$ with a threshold of $E_\mu > 500$ GeV Nusex[14] reports a flux of $2.5 \times 10^{-12} \mu/cm^2/sec$ with a threshold of $E_\mu > 2$ TeV. These results have not been confirmed by other groups. Kamioka gives a flux limit of $2.2 \times 10^{-12} \mu/cm^2/sec$ with a threshold near 600 GeV. IMB has searched [16] but not had enough sensitivity to confirm the effect. Frejus [23] has also failed to confirm the effect and quotes a bound of $1.9 \times 10^{-13} \mu/cm^2/sec$ at a threshold comparable to the Nusex group.

One problem with comparing different experiments is that Cygnus X3 is known to be a variable source. It is possible that different groups have observed at different times and in spite of significant overlap have failed to get consistent results.

6   THE FUTURE

In the near future we can look forward to more data, in particular the Frejus detector which is the first of a new generation of massive fine grain detectors. Background estimates should improve. At present background subtractions are still statistics limited. I expect the systematic effects associated with the background estimates to decrease at least as fast as the statistical errors on the data.

Kamioka has completed a major upgrade and will attempt to observe a solar neutrino signal via ν_e scattering. This first attempt to observe a prompt, rather than integrated signal could shed some new light on this old question. IMB will finish an upgrade shortly. The upgrade was meant to increase their sensitivity to low light decay modes by increased light collection. It may yield better muon decay sensitivity and better event reconstruction. The Soudan II detector, a second generation tracking calorimeter is in the construction stage. Completion is anticipated in two years. This should greatly extend sensitivity into the rare decay modes where good tracking of high multiplicities and particle identification could make a significant difference.

There will be an increased interest in astrophysical problems as detectors are improved. Neutrinos and muons are still very good probes of high energy phenomena. These are the source of the only signals in underground detectors.

7 ACKNOWLEDGEMENT

This work was supported in part by the National Science Foundation and the U.S. Department of Energy. I would like to thank the organizers for their hospitality at Lake Louise.

8 REFERENCES

1. G. Blewitt, <u>et al.</u>, Phys. Rev. Lett., <u>27</u>, 2114 (1985)
   G. Blewitt, <u>et al.</u>, "Search for Two Prong Proton Decays" CALT68-1265

2. K. Arisaka, <u>et al.</u>, Journal of the Physical Society of Japan <u>54</u>, 3213 (1985)

3. M.R. Krishnaswamy, et al., Proc. of the 19th International Cosmic Ray Conference, La Jolla (1985) Vol. 8, 261-264

4. G. Battistoni,et al., Proc. of the 19th International Cosmic Ray Conference, La Jolla (1985) Vol. 8, 271-274

5. Aachen, Orsay, et al. Collaboration, Proc. of the 19th International Cosmic Ray Conference, La Jolla (1985) Vol. 8, 257-260

6. T. Stanev, "Muons and Neutrinos" MAD/PH/270, Proc. of the 19th International Cosmic Ray Conference, La Jolla (1985) (to be published)

7. J. LoSecco, Comments on Nuclear and Particle Physics, XV, 23 (1985)

8. J. LoSecco, et al., Phys. Rev. Lett. $\underline{54}$, 2299 (1985)
J. LoSecco, et al., Proc. of the 19th International Cosmic Ray Conference, La Jolla (1985) Vol. 8, 116-119

9. P. B. Price and M. H. Salamon, Proc. of the 19th International Cosmic Ray Conference, La Jolla (1985) Vol. 8, 242-245
P. B. Price and M. H. Salamon, Phys. Rev. Lett. $\underline{56}$, 1226 (1986)

10. S. Errede, "Experimental Limits on Monopole Catalysis", Monopole '83, 251, (J. L. Stone ed.) (Plenum, New York)

11. K. Freese, M. Turner and D. Schramm, Phys. Rev. Lett. $\underline{51}$, 1625 (1983)

12. M. R. Krishnaswamy, et al., Proc. of the 19th International Cosmic Ray Conference, La Jolla (1985) Vol. 8, 4-5

13. M. L. Marshak, et al., Phys. Rev. Lett. $\underline{54}$, 2079 (1985), and Phys. Rev. Lett. $\underline{55}$, 1965 (1985)

14. G. Battistoni, et al., Phys. Lett. $\underline{155B}$, 263 (1985)

15. Y. Oyama, et al., Phys. Rev. Lett. $\underline{56}$, 991 (1986)

16. J. LoSecco, et al., "Neutrino Physics with a Massive Underground Detector", Proceedings of the XXI Recontre de Moriond, Workshop on Massive Neutrinos, January 1986 (to be published) UND-PDK-86-2

17. C. Dover, et al., Phys. Rev. $\underline{D27}$, 1090 (1983)

18. M. Koshiba, Talk presented at the 1986 Aspen Winter Conference on Particle Physics (to be published)

19. L. Wolfenstein, Phys. Rev. $\underline{D17}$, 2369 (1978)

20. J. LoSecco, et al., "Contained Neutrino Interactions in an Underground Water Detector" UND-PDK-86-1 (1986)

21. R. Svoboda, "A Search for Astrophysical Point Sources of Neutrinos Using a Large Underground Water Cerenkov Detector", Ph.D. thesis, University of Hawaii, 1985

22. R. Bionta, et al., Proceedings of the DPF Santa Fe Meeting (1984) page 255, (T. Goldman and M. M. Nieto editors)

23. G. Chardin private communication and Proceedings of the XXI Recontre de Moriond, Workshop on Massive Neutrinos January 1986 (to be published)

24. H. Bethe, Phys. Rev. Lett., $\underline{56}$, 1305 (1986)

25. G. Blewitt "A Search for Free Proton Decay and Nucleon Decay in O^{16}, Using the Invariant Mass and Momentum of Exclusive Final States", PhD. Thesis, California Institute of Technology, 1985

26. J. LoSecco, "Comment on 'Possible Explanation of the Solar Neutrino Puzzle' ", UND-PDK-86-5 (to be published)

Dark Matter as a Probe of the Early Universe

Keith A. Olive
School of Physics and Astronomy
University of Minnesota
Minneapolis, Minnesota 55455 USA

Abstract

The role of dark matter in the evolution of the
Universe is reviewed. Observational consequences
of cold dark matter in the galactic halo is also
discussed.

A key element to our understanding of the nature of the Universe
centers heavily on knowing how much stuff there is and what the
identity of that stuff is. Indeed, there is a growing body of
evidence (which will be briefly reviewed below) that the Universe is
filled with dark stuff (dark matter). The quantity and quality of the
dark matter profoundly affects the present state of affairs in the
Universe from the distribution of galaxies and clusters of galaxies to
the eventual fate of the Universe: a terminal collapse or an unending
expansion. Particle physics models are known to give a variety of
candidates including neutrinos and photinos. The masses of the
particles are, however, in almost all cases subject to cosmological
constraints. In this review, I will look at some of the reasons for
expecting the existence of dark matter and look at implications coming
from our knowledge of the early Universe. In addition, if the dark
matter is nonbaryonic and somewhat weakly interacting, the sun and
proton-decay detectors may be able to either rule out or "see" certain
candidates.

In the early Universe, the first prediction (chronologically) of
dark matter comes from inflation. Before discussing inflation,

however, it will be necessary to review briefly the essentials in the standard Big Bang model. One starts with the assumption that the Universe is very nearly homogeneous and isotropic and described by the Friedmann-Robertson-Walker metric

$$ds^2 = -dt^2 + R^2(t) \left[\frac{dr^2}{1-kr^2} + r^2(d\theta^2 + \sin^2\theta \ d\phi^2) \right] \tag{1}$$

where $R(t)$ is the R-W scale factor and k is the curvature constant describing the overall geometry ($k=0,+1,-1$ for a flat, closed or open Universe). Einstein's equations lead to the Friedmann equation which relates the expansion rate to the energy density and curvature,

$$H^2 \equiv (\dot{R}/R)^2 = 8\pi\rho/3M_p^2 - k/R^2 + \Lambda/3 \tag{2}$$

where H is the Hubble parameter, ρ is the total mass-energy density, Λ is the cosmological constant and $M_p = G_N^{-1/2} = 1.2 \times 10^{19}$ GeV is the Planck mass.

When combined with the equation for energy conservation

$$\dot{\rho} = -3 \left(\frac{\dot{R}}{R} \right) (\rho + p) \tag{3}$$

where p is the isotropic pressure, the Freidmann equation (2) leads to several typical expansion stages in the early Universe. At early times, the Universe is thought to have been dominated by radiation so that the equation of state was just $p = \rho/3$ and at early times we can assume that the contribution to H from k and Λ were negligible. From eq. (3) we have that $\dot{\rho} = -4\rho \ \dot{R}/R$ or $\rho \sim R^{-4}$ and hence $\dot{R} \sim R^{-1}$ so that

$$R \sim t^{1/2} \tag{4}$$

for a radiation dominated Universe. One then also finds the time temperature relation through

$$t = (3/32\pi G\rho)^{1/2} + \text{constant} \qquad (5)$$

and realizing that $\rho \sim T^4$. Similarly for a matter dominated Universe (late times) we take $p = k = \Lambda = 0$ and find $\dot{\rho} = -3\rho \, \dot{R}/R$ or $\rho \sim R^{-3}$ and

$$R \sim t^{2/3} \quad . \qquad (6)$$

If we maintain that $\Lambda = 0$, we can define a critical energy density ρ_c such that $\rho = \rho_c$ for $k = 0$

$$\rho_c = \frac{3H^2}{8\pi G_N} \qquad (7)$$

In terms of the present value of the Hubble parameter

$$\rho_c = 1.88 \times 10^{-29} \, h^2_0 \, \text{g cm}^{-3} \, , \qquad (8)$$

where

$$h_0 = H_0/(100 \text{ km Mpc}^{-1}\text{s}^{-1}) \qquad (9)$$

is the present value of the Hubble parameter in units of 100 km Mpc^{-1} s^{-1}. The cosmological density parameter is then defined as the ratio of the present energy density to the critical density

$$\Omega \equiv \rho/\rho_c \qquad (10)$$

Furthermore, the value of Ω will determine the sign of k. For $\Omega > 0$ we have $k = +1$, $\Omega = 1$ corresponds to $k = 0$ and $\Omega < 0$ to $k = -1$. In terms of Ω the Friedmann equation can be rewritten as

$$(\Omega - 1) \, H^2_0 = \frac{k}{R^2} \qquad (11)$$

The observational limits on h_0 and R are[1]

$$1/2 < h_0 < 1 \qquad (12a)$$

$$0.1 < \Omega < 4 \qquad (12b)$$

Included in Ω, is the total mass density of dark matter. If Ω_L is the total contribution to Ω from luminous matter $\Omega_{DM} = \Omega - \Omega_L$.

The density of the Universe is generally determined by means of a mass-to-light ratio. The mass of a galaxy or gravitational system if in gravitational equilibrium can be computed via the virial theorem from measured rotational velocities. The total mass of the system is then compared with its absolute luminosity which is derived from the measured apparent luminosity. The total density ρ is then

$$\rho = \left(\frac{M}{L} \right) \mathcal{L}, \qquad (13)$$

where (M/L) is the above described mass-to-light ratio and $\mathcal{L}$ is the total luminosity density of the night sky

$$\mathcal{L} \approx 2 \times 10^8 \ h_0 \ L_\odot \ Mpc^{-3}, \qquad (14)$$

where $L_\odot$ is the solar luminosity $L_\odot = 3.9 \times 10^{33}$ erg s^{-1}. We can now define a critical mass-to-light ratio

$$(M/L)_c = \rho_c / \mathcal{L} \approx 1200 \ h_0 \qquad (15)$$

and the cosmological density parameter is given by

$$\Omega = (M/L)/(M/L)_c . \qquad (16)$$

In principle this could give us an accurate determination of Ω. The problem is that the derived value of Ω seems to depend on what scale we measure (M/L). For example, the following four systems all give different values[2] of Ω

$$1) \text{ solar neighborhood}$$

$$(M/L) \approx 2 \pm 1 \implies \Omega \approx (0.0016 \pm .0008)/h_0$$

$$2) \text{ central parts of galaxies}$$

$$(M/L) \approx (10\text{-}20) \ h_0 \implies \Omega \approx (0.008 - 0.017)$$

$$3) \text{ binaries and small groups of galaxies}$$

$$(M/L) \approx (60\text{-}180) \ h_0 \implies \Omega \approx (0.05 - 0.15)$$

$$4) \text{ clusters of galaxies}$$

$$(M/L) \approx (300\text{-}1000) \ h_0 \implies \Omega \approx (0.25 - 0.8) \ .$$

The dependence on h_0 of the last three mass-to-light ratios is due to the uncertainties in estimating the mass and absolute luminosities of distant objects. It is evident that as we look on larger and larger scales the value of Ω seems to be increasing. This is known as the missing mass problem. In particular, it seems to indicate that there is dark matter present in the Universe on large scales.

As there are already several reviews[3] about inflation[4], I will try to be brief here. However, any review of the very early Universe would be incomplete if it did not at least touch upon inflation. In short what is meant by inflation, is the effect of exponential expansion due to a supercooled phase transition in order to resolve several finetunings regarding the intial conditions in the standard big bang model. As examples of these problems, I will briefly describe what is known as the horizon problem and the curvature problem. The horizon volume or causally connected volume today, is just related to the age of the Universe $V_0 \propto t_0$. The microwave background radiation with the temperature $T_0 \sim 3$ K has been decoupled from itself since the epoch of recombination at $T_d \sim 10^4$ K. The horizon volume at that time was $V_d \sim t_d$. Now the present horizon volume scaled back to the period of decoupling will be $V' = V_0 (T_0/T_d)^3$ and the ratio of this volume to the horizon volume at decoupling is

$$V_0'/V_d \sim (V_0/V_d)(T_0/T_d)^3 \sim (t_0/t_d)^3 \ (T_0/T_d)^3 \sim 10^5, \tag{17}$$

where I have used $t_d \sim 3 \times 10^{12}$ sec and $t_0 \sim 5 \times 10^{17}$ sec. The ratio (17) corresponds to the number of regions that were casually disconnected at recombination which grew into our present visible Universe. Because the anisotropy of the microwave background is so small,[5], $\delta T/T \lesssim$ few x 10^{-5}, the horizon problem, therefore, is the lack of an explanation as to why 10^5 causally disconnected regions at t_d all had the same temperature to within one part in 10^4!

The curvature problem (also known as the flatness or oldness problem) stems from the fact that although the Universe is very old, we still do not know whether it is open or closed. If we look at the Freidmann equation (2) for the expansion of the Universe and use the limits $\Omega < 4$ and $H_0 < 100$ km s^{-1} Mpc^{-1} we can form a dimensionless constant

$$\hat{k} = k/R^2 T^2 = (\Omega - 1) H_0^2/T^2 \quad < 3H_0^2/T_0^2 < 2 \times 10^{-58} \tag{18}$$

where I have used $T_0 > 2.7$ K. In an adiabatically expanding Universe, $\hat{k}$ is absolutely constant $(R \sim T^{-1})$ and thus the limit (18) represents an initial condition which must be imposed so that the universe will have lived this long looking still so flat.

A more natural initial condition might have been $\hat{k} \sim 0(1)$. In this case the universe would have become curvature dominated at $T \sim 10^{-1} M_p$. For $k = +1$, this would signify the onset of recollapse. Even for $\hat{k}$ as small as $0(10^{-40})$ the Universe would have become curvature dominated when $T \sim 10$ MeV or when the age of the Universe was only $0(10^{-2})$ sec. Thus not only is (18) a very tight constraint, it must also be strictly obeyed. Of course, it is also possible that $k = 0$ and the Universe is actually spatially flat.

These are the two main problems that led Guth[4] to consider inflation. In the problems that were just discussed it was assumed that the Universe has always been expanding adiabatically. During a phase transition, however, this is not necessarily the case. If we consider a scalar potential describing a 1st order phase transition from a symmetric false vacuum state $\langle \Sigma \rangle = 0$ to the broken true vacuum at $\langle \Sigma \rangle = v$, and we suppose that because of a barrier separating the two minima the phase transition was a supercooled transition. If in addition, the transition takes place at T_c such that $T_c^4 < V(0)$, the energy stored in the form of vacuum energy will be released. If released fast enough, it will produce radiation at a temperature $T_R^4 \sim V(0)$. In this reheating process entropy has been created and

$$(RT)_f \sim (T_R/T_c) \; (RT)_i \tag{19}$$

provided that T_c is not too low. Thus we see that during a phase transition the relation $RT \sim$ constant need not hold true and thus our dimensionless constant $\hat{k}$ may actually not have been constant.

The inflationary Universe scenario[4], is based on just such a situation. If during some phase transition, such as $SU(5) \rightarrow SU(3) \times SU(2) \times U(1)$ the value of RT changed by a factor of $O(10^{29})$, these two cosmological problems would be solved. The isotropy would in a sense be generated by the immense expansion; one small causal region could get blown up and hence our entire visible Universe would have been at one time in thermal contact. In addition, the parameter k could have started out $O(1)$ and have been driven small by the expansion.

If, in an extreme case, the barrier between the minima caused a lot of supercooling such that $T_c^4 \ll V(0)$, the dynamics of the expansion would have greatly changed. In this example the energy density of the symmetric vacuum, $V(0)$ acts as a cosmological constant with

$$\Lambda = 8\pi \; V(0)/M_p^2 \; . \tag{20}$$

If the Universe is trapped inside the false vacuum with $\Sigma = 0$, eventually the energy density due, to say, radiation will fall below the vacuum energy density, $\rho \ll V(0)$. When this happens, the expansion rate will be dominated by the constant $V(0)$ and we will get the De Sitter-type expansion and from Eq. (2)

$$R \sim \exp[Ht]. \tag{21}$$

where

$$H^2 = \Lambda/3 = 8\pi \, V(0) \, /3M_p^2 \, . \tag{22}$$

The cosmological problems could be solved if

$$H\tau \gtrsim 65, \tag{23}$$

where τ is the duration of the phase transition and the vacuum energy density was converted to radiation so that the reheated temperature is found by

$$\frac{\pi^2}{30} \, N(T_R) \, T_R^4 = V(0) \tag{24}$$

where $N(T_R)$ is the number of degrees of freedom at T_R.

In the original inflationary scenario, the phase transition given by a potential with a large barrier proceeds via the formation of bubbles[6]. The Universe would reheat, i.e., the release of entropy must occur through bubble collisions and the transition is completed when the bubbles fill up all of space. It is now known[7], however, that the requirement for a long timescale τ is not compatible with the completion of the phase transition. The Universe as a whole remains trapped in the exponentially expanding phase containing only a few isolated bubbles of the broken SU(3) x SU(2) x U(1) phase.

The well-known solution to this dilemma is called the new inflationary scenario[8]. If the shape of the potential V (ϕ) is such that following the barrier (if there is one) there is a long flat region out to $\langle\phi\rangle = V$, the phase transition would proceed not through the formation of bubbles, but rather, by the long rollover in which ϕ picks up a vacuum expectation value. (ϕ may be the adjoint in the case of SU(5) or any other field with potential V.) During the

rollover, the vacuum energy density would remain essentially constant
for a long period of time triggering the exponential expansion.
Completion of the transition is thus guaranteed and reheating occurs
through the dissipation of energy due to field oscillations about the
global minimum. Early models for new inflation utilized Coleman-
Weinberg[9] type symmetry breaking for SU(5). These too, turned out to
be problamatic.

New inflation makes two important predictions: $\Omega = 1$ and $\delta\rho/\rho \neq 0$.
The way inflation cures the curvature problem, is by driving R to
exponentially large values. As one can see from eq. (18) $\hat{k}$ is driven
from O(1) to very small values. Indeed, without additional fine
tuning it is not feasible for $\hat{k}$ to settle at the relatively large
value of $O(10^{-58})$. In general, it may easily go hundreds of orders of
magnitude below this, thus ensuring $\Omega = 1$ to a very high accuracy. In
addition, as inflation occurs, the field ϕ will not roll over
uniformly over all space. There will be a time spread δt (computable
in terms of V) in which ϕ rolls down faster or slower in certain
spatial regions than in others. Density perturbations have been
calculated[10] in terms of δt; $\frac{\delta\rho}{\rho} = H\delta t$ where $\frac{\delta\rho}{\rho}$ is the magnitude of
the perturbation as it enters the horizon. One of the major problems
with the Coleman-Weinberg type of inflation in SU(5) is that it
predicts $\frac{\delta\rho}{\rho} \sim 50$ whereas limits from the microwave background
anisotropy require $\frac{\delta\rho}{\rho} \lesssim 10^{-4}$.

A varient of the new inflationary scenario is called primordial
inflation[11]. Primordial inflationary models are simply those in
which the scalar field responsible for inflation is no longer
associated with the 24, of SU(5), but rather with some other field ϕ ,
dubbed the inflaton, which picks up a vacuum expectation value $\langle\phi\rangle \gg$
$\langle\Sigma\rangle$, thus allowing for a longer rollover timescale.

Supersymmetry offers the possibility for producing flat
potentials necessary for a long roll-over time scale [11,12] and
density perturbations of the right order of magnitude [13]. In the
limit that $\langle\phi\rangle = M_p$, it is useful to consider inflation in the context
of N=1 supergravity[14].

In N=1 supergravity, the scalar potential $V(\phi)$ is determined in terms of the superpotential f by[15].

$$V(\phi) = e^G \left[\frac{\partial G}{\partial \phi} \frac{\partial G}{\partial \phi*} \left(\frac{\partial G}{\partial \phi \partial \phi*} \right)^{-1} -3 \right]$$ (25)

where

$$G = \phi \phi* + \ln|f|^2$$ (26)

is the Kahler potential in minimal N = 1 supergravity, f must then be chosen so that $V(\phi)$ complies with the conditions for inflation discussed earlier. The simplest example for f, satisfying the inflationary requirements, is[16]

$$f = m^2 \left(1 - \frac{\phi}{m_P} \right)^2 M_P$$ (27)

The mass scale m is determined to be $O(10^{-4})$ from the calculation of $\delta\rho/\rho$.

Because the inflaton is generally taken to be a gauge non-singlet, there is an additional concern regarding initial conditions. Normally, one expects that symmetries are restored at high temperatures so that a natural intial condition for $\langle\phi\rangle$ is $\langle\phi\rangle = 0$. Inflation then takes place as $\langle\phi\rangle$ moves from 0 to $v \sim M_P$. In these models "symmetry restoration" must be added as an additional constraint[17]. Namely, we must also require that at very high temperatures there exists a minimum at $\langle\phi\rangle = 0$. It has been shown[18], however, that no choice of a superpotential can simultaneously satisfy the inflationary constraints and the thermal constraint.

Another constraint on the initial conditions comes from the fact that at high temperatures ϕ is not localized near $\phi \sim 0$ but rather, $\phi \sim T$ and hence, as the Universe cools ϕ may fall directly to the global minimum at $\phi = v$[19]. Recently, it has been shown that these difficulties are most easily avoided in models of primordial inflation[20].

A route to tackle the problem regarding the thermal constraint and finding a model with a suitable superpotential is to look at non-minimal supergravity models, i.e., those in which

$$\partial^2 G/\partial\phi\partial\phi* = 1 \tag{28}$$

If, instead, one considered[21] a general form for G

$$G = g_\phi(\phi + \phi*) + g_\phi^\phi(\phi\phi*) + g_{\phi\phi}(\phi^2 + \phi*^2) + g_{\phi\phi}^\phi (\phi\phi*)(\phi + \phi*) \tag{29}$$

where g_ϕ , g_ϕ^ϕ etc. are couplings. It is then possible to derive relations between the g_ϕ's to satisfy the inflationary constraints as well as the thermal constraint. The extra degrees of freedom in G allowed the constraints to be satisfied without the inclusion of several scales.

There are also specific non-minimal models based on SU(n,1) supergravity[22] which satisfy all constraints[23]. In these models the Kahler potential has the form

$$G = -3 \ln (z + z* - \phi\phi*/3) + \ln |f|^2 \tag{30}$$

where z is the field which breaks supergravity (see references[24] for a discussion on the z field in minimal supergravity theories) ϕ is the inflaton and f is the superpotential. In these theories it is possible to write down a superpotential as in Eq. (27) which in addition satisfies the thermal constraint[23]

$$f = m^2(\phi-\phi^4/4M_P^3) \quad . \tag{31}$$

I note here that SU(n,1) models have also been of interest in recent superstring models[25] and that inflationary models have been attempted in that context as well[26]. In summary, inflation's impact on dark matter is the prediction that $\Omega = 1$. If this is so, then $\Omega \neq \Omega_L$ and $\Omega_{DM} \neq 0$.

Big Bang Nucleosynthesis also gives us important clues as to the nature of the dark matter. What we will find is that if inflation had occurred, then not only $\Omega_{DM} = 0$, but a large portion of Ω is also nonbaryonic. Again, I refer the reader to a more extensive treatment[27,28] and review[29]. I will only highlight on the implications toward dark matter.

Big Bang Nucleosynthesis is a fundamental part of the standard Big Bang model, and it offers the only real evidence justifying an extrapolation to early times and high temperatures (T~1MeV) in the Universe. Starting with only a minimal set of assumptions (e.g., the early Universe was hot and all chemical potentials, aniostropies or inhomogeneities were all zero) and known nuclear cross sections, it is possible to calculate the abundances of the light isotopes, D, ^{3}He, ^{4}He and ^{7}Li. That the calculated abundances agree very well with the observed abundances, is an a posteori justification of the early time, high temperature extrapolation.

There are, however, three key parameters on which the calculated abundances are sensitive.[27,28] They are 1) the baryon to photon ratio η; 2) the number of light (m<1MeV) particle species (or neutrino flavors N_ν; 3) the neutron half-life t_n. Because the dependence of the abundances on t_n is rather weak, I will not discuss it any further here. Of the isotopes produced in the Big Bang, ^{4}He is the most abundant (about 25% by mass). Because the primordial ^{4}He abundance Y_p, is an increasing function of both η and N_ν, the observational upper limit $Y_p \leq 0.254$[28,29,30] implies that $\eta \leq 10^{-9}$ for $N_\nu = 3$. The upper limit to N_ν, however, depends on a lower limit to η.[27]

By considering D and ^{3}He it is possible to find stronger limits on η.[28] To begin with, D is very easily destroyed in stars. Hence it is generally believed that the Big Bang production of D must at least be as large as the observed abundance $D/H \sim 10^{-5}$ (by number). Because D (and ^{3}He) is a decreasing function of η, the lower limit on D/H implies that $\eta \leq 7 \times 10^{-10}$. When D is destroyed by stars, ^{3}He is produced, which is in turn rather difficult to destroy in stars. Stellar evolution calculations[30a] indicate that at least 1/4 of the ^{3}He abundance survives stellar processing. Thus the Big Bang production of the sum $(D + {}^3He)/H$ must not exceed the observed abundance by more than a factor of ~ 4 so that $(D + {}^3He)/H < 10^{-4}$ and gives $\eta > 3 \times 10^{-10}$. Given this as a lower limit to η, it is now possible to conclude an upper limit to N_ν, $N_\nu \leq 4$, with the equality being at best marginal.

This limited range on η, makes a powerful statement as to the nature of the dark matter. The mass density in baryons corresponding to a given value of η is given by

$$\rho_B = m_N n_\gamma \eta \tag{32}$$

and

$$\Omega_B = \rho_B/\rho_c = 3.56 \times 10^7 \eta \, h_o^{-2} (T_o/2.7k)^3 \tag{33}$$

where m_N is the nucleon mass and T_o is the present temperature of the microwave background. The limits

$$3 \times 10^{-10} \leq \eta \leq 7 \times 10^{-10} \tag{34}$$

imply that (for $1/2 < h_o < 1$, $2.7 < T_o < 2.8$)

$$0.01 \leq \Omega_B < 0.15 \tag{35}$$

Thus we can conclude the following: 1) the Universe is not closed by baryons, and 2) if $\Omega=1$, the Universe must be dominated by nonbaryonic matter.

One of the strongest arguments for dark matter comes from galaxy formation and the growth of perturbations. Again, without attempting a comprehensive review[31], I will only highlight the key arguments. The type of perturbations produced by inflation are adiabatic perturbations ($\delta\rho/\rho \propto \delta T/T$) and I will restrict my attention to these. Indeed, these perturbations also have the very nearly scale-free spectrum described by Harrison[32] and Zeldovich.[33] When produced scale-free perturbations fall off as $\frac{\delta\rho}{\rho} \propto \ell^{-2}$ (increase as the square of the wave number). At early times $\delta\rho/\rho$ grows as t so that at the time when the horizon scale ($\propto t_u$) is comparable to ℓ, the growth halts (the mass contained with the volume ℓ^3 has become smaller than the Jean's mass) and $\frac{\delta\rho}{\rho} \sim \delta$ independent of scale. When the Universe becomes matter dominated, the Jean's mass drops dramatically and growth continues as $\frac{\delta\rho}{\rho} \propto R \sim 1/T$. (For baryons, growth is delayed until the recombination of electrons and protons to form neutral hydrogen at $T_R \sim$ few x 10^3 K.)

Because we are considering adiabtic perturbations, there will be anisotropics produced in the microwave background radiation on the order of $\delta T/T \sim \delta/4$. Small angular scale observations[34] of the microwave background indicate that $\delta T/T \leq (2.5)$ x 10^{-5} implying that $\delta \lesssim 10^{-4}$. Without the existence of dark matter, $\delta\rho/\rho$ in baryons could then achieve only a maximum value $\delta\rho/\rho \sim \delta(T_R/T_o) \sim 10^{-1}$. This is too small to argue that growth has entered a nonlinear regime needed to explain the large value (10^5) of $\delta\rho/\rho$ in galaxies.

Dark matter easily remedies this dilemma in the following way. The transition to matter dominance is determined by setting equal to each other the energy densities in radiation (photons and any massless neutrinos) and matter (baryons and any dark matter). For three massless neutrinos and only baryons, matter dominance begins at

$$T_m = 0.22 \, m_B \eta \qquad (36)$$

while if we suppose that there exists dark matter with an abundance $Y_x = n_x/n_\gamma$ (the ratio of the number density x's to photons) then

$$T_m = 0.22 \ m_x \ Y_x \tag{37}$$

If we now consider neutrinos with small masses for the candidate x, $Y_\nu = 4/11$ so that

$$T_m \simeq 0.1 \ (\frac{g_\nu}{2}) \ m_\nu \tag{38}$$

($g_\nu = 2(4)$ for a Majorana (Dirac) mass neutrino) and could easily be a factor $0(10)$ greater than T_R and lead to $\delta\rho/\rho \sim 1$. Clearly, other forms of dark matter would have a similar effect. The baryons, although still bound to the radiation until T_R, now see deep potential wells formed by the dark matter perturbations to fall into and are no longer bound by the growth rate $\delta\rho/\rho \propto R$.

With regard to dark matter and galaxy formations, all forms of dark matter are not equivalent. They can be distinguished by their relative temperature at T_M.[35] Particles which are still largely relativistic at T_M (like neutrinos or other particles with $m_x < 100$ eV) have the property[36] that they have erased perturbations due to free streaming out to very large scales

$$M_J = 3 \times 10^{18} \ M_\Theta/M_\nu^{\ 2} \ (eV) \tag{39}$$

Thus, very large scale structures form first and galaxies are expected to fragment out later. Particles with this property are termed hot particles. Cold particles (m>1 meV) have the opposite behavior. Small scale structure forms first aggregating to form larger structures later. Neither candidate is completely satisfactory when the resulting structure is compared to the observations. For more details, I refer the reader to reviews in refs. 31) and 37).

408

There is also good evidence for dark matter on the scale of galactic halos. The strongest evidence comes from the rotation curves of spiral galaxies which measure the rotational velocity as a function of distance from the galactic center. In equilibrium, one expects that $GmM/r^2 \propto mv^2/r$ or $GM \propto v^2 r$ for a body of mass m rotating with velocity v in a galaxy of mass M. Without the presence of dark matter, at distances beyond the bulk of the luminous matter one would expect $v^2 \sim 1/r$. Instead, one finds [38,39] $v^2 \sim$ constant indicating that $M \sim r$ and hence implying the existence of an additional non-luminous component.

As was indicated earlier, the value of Ω on the scale of galactic halos is[2] $\Omega_H \sim 0.05 - 0.15$ which is in concordance with nucleosynthesis if the dark matter in halos was made of baryons. There are, however, numermous arguments[40] against a baryonic halo. Put briefly, it is very difficult to have a large baryon density in such a way that it is unobservable. In the form of gas the baryons would heat up and emit X-rays in violation of observed limits. To put the baryons in non-nuclear burning stars (Jupiters) would require an extrapolation of the stellar mass distributing which is very different from what is observed. Dust or rocks along with dead remnants such as neutron stars or black holes would require a metal abundance in great excess of the galactic metallicity. Very massive (≥ 100 $M_\odot$) black holes remain a possibility.

There are, of course, many candidates for non-baryonic dark matter. In the remainder of this contribution, I will concentrate on cold dark matter candidates which include massive (> few GeV) neutrinos[41], and the supersymmetric candidates: photinos,[42,43] higgsinos[43] and sneutrinos.[44] With the exception of sneutrinos, the overall cosmological density places lower limits on the dark matter candidate's mass. For example, the relic neutrino density is $\rho_\nu \propto \sigma_A^{-1} \propto m_\nu^{-1}$ where σ_A is the annihilation cross section. While the neutrino limit can be explicitly calculated,[41] the supersymmetric mass limits[42,43] are given in terms of the unknown scalar quark and lepton masses. The sneutrino is an exception because σ_A is independent of m_ν.[44] (If sneutrinos make up the dark matter in the galactic halo, their effects in white dwarfs[45] lead to some constraints on their mass).

The eventual verification of the existence of a dark matter in galactic halo obviously depends on some kind of signature or signal. One interesting suggestion[46] has been to use a very cold detector with superconducting grains which would flip as the DM passes through. What I will discuss here is some possible signatures due to DM annihilations in the halo and in the sun. I will throughout assume that there is on type of DM with $\Omega \simeq 1$ and a local density $n_x = (0.3/m_x)$ cm^{-3} with velocities $v \simeq 300$ km s^{-1}.

The possible observation of annihilation of DM in the galactic halo was recently examined[47] in the case where photinos of mass $m_\gamma \simeq 3$ GeV were the DM. The annihilations of photinos can lead to appreciable fluxes of cosmic rays. Although the γ-ray flux from these annihilations is well below the established backgrounds the predicted flux of positrons and antiprotons is significant. The predicted fluxes are $F_{e^+} \approx 5\times10^{-4}$ cm^{-2} s^{-1}sr^{-1} and $F_{\bar{p}} \approx 1.5\times10^{-6}$ cm^{-2}s^{-1}sr^{-1} while the observed fluxes are [48] $F_{e^+} \approx 10^{-3}$ cm^{-2}s^{-1}sr^{-1} and $F_{\bar{p}} \approx 3\times10^{-6}$ cm^{-2}s^{-1}sr^{-1}. This is particularly important for the case of the antiproton flux because these are low energy $\bar{p}$'s $(0.6 \leq E \leq 1.2$ GeV$)$ whose origins are otherwise difficult to explain.

The effects of cold DM in the sun were first discussed by Steigman et al.[49] and more recently in the context of the solar neutrino problem.[50-52] As dark matter particles from the halo pass through the sun, they will become trapped due to their elastic scatterings with protons in the sun. Annihilations of dark matter in the sun[53] can however lead to an appreciable flux of high energy neutrinos. The flux of neutrinos depends primarily on the rate of capture by the sun of the DM which I assume to be a photino here for definiteness. The capture rate is computable[51,53] in terms of the photino-proton elastic scattering cross-section σ_E

$$\Gamma_c \simeq 10^{29} \sigma_{E,36}/m_\gamma \quad s^{-1} \tag{40}$$

where $\sigma_{E,36} = \sigma_E/10^{-36}$ cm.^{2}

In equilibrium, the rate of annihilations is just equal to the capture rate Γ_c. In addition, annihilations lead to a spectrum of particles, of which only neutrinos escape. The total flux of neutrinos of type i arriving at earth is

$$\phi_{\nu_i} = \frac{1}{2} \Gamma_c N_i/4\pi d^2 \tag{41}$$

where N_i is the number of neutrinos of type i produced in the annihilation and $d = 1$ A.U. (the factor $\frac{1}{2}$ corresponds to the fact that each annihilation destroys two captured particles). Numerically Eq. (41) becomes

$$\phi_{\nu_i} \simeq 16 N_i \sigma_{E,36}/m_x \quad cm^{-2} s^{-1} \tag{42}$$

The spectrum of neutrinos will be very broad, but we can estimate the average neutrino energy to be $m_x/\Sigma_i N_i$, and use this to compare the solar flux to the atmospheric background of neutrinos.[54] For example, for 10 GeV photinos, it is found[53] that $N_{\nu_e} \simeq 2.5$, $\sigma_{E,36}$ 0.01 so that $\phi_{\nu_e+\bar{\nu}_e} \simeq .08$ cm^{-2} s^{-1}. With $E_\nu = 500$ MeV one finds that the atmospheric flux is in the range 0.06 - 0.14 cm^{-2} s^{-1} (the range depends on geomagnetic effects). For the case of sneutrinos, annihilations lead to a monochromatic flux of neutrinos with $E_\nu = m_\gamma$. For $m_\nu = 5$ GeV, $N_{\nu_e} = 1$, $\sigma_{E,36} \simeq 0.025$ so that $\phi_{\nu_e+\bar{\nu}_e} \simeq 0.1$ cm^{-2} s^{-1} whereas the background at 5 GeV is only $\simeq 10^{-4}$ cm^{-2} s^{-1}. An additional test to compare the solar flux from dark matter annihilations to the atmospheric background is to look at only the prompt neutrinos produced by the decay of an annihilation product,[55] e.g., $\gamma + \gamma \to \tau^+\tau^-$; $\tau^+ \to e^+ + \nu_e + \bar{\nu}_\tau$. By comparing the computed spectrum to the atmospheric background, one finds that for $E_\nu > 1\,\text{GeV}$, the solar flux is comparable to the background.

At this point it is worthwhile to note that I have as yet to discuss superstrings. In keeping with the overwhelming emphasis of this meeting, I will conclude by discussing the implications on dark matter[56] by the superstring phenomonological models[57] discussed by J. Ellis. Similar to the situation in ordinary supersymmetric theories,[43] low energy superstring models based on E_6 will contain one stable particle (the lightest supersymmetric particle, LSP) which becomes a candidate for the dark matter. However, instead of the neutral fermion sector containing the wino, bino (photino and zino) and two higgsinos, the LSP now may be a mix of these plus an additional zino and an additional neutral fermion N which is an SU(3)xSU(2)xU(1) singlet in the 27 of E_6. Performing a similar analysis to that in ref. 43) and using the parameters derived in ref. 57), it was found that for the allowed gaugino mass of 100-1000 GeV

$$\Omega_{LSP}\,(h_o/0.5)^2 = 2^{0\pm1} \tag{43}$$

Thus making the superstring LSP an excellent candidate for the dark matter.

Because superstring dark matter will also become trapped in the sun, we can get additional constraints on parameters such as the gaugino mass. Indeed only for $M_{LSP} < m_t$ (the top quark mass) was the solar neutrino flux sufficiently below the atmospheric background flux. If we now assume that $\Omega_{LSP} = 1$, the hubble parameter h_o is determined by the gaugino mass $M_{1/2}$. But M_{LSP} is also determined by $M_{1/2}$ as well so that $M_{LSP} < M_{top}$ places an upper limit on $M_{1/2}$ as well as h_o. In the figure below the upper bound on h_o as a function of M_t is shown. If we want to go even further, it is amusing to note that phenomonological considerations[57] and cosmological arguments based on nucleosynthesis[58] prefer larger values of $M_{1/2}$ so that perhaps we have a direct relation between h_o and m_t. Thus can we conclude that in the context of E_6 superstring theories a measurement of m_t will determine h_o?

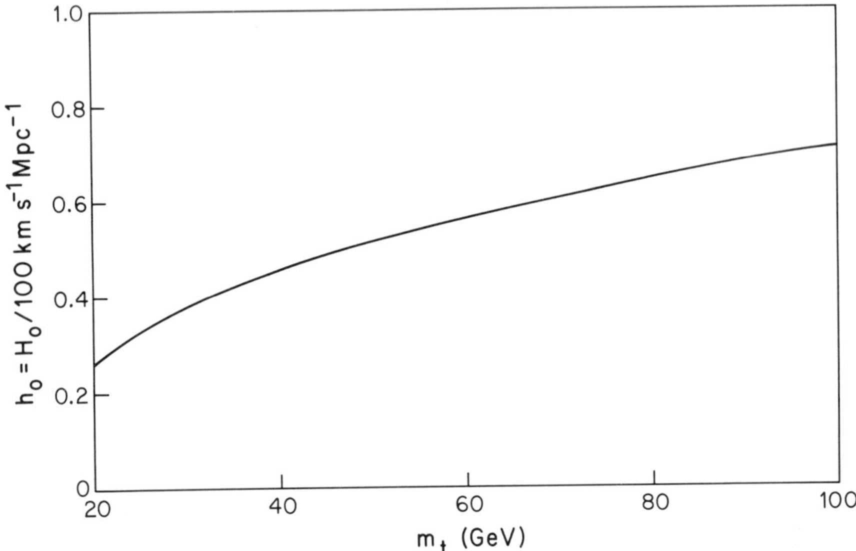

Surely this last remark was not meant to be taken too seriously. With that, I will conclude with the following observations. All observations to date are consistent with $\Omega \simeq 0.1$-0.2. Big Bang Nucleosynthesis is also consistent with $\Omega_B \simeq 0.1$. In addition, galaxy formation models can even be made to work if $\Omega_\Lambda = 1$-$\Omega = 0$.[59] In this case, inflation is also consistent because $\Omega_{total} = 1$. Unless there is some way of hiding baryons, galactic halos still require nonbaryonic dark matter.[40] Finally, searches for dark matter either using superconducting detectors[46] or the sun and searching for high energy neutrinos[53] are essentially unchanged whether $\Omega = 0.1$ or 1 as the local density must be that which is determined by rotation curves. But because not all forms of dark matter yield the same signature,[53,55] these searches may eventually yield the identity of the dark matter.

References

1) See e.g., G. A. Tammann, A. Sanduge and A. Yahill, in <u>Physical Cosmology</u>, R. Balian, J. Audouze and D. N. Schramm, eds. (North-Holland Pub. Co., Amsterdam) 1980.

2) S. M. Faber and J. J. Gallagher, Ann. Rev. Astron. Astrophys. 17 (1979) 135.

3) R. Brandenberger, Rev. Mod. Phys. 57 (1985) 1; A. D. Linde, Rep. Prog. Phys. 47 (1984) 925.

4) A. H. Guth , Phys. Rev. D23 (1981) 347.

5) J. Uson and D. Wilkinson, in <u>Inner Space/Outer Space</u>, E. Kolb, M. S. Turner, D. Lindley, K. A. Olive and D. Seckel eds. (Univ. of Chicago Press) 1986, p. 126.

6) S. Coleman, Phys. Rev. D15 (1977) 2929; C. Callan and S. Coleman, Phys. Rev. D16 (1977) 1762.

7) A. H. Guth and E. Weinberg, Phys. Rev. D23 (1981) 826; Nucl. Phys. B212 (1983) 321.

8) A. D. Linde, Phys. Lett. 108B (1982) 389; A. Albrecht and P. Steinhardt, Phys. Rev. Lett. 48 (1982) 1220.

9) S. Coleman and E. Weinberg, Phys. Rev. D7 (1973) 1888.

10) W. H. Press, Phys. Scr. 21 (1980) 702; S. W. Hawking, Phys. Lett. 115B (1982) 295; A. H. Guth and S. - Y. Pi, Phys. Rev. Lett. 49 (1982) 1110; A. A. Starobinski, Phys. Lett. 117B (1982) 175; J. M. Bardeen, P. J. Steinhardt and M. S. Turner, Phys. Rev. D28 (1983) 679.

11) J. Ellis, D. V. Nanopoulos, K. A. Olive and K. Tamvakis, Nucl. Phys. B221 (1983) 224.

12) J. Ellis, D. V. Nanopoulos, K. A. Olive and K. Tamvakis, Phys. Lett. 118B (1982) 335.

13) J. Ellis, D. V. Nanopoulos, K. A. Olive and K. Tamvakis, Phys. Lett. 120B (1983) 331.

14) D. V. Nanopoulos, K. A. Olive,M. Srednicki and K. Tamvakis, Phys. Lett. 123B (1983) 41.

414

15) E. Cremmer, B. Julia, J. Scherck, L. Girardello and
     P. Van Nieuwenhuizen, Phys. Lett. 79B (1978) 231 and Nucl. Phys.
     B147 (1979) 105; E. Cremmer, S. Ferraro, L. Girardello and
     A. Von Proeyen, Phys. Lett. 116B (1982) 231 and Nucl. Phys.
     B212 (1983) 431.

16) R. Holman, P. Ramond and G. G. Ross, Phys. Lett. 137B, (1984)
     343.

17) G. Gelmini, D. V. Nanopoulos and K. A. Olive, Phys. Lett. 131B,
     (1983) 53.

18) B. Ovrut and P. Steinhardt, Phys. Lett. 133B, (1983) 161.

19) G. Mazenko, W. Unruh and R. Wald, Phys. Rev. D31, (1985) 273.

20) A. Albrecht and R. Bradenberger, Phys. Rev. D31, (1985) 1225;
     Coughlan and G. G. Ross, Phys. Lett. 157B (1985) 151; L. Jensen
     and K. A. Olive, Phys. Lett. B159B (1985) 99; K. Engvist,
     D. V. Nanopaulos, E. Papantonopoules and K. Tamvakis, CERN
     Preprint Th-4274 (1985).

21) L. Jensen and K. A. Olive, Nucl. Phus. B263 (1986) 731.

22) E. Cremmer, S. Ferraro, C. Kounnas and D. V Nanopoulos, Phys.
     Lett. 133B, (1983) 287; J. Ellis, C. Kounnas, D. V. Nanopoulos,
     Nucl. Phys. B241, (1984) 406 and Phys. Lett. 143B, (1984) 410
     and Nucl. Phys. B247 (1984) 373.

23) J. Ellis, K. Enquist, D. V. Nanopoulos, K. A. Olive and
     M. Srednicki, Phys. Lett. 152B, (1985) 175.

24) G.D. Coughlan, W. Fischler, E.W. Kolb, S. Raby and G. G. Ross,
     Phys.Lett. 131B (1983) 59; M. Dine, W. Fischler and
     D. Nemeschansky, Phys. Lett. 136B (1984) 169; G. D. Coughlan,
     R. Holman, R. Raymond and G. G. Ross, Phys. Lett. 140B (1984)
     44.

25) E. Witten, Phys. Lett. 155B (1985) 151.

26) J. Ellis, K. Enquist, D. V. Nanopoulos and M. Quiros, CERN
     preprint TH-4325 (1985).

27) K. A. Olive, D. N. Schramm, G. Steigman, M. S. Turner and
     J. Yang, Ap. J. 246 (1981) 547.

28) J. Yang, M. S,. Turner, G. Steigman, D. N. Schramm and
     K. A. Olive, Ap. J. 281 (1984) 493.

29) A. Bosegard and G. Steigman, Ann. Rev. Astron. Astrophys.

30) See e.g. Proceedings of the ESO Workshop on Primordial Helium,
     eds. P. S. Shaver , D. Kurth and K. Kajar, (Garchiy, Germany)
     1983.

30a) D.S.P. Dearborn, D. N. Schramm and G. Steigman, Ap. J. (in press)
     1986.

31) J. Primack, SLAC preprint 3387 (1984).

32) E. R. Harrison, Phys. Rev. D1 (1970) 2729.

33) Ya. B. Zeldovich, Mon. Not. R. Ast. Soc. 160 (1972) 1P.

34) J. Uson and D. Wilkinson in Innner Space/Outer Space eds.
     E.W. Kolb, M. S. Turner, D. Lindley, K. A. Olive and D. Seckel,
     (Univ. of Chicago Press) 1986, p. 126.

35) J. R. Bond and A. S. Szalay, Ap. J. 274 (1983) 443.

36) J. R. Bond, G. Efstathiov and J. Silk, Phys. Rev. Lett. 45 (1980)
     1980; Ya. B. Zeldovich and R.A. Sunyaev, Pisma Astr. Zh. 6 (1980)
     457.

37) J. Silk, Proc. 1985 Int. Symposium on Lepton and Photon
     Interactions at High Energies eds. M. Konume and K. Takahashi
     (Kyoto Univ.) 1986.
38) V. C. Rubin,W. K. Ford and N. Thonnard, Ap. J. 255 (1978) L107.

39) D. Burstein and V. C. Rubin, Ap. J. 297 (1985) 423.

40) D. J. Hegyi and K. A. Olive, Phys. Lett. 126B (1983) 28; Ap. J.
     303 (1986) 56.

41) P. Hut, Phys. Lett. 69B (1977) 85; B. W. Lee and S. Weinberg,
     Phys. Rev. Lett. 39 (1977) 165; L. Krauss, Phys. Lett. 128B
     (1983) 37, E. W. Kolb and K. A. Olive, Phys. Rev. D33 (1986)
     1202.

42) H. Goldberg, Phys. Rev. Lett. 50 (1983) 1419, L. Krauss, Nucl.
     Phys. B227 (1983) 556.

43) J. Ellis, J. Hagelin, D. V. Nanopoulos, K. A. Olive and
     M. Srednicki, Nucl. Phys. B238 (1984) 453.

44) L. E. Ibanez, Phys. Lett. 137B (1984) 160; J. Hagelin, G. Kane
     and S. Ruby, Nucl. Phys. B241 (1984) 638.

45) R. Mochkovitch, K. A. Olive and J. Silk, Ap. J. Lett (submitted)
     1986.

46) A. Drukier and L. Stodolsky Phys. Rev. D30 (1984) 2295;
    M. Goodman and E. Witten, Phys. Rev. D31 (1984) 3059; A. Drukier,
    K. Freese and D. Spergel, Harvard CFA Preprint 1985.

47) J. Silk and M. Srednicki, Phys. Rev. 53 (1984) 624; J. Hagelin
    and G. Kane, Nucl. Phys. B263 (1986) 399; S. Rudaz, T. F. Walsh
    and F. Stecker, Phys. Rev. Lett. 55 (1985) 2622.

48) A. Buffington, S. Schindler and C. Pennypacker, Ap. J. 248
    (1981) 1179; R. Protheroe, Ap. J. 254 (1982) 391.

49) G. Steigman, C. L. Sarazin, H. Quintana and J. Faulkner, Ap. J.
    83 (1978) 1050.

50) D. N. Spergel and W. H. Press, Ap. J. 294 (1985) 663.

51) W. H. Press and D. N. Spergel, Ap. J. 296 (1985) 679.

52) L. Krauss, K. Freese, D.N. Spergel and W.H. Preess, Ap. J. 299
    (1986) 1001.

53) J. Silk, K. A. Olive and M. Srednicki, Phys. Rev. Lett. 55, 257
    (1985); M. Srednicki, K. A. Olive and J. Silk, Nucl. Phys. B
    (submitted) (1986).

54) D. H. Perkins, Ann. Rev. Nucl. Part. Sci. 34 (1984) 1.

55) J. Hagelin, K.-W. Ng and K. A. Olive, (in preparation) 1986.

56) B. Campbell, J. Ellis, K. Enquist, J. Hagelin, D. V. Nanopoulos
    and K. A. Olive, Phys. Lett. B (in press) (1986).

57) J. Ellis, K. Enquist, D.V. Nanopoulos and F. Zwirner, CERN
    Preprints TH-4323/85 and TH-4350/86; J. Ellis, these proceedings.

58) J. Ellis, K. Enquist, D. V. Nanopoulos and S. Sarkar, Phys. Lett.
    167B (1986) 457; G. Steigman, K. A. Olive, D. N. Schramm and
    M. S. Turner (in preparation) 1986.

59) N. W. Vittorio and J. Silk, Ap. J. 297 (1985) L1.

NEW PARTICLE SEARCHES BY DOUBLE BETA DECAY

David O. Caldwell
Physics Department, University of California
Santa Barbara, CA 93106
U.S.A.

ABSTRACT

So far, neutrinoless double beta decay has not been observed, but the lifetime limit (3×10^{23} years) is such a stringent one that its non-observation can set the best upper limit
for the mass of the electron neutrino (if it is a Majorana
particle) and the best lower limit on the mass of a heavy
Majorana neutrino for a given coupling to the electron neutrino, ν_e. In addition it provides the best limit on the
coupling of light bosons, such as Majorons, to ν_e, and also
the best limit on the existence of right-handed currents in
the case in which all right-handed Majorana neutrinos are
heavier than all left-handed leptons. The apparatus has to
have such low backgrounds even in the keV energy region that
it also can be used to set the best terrestrial limits on the
mass of solar axions and of other dark matter candidates.

1. INTRODUCTION TO DOUBLE BETA DECAY

Double beta decay, although in the domain of nuclear physics, is
an extremely sensitive probe of the new frontiers in particle physics.
In particular, it provides an excellent way to set limits on the existence of light Majorana neutrinos, the right-handed W boson, heavy
Majorana neutrinos, and light bosons, such as Majorons or familons.
In addition, because the apparatus has to have lower backgrounds in the
keV energy region than any previous detection system by orders of magnitude, it also provides a means of searching for solar axions and for
dark matter. The limitations this type of experiment places on the

418

existence of the various particles will be presented after an introduction to double beta decay, which unfortunately still remains
relatively unfamiliar.[1]

Since the pairing energy makes nuclei with even numbers of protons
and neutrons more tightly bound than their odd-odd neighbors, it often
occurs that the only decay energetically allowed for an even-even nucleus is one that would require its changing charge by two units. This
can in principle occur via the second-order weak interaction

$$(A,Z) \rightarrow (A,Z+2) + 2e^- + 2\bar{\nu}, \tag{1}$$

which would have a very long lifetime. A decay which would be favored
by a huge ($\sim 10^8$) phase space factor would be

$$(A,Z) \rightarrow (A,Z+2) + 2e^-, \tag{2}$$

but this violates lepton number and requires a helicity reversal for
the virtual neutrino exchanged between the decaying neutrons. Another
type of neutrinoless decay is possible if a Majoron or similar very
light or massless boson, χ°, existed:

$$(A,Z) \rightarrow (A,Z+2) + 2e^- + \chi^\circ. \tag{3}$$

These three decays can be distinguished in a detector which measures the sum of the electron energies, since decay (2) (designated
$\beta\beta_{0\nu}$) gives a single peak, (1) (or $\beta\beta_{2\nu}$) has a four-body spectrum with
a maximum at about one-third of the total energy, and (3) (symbolized
by $\beta\beta_{0\nu,\chi}$) has a three-body spectrum with a maximum at about 0.8 of
the total energy, as shown in Fig. 1.

Since we shall devote the most attention to decay (2), a little
more explanation is in order to understand how it could occur. In normal β decay a neutron emits an e^- and an antineutrino, conserving lepton number, but for $\beta\beta_{0\nu}$ the neutral particle is virtual and must be
reabsorbed by the second neutron as a neutrino, in order to conserve
lepton number and permit a second e^- to be emitted. This could occur
only if neutrino and antineutrino are identical, which is the case if
the neutrino is a Majorana and not a Dirac particle. Besides this requirement of overall violation of lepton number, there is also the

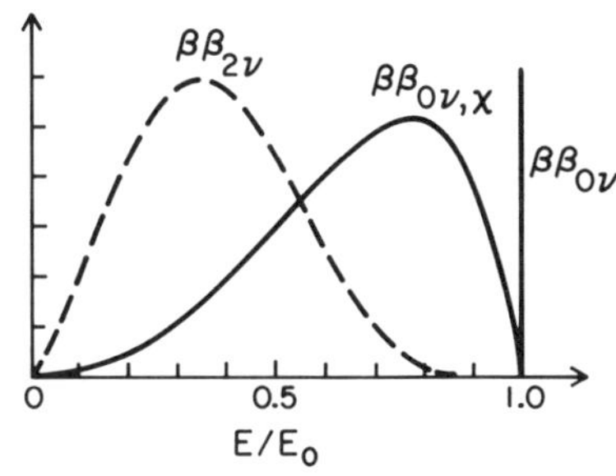

Fig. 1. The summed kinetic energies of the two electrons for $\beta\beta_{2\nu}$, $\beta\beta_{0\nu}$, and $\beta\beta_{0\nu,\chi}$ decay modes (arbitrary scales).

necessity to achieve a helicity reversal. If the neutrino were massless this could be achieved if there were a small admixture of right-handed currents (RHC), as shown in Fig. 2. Alternatively, if the neutrino has mass, then helicity reversal is possible, as in the case of π decay, but it is very costly in rate, the lifetime being proportional to the square of the neutrino mass. For a decay induced by neutrino mass, two normal left-handed W bosons can be involved so that two left-handed e^- are emitted back to back, canceling their helicities, making S-wave emission possible, so that the selection rule for the transition is $\Delta J=0$. The decay then occurs from the 0^+ ground state of one even-even nucleus to the 0^+ ground state of the next even-even nucleus. As shown in Fig. 2, for the case of RHC, one left-handed and one right-handed e^- can be emitted, requiring a p-wave to supply the angular momentum of one from the helicity addition. This permits $\Delta J=2,1$, or 0, so that the transition can go from the 0^+ state to the 0^+ ground state or to an excited state, the first one of which is generally 2^+. Thus the $0^+\rightarrow2^+$ transition can occur only if RHC exist, and hence this is useful for separating neutrino mass and RHC effects, if the more probable $0^+\rightarrow0^+$ transition is ever observed.

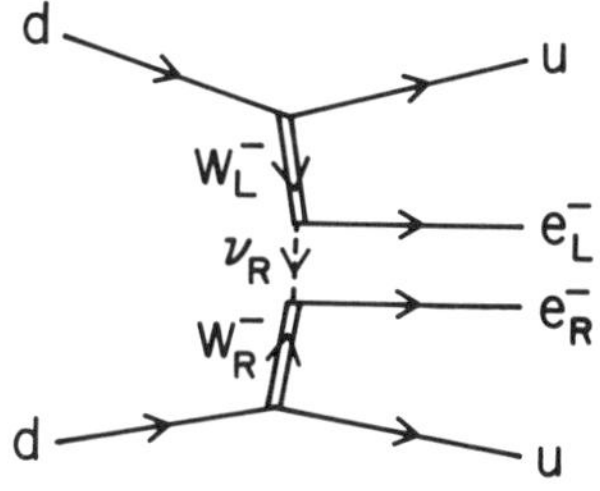

Fig. 2. An example of neutrinoless double beta decay mediated by right-handed currents.

2. NEUTRINOLESS DOUBLE BETA DECAY EXPERIMENTAL CONSIDERATIONS

A second-order weak interaction inhibited by the requirement of helicity reversal is clearly going to be a very slow process with a lifetime ($>10^{23}$ years) longer than that of any conjectured process except proton decay. Unlike proton decay, the $\beta\beta$ source is not prevalent, and acquiring enough of an isotope capable of displaying $\beta\beta$ decay is expensive. Also proton decay releases $\sim10^3$ MeV, whereas $\beta\beta$ energies are $\sim$ MeV, which means that the decay electrons have a short range. Then the source, which has to be massive to get sufficient rate, must be quite thin to get the electrons out. One solution to this problem is to use ^{76}Ge as a source, since it constitutes 7.76% of normal Ge, which can be made into a very high resolution detector of electrons. Then the unified source and detector can be made as massive as can be afforded. Actually, Ge detectors have not been made larger than 200 cm^3, so one has to go to multidetector arrays.

Even with a quantity of Ge costing $\sim10^5$\$ one could hope to get only a signal of the order of a count or a very few counts per year in a peak at 2.041 MeV which would typically have a full width of about 3.5 keV. Clearly background is then the essential issue. Direct effects of cosmic rays can be eliminated by going deep underground and/or providing surrounding anticoincidence counters, but indirect effects of the production of neutrons in surrounding rock is difficult to eliminate at these levels. More difficult to deal with is the pervasive radioactivity which one finds in all materials. Most of the effort experimenters have put into making these low-background systems has been in finding materials of sufficiently low radioactivity. Certain materials are better than others, but even within the same type of material there can be large differences in radioactivity, and hence each component must undergo extensive testing. Some experiments are installed very deep underground and have a surrounding shield against local radioactivity made of materials with as little radioactivity as can be obtained. Other experiments need to go less deep and can be less selective about passive shielding by using an active shield of NaI scintillator. Such a shield not only reduces effects of external sources but more importantly reduces the effect of residual radioactivity internal to the shield by vetoing Compton scatters in the Ge. A γ can deposit 2.04 MeV in a Ge detector by Compton scattering, but if it (or one of the cascade γ's usually associated with it) then is counted in the NaI, that background is eliminated. Thus although the

NaI, having a greater radioactivity than the best passive shielding materials (even allowing for its self-vetoing), may display higher photon peak backgrounds than is seen in the experiments without an active shield, there may be a lower Compton continuum using the NaI. Indeed, the best backgrounds achieved around 2 MeV are about the same in three experiments, one with NaI and two without. The source of that background is still unknown, although some of it is due to radioactivity induced in the material by cosmic rays before it was put underground.

3. TWO-NEUTRINO DOUBLE BETA DECAY EXPERIMENTS

Except toward the end of this presentation when future directions will be mentioned, the only type of experiment discussed for $\beta\beta_{0\nu}$ will be the Ge detector; results from other techniques are not competitive, with one exception which will be noted shortly. However, for observing $\beta\beta_{2\nu}$ quite different approaches have so far been more successful. In particular, the geochemical method, which has the advantages of a long integration time and a large source sample, has provided the only certain observation of double beta decay. The noble gas decay products which could come only from double beta decay have been found in billion-year-old rocks. There is no way to distinguish experimentally whether these products come from $\beta\beta_{2\nu}$ or $\beta\beta_{0\nu}$, but the former should have a shorter lifetime, and hence one can safely presume that the resulting lifetime measurements are for $\beta\beta_{2\nu}$. As can be seen from Table I the geochemical measurements[2] of lifetimes for ^{82}Se and ^{130}Te, as well as the limit for ^{128}Te, are one to two orders of magnitude longer than the numerous calculations for this second-order weak process. Note that the completeness of the calculation does not necessarily correspond to the closeness of the result to experiment. The source of this discrepancy is completely unknown at this time. That the geochemical measurements are not likely to be that incorrect is given support by the one laboratory measurement[3] which may actually be observing $\beta\beta_{2\nu}$ in ^{82}Se. The Irvine group uses a thin source of enriched ^{82}Se and observes the two electrons in a Time Projection Chamber. Although this work is still in progress, they have candidate events and their results are consistent with the geochemical lifetime as shown in

Table I. The problem is then a theoretical one, and while the discrepancy is probably in the wrong direction to indicate new particle physics, it may well be the harbinger of new nuclear physics.

Table I: Half-Life Values in Years for $\beta\beta_{2\nu}$ Decays

Decay	Geochemical[2]	Laboratory	Theory
^{82}Se	$(1.45\pm0.15)\times10^{20}$	$(1.7\pm0.6)\times10^{20}$ [3]	$(1.5^{[4]},2.6^{[5]}3.3^{[6]})\times10^{19}$
^{130}Te	$(2.6\pm0.3)\times10^{21}$		$(1.7^{[5]},2.3^{[6]},12^{[4]})\times10^{19}$
^{128}Te	$>7.9\times10^{24}$ (2σ)		$(0.88^{[5]},1.1^{[6]},5.7^{[4]})\times10^{23}$
^{150}Nd		$>2.4\times10^{19}$ [7]	$(0.48^{[8]},2.2^{[6]})\times10^{17}$
^{76}Ge		$>8\times10^{19}$ [9]	$(2.1^{[10]},2.2^{[4]},4.2^{[5]},18^{[6]})\times10^{20}$

4.   NEUTRINOLESS DOUBLE BETA DECAY EXPERIMENTAL RESULTS

Following a suggestion by Pontecorvo,[11] one can utilize the geochemical results for ^{128}Te and ^{130}Te to look for the $\beta\beta_{0\nu}$ process. The phase space for $\beta\beta_{2\nu}$ [lifetime $T\sim(E_0/m_e)^{11}$] and $\beta\beta_{0\nu}$ [$T\sim(E_0/m_e)^5$] is quite different, and this coupled with the very different energy release, E_0, for the two isotopes ($E_0/m_e=5.0$ for ^{130}Te and 1.7 for ^{128}Te) makes the lifetime ratio a test for $\beta\beta_{0\nu}$. More specifically, the expected[1b] ratios are $T^{2\nu}(128)/T^{2\nu}(130)=5130$ and $T^{0\nu}(128)/T^{0\nu}(130)=25$ (for no right-handed currents). The results in Table I then give[1b] $m_\nu<6$ eV under the assumption that the nuclear matrix elements cancel. The lack of understanding of the $\beta\beta_{2\nu}$ rates raises doubts[1b] about this assumption, and more specifically it has been argued[4] that the $\beta\beta_{2\nu}$ rates depend critically on low-lying 1^+ intermediate states which can be quite different for the two isotopes. On the other hand, the limit on the ^{128}Te lifetime, even if its decay were all $\beta\beta_{0\nu}$ could provide a more stringent limit on m_ν.[12] However, in this case the experimental uncertainties in ore age and diffusive loss of xenon, as well as the theoretical uncertainties, make conclusions less clear.

As stated above, the rest of the competitive results for the $\beta\beta_{0\nu}$ process come from ^{76}Ge experiments. Recently these have improved both in background reduction--all experiments are now underground and radioactivities have been reduced substantially--and also in quantity of Ge used. A brief status report will be given on the work of the various groups, starting with Milan, which was the first[13] to use this technique. Their 115 cm^3 detector in the Mt. Blanc tunnel reached a limit of 1.2×10^{23}y after 2.4y of counting time, while a lower background 138 cm^3 detector achieved 0.7×10^{23}y after only 0.84y.[14] Their main efforts are now turning to a different technique, the use of ^{136}Xe in a multiwire proportional chamber. The Pacific Northwest Laboratory, University of South Carolina group has used a 135 cm^3 detector for 0.92y and achieved a limit of 1.4×10^{23}y because of good background reduction.[15] They are continuing to work on this issue before implementing a multidetector array. The Guelph-Aptec-Kingston group has published[16] a 0.3×10^{23}y result with a 194 cm^3 detector but now has improved backgrounds and is using three detectors of this size. The Cal Tech group had a result of 0.2×10^{23}y using a 90 cm^3 detector above ground,[17] but they have moved into the St. Gotthard tunnel, greatly improved radioactivity backgrounds, and are now installing eight 140 cm^3 detectors. They are also working on a ^{136}Xe Time Projection Chamber. The Osaka group, using a 164 cm^3 detector inside a NaI anticoincidence shield achieved 0.6×10^{23}y in 0.52y[18] and is continuing to count in the Kamioka mine. They are also doing a ^{100}Mo experiment. Finally, the UCSB-LBL group,[10] also using NaI, reached 0.5×10^{23}y in 0.18y above ground with two detectors (336 cm^3), 2.5×10^{23}y below ground with four detectors (658 cm^3) after 0.40y and 3.4×10^{23}y with six detectors (986 cm^3) for an additional 0.45y. They now have eight detectors (1.3 liters) operating.

5. INTERPRETATION OF THE RESULTS

The $\beta\beta_{0\nu}$ results for ^{76}Ge can be used to set a number of important limits, but there are caveats. If there is a mixture of light neutrinos, these can have opposite CP eigenstates,[19] producing a cancellation which makes the effective $<m_\nu>$ less than it might be for, say, single β decays. Similarly the phase of a heavy neutrino mixing with the light one could produce a cancellation, the effect of which would vary from one nucleus to another.[20] There is also the uncertainty introduced by the disagreement between calculated and experimental $\beta\beta_{2\nu}$

rates. One can conservatively scale all the limits given below by $\sim\sqrt{10}$, since in the ^{82}Se case the disagreement is an order of magnitude. However, it can also be argued[4] that the problem with the $\beta\beta_{2\nu}$ decays relates to this being a long-range process, and that the exchange of a high-momentum virtual neutrino makes $\beta\beta_{0\nu}$ a short-range interaction for which the calculations are more reliable. Finally, there is the statistical uncertainty resulting from the convention of quoting 68% confidence level limits.

These problems aside, the calculation of this process involves approximations, and some indication of the uncertainties in these is shown by the difference in results obtained by different groups. In Table II below results are given with more significant figures than is warranted just to show these differences. For example, the Table shows the range of neutrino mass, $<m_\nu>$, based on the UCSB-LBL lifetime limit of 3.4×10^{23}y, and assuming no RHC. Also given are limits for two RHC parameters assuming $<m_\nu>=0$ and that only the one RHC parameter contributes. It is more likely that two or three of these would be involved. To understand the notation for the RHC parameters, one can use a simplified form of the Hamiltonian: $H_W=-2^{-1/2}G_F\cos\theta_c[j_L^\mu(J_{L\mu}+\eta_{LR}J_{R\mu}^\dagger) + j_R^\mu(\eta_{RL}J_{L\mu}^\dagger+\eta_{RR}J_{R\mu}^\dagger)] + h.c.$ The j (J) are components of the leptonic (hadronic) current, with L and R designating handedness, and G_F is the weak coupling constant, with θ_c the Cabbibo angle. Note that the η's are defined so that the first (second) subscript refers to the leptonic (hadronic) current.

Table II: Neutrino Mass and RHC Limits for $T_{1/2}>3.4\times10^{23}$y

Parameter	Los Alamos[1b]	Tübingen[9]	Heidelberg[12]	Osaka[1c]
$<m_\nu>$ eV	2.2	1.6	0.9	--
$<\eta_{RL}>$	5.4×10^{-6}	3.9×10^{-8}	--	3.9×10^{-7}
$<\eta_{RR}>$	3.6×10^{-6}	3.1×10^{-6}	--	4.2×10^{-6}

For $<\eta_{RL}>$ the calculation of Ref. 1b does not include the relativistic P-wave effect or the nucleon recoil term, so the more stringent limits apply. If heavy as well as light Majorana neutrinos exist--the usual mechanism to give neutrinos a mass--then these limits are 10^4 to 10^6 times better than those obtained from μ decay, and hence set stringent limits on a product of right-handed W boson mass times the coupling angle between left- and right-handed W bosons. Note that the RHC limits from the $0^+ \rightarrow 0^+$ decay are stronger than those that can be obtained from the best limit for the $0^+ \rightarrow 2^+$ transition, which is 1.0×10^{23}y from the UCSB-LBL experiment.

That experiment has also set the best limit on the lifetime for $\beta\beta_{0\nu,\chi}$ of 6×10^{20}y (90% confidence level). From the evaluation of Ref. 1c, this result gives a limit on the coupling of the Majoron to the electron neutrino of $< 10 \times 10^{-4}$, using the nuclear matrix elements of Ref. 1b and $< 4.1 \times 10^{-4}$ using those of Ref. 12.

Searching for heavy neutrinos has become more and more important, not only to determine whether there is a fourth generation of leptons, in which the neutrino could be either a Dirac or Majorana particle, but also to find extra neutrinos--most frequently Majorana particles--such as are usually required in left-right symmetric models and particularly in the low-energy limit (E_6) of superstring theories. If the heavy neutrino of mass M_ν is a Majorana particle it can couple to the light electron neutrino with a mixing coefficient U_e^L and induce $\beta\beta_{0\nu}$ decay. The UCSB-LBL limit of 3.4×10^{23}y then imposes in the analysis of Ref. 1b the limit on M_ν and U_e^L shown by the line in Fig. 3. Values below and to the right of the line are allowed. The limit is seen from the figure to be much more stringent than those[21] from other experiments and universality, although those limits apply to Dirac particles as well.

Finally, brief mention should be made of a new use for $\beta\beta$ apparatus to search for the unknown dark matter which probably makes up about 90% of the mass of the universe. The $\beta\beta$ detector can be used in a general way to look for the recoil energy of a cold (i.e., heavy) dark matter candidate, with a lower limit being set on the mass the lower the energy one can measure in the detector without being swamped by background. One can already reach mass limits ~ 10 GeV for heavy Dirac neutrinos or scalar neutrinos, but considerable improvement in the Ge detectors is possible, since they were configured originally to have good detector characteristics and low backgrounds in the MeV region, whereas

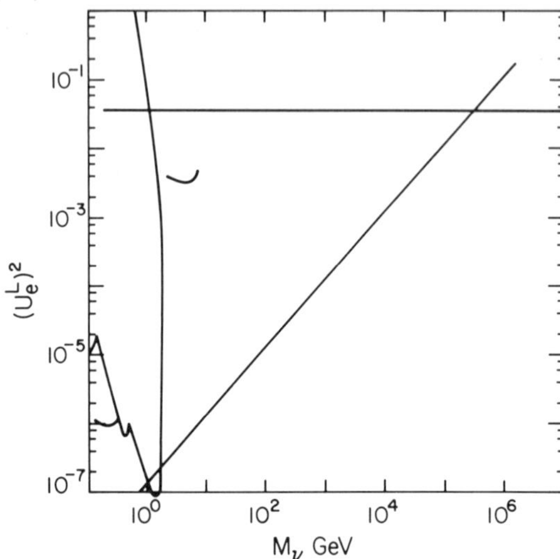

Fig. 3. Limit on the mass of a heavy Majorana neutrino as a function of its probability for mixing with an electron neutrino.

in this case counting must be done in the keV region. The axion is a promising dark matter candidate, and one could hope to see solar axions because of atomic enhancements in their detection.[22] At the limit of $\sim$5 keV reached by present detectors no meaningful limit can be set on axion mass, since the background rate detected gives a limit comparable to that required to keep the sun from cooling too fast. However, an order of magnitude can be gained by going to 1.5 keV,[22] and that is possible with reconfigured detectors.

Thus the remarkable sensitivity of this second-order weak process can tell us a lot about the necessary characteristics of many existing and conjectured particles, and the equally remarkable sensitivity of the apparatus to detect it may add much more information.

6.   REFERENCES

1)   Very complete reviews of double beta decay are given by (a) H. Primakoff and S.P. Rosen, Ann. Rev. Nucl. Part. Sci. 31, 145 (1981); (b) W.C. Haxton and G.J. Stephenson, Jr., Progr. Part. Nucl. Phys. 12, 409 (1984); and (c) M. Doi, T. Kotani, and E. Takasugi, Progress of Theor. Phys. Supp. 83 (1985).

2)   T. Kirsten, H. Richter, and E. Jessberger, Phys. Rev. Lett. 50, 474 (1983); AIP Conf. Proc. 96, 396, Neito, et al. (New York, 1983).

3)   S.R. Elliott, A.A. Hahn, and M.K. Moe, UCI Preprint (Feb. 1986, unpublished).

4)   K. Grotz and H.V. Klapdor, Phys. Lett. 142B, 323 (1984).

5)   W.C. Haxton, G.J. Stephenson, Jr., and D. Strottman, Phys. Rev. Lett. 47, 153 (1981) and Phys. Rev. D25, 2360 (1982); see also ref. 1b) for more recent results in some cases.

6)   P. Vogel and P. Fisher, Phys. Rev. C32, 1362 (1985).

7)   A.A. Klimenko, A.A. Pomansky, and A.A. Smolnikov, Proc. "Neutrino-84", ed. by Kleinknecht and Paschos (World Scientific Pub., Singapore, 1984), p. 161.

8)   K. Grotz and H.V. Klapdor, Phys. Lett. 157B, 242 (1985).

9)   T. Tomoda, et al., Nucl. Phys. 153B, 1 (1985).

10) D.O. Caldwell, et al., Phys. Rev. Lett. 54, 281 (1985) and Phys. Rev. D (to be published).

11) B. Pontecorvo, Phys. Lett. 26B, 630 (1968).

12) K. Grotz and H.V. Klapdor, Phys. Lett. 153B, 1 (1985).

13) E. Fiorini, et al., Phys. Lett. 25B, 602 (1967) and Nuovo Cimento 13A, 747 (1973).

14) E. Bellotti, et al., Phys. Lett. 146B, 450 (1984).

15) F.T. Avignone, et al., Phys. Rev. Lett. 54, 2309 (1985) and PNL-USC Preprint (Nov. 1985, unpublished).

16) J.J. Simpson, et al., Phys. Rev. Lett. 53, 141 (1984).

17) A. Forster, et al., Phys. Lett. 138B, 301 (1984).

18) H. Ejiri, et al., Proc. of the Santa Fe Meeting of APS, ed. by Goldman and Nieto (World Scientific Pub., Singapore, 1984), p.252.

19) M. Doi, et al., Phys. Lett. 102B, 323 (1981) and Prog. Theo. Phys. 69, 602 (1983); L. Wolfenstein, Phys. Lett. 107B, 77 (1981); B. Kayser and A.S. Goldhaber, Phys. Rev. D28, 2341 (1983).

20) A. Halprin, S.T. Petcov, and S.P. Rosen, Phys. Lett. 125B, 335 (1983).

21) F.J. Gilman and S.H. Rhie, Phys. Rev. D32, 324 (1985) and F.J. Gilman, SLAC-PUB-3898 (1986), to be published in Comments in Nuclear and Particle Physics.

22) S. Dimopoulos, G.D. Starkman, and B.W. Lynn, SLAC-PUB-3850 (1985), submitted to Phys. Lett. B.

CONSTRAINTS ON MASSIVE NEUTRINOS FROM $\pi \to e\nu$ DECAY[*]

D.I. Britton

TRIUMF
4004 Wesbrook Mall, Vancouver, B.C., Canada, V6T 2A3

ABSTRACT

The positron energy spectrum from $\pi \to e\nu$ decay has
been searched looking for evidence of massive neutrinos.
A new technique, developed to suppress the dominant π-μ-e
decay chain background, has resulted in an order of magnitude improvement in the limits for the neutrino mass
range 60-130 MeV.

If all neutrinos are massless then mixing amongst them is
physically unobservable. This effect manifests itself as conservation
of lepton family number. Alternatively, if neutrinos have mass the
mixing matrix is no longer necessarily diagonal. In the quark sector
mass and weak eigenstates are related by the Kobayashi-Maskawa matrix
and in the lepton sector a similar structure may be used:

$$\nu_\ell = \sum_i U_{\ell i}\nu_i$$

Here, the weak eigenstates ν_ℓ are related to the mass eigenstates ν_i
by a mixing matrix $U_{\ell i}$. The experimental problem is to determine, or
set limits upon, both the masses and the mixing matrix elements.
Neutrino mass may be looked for in a variety of ways including the
production or decay of massive neutrinos, neutrino oscillations, and
neutrinoless double beta decay. Production or decay experiments
require relatively large neutrino masses, of the order of MeV, but are
sensitive to very small mixing. Oscillation experiments are sensitive
to small neutrino masses, eV and less, but require significant mixing.
Neutrinoless double beta decay is sensitive to Majorana mass.

[*]The $\pi e\nu$ collaboration at TRIUMF includes G. Azuelos, D.A. Bryman,
E.T.H. Clifford, P. Kitching, J.A. Macdonald, T. Numao, B. Olaniyi,
A. Olin, and J-M. Poutissou; and M.S. Dixit, NRC, Ottawa, Ont, K1A OR6

At TRIUMF, the decay $\pi \to e\nu$ has been studied to measure its branching ratio as a test of electron-muon universality.[1] The data has also been searched for evidence of massive neutrinos.[2] In the case of massless neutrinos the decay $\pi \to e\nu$ is kinematically simple and the positron is emitted with a unique energy of 69.3 MeV. In general, for finite neutrino masses, mixing occurs. Therefore, $\pi \to e\nu$ decay could show evidence of all the kinematically accessible mass eigenstates. More specifically,[3] examination of the energy spectrum of emitted positrons in the region below the main peak could reveal additional peaks with the following branching ratios:

$$R_{ei} = \frac{\Gamma\ (\pi^+ \to e^+\nu_i)}{\Gamma\ (\pi^+ \to e^+\nu_1)} = |U_{ei}|^2\ \rho_e(\delta_i) \tag{1}$$

Where ν_1 refers to the conventional massless neutrino and $\rho_e(\delta_i)$ is the product of a phase space and a helicity suppression factor:

$$\rho_e(\delta_i) = \rho_e\left(m(\nu_i)^2/m_\pi^2\right) = \frac{[1+\delta_e^2+\delta_i^2-2(\delta_i+\delta_e+\delta_i\delta_e)]^{1/2}[\delta_i+\delta_e-(\delta_i-\delta_e)^2]}{\delta_e(1-\delta_e)^2}$$

where $\delta_e = m_e^2/m_\pi^2$.

Since the observed positron energy spectrum in $\pi \to e\nu$ decay consists of a single peak at 69.3 Mev the electron neutrino couples predominantly to a light or zero mass eigenstate. However, data from this experiment allows limits to be set on weak coupling to more massive eigenstates which would result in additional peaks buried under the background in the positron energy region below 70 Mev.

An important asset in looking for massive neutrinos in this decay is the helicity suppression factor $\sim 10^4$ contained in the kinematical term above. In the case of massive neutrinos the helicity suppression is reduced and the extra peaks due to decays to heavy neutrinos are enhanced relative to decays to massless neutrinos. Consequently this experiment provides a sensitive test of mixing.

The experiment was performed at TRIUMF using a 77 MeV/c beam of positive pions. Referring to Fig. 1, the beam was defined by the

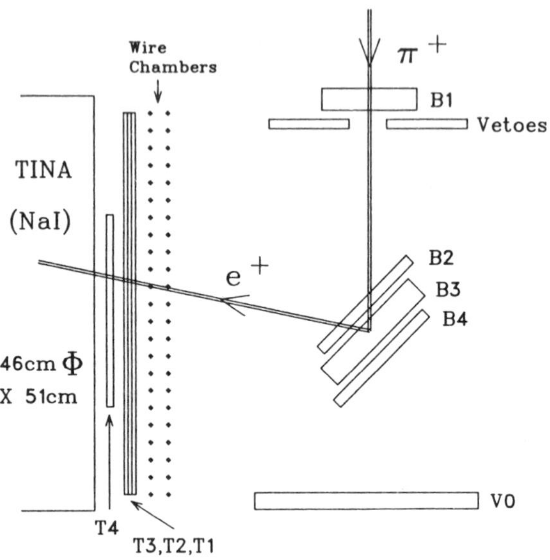

Fig. 1. Experimental set-up

scintillators B1 and the two veto counters, and then stopped in the target counter B3 at a rate of 10^5 s^{-1}. Following a stop, the pion decays to a positron, either directly in the decay $\pi \to e\nu$, or much more often, via a muon in the chain $\pi \to \mu\nu$ followed by $\mu \to e\nu\bar{\nu}$. In order to minimise the energy deposited by the exiting positrons, the target counters were oriented at 45° and the beam momentum was chosen so that the pions stopped about 3 mm from the front edge of the 13 mm thick target counter, B3.

The positrons were detected in a 51 cm × 46 cm Φ NaI(Tℓ) detector (TINA) which was positioned 20 cm from the target. The solid angle acceptance ~ 5% was defined by a 178 mm diameter scintillator T4 directly preceding TINA. Three other thin scintillators T_1, T_2, T_3 provided energy loss information to discriminate against protons. There was also a pair of wire chambers which gave the capability of further restricting the acceptance.

The energy spectrum detected in TINA, shown in Fig. 2, has two components: A distribution from 0–53 MeV due to the positrons from the

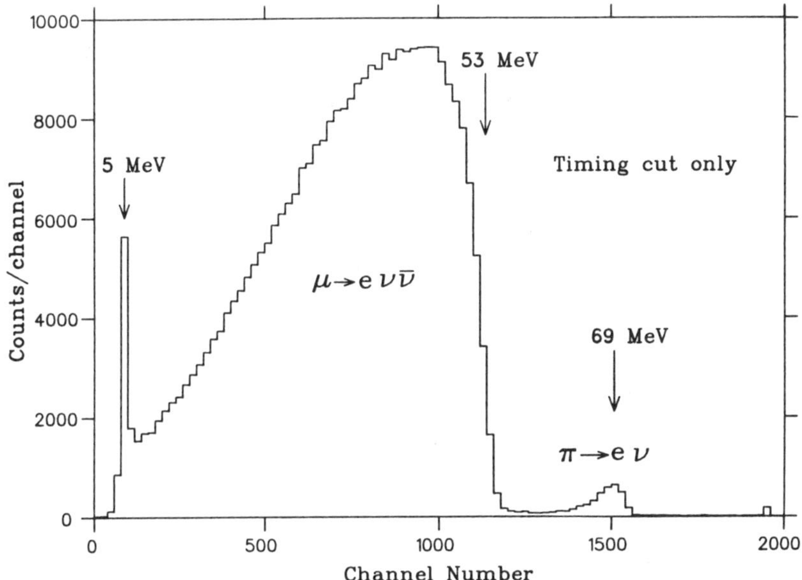

Fig. 2. Positron energy spectrum in TINA.

3-body muon decay, and a monoenergetic peak at 69 MeV which are the
positrons from the direct 2-body pion decay. The problem in looking
for massive neutrinos is immediately obvious: The region below 70 MeV,
to be searched for small additional peaks, is swamped by the huge background of positrons from muon decay. In fact, the muon decay component
in Fig. 2 has already been suppressed by a factor of 100 by selecting
only those events that occur between 2 and 30 ns after a pion stop.
The positrons from muon decay are thus suppressed due to the relatively
long lifetime of the muon.

A technique for obtaining additional suppression of the π-μ-e
chain background has been developed, based upon measurement of the
total energy deposited in the target counter B3. The principle is
illustrated in Fig. 3. If a pion stopping in the target decays
directly to a positron then the energy deposited in the target scintillator is the kinetic energy of the pion T_π, 20 MeV, along with a small
portion of the positron's kinetic energy ($\varepsilon \sim 1$ MeV). Alternatively, if
the pion decays first to a muon then an additional 4.2 MeV is detected

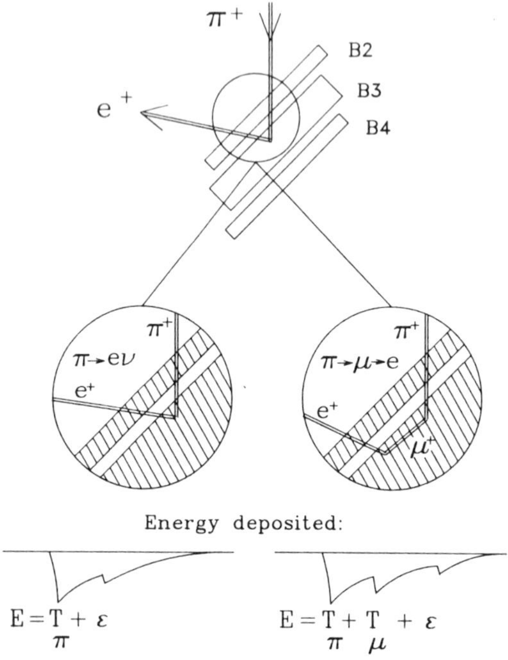

Fig. 3.   Target energy technique.

corresponding to the kinetic energy of the muon. The mean range of
4 MeV muons in plastic scintillator is less than 1.4 mm, so by stopping
the pions on average 3 mm into the target, all the muons are contained.
Two distinct peaks are seen in the spectrum of target energy as shown
in Fig. 4. By selecting only those events that fall in the lower
energy peak the background, due to the predominant decay involving
muons, is strongly suppressed.

Additional suppression may be obtained from this technique by
using 2 ADCs to digitise the signal from the target. The first ADC is
gated with a 200 ns wide pulse that starts 5 ns before the pion stop in
the target and the second ADC has a narrow gate that only views the
rising edge of the analogue pulse. Using the difference between the
signals from these two ADC's, an energy dependent cut can be applied to
identify those events in which a second pulse has occurred. The result
of both these target energy cuts is a further reduction in the

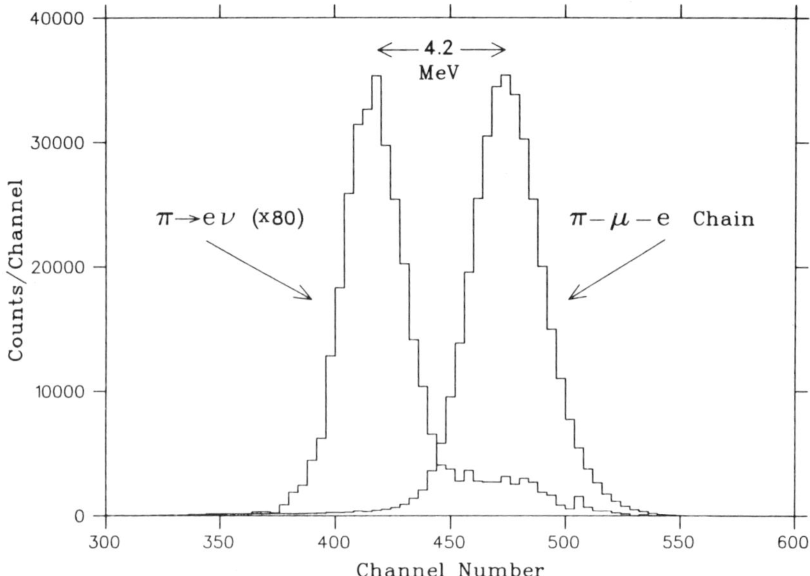

Fig. 4. Target energy spectrum.

background due to muon decay by a factor of 160, at the cost of a 34%
loss in efficiency. The background suppressed data is shown in Fig. 5.
There are 2.7×10^4 events in the $\pi \rightarrow e\nu$ peak, and the FWHM resolution is
$\Delta E/E = 7\%$.

Having beaten down the background by more than a factor of 10,000
the residual data is suitable for the search for evidence of massive
neutrinos. The remaining muon decay component, which arises mostly
from pion decay in flight near the target, was fitted with a normalised
distribution of events obtained from analysing those events associated
with the higher energy peak in the target energy spectrum (Fig. 4).
The $\pi \rightarrow e\nu$ peak in the TINA energy spectrum of Fig. 5 has a tail that
extends down to zero energy due to the radiative decays $\pi \rightarrow e\nu\gamma$, and
due to the detector response function. This distribution was fitted
with an analytic function consisting of a delta function with linear
and exponential tail components, folded with a Gaussian distribution.

The search for additional peaks was carried out using two different approaches. The first method duplicated the analysis used in
earlier experiments[4]) where the residual spectrum obtained after

434

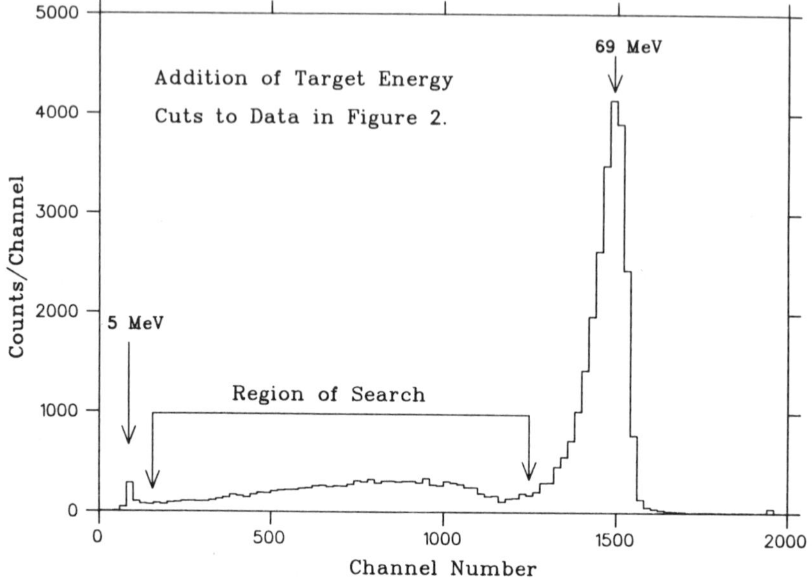

Fig. 5. Background suppressed position energy spectrum in TINA.

subtracting the muon decay, and direct pion decay, components, was
smoothed using a Fourier transform technique. This spectrum was then
searched for maxima and minima; the minima were used to define a background and then the peaks were examined for statistical significance.
The second method of analysis was to try and fit additional peaks whose
shape was identical to the main $\pi \to e\nu$ peak but scaled to approximate
the response function of TINA at any given energy. This peak, with a
free amplitude, was then stepped through the energy range from 8-57 MeV
to find the most probable peak amplitude at each energy. After a
smooth background was fitted, the sample peak areas were consistent
with statistical fluctuations about zero. The largest peak being 1.9
standard deviations from zero at M_ν=115 MeV. From both methods it
is concluded that there is no evidence of massive neutrinos in the
region investigated. The results of these analyses are presented in
Fig. 6.

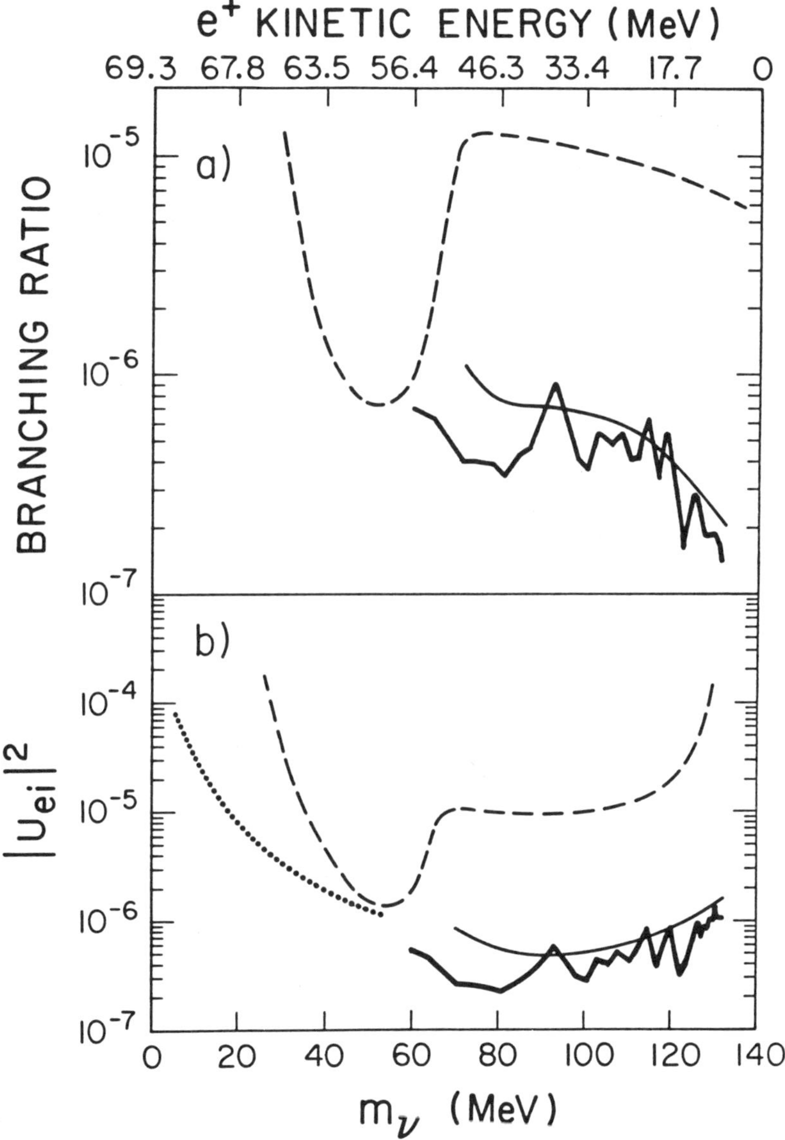

Fig. 6. Limits on (a) the branching ratio R_{ei} (Equation 1) and (b) the corresponding mixing parameter $|U_{ei}|^2$.

The dashed lines in Fig. 6 show the limits from earlier experiments[4]) and the thin and thick solid lines are the results of the two methods of analysis described above, respectively. In the positron energy region above the background from muon decay, the extra suppression is of little advantage. This corresponds to neutrino masses of less than 60 MeV and the search has not been extended below this. In this region measurement of the branching ratio provides a better limit on massive neutrinos than does the peak search. This arises because when the branching ratio is calculated, peaks due to extra neutrino would be counted as part of the main $\pi \to e\nu$ peak rather than as part of the $\mu \to e\nu\bar{\nu}$ distribution. Since the branching ratio is $\sim 10^{-4}$ it is sensitive to such extra counts. The dotted line shows the limits obtained from the branching ratio measurements of the earlier experiments.[4])

The new background suppression technique is also of importance in the new branching ratio measurements because the dominant uncertainty has previously been the limited knowledge of the tail of the main $\pi \to e\nu$ peak under the π-μ-e background. Current experiments at TRIUMF, designed to measure the branching ratio to the level of 0.3%, should also provide improved limits for coupling to massive neutrinos in the range 5-60 MeV.

REFERENCES

1. Bryman, D.A. et al., Phys. Rev. D33, 1211 (1986).
2. Azuelos, G. et al., TRIUMF Preprint TRI-PP-86-8.
3. Shrock, R.E., Phys. Rev. D24, 1232 (1981).
4. Bryman, D.A. et al., Phys. Rev. Lett. 50, 1546 (1983).

EVIDENCE OF RADIATIVE CORRECTION EFFECTS
IN ν-INDUCED DEEP INELASTIC SCATTERING

G.L. Fogli

Dipartimento di Fisica, Università di Bari, Bari, Italy
Istituto Nazionale di Fisica Nucleare, Sezione di Bari, Bari, Italy

ABSTRACT

An updated analysis of the structure of the hadronic neutral current is reported, inclusive of radiative correction effects, starting from the ν-induced deep-inelastic scattering. The comparison with the vector boson masses as measured in $p\bar{p}$ collider gives a first evidence, not conclusive but rather persuasive, that radiative corrections are indeed "active" in the standard model.

1. AN UPDATED ANALYSIS OF ν-INDUCED DEEP-INELASTIC SCATTERING

The results of a model-independent analysis of the hadronic neutral current structure, starting from the ν-induced deep-inelastic scattering, and including radiative corrections, are presented here and compared with the standard model predictions.

The approach is similar to that described in [1], where all details concerning method, parametrization, sources of theoretical uncertainty, etc., are reported. However, with respect to [1]:

a) An impressive amount of data is now available, since new data coming from recent, high precision, experiments have been included, as it can be verified from Table I, where all experimental data used in the present analysis are collected.

b) Even all n, p target data are now parametrized, as those coming from the isoscalar target experiments, by inserting the explicit form of the nucleon structure functions, their Q^2-dependence provided by QCD, the effects of the experimental cuts, the estimate of the theoretical uncertainties, and so on.

TABLE I - Experimental data used in the present analysis ((a) previously used in [1], (b) used in [1] but improved in statistics, (c) "new entries", (d) unpublished results presented very recently). Statistical and systematic errors are added quadratically.

experiment	ref.	measured quantities		
GGM	[2]	$R_\nu = 0.25 \pm 0.04$	$R_{\bar\nu} = 0.56 \pm 0.08$	(c)
CDHS I	[3]	$R_\nu = 0.293 \pm 0.010$	$R_{\bar\nu} = 0.350 \pm 0.030$	(c)
HPWF	[4]	$R_\nu = 0.30 \pm 0.04$	$R_{\bar\nu} = 0.33 \pm 0.09$	(a)
CITF	[5]	$R_\nu = 0.28 \pm 0.03$	$R_{\bar\nu} = 0.35 \pm 0.11$	(a)
CDHS II	[6]	$R_\nu = 0.301 \pm 0.007$	$R_{\bar\nu} = 0.363 \pm 0.015$	(b)
CHARM '79	[7]	$R_\nu = 0.320 \pm 0.010$	$R_{\bar\nu} = 0.377 \pm 0.020$	(a)
ABCDLOS	[8]	$R_\nu = 0.345 \pm 0.018$	$R_{\bar\nu} = 0.364 \pm 0.030$	(a)
BEBC D	[9]	$R_\nu = 0.328 \pm 0.027$	$R_{\bar\nu} = 0.353 \pm 0.043$	(a)
MB-ITEP	[10]	$R_\nu = 0.330 \pm 0.040$	$R_{\bar\nu} = 0.405 \pm 0.028$	(c)
FMM	[11]	$R_\nu = 0.307 \pm 0.011$	$R_{\bar\nu} = 0.384 \pm 0.023$	(c)
CDHS III	[12]	$R_\nu = 0.3059 \pm 0.0032$	$R_{\bar\nu} = 0.380 \pm 0.016$	(d)
CHARM '84	[13]	$R_\nu = 0.3174 \pm 0.0060$		(c)
BEBC TST	[14]	$R_\nu^p = 0.455 \pm 0.038$	$R_{\bar\nu}^p = 0.320 \pm 0.038$	(a)
MB	[15]	$R_\nu^{n/p} = 1.08 \pm 0.19$		(b)
SIMMT	[16]	$R_\nu^{n/p} = 1.01 \pm 0.14$		(a)
ITEP	[10]		$R_{\bar\nu}^{n/p} = 0.88 \pm 0.17$	(b)
FNAL	[17]	$R_\nu^p = 0.40 \pm 0.14$		(a)
ABCMO	[18]	$R_\nu^p = 0.51 \pm 0.04$		(a)
SIMMT	[16]	$R_\nu^p = 0.49 \pm 0.06$		(a)
FNAL	[19]		$R_{\bar\nu}^p = 0.35 \pm 0.05$	(a)
BEBC D	[9]	$r_\nu = 0.06 \pm 0.06$	$r_{\bar\nu} = 0.02 \pm 0.09$	(a)
BBCLMO	[20]	$R_\nu^p = 0.384 \pm 0.028$	$R_{\bar\nu}^p = 0.388 \pm 0.021$	(d)

c) Radiative corrections, included following the renormalization framework proposed by Marciano and Sirlin [21,22], are now introduced at the level of double differential cross-sections with their explicit dependence on the kinematical variables: in this way the effects of the experimental cuts are taken into account also at the level of their influence on radiative correction contributions.

d)   The effect of varying the c-quark mass, relevant because of the threshold effect in c-quark excitation, is explicitly considered. The value m_c = 1.5 Gev/c , adopted in [1], is maintained, but the shift induced by a different choice of m_c (1.35 and 1.20 GeV/c^2) is introduced.

The model-independent solution inclusive of radiative corrections (RC solution) is obtained analytically through the two-step iterative procedure described in [1]. In terms of the sums and differencies of the four squared chiral couplings, it is

$$\begin{cases} g_L^2 = u_L^2 + d_L^2 = 0.3070 \pm 0.0036 \ (\pm 0.0019) \\ g_R^2 = u_R^2 + d_R^2 = 0.0283 \pm 0.0032 \ (\pm 0.0019) \\ \delta_L^2 = u_L^2 - d_L^2 = -0.0426 \pm 0.0313 \ (\pm 0.0027) \\ \delta_R^2 = u_R^2 - d_R^2 = 0.0181 \pm 0.0164 \ (\pm 0.0049) \end{cases} \tag{1}$$

with correlation coefficients

$$\{\rho\}_{exp} = \begin{pmatrix} 1 & -0.766 & -0.145 & 0.208 \\ & 1 & 0.076 & 0.302 \\ & & 1 & -0.263 \\ & & & 1 \end{pmatrix} \quad \{\rho\}_{par} = \begin{pmatrix} 1 & -0.768 & -0.901 & 0.682 \\ & 1 & 0.477 & -0.415 \\ & & 1 & -0.752 \\ & & & 1 \end{pmatrix} \tag{2}$$

In eqs. (1) the first error comes from the experimental errors given in Table I, whereas in bracket the "theoretical" error is reported, derived from the different uncertainties affecting the theoretical approach. The correlation coefficients in (2) refer to the experimental and to the theoretical errors, respectively.

In terms of the squared chiral couplings the above solution corresponds to

$$\begin{cases} u_L^2 = 0.1322 \pm 0.0155 \ (\pm 0.0006) \\ d_L^2 = 0.1748 \pm 0.0160 \ (\pm 0.0023) \\ u_R^2 = 0.0232 \pm 0.0088 \ (\pm 0.0023) \\ d_R^2 = 0.0051 \pm 0.0079 \ (\pm 0.0030) \end{cases} \tag{3}$$

with correlation coefficients

$$\{\tau\}_{exp} = \begin{pmatrix} 1 & -0.974 & -0.227 & 0.249 \\ & 1 & 0.231 & -0.325 \\ & & 1 & -0.931 \\ & & & 1 \end{pmatrix} \quad \{\tau\}_{par} = \begin{pmatrix} 1 & -0.968 & -0.228 & 0.240 \\ & 1 & 0.237 & -0.341 \\ & & 1 & -0.902 \\ & & & 1 \end{pmatrix} \tag{4}$$

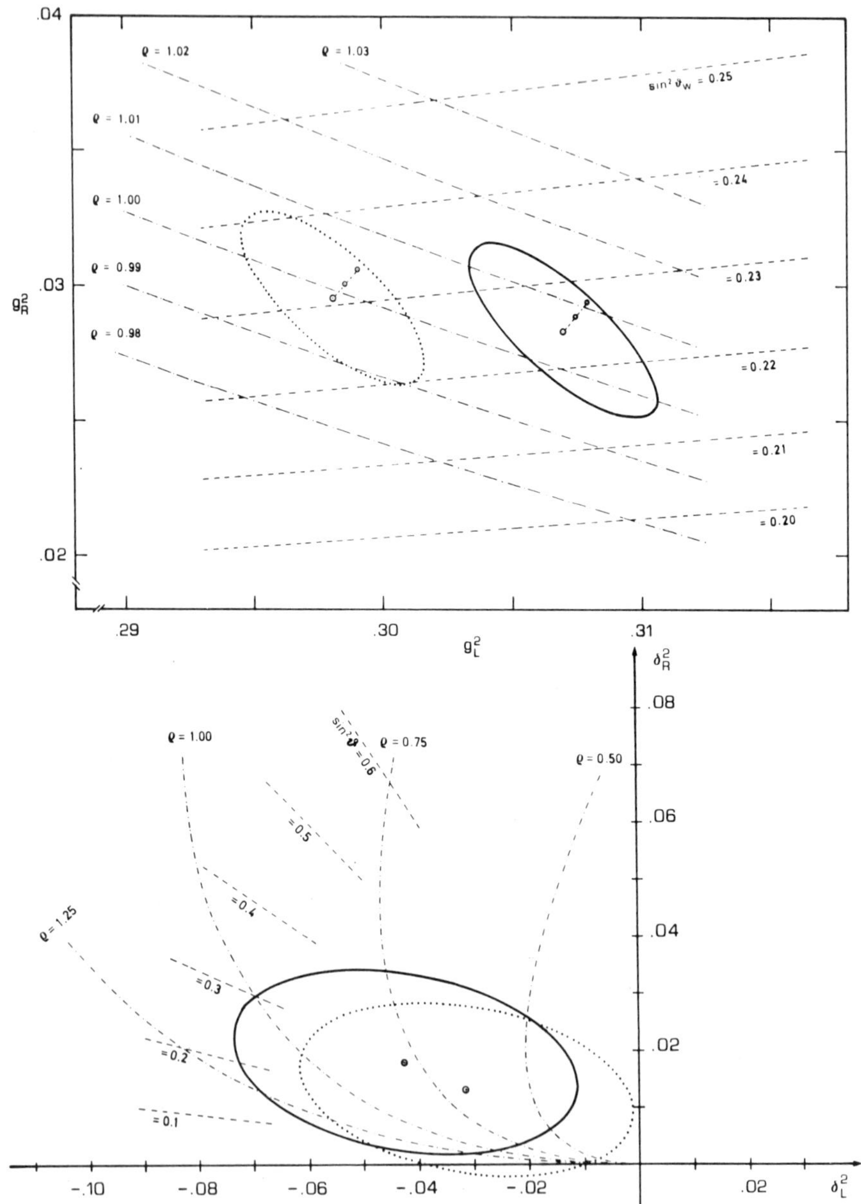

Fig. 1 - The two model-independent RC and no RC solutions projected in the planes (g_L^2, g_R^2) and (δ_L^2, δ_R^2) (heavy and dotted ellipses, respectively). The shift induced by the change of m_c from the adopted value 1.5 to 1.35 and 1.20 GeV/c^2 is shown.

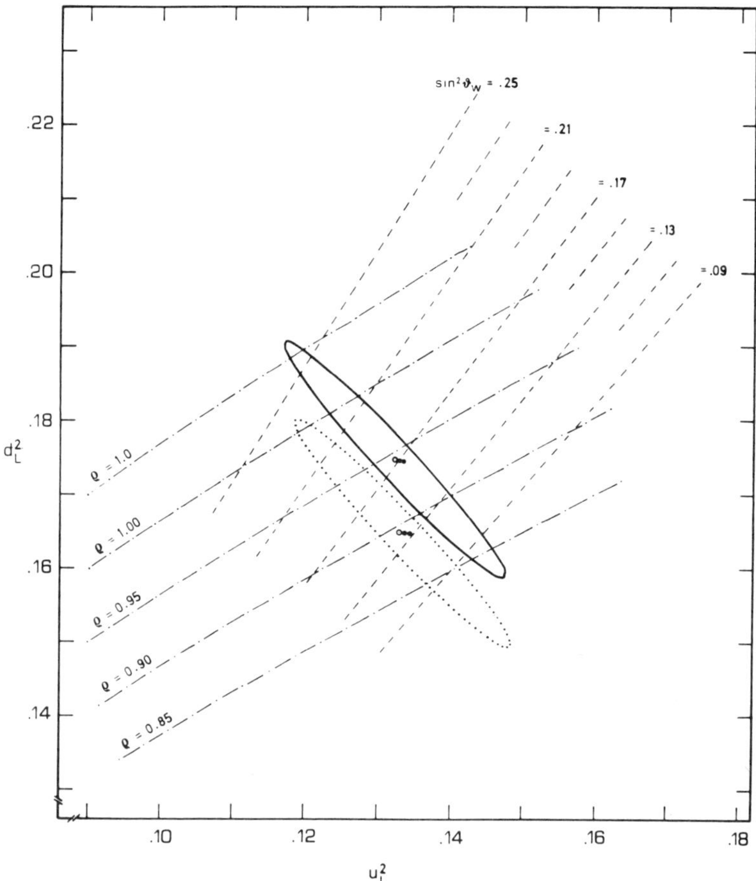

Fig. 2 - The two model-independent RC and no RC solutions projected in the plane (u^2_L, d^2_L) (heavy and dotted ellipses, respectively). The shift induced by the change of m_c from the adopted value 1.5 to 1.35 and 1.20 GeV/c² is shown.

If radiative corrections are neglected, the following solution (no RC solution) is obtained:

$$\begin{cases} (g^2_L)^\circ = 0.2981 \pm 0.0035 \ (\pm 0.0020) \\ (g^2_R)^\circ = 0.0295 \pm 0.0032 \ (\pm 0.0014) \\ (\delta^2_L)^\circ = -0.0315 \pm 0.0302 \ (\pm 0.0025) \\ (\delta^2_R)^\circ = 0.0133 \pm 0.0153 \ (\pm 0.0038) \end{cases} \tag{5}$$

corresponding, in terms of the squared chiral couplings, to

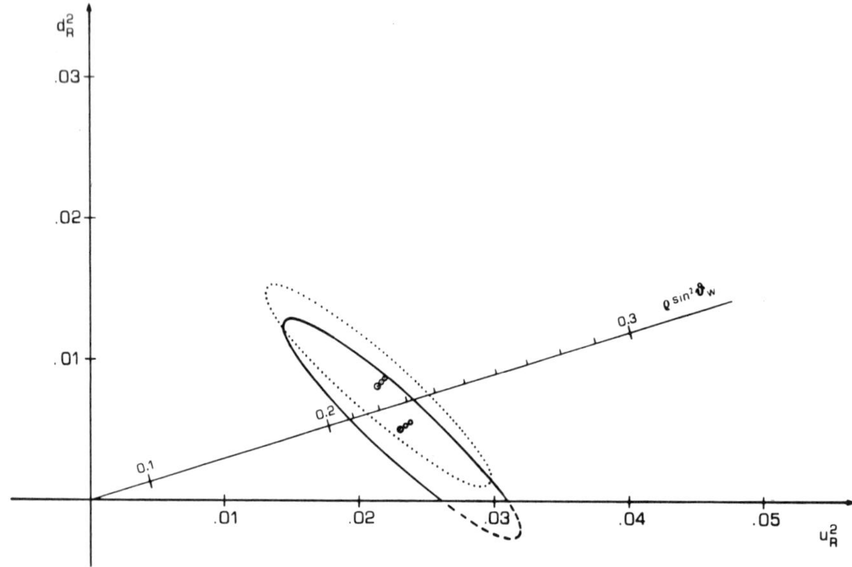

Fig. 3 - The two model-independent RC and no RC solutions projected in the plane (u_R^2, d_R^2) (heavy and dotted ellipses, respectively). The shift induced by the change of m_c from the adopted value 1.5 to 1.35 and 1.20 GeV/c² is shown.

$$\left[\begin{array}{l} (u_L^2)^\circ = 0.1333 \pm 0.0150 \ (\pm 0.0006) \\ (d_L^2)^\circ = 0.1648 \pm 0.0154 \ (\pm 0.0022) \\ (u_R^2)^\circ = 0.0214 \pm 0.0084 \ (\pm 0.0019) \\ (d_R^2)^\circ = 0.0082 \pm 0.0071 \ (\pm 0.0021) \end{array}\right. \tag{6}$$

the relative correlation coefficients being very similar to those given in (2) and (4), respectively.

The two solutions are drawn in Figs. 1-3, projected in the planes (g_L^2, g_R^2), (δ_L^2, δ_R^2), (u_L^2, d_L^2) and (u_R^2, d_R^2). All allowed elliptical regions correspond to 1-standard deviation on each variable separately.

2. COMPARISON WITH THE STANDARD MODEL

The comparison is performed through a least squares method (solved exactly) applied to the model independent solution. The estimated χ^2/DOF represents a reliable "measure" of the "goodness" of the standard model. Starting from the RC solution, if $\rho = 1$ is assumed $(x = \sin^2\theta_W)$

$$x = 0.2194 \pm 0.0031 \ (\pm 0.0017) \qquad \chi^2/\text{DOF} = 0.42/3 \tag{7}$$

whereas the two-parameters analysis gives (τ being the correlation, with in bracket the total correlation)

$$x = 0.2236 \pm 0.0103 \; (\pm 0.0072)$$
$$\rho = 1.0036 \pm 0.0080 \; (\pm 0.0069) \qquad \chi^2/DOF = 1.73/2 \qquad (8)$$
$$\tau = 0.9536 \; (0.9575)$$

The above determinations show the impressive agreement of the data with the standard model, and lead to the best estimate available at present of the parameters $\sin^2\theta_W$ and ρ. It is worthwhile to note that the errors coming from the experiments are now comparable with (but still slightly larger than) the "theoretical" errors due to the parametrization.

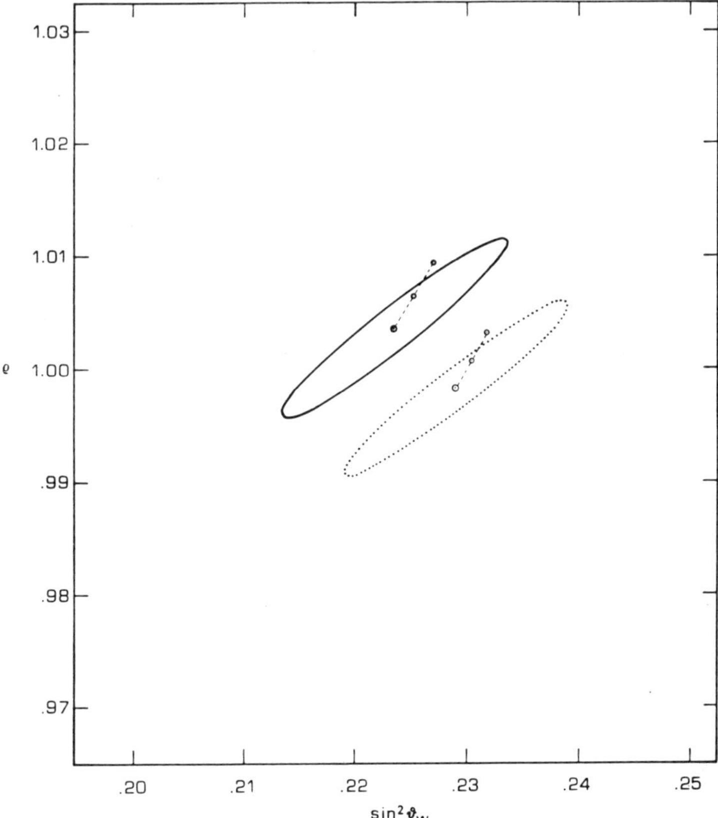

Fig. 4 – The two model-independent RC and no RC solutions projected in the plane $(\sin^2\theta, \rho)$ (heavy and dotted ellipses, respectively). The shift induced by the change of m_c from the adopted value 1.5 to 1.35 and 1.20 GeV/c^2 is shown.

If radiative corrections are not included (no RC solution), then in the case $\rho = 1$

$$(x)^\circ = 0.2311 \pm 0.0029 \; (\pm 0.0018) \qquad \chi^2/DOF = 0.85/3 \qquad (9)$$

and the simultaneous determination of both the two parameters leads to

$$\begin{cases} (x)^\circ = 0.2290 \pm 0.0102 \; (\pm 0.0073) \\ (\rho)^\circ = 0.9983 \pm 0.0078 \; (\pm 0.0070) \qquad \chi^2/DOF = 1.19/2 \qquad (10) \\ (\tau)^\circ = 0.9614 \; (0.9612) \end{cases}$$

The effect of the radiative corrections can be seen in Fig. 4, where both solutions (8) and (10) are reported.

3. ARE THERE INDICATIONS ABOUT RADIATIVE CORRECTIONS? A COMPARISON WITH THE VECTOR BOSON MASSES MEASURED AT $p\bar{p}$ COLLIDER

Even if the accuracy reached in the determination of the squared chiral couplings distinguishes in a rather clean way the two (RC and no RC) solutions, however very poor indications, if any, can be extracted in favour of radiative corrections within the ν-induced deep-inelastic scattering itself. Both solutions show an impressive agreement with the standard model (even though slightly better if RC are included) and, within a minimal model scheme, ρ is very near to 1 in both cases.

Indications about RC can be obtained only through a comparison of the above results with those coming from a different process. In principle, a comparison with the ν-induced leptonic scattering would allow to extract some indication about RC. But, waiting for CHARM II [23], at present [24] leptonic processes are still far from the precision required.

At present, the only process which allow a comparison at the level of radiative correction effects is the direct measurement of the vector meson masses at the $p\bar{p}$ collider. Here the most recent results of UA1 [25] and UA2 [26] (see also [27]), will be used. With masses expressed in GeV/c^2

$$UA1 \begin{bmatrix} M_W = 83.5 \pm 1.1 \; (\pm 2.7) \\ M_Z = 93.0 \pm 1.4 \; (\pm 3.0) \end{bmatrix} \qquad UA2 \begin{bmatrix} M_W = 81.2 \pm 1.1 \; (\pm 1.3) \\ M_Z = 92.5 \pm 1.3 \; (\pm 1.5) \end{bmatrix} \qquad (11)$$

where the first error comes from statistics and the second reflects the instrumental uncertainty on the absolute mass scale due to the calorimeter calibration (it is usually denoted as "systematic", even though strictly speaking it is not so).

3.1   Comparison with the measured ratio M_W/M_Z

The ratio M_W/M_Z is not affected by "systematic" errors. More-over, being related only to the direct measurement of the physical masses, it does not require to be corrected because of RC effects. By combining the results coming from UA1 and UA2

$$\frac{M_W^2}{M_Z^2} = \rho(1-\sin^2\theta_W) = 0.787 \pm 0.022 \tag{12}$$

which defines a band in the plane $(\sin^2\theta_W,\rho)$. Even if the relatively low statistics limits the result, however the RC solution is clearly preferred, in particular in the minimal model $(\rho=1)$ case, as it can be seen by the compatibility check reported in Table II, where the confidence levels (CL) obtained from the χ^2 distribution of a last squares method which compares RC and no RC solutions to the $p\bar{p}$ result (12) are given.

TABLE II – Compatibility check of RC and no RC solutions of deep-inelastic scattering (DIS) with the $p\bar{p}$ result (12). Confidence levels derived from the χ^2 distribution of a least squares method.

		CL with only statistical errors in DIS	CL with stat. and theor. errors in DIS added in quadrature
ρ arbitrary	RC solution	72.2%	72.3%
	no RC solution	42.9%	43.1%
$\rho = 1$ fixed	RC solution	92.3%	92.3%
	no RC solution	40.5%	40.7%

3.2   Comparison in terms of $\sin^2\theta_W$ and ρ

Further comparisons are possible, but require to add UA1 and UA2 by taking into account also "systematic" errors. Here a weighted mean is taken in which total errors (statistical and systematic errors added in quadrature) are used as weights, but systematic errors are added linearly in order to account for their specific character. It follows

$$\text{UA1 + UA2} \begin{cases} M_W = 81.79 \pm 0.78 \ (\pm 1.66) \quad \text{GeV/c}^2 \\ M_Z = 92.63 \pm 0.95 \ (\pm 1.90) \quad \text{GeV/c}^2 \end{cases} \tag{13}$$

The above estimate allows two different determinations of the variables $\sin^2\theta_W$ and ρ, depending on the possibility of introducing (or not) RC effects ($(\pi\alpha/\sqrt{2}G_\mu)^{1/2} = 37.2810 \pm 0.0003$ GeV, $r = 0.0696 \pm 0.0020$ [22] is

TABLE III – Compatibility check of DIS and $p\bar{p}$ analyses in terms of $\sin^2\theta_W$ and ρ (see Fig. 5). CL derived from the χ^2 distribution of a least squares method applied to the two pairs of solutions.

	statistical errors only	stat. and syst. errors added in quadrature	central value varied along the straight line of syst. error
RC solutions	99.9%	99.9%	99.9%
no RC solutions	22.7%	54.9%	79.5%

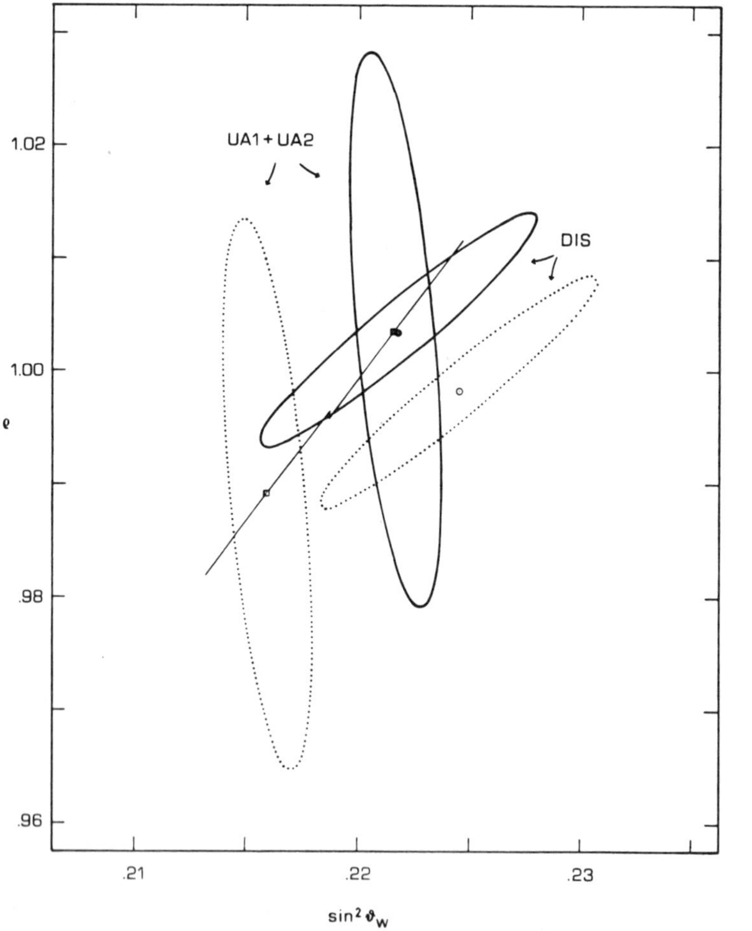

Fig. 5 – Comparison of DIS and $p\bar{p}$ in the plane $(\sin^2\theta_W, \rho)$. Heavy (dotted) ellipses correspond to RC (no RC) solutions. Systematic errors given as straight lines.

used). The results are reported in Fig. 5, with systematic errors represented by straight lines (along which, in principle, the central value is allowed to vary). The compatibility check of the two pairs of solutions (RC and no RC) is given in Table III: with a rather clear indication in favour of RC.

If a weighted mean of the two RC solutions is taken, a very accurate estimate of the two standard model parameters is derived (because of the different structure of the two allowed regions)

$$p\bar{p} + DIS \begin{cases} x = 0.2234 \pm 0.0032 \ (\pm 0.0071) \\ \rho = 1.0035 \pm 0.0032 \ (\pm 0.0061) \\ \tau = 0.6727 \ (0.9001) \end{cases} \tag{14}$$

where in bracket the non-statistical error comes from a quadratic addition of the "theoretical" error affecting the DIS solution with the "systematic" error which characterizes the $p\bar{p}$ solution.

3.3   Comparison in terms of the vector boson masses

The vector boson masses are directly measured in $p\bar{p}$, but they can be also determined starting from the values of $\sin^2\theta_W$ and ρ found in DIS. Since this determination depends on the possibility of including or not RC, two solutions are derived. Their comparison with the direct estimate of $p\bar{p}$ is drawn in Fig. 6. The compatibility check clearly favours RC solution and is reported in Table IV.

TABLE IV - Compatibility check of RC and no RC solutions of DIS in terms of the vector boson masses compared with the direct measure of $p\bar{p}$ (see Fig. 6). CL derived from the χ^2 distribution of a least squares method applied to pairs of solutions.

	statistical errors only	stat. and syst. errors added in quadrature	central value varied along the straight line of syst. error
RC solution	99.9%	99.9%	99.9%
no RC solution	10.7%	38.3%	61.5%

An accurate estimate of the mass difference $M_Z - M_W$ versus M_Z is derived from the weighed mean of the $p\bar{p}$ estimate with the RC solution of DIS: adding also in this case non-statistical errors in quadrature

$$p\bar{p} + DIS \begin{cases} M_Z = 92.63 \pm 0.58 \ (\pm 1.32) \ \text{GeV/c}^2 \\ M_Z - M_W = 10.86 \pm 0.11 \ (\pm 0.09) \ \text{GeV/c}^2 \\ \text{corr} = 0.094 \ (0.190) \end{cases} \tag{15}$$

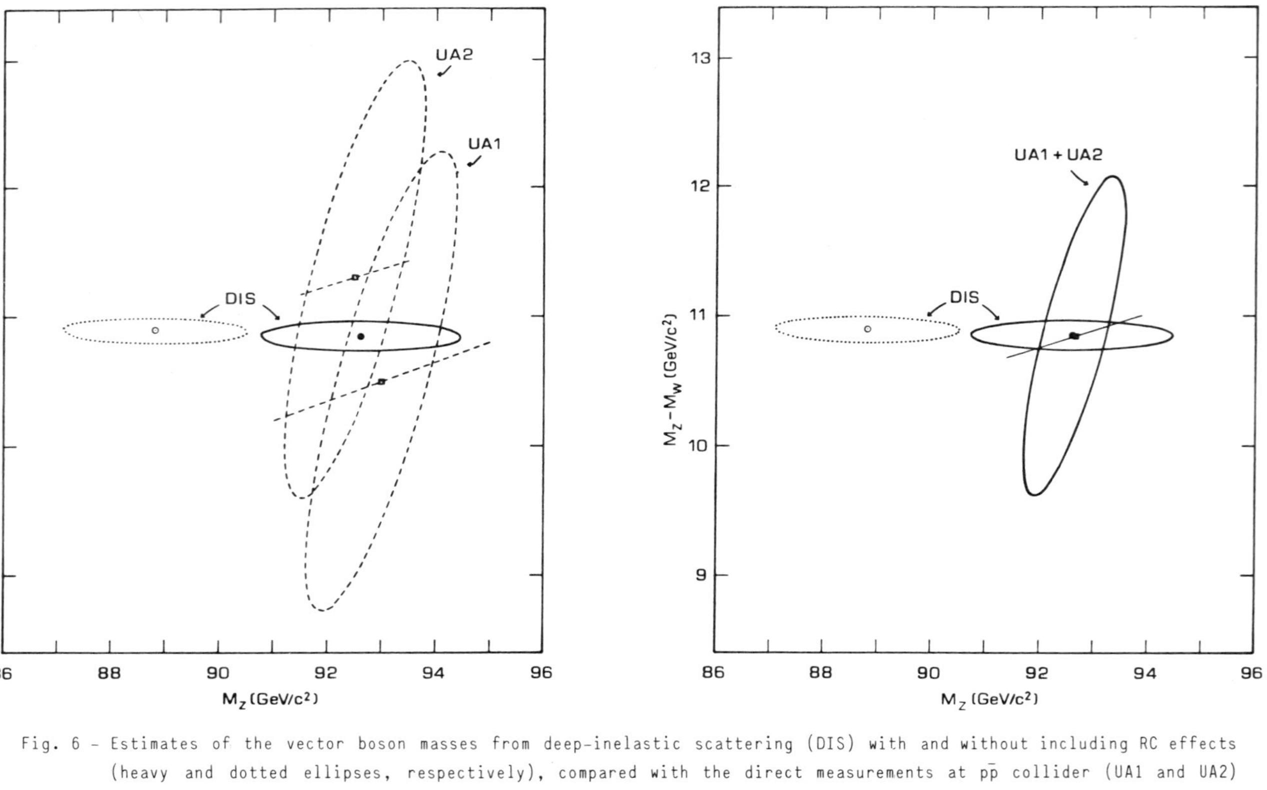

Fig. 6 – Estimates of the vector boson masses from deep-inelastic scattering (DIS) with and without including RC effects (heavy and dotted ellipses, respectively), compared with the direct measurements at $p\bar{p}$ collider (UA1 and UA2) and with their weighted mean.

3.4 Comparison assuming $\rho = 1$

A further comparison is possible, within the minimal model scheme requiring $\rho = 1$ (very well verified by both DIS and $p\bar{p}$ solutions). RC and no RC solutions are compared in Fig. 7, the compatibility check being reported in Table V: the two no RC solutions are incompatible, whereas, if RC are included, the compatibility is rather good also when only statistical errors are considered. A weighted mean of the two RC solutions (non-statistical errors combined in quadrature) gives the best estimate available at present for $\sin^2\theta_W$

$$p\bar{p} + DIS \quad \sin^2\theta_W = 0.2199 \pm 0.0024 \; (\pm 0.0024) \tag{16}$$

TABLE V – Compatibility check of the values of $\sin^2\theta_W$ as derived from DIS and $p\bar{p}$ (see Fig. 7). CL derived from the χ^2 distribution of a least squares method applied to the different pairs of solutions.

	statistical errors only	stat. and syst. errors added in quadrature	central value varied along the straight line of syst. error
RC solutions	41.3%	74.8%	99.9%
no RC solutions	incompatible	always less than 1%	

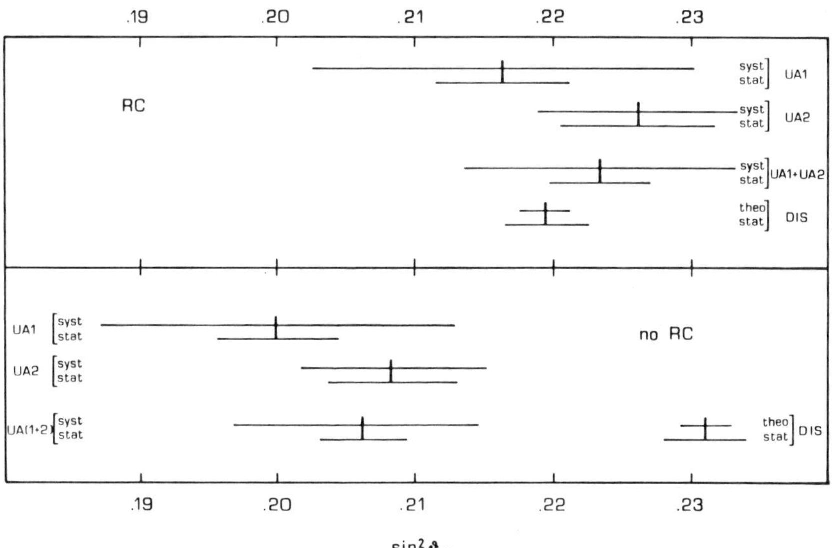

Fig. 7 – Comparison of the estimate of $\sin^2\theta_W$ assuming $\rho = 1$ from deep-inelastic scattering (DIS) and $p\bar{p}$ collider (UA1 and UA2) with and without the inclusion of radiative corrections (RC) effects.

4.   CONCLUSIONS

Deep-inelastic scattering, with the considerable contribution of the last, high precision experiments, is seen to have a fundamental role in the study of the structure of the hadronic neutral current: it allows a precise determination of the NC couplings and shows a spectacular agreement with the standard model.

As far as radiative corrections are concerned, the two solutions obtained with and without the inclusion of RC effects can now be distinguished in a clean way. However, deep-inelastic scattering is not able, by itself, to provide an indication in favour of RC: only if compared with other processes, the indication, at least in principle, can be achieved.

The only significant comparison which can be performed at present is that with the direct measurement of the vector boson masses at the $p\bar{p}$ collider. The comparison leads, despite the large "systematic" errors affecting the $p\bar{p}$ measurements, to the first indications, not conclusive, but rather persuasive, that RC are "active". In particular, within the minimal version of the standard model, RC effects are confirmed on the basis of the compatibility of the experimental results.

REFERENCES

[1]   G.L. Fogli, Phys. Lett. 158B, 66 (1985);
      Nucl. Phys. B260, 593 (1985).
[2]   J. Blietschau et al., Nucl. Phys. B118, 218 (1977).
[3]   M. Holder et al., Phys. Lett. 71B, 222 (1977).
[4]   P. Wanderer et al., Phys. Rev. D17, 1679 (1978).
[5]   F.S. Merritt et al., Phys. |17, 2199 (1978).
[6]   H. Abramowicz et al., Z. Phys. C24, 51 (1985).
[7]   M. Jonker et al., Phys. Lett. 99B, 265 (1981).
[8]   P. Bosetti et al., Nucl. Phys. B217, 1 (1983).
[9]   D. Allasia et al., Phys. Lett. 133B, 129 (1983).
[10] J. Marriner et al., Phys. Rev. D27, 2569 (1983);
      P.A. Gorichev et al., Sov. J. Nucl. Phys. 39, 396 (1984).
[11] D. Bogert et al., Phys. Rev. Lett. 55, 1969 (1985).
[12] CDHS Collab., A. Blondel, presented in: Proc. Intern. Conf. on High energy physics (Bari, 1985) p. 252. Used here more recent data presented by F. Perrier, talk given at CERN (1986).
[13] F. Bergsma et al., CERN-EP/85-113. (1985).
[14] N. Armenise et al., Phys. Lett. 122B, 448 (1983);
      J. Moreels et al., Phys. Lett. 138B, 230 (1984).
[15] J. Marriner et al., Phys. Rev. D27, 2569 (1983).
[16] T. Kafka et al., Phys. Rev. Lett. 48, 910 (1982).

[17] F.A. Harris et al., Phys. Rev. Lett. 39, 437 (1977).

[18] J. Blietschau et al., Phys. Lett. 88B, 381 (1979).

[19] D.D. Carmony et al., Phys. Rev. D26, 2965 (1982).

[20] BBCLMO Collab., A. Pain, presented in: Proc. Intern. Conf. on High energy physics (Bari, 1985) p. 249. Used here more recent data presented by M.M. Mobayyen at XXI Rencontre de Moriond on "Perspectives in Electroweak Interactions" (La Plagne, 1986), to appear.

[21] A. Sirlin, Phys. Rev. D22, 971 (1980);
W.J. Marciano and A. Sirlin, Phys. Rev. D22, 2695 (1980);
A. Sirlin and W.J. Marciano, Nucl. Phys. B189, 442 (1981).

[22] W.J. Marciano and A. Sirlin, Phys. Rev. D29, 945 (1984).

[23] CHARM II proposal CERN/SPSC/83-24 and 83-37 (1983).

[24] L.A. Ahrens et al., Phys. Rev. Lett. 55, 1814 (1985);
F. Bergsma et al., Phys. Lett. 147B, 481 (1984).

[25] UA1 Collab., G. Arnison et al., CERN-EP/85-185 (1985).

[26] UA2 Collab., J.A. Appel et al., CERN-EP/85-166 (1985).

[27] L. Di Lella, Proc. Intern. Conf. on High energy physics (Bari, 1985) p. 761.

A NEW SOLUTION TO THE SOLAR NEUTRINO PROBLEM ?

M. K. Sundaresan
Physics Department
Carleton University
Ottawa, K1S5B6
CANADA

ABSTRACT

Recent work by Mikheyev and Smirnov on the possibility of resonant neutrino oscillations in the sun is reviewed. The application of this mechanism, as discussed by Bethe, and by Rosen and Gelb, to explain the solar neutrino problem is also reviewed. Implications of the Mikheyev-Smirnov work for the proposed Sudbury neutrino detector are worked out.

1.   INTRODUCTION

Considerable excitement has been caused by the work of Mikheyev and Smirnov[1] in the USSR who have shown that neutrino oscillations in the sun can be greatly amplified through the Wolfenstein[2] matter oscillation mechanism. Two further papers have appeared, one by Bethe[3] and the other by Rosen and Gelb[4] which go into details of the MS mechanism and its application to the solar neutrino problem[5]. In what follows, we review the work of MS as extended by Bethe and by Rosen and Gelb. Recently, a proposal to establish a deep underground D_2O detector for neutrinos in a mine in Sudbury[6], has been made to NSERC by a collaboration involving a number of universities in Canada, USA, and UK. C. K. Hargrove and I have calculated the expected number of events in the Sudbury detector based on the Rosen-Gelb calculations. The result of these calculations are presented.

2.   BRIEF DESCRIPTION OF THE SOLAR NEUTRINO PROBLEM

The expected flux of neutrinos from the sun according to the standard solar model is shown in Fig 1. Tha abscissa is the neutrino energy and the ordinate is the solar flux at the earth's surface. The neutrinos produced in the different nuclear reactions are labelled in the diagram. The units used for the continuum neutrinos are

$cm^{-2}\ s^{-1}MeV^{-1}$ while those from the PeP and ^{7}Be reactions are in $cm^{-2}s^{-1}$.

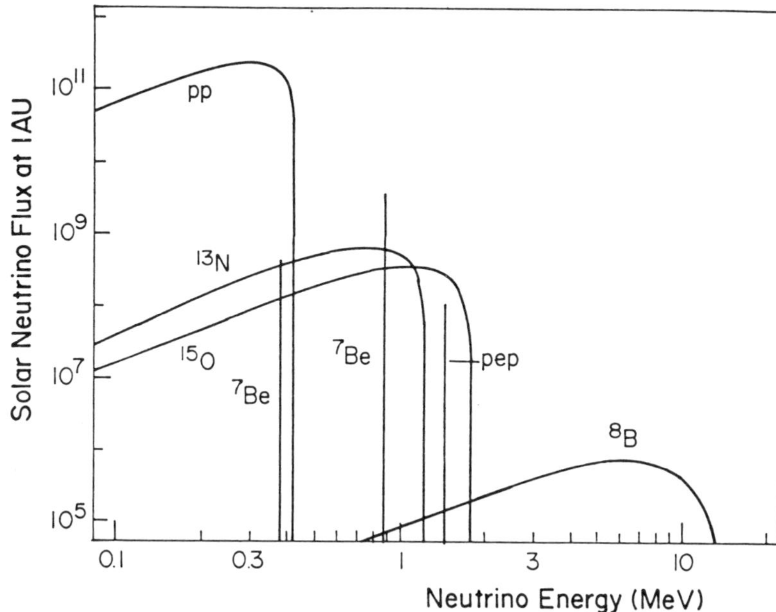

Fig. 1. Solar neutrino spectrum according to the standard solar model.

Davis' experiment is designed to detect high energy neutrinos produced in a side branch reaction involving $^8_5$B. These neutrinos are spread in energy between about 1 MeV and 14 MeV with the bulk of them having energies about 5 MeV. For detecting the $^8_5$B neutrinos, Davis set up 100,000 gallons of C_2Cl_4 one mile underground in the Homestake mine and looked for captures in $^{37}_{17}$Cl,

$$^{37}_{17}Cl + {}^0_0\nu_e \rightarrow {}^{37}_{18}Ar + {}^0_{-1}e \ .$$

The radioactive argon atoms produced by this capture were flushed out of the fluid and were counted. In this manner Davis could get a measure of the solar $^8_5$B neutrino flux. In quantitative terms, the results of the Davis' experiment can be expressed in terms of SNU's, where

$$1\ SNU = 10^{-36}\ events/s/scatterer.$$

In an analysis of the Davis experiment, Bahcall et al.,[7] give results which we state in the following form :

Number of SNU's detectable in ^{37}Cl :

$$\text{from } {}^8_5B \ : \quad 5.8 \pm 2.2\ SNU$$

$$\underline{\text{from PeP, } {}^7Be \ldots (E_\nu < 2.8\ MeV) \ : \quad 1.6 \pm 0.2\ SNU}$$

$$\text{Total} \qquad : \quad 7.4 \pm 2.4\ SNU$$

454

        Number observed by Davis   :   2.1 ± 0.3 SNU.
Where are the other (5.3 ± 2.7) SNU's ? This in brief is the solar
neutrino problem.

3.   BRIEF SUMMARY OF THE MS WORK

     If we take the result of Davis' experiment at its face value, an
explanation for the missing neutrinos in terms of the possibility of
neutrino oscillations between different flavours has been attempted.
The "vacuum" and "matter" oscillations of neutrinos between different
flavours was studied by Wolfenstein[2] some time ago. What MS have
shown is that "matter" oscillations of Wolfenstein can act upon
"vacuum" oscillations in a resonant manner and selectively amplify
the oscillations in a part of the neutrino spectrum from the sun. It
is possible to have complete oscillation of, for example, ν_e into ν_μ
for $E_\nu > E_{min}$, where E_{min} is some minimum energy.

     Assuming this is the reason why Davis sees far fewer ν_e's ,
Bethe[3] calculates E_{min} from the Wolfenstein-MS mechanism and shows
that it is about 5 MeV. Bethe shows further that Δm^2, which represents the difference between the square of ν_μ mass and the square of
the ν_e mass, is 9 x 10^{-5} $(eV)^2$, and there is a minor restriction on
the mixing angle, $\tan^2 2\theta > 10^{-4}$.

     Rosen and Gelb[4] have carried out detailed calculations with the
Wolfenstein-MS mechanism to determine the parameters Δm^2 and $\sin^2 2\theta$
for which the rate in the Cl detector is reduced by factors of 2 to 4.
They also give results for ^{71}Ga detector (for the lower energy
neutrinos). To quote Rosen : " Resonant amplification of neutrino
oscillation in the sun discussed by Mikheyev and Smirnov is an effect
so elegant that nature ought to use it."

4.   DETAILS ON NEUTRINO OSCILLATIONS [2].

     a) Vacuum Oscillations

     The prerequisites for vacuum oscillations are, (i) neutrino
masses must not all be degenerate, with at least one being non-zero,
and (ii) weak current eigenstates should be mixtures of mass matrix
eigenstates and vice versa (non-conservation of lepton number).
     For simplicity, considering only two flavours, let ν_1 and ν_2
be the mass eigenstates, and let ν_e, ν_μ be the weak current eigenstates. Then from (ii) we have

$$\nu_e = \nu_1 \cos\theta + \nu_2 \sin\theta , \quad \nu_\mu = -\nu_1 \sin\theta + \nu_2 \cos\theta, \qquad (1)$$

where θ is the mixing angle (assumed < 45°). The time behaviour of
ν_1, ν_2 is given by

$$\nu_j \quad \exp\{ -it(p + (m_j^2/2p)) \} , \quad j=1,2.$$

Thus ν_1, ν_2 develop a phase difference for the same p if $m_1 \neq m_2 \neq 0$

(condition (i) above). This difference in phase leads to $\nu_e \leftrightarrow \nu_\mu$ oscillations and thus leads to a reduction of the flux of ν_e. M, the square of the mass matrix , in the weak current basis can be written as

$$M = \frac{1}{2}\,(m_1^2 + m_2^2)\begin{pmatrix} 1 & 0 \\ 0 & 1 \end{pmatrix} + \frac{1}{2}\,(m_2^2 - m_1^2)\begin{pmatrix} -\cos 2\theta & \sin 2\theta \\ \sin 2\theta & \cos 2\theta \end{pmatrix}, \quad (2)$$

and the equation for the time evolution of ν_e , ν_μ is

$$i\,\frac{\partial}{\partial t}\begin{pmatrix} \nu_e \\ \nu_\mu \end{pmatrix} = \frac{1}{2p}\,M\begin{pmatrix} \nu_e \\ \nu_\mu \end{pmatrix} . \quad (3)$$

From this one gets for the probability that ν_e remains ν_e after propagating through a distance R, the expression,

$$P(\nu_e \leftrightarrow \nu_e\,;\,R) = 1 - \sin^2 2\theta\,\sin^2\!\left(\frac{\pi R}{L_v}\right), \quad (4)$$

where L_v = vacuum oscillation length = $(4\pi\,p/\,\Delta m^2)$, $\quad (5)$

and $\quad\Delta m^2 = m_2^2 - m_1^2. \quad (6)$

When p is measured in MeV, Δm^2 in $(eV)^2$, a convenient expression that gives L_v in meters is

$$L_v = 2.5\,\left\{\frac{(p/\mathrm{MeV})}{(\Delta m^2/\,(eV)^2)}\right\}\ \mathrm{meters}. \quad (7)$$

For the sun, taking $p(\nu_e) \simeq 0.25 - 14$ MeV , $L_v \simeq R_\odot = 1.5 \times 10^{11}$ m, one gets $\Delta m^2 \simeq (0.4 - 25\,)\ \times\ 10^{-11}\,(eV)^2$. Such a small Δm^2 (or large L_v) poses problems for performing laboratory experiments to discover neutrino oscillations. Nevertheless evidence for neutrino oscillations have been sought at reactors and accelerators. Early claims for evidence of oscillations with $\Delta m^2 \simeq 0.2\ (eV)^2$ and $\sin^2 2\theta \simeq 0.2$ have not been supported. In Fig 2 we present the region of

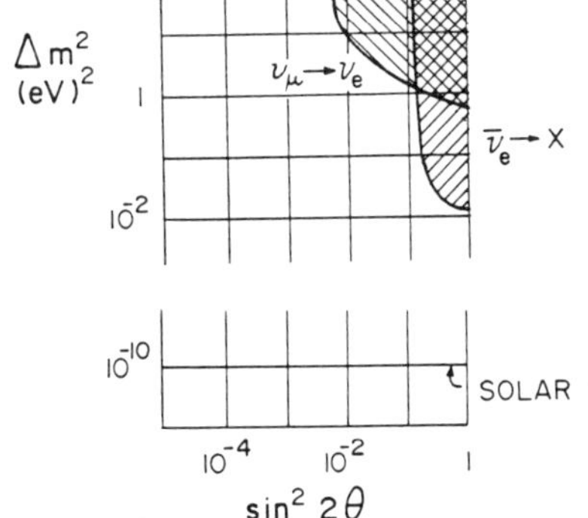

Fig.2. Regions of Δm^2, $\sin^2 2\theta$ excluded by experiments on neutrino oscillations.

Δm^2, $\sin^2 2\theta$ plane for neutrino oscillations excluded by reactor and accelerator experiments.

(b) INFLUENCE OF MATTER ON NEUTRINO OSCILLATIONS[2]

To understand how the presence of matter influences neutrino oscillations, it is convenient to define the refractive index n_e, n_μ for the propagation of ν_e, ν_μ respectively through matter :

$$n_{e,\mu} = 1 + \frac{2\pi N}{p^2} f_{e,\mu}(0) \tag{8}$$

where N is the number density of scattering centers and $f(0)$ is the forward scattering amplitude for the appropriate neutrino. Due to the refractive index, additional phase differences will arise in the propagation of ν_e, ν_μ if $f_e(0) \neq f_\mu(0)$, so that $n_e \neq n_\mu$. It is easy to see that $f_e(0) \neq f_\mu^\mu(0)$. This is due to the fact that the weak charged current contributes only to $f_e(0)$ in the scattering of ν_e on e; the weak neutral current on the other hand, contributes equally to $f_e(0)$ and $f_\mu(0)$. The total contribution of weak charged and neutral currents leads to $f_e(0) \neq f_\mu(0)$.

The presence of $(\nu_e\text{-e})$ scattering gives rise to a modification of the matrix M to M' where [3]

$$M' = \frac{1}{2}(m_1^2+m_2^2+A)\begin{pmatrix} 1 & 0 \\ 0 & 1 \end{pmatrix} + \frac{1}{2}\begin{pmatrix} A-\Delta m^2\cos2\theta & \Delta m^2\sin2\theta \\ \Delta m^2\sin2\theta & -A+\Delta m^2\cos2\theta \end{pmatrix}. \tag{9}$$

Here the quantity A is

$$A = \frac{2\sqrt{2}\,G_F}{m_n} Y_e \, \rho \, E, \tag{10}$$

where m_n is the nucleon mass, ρ is the density, Y_e is the number of electrons per nucleon, E $(\simeq p)$ is the neutrino energy, and G_F is the Fermi constant for weak interaction. For normal matter $Y_e = 1/2$, and if ρ is measured in (gm/cm^3) and E in MeV, then A in $(\text{eV})^2$ is

$$A = 1.33 \times 10^{-7} \{ (\rho/(\text{gm/cm}^3)). (E/\text{MeV})\} (\text{eV})^2. \tag{11}$$

The eigenvalues of M' are

$$m_\nu^2 = \frac{1}{2}(m_1^2+m_2^2+A) \pm \sqrt{\{(\Delta m^2\cos2\theta-A)^2+\Delta m^4\sin^2 2\theta\}}. \tag{12}$$

The splitting of the eigenvalues is given by the square root. The splitting as a function of A is a minimum if $\Delta m^2 > 0$ (that is $m_2 > m_1$)

when

$$A = \Delta m^2 \cos 2\theta \equiv A_o \ . \tag{13}$$

This is the resonance condition. In Fig 3, the mass of the two neutrino flavours is plotted as a function of the density.

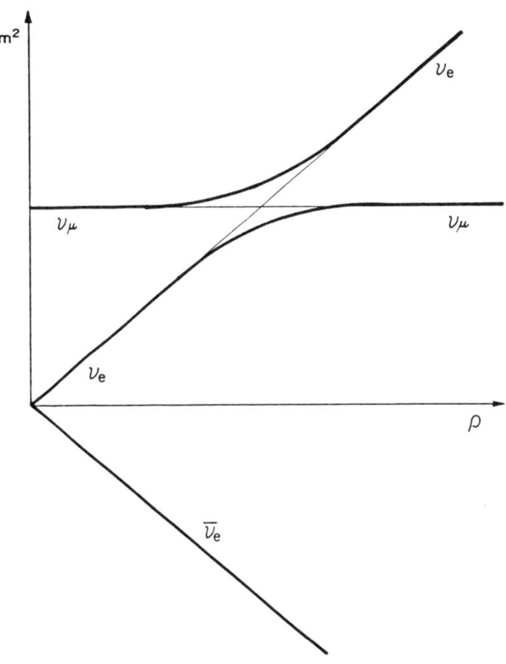

Fig. 3. The masses of two neutrino flavours plotted against density.

Note from the figure that for small ρ (A small), ν_e has the lower mass. As ρ increases, A sweeps past the value A_o and the two curves would cross if the term $\Delta m^2 \sin 2\theta$ was absent in the square root. For $A > A_o$, it is ν_e that is larger. The near crossing point in this figure is the resonance of MS.

From (11), for a given energy of the neutrino, resonance occurs at a definite density ρ_E , where

$$E \rho_E = \frac{3}{4} \times 10^{7} \Delta m^2 \cos 2\theta \equiv \Lambda. \tag{14}$$

There is thus a critical energy $E_c = (\Lambda/\rho_c)$, where ρ_c is the central density in the sun. All neutrinos ν_e with $E > E_c$ go through the resonance and oscillate into ν_μ while the neutrinos with $E < E_c$ emerge unchanged as ν_e. In Fig 4, the solar density is plotted as a function of the distance from the center of the sun.

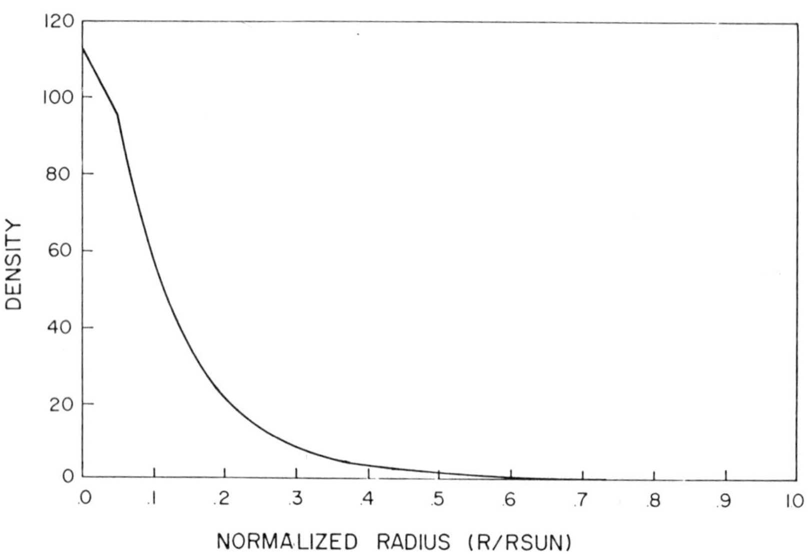

Fig. 4. Solar Density as a function of the normalized radius R/R_{sun}.

If the MS resonance is to explain the result of Davis' experiment, we may say that Davis observes only those neutrinos ν_e with $E < E_c$. Davis gets a total of (2.1 ± 0.3) SNU's of which (1.6 ± 0.2) SNU's is contributed by PeP, ^{7}Be etc. Thus the remaining (0.5 ± 0.5) SNU's can be attributed to neutrinos of $E < E_c$ from $^8_5$B . This represents $(8 \pm 8)\%$ of all the SNU's expected from $^8_5$B. This fact is sufficient to determine the value of E_c and Bethe[3] gives $E_c \simeq 5 - 6$ MeV. This E_c is well above the energy of all electron neutrinos other than from ^{8}B. To determine Δm^2, taking the density from the central portion of the sun, $\rho_c = 130$ gm/cm^3, Bethe gets

$$\Delta m^2 \cos 2\theta = 9.2 \times 10^{-5} \ (\text{eV})^2 \ . \tag{15}$$

If $m_1 = 0$ and $\cos 2\theta \simeq 1$, (15) gives $m_2 \simeq 0.01$ eV. Bethe remarks that with such a small value of Δm^2, it may be difficult to confirm this result by laboratory experiments, and it may very well be that the astrophysical evidence in the form of deficit of solar ν_e flux is the only way to determine the important particle physics parameter Δm^2. MS and Bethe further show that for the resonance to occur under the varying density conditions in the sun, the mixing angle must be restricted to the region $\tan^2 2\theta > 2 \times 10^{-4}$.

That the results obtained by MS are an intersting limit of the results obtained earlier by Wolfenstein[2] may be seen as follows. Wolfenstein showed that in the presence of matter, one gets a new mixing angle θ_m, and a new oscillation length L_m :

$$\sin^2 2\theta_m = \frac{\sin^2 2\theta}{1 + \left(\frac{L_v}{L_\rho}\right)^2 - 2 \cos 2\theta \left(\frac{L_v}{L_\rho}\right)} \ , \tag{16}$$

$$L_m = \frac{L_v}{\sqrt{\{1 + \left(\frac{L_v}{L_\rho}\right)^2 - 2 \cos 2\theta \left(\frac{L_v}{L_\rho}\right)\}}} \ , \tag{17}$$

where L_v is the vacuum oscillation length given in (5) and L_ρ is given by

$$L_\rho = \frac{2\pi}{\sqrt{2} \ G_F \ N} \ . \tag{18}$$

where again G_F = Fermi constant and N the electron density in the material. Wolfenstein considered the limiting cases (i) $L_v < L_\rho$, in which case $\theta_m \simeq \theta$ and $L_m \simeq L_v$, and (ii) $L_v > L_\rho$, in which case $L_m \simeq L_\rho$ independent of the angle θ. He did not consider the case $L_v \simeq L_\rho$. This is the limit that was considered by MS who pointed out that if

$$\frac{L_v}{L_\rho} = \cos 2\theta, \tag{19}$$

then from (16) one sees that $\sin^2 2\theta_m \to 1$ (resonant amplification of the mixing angle). What is interesting is that under the conditions prevailing in the sun, it is possible to satisfy (19) (which is the same as (13)). Note that the width of the resonance (19) or (13), is proportional to $\sin 2\theta$, hence it is very narrow for very small θ.

Rosen and Gelb[4] have used the computer to solve the time evolution equations for ν_e through the sun and obtained the parameters Δm^2 and $\sin^2 2\theta$ which give a factor (1/3) reduction of the solar ν_e flux at the earth's surface. Figure 5 shows their results for these parameters for both the ^8_5B and the PP neutrinos.

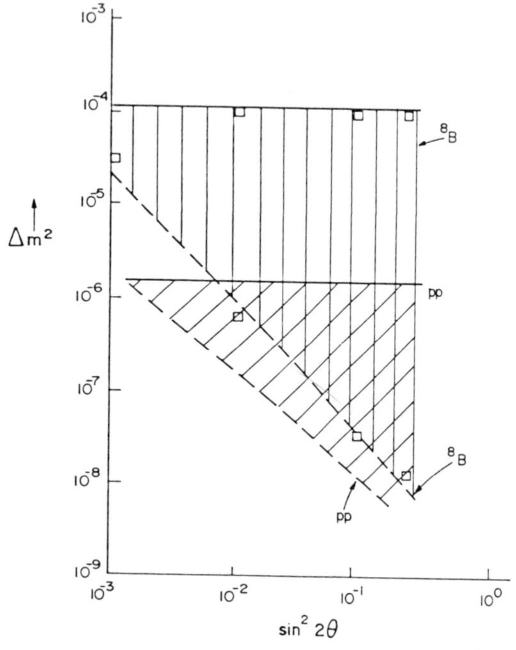

Fig. 5. Δm^2 and $\sin^2 2\theta$ which give 1/3 reduction of the solar neutrino flux at the earth.

The horizontal lines in the figure, where Δm^2 is relatively independent of $\sin^2 2\theta$, results if the resonance occurs in the core of the sun. On the other hand, Δm^2 falls almost linearly with $\sin^2 2\theta$, if the resonance is outside the core (the dashed lines in the figure). Note the different values of Δm^2 to which one is sensitive in the case of ^8_5B and in the case of PP neutrinos. The results of the detailed calculations for the flux of electron neutrinos at the earth for different Δm^2 and $\sin^2 2\theta$ which gave a reduction of 1/3 in the flux were kindly supplied to us by Drs. Rosen and Gelb. These we used in obtainig our results for the expected rates in the

Sudbury neutrino detector.

5.   PROPOSAL FOR A NEUTRINO OBSERVATORY AT SUDBURY[6]

A collaboration consisting of physicists from a number of institutions (Queen's, Univ. of California(Irvine), Oxford, NRC, CRNL, Guelph, Laurentian, Princeton and Carleton) have made a proposal to the Canadian government for funding the construction of a large volume (1000 tonnes) D_2O detector to be located 2070 m below ground in a mine at Creighton near Sudbury. This large volume will be looked at by a large number of Cerenkov detectors. One of the objectives of this detector is to study the ^8_5B neutrinos from the sun by a direct counting method that will measure their energy and direction. The detector is sensitive to both ν_e and ν_μ provided $E_\nu > 2.2$ MeV, through the reactions

$$\nu_e + D \rightarrow P + P + e \quad \text{(charged current reaction)},$$

$$\nu + D \rightarrow P + N + \nu \quad \text{(neutral current reaction)}.$$

In Fig 6 the cross sections for these reactions in the energy region of interest are shown.

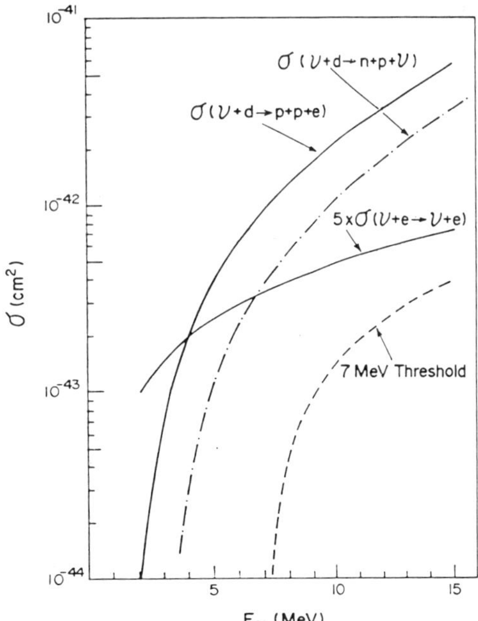

Fig. 6. Neutrino cross sections as a function of the energy.

The charged current reaction gives an electron in the final state
which can be detected through its Cerenkov radiation in D_2O. In the
case of the neutral current reaction, following the breakup of the
deuteron, the neutron wanders and is eventually captured by a deuteron
and produce 6.25 MeV gamma rays. The cross section for the breakup
reaction is independent of the type of neutrino; hence it is sensitive
to the total flux of neutrinos.

Using the results of Rosen and Gelb[4], we have calculated the
expected number of electron neutrino events for different values of
$\sin^2 2\theta$ and Δm^2 (all of which 1/3 reduction of ν_e flux at the earth)
for the Sudbury detector, and these results are shown in Fig 7. These
are based on the standard solar model and the left hand set of curves
are for resonance occurring in the core, ($\Delta m^2 \simeq 10^{-4}$), while the right
hand set of curves are for resonance occurring outside the core, ($\Delta m^2 \simeq$
10^{-6} or smaller) for different $\sin^2 2\theta$. Note the sensitivity to the
values of the different parameters.

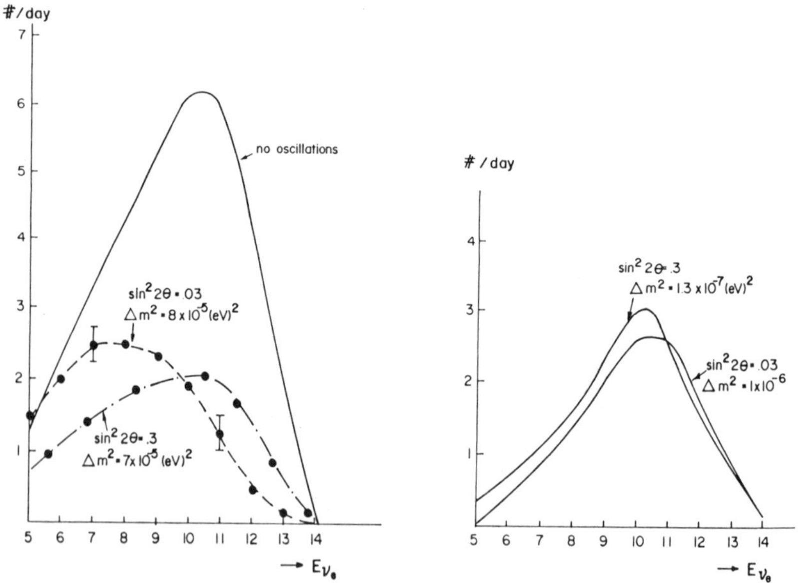

Fig. 7. Expected rates in the Sudbury detector for different Δm^2,
$\sin^2 2\theta$.

It is also clear that one can obtain predictions similar to those shown in Fig 7 by changing the solar model. Such a detector as the one proposed in Sudbury, if constructed, will clearly able to test the solar model independently of whether neutrino oscillations occur or not.

6. REFERENCES

1. Mikheyev, S. P. and Smirnov, A. Yu., Institute for Nuclear Research of Academy of Sciences of the USSR Preprint, 60th Anniversary Prosp.7a, Moscow 117342, USSR (hereafter referred to as MS)
2. Wolfenstein, L., Phys. Rev. D $\underline{17}$, 2369 (1978).
3. Bethe, H. A., Phys. Rev. Letts., $\underline{56}$, 1305, (1986).
4. Rosen, S. P. and Gelb, J. M., Los Alamos Preprint, "Mikheyev-Smirnov-Wolfenstein Enhancement of Oscillations as a Possible Solution to the Solar Neutrino Problem," March 1986.
5. Davis, R. et al., Science Underground AIP Conference Proceedings No.96, American Institute of Physics, N. Y., (1983). Series Editor: H. C. Wolfe.
6. Ewan, G. T. et al., "Sudbury Neutrino Observatory- Feasibility Study for a Neutrino Observatory Based on a Large Heavy Water Detector Deep Underground", SNO-85-3, July (1985)
7. Bahcall, J. N. et al.,Rev. Mod. Phys. $\underline{54}$, 767, (1982).

NEW RESULTS ON DECAYS OF THE B($b\bar{u}$/$b\bar{d}$) MESON

M. S. ALAM

State University of New York at Albany, Albany, NY12222

ABSTRACT

On behalf of the CLEO Collaboration, new and updated results on inclusive decays of B mesons into D^o ,D^{*+},F^+ , ψ ,and ψ' mesons besides decays into few-body states involving them, are presented.

INTRODUCTION

$B^o\bar{B}^o$ AND B^+B^- ,pairs of mesons with beauty quantum number are produced from the decays of $\Upsilon(4S)$ created in electron-positron collisions at centre-of-mass energy equal to the $\Upsilon(4S)$. In fact,the $\Upsilon(4S)$ has turned out to be a veritable B meson factory. The study of inclusive and exclusive decays of neutral and charged B mesons provides information on the weak decays of the b-quark. It is therefore a very useful tool for testing the data against the predictions of the so-called " Standard Model "[1] .

Results on inclusive decays of B mesons into D^o ,D^{*+} , and ψ besides decays into few-body final states have been reported earlier[2]. These results were based on data recorded in 82 and 83. Results presented here include old data as well as new data recorded in 84 and 85. The old CLEO[3] detector has been improved by adding a ten-layer, high precision vertex tracking chamber with a thiner Beryllium pipe (1% r.l.) besides instrumenting all 17 layers of the drift chambers to record dE/dx information. A comparision of the old and new data sample is presented:

	Old Data	New Data
Integrated Luminosity on $\Upsilon(4S)$	41 pb^{-1}	77 pb^{-1}
Integrated Luminosity Contin.	17 ..	38 ..
Number of $B\bar{B}$ Pairs Produced	42 K	89 K
$\delta p/p$ (large momentum) GeV/c	.012 p	.007 p
Lower Mom. Cutoff for Kaons	.450 GeV/c	.200 GeV/c
Lower Mom. Cutoff for protons	.650 GeV/c	.300 GeV/c

Results are presented on (1)Inclusive decays of B->FX, (2)Inclusive decays of B->ψX and ψ'X, (3)Inclusive decays of B->D^oX and B->D^{*+}X besides (4)Exclusive decays of B mesons into few-body final states involving charm and charmless particles and finally (5)Measurements of the B^o-$\bar{B}^o$ mixing parameter y_B in terms of measured same sign and opposite sign dileptons. Finally the conclusion and references are presented.

1. INCLUSIVE DECAYS OF B --> FX

B mesons are expected to decay weakly into F mesons according to the spectator diagrams shown in Fig. 1(a and b).

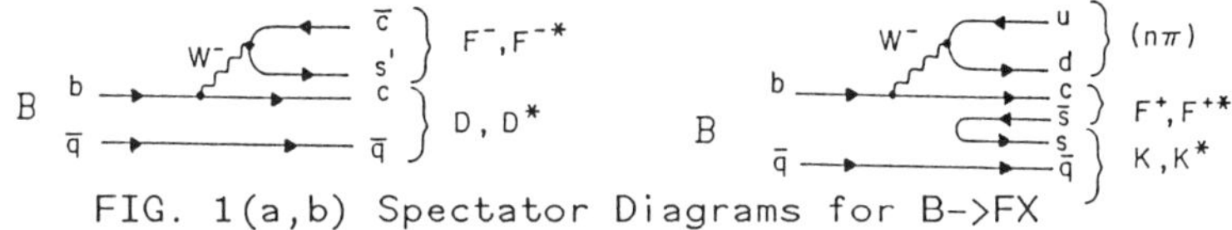

FIG. 1(a,b) Spectator Diagrams for B->FX

Non-spectator diagrams shown in Fig. 2(a and b) are also expected to contribute if helicity-supression in these diagrams could be overcome by initial-state soft-gluon emission [4].

FIG. 2(a,b) Non-Spectator Diagrams for B->FX

Spectator diagrams of the first kind will lead to final states of the type FD, F*D, FD* and F*D* in the two-body mode or with extra pions in multibody mode. The non-spectator annihilation diagram, which is only possible for charged B's, is suppressed both by helicity violation and the smallness of buW^-coupling. The non-spectator exchange diagram, which is only possible for the neutral B's, gives rise to two-body states like F^+K^-, F^+K^{*-}, $F^{*+}K^-$, $F^{*+}K^{*-}$ and more complicated multibody states with extra pions. Observation of such final states would be interpretted as evidence for exchange diagrams in B decays. Suzuki[5] estimates the branching ratio of B -> FX to be about 6 per cent. CLEO has observed the inclusive decay B->FX with the F detected in the $\phi\pi$ decay mode. The analysis is based on 77 pb^{-1} data at the Υ(4S) and 38 pb^{-1} of data on the continuum below the Υ(4S). F mesons produced from the decay of the B mesons have momenta less than 2.5 GeV/c . Fig. 3 shows the invariant mass distributions for all $\phi\pi$ combinations at the Υ(4S) (data points) and the continuum sca-

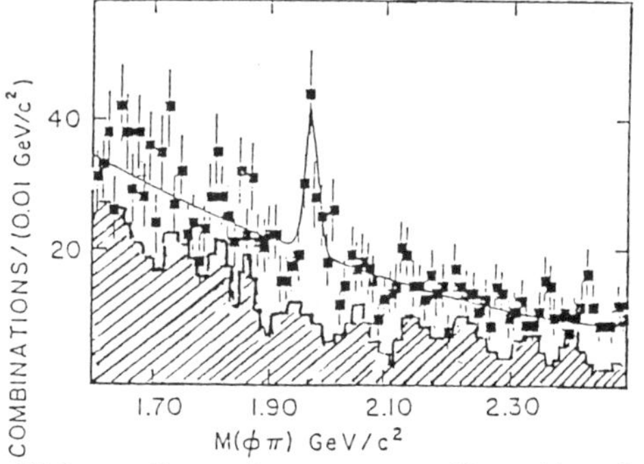

FIG. 3 Invariant mass distribution for $\phi\pi$ combinations with p>2.5 GeV/c at the Υ(4S) (data points) and on the continuum (dashed hsitogram) scaled by the luminosity factor.

led by the lumnosity factor(dashed histogram). ϕ^o 's are detected in the decay mode K^+K^- where oppositely charged pairs of tracks consistent with being kaons were used. Only those $\phi\pi$ combinations were used, whose momenta were less than 2.5 GeV/c and the decay angle θ_K of the kaons with respect to the F-direction in the ϕ restframe satisfies $|\cos(\theta_K)| > 0.6$. An excess of $(60 + 15.6)$ events corresponding to the direct decay of $B\text{->}FX$ is obtained after subtraction of the continuum contribution. The corrected momentum distribution for direct $B\text{->}FX$ decays is shown in Fig. 4. The spectrum is peaked in the region between 1.0 and 2.0 GeV/c and is consistent with two-body decays of the type DF, D^*F, DF^* ,and D^*F^*. CLEO measures the following product branching ratio:

$B(B\text{->}FX)*B(F\text{->} \phi\pi) =$
$.0038 \pm .001$

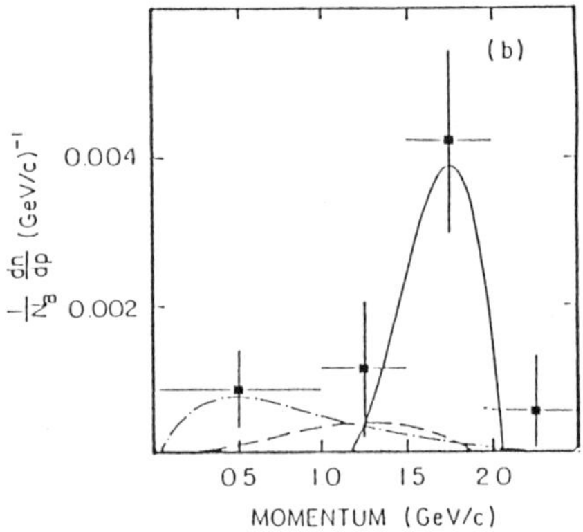

If we assume the semi-measured value[6] of 4.4 percent for $B(F\text{->} \phi\pi)$, we measure:

$B(B\text{->}FX) = .086 \pm .023$

A fit to the momentum spectrum with two-body final states of the type DF and three-body final states of the type $FD\pi$ and FKW^- gives the two-body component of the decay to be :

$B[B\text{->}FX(2body)]/B[B\text{->}FX(all)] = .64 \pm .22$

FIG. 4 Momentum of F mesons from direct $B\text{->}FX$ decays. Monte Carlo predictions for BF type two-body decays(solid), $B\text{->}FD$ type three-body decays (- - -) and FKW^- type three-body decays($\_.\_.\_$) are shown.

2. INCLUSIVE DECAYS OF $B\text{->} \psi X$ and $B\text{->} \psi' X$

Inclusive decays of B mesons into final states with $\psi(c\bar{c})$, $\psi'(c\bar{c})$ and $\chi_c(c\bar{c})$ are expected to decay according to the spectator diagram shown in Fig. 5 . The probability that the color of the c quark from the W is the same as that of the original b quark to form a color-singlet $c\bar{c}$ state (e.g. ψ, ψ', χ_c) is one out of three so that the decay processes of the above kind are expected to be color-supressed by a factor of 9 relative to decays that do not involve color-matching. Theorists have argued that the color-suppression can be overcome by initial-state soft gluon radiation. Ruckl[6] estimates the branching ratio $B(B \rightarrow \psi_{tot} X)$ to be between 1.6 and 2.5 percent where the color-suppression factor of 9 has been included and ψ_{tot} includes ψ's produced from higher excited states.

CLEO[8] and ARGUS[9] have both already reported the observation of $B \rightarrow \psi X$. Preliminary results on a larger sample of data ('83 and '85) are reported here. The ψ's are detected through their decays to e^+e^- and $\mu^+\mu^-$ pairs. From a sample of 118 pb^{-1} of data ('83 and '85) at the $\Upsilon(4S)$, the mass distribution for all $\mu^+\mu^-$ pairs with one muon cleanly identified and the other loosely identified is shown in Fig. 6 . A clear signal of (53 $\pm$ 9) events above background ,corresponding to $\psi \rightarrow \mu^+\mu^-$ is visible. A smaller signal of (26 $\pm$ 8) events corresponding to $\psi \rightarrow e^+e^-$ is also observed using only the '85 data sample of 78 pb^{-1} . The corrected ψ momentum is shown in Fig. 7 below.

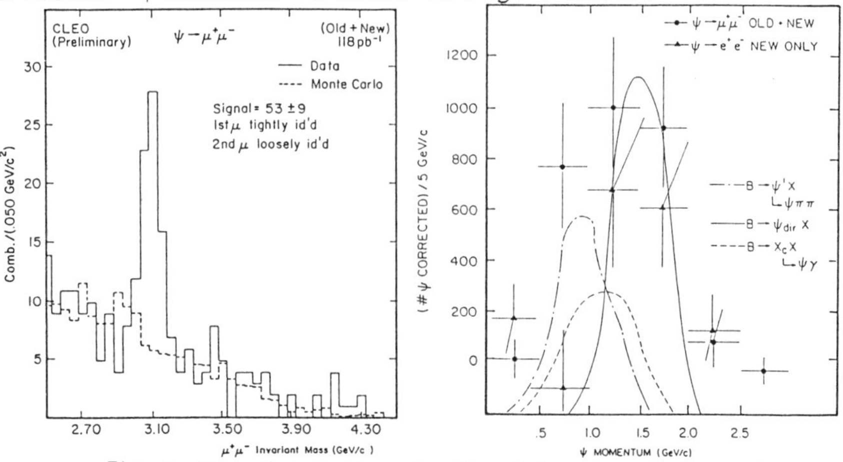

Fig. 6 Mass distribution of $\mu^+\mu^-$ pairs at the $\Upsilon(4S)$.

Fig. 7 Momentum of ψ's produced from direct $B \rightarrow \psi X$ decays.

The inclusive branching ratio is measured to be

$$B(B \rightarrow \psi X) = (0.94 \pm 0.30 \pm 0.26)\% \quad \text{(only } e^+e^- \text{ used)}$$

$$B(B \rightarrow \psi X) = (1.07 \pm 0.18 \pm 0.26)\% \quad \text{(only } \mu^+\mu^- \text{ used)}$$

and combining the two results, CLEO reports

$$B(B \rightarrow \psi X) = (1.03 \pm 0.17 \pm 0.26)\%$$

to be compared with the ARGUS [9] result of (1.37 $\pm$ 0.6)% .

A look at the ψ momentum spectrum suggests a substantial contribution from secondary production of ψ's from ψ''s and χ_c's higher mass charmonium states or a softening of the momentum spectrum due to initial-state soft-gluon radiation. CLEO has searched for the inclusive process B-> ψ'X where the direct decay of the ψ' to e^+e^- and $\mu^+\mu^-$ pairs and the cascade decay to $\psi\pi^+\pi^-$ have both been used to search for ψ'. Using the decay $\psi \to \psi\pi^+\pi^-$, CLEO places an upper limit of 0.8 percent at the 90% C.L. on the branching ratio B(B -> ψ' X). The corresponding upper limit on B(B -> ψ'X) is 1.1 percent if the direct decay of the $\psi' \to e^+e^-$ or $\mu^+\mu^-$ is used.

3. INCLUSIVE DECAYS OF B-> D^0X AND D^{*+}X

CLEO has already reported the inclusive decays of B mesons to the charmed mesons D^0 and D^{*+} [2]. Updated results using data recorded in '83 and '85 are presented. The results on inclusive D^0's are based on 89 pb^{-1} of data at the Υ(4S) and 38 pb^{-1} of data on the continuum at energies below the Υ(4S). The results on inclusive decays into D^{*+}'s are based on 118 pb^{-1} at the Υ(4S) and 53 pb^{-1} below the Υ(4S). D^0's are detected through the decay D^0 ->$K^-\pi^+$. D^{*+}'s are detected using the double decay D^{*+} ->$D^0\pi^+$ and D^0 ->$K^-\pi^+$, where we use the excellent resolution of 2 MeV/c^2 on the mass difference $\Delta M = M(K^-\pi^+\pi^+) - M(K^-\pi^+)$. Fig. 8 shows the mass distribution for $(K^-\pi^+/K^+\pi^-)$ combinations for which the magnitude of the mass difference (M - 145.4) MeV/c^2 is less than 2.5 MeV/c^2. A clear signal for D^0's from D^{*+} decays can be seen from both the Υ(4S) and the continuum. However, a clear excess corresponding to D^{*+}'s from B decays is evident above the continuum contribution. The momentum of the D^{*+}'s is restricted between 1.0 and 2.5 GeV/c. The lower momentum cut is used to reduce background. The upper momentum cut is the kinematic limit for D momenta from B decays. For the D^0 analysis, the full momentum spectrum up to the limit of 2.5 GeV/c is used.

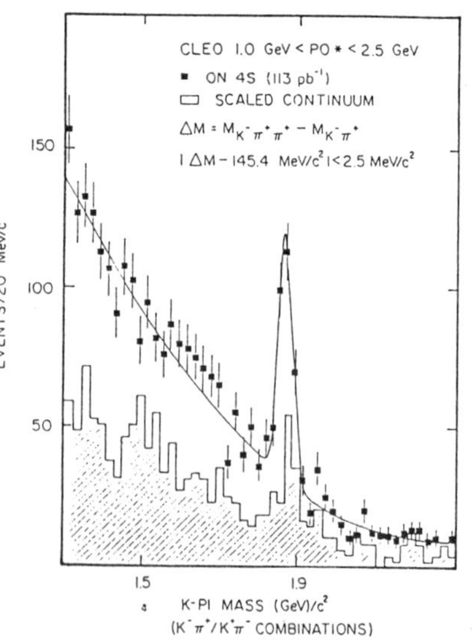

FIG. 8 Mass distribution of ($K^-\pi^+$/$K^+\pi^-$) combinations from the decay D^{*+} -> $D^0\pi^+$ and D^0 ->$K^-\pi^+$ at the Υ(4S) and on the cont. properly scaled

The corrected momentum spectra of D^0's and D^{*+}'s from the decays of the B mesons, after subtraction of the continuum contributions, are shown in Fig. 9 . The D^{*+} spectrum has been scaled by a factor of 2 for comparison with the D^0 spectrum and also on the assumption that $B(B\to D^{*+}X)$ equals $B(B\to D^{*0}X)$. Using the MARKIII measurements of $B(D^0\to K^-\pi^+)=.056 +.005$ and $B(D^{*+}\to D^0\pi^+)=0.6\ ^{+.08}_{-.06}$, CLEO measures the inclusive decays of the B mesons into D^0's and D^{*+}'s to be

$$B(B\to D^0 X)= 0.40 \pm 0.04 \pm .04$$
and
$$B(B\to D^{*+} X)= 0.27 \pm 0.04\ ^{+.07}_{-.04}$$

Fig. 9 D^0 and D^{*+} momentum spectra from the direct B decays and predictions from a $b\to cW^-$ Monte Carlo are shown.

where the systematic errors essentialy reflect the error on the D^0 and D^{*+} branching ratios . If we assume $B(B\to D^{*+}X)= B(B\to D^{*0}X)$, it appears that most of the D^0's are produced from the decay of the D^{*+}'s and D^{*0}'s. In fact, CLEO places the 90% C.L. upper limit :
$$B(B \to D^0X)\text{direct}/B(B \to D^{*+} X + D^{*0}X) < .2$$

4. EXCLUSIVE DECAYS OF B MESONS

A. Decays Involving Charm

B mesons are expected to decay into final states involving charmed mesons and baryons. In fact, if $b\to cW^-$ dominates, D^0, D^+, D^{*0} or D^{*+} mesons together with one or more pions will dominate as shown in the spectator diagrams for B decay in Figure 10. Charmed baryons together with a proton or neutron and pions could also be formed. Non-spectator

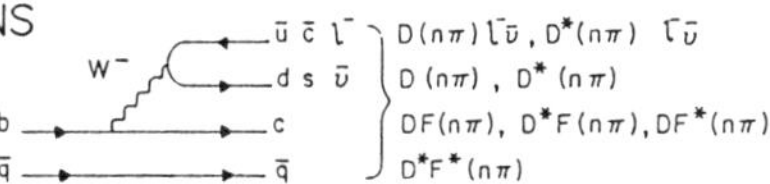

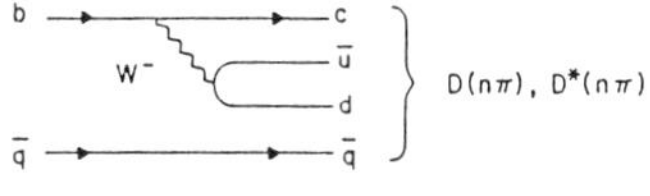

Fig. 10 Spectator Diagrams for B decays. The light quark behaves as a spectator.

diagram as shown in Fig. 11 may also contribute significantly if the helicity suppression associated with the diagram is overcome by initial-state gluon radiation. The first evidence for B meson decays into exclusive final states involving charmed mesons D^0 and D^{*+} and one or more pions was presented in '84[10]. The charge conjugate states are also implied but omitted for the sake of brevity. Using the 42 pb^{-1} of data recorded in '84 and 78 pb^{-1} of data recorded in '85, updated results on B decays into exclusive states involving D^0, D^+, and D^{*+} and ψ are presented. D^0's are detected in the $D^0 \to K^-\pi^+$ and $K^-\pi^+\pi^-\pi^+$ decay modes. D^{*+}'s are detected in the decay mode $D^{*+} \to D^0\pi^+$ using the mass difference technique discussed earlier. D^+'s are detected in the $K^-\pi^+\pi^+$ decay mode . ψ's are detected in the $\psi \to e^+e^-$ and $\mu^+\mu^-$ decay modes. Particle id was used when available.

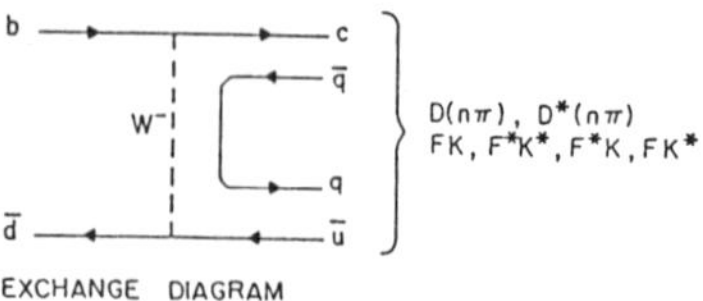

Fig. 11 Exchange diagram for B decays.

Instead of plotting the invariant mass of the tracks involved in the the exclusive final state e.g. $\bar{B}^0 \to D^0\pi^-$ defined as

$$e^+ e^- \to \Upsilon(4S) \to B^+ B^-$$

$$B^- \to D^0 \pi^- \quad \text{,and}$$
$$B^+ \to \bar{D}^0\pi^+ \quad \text{, and}$$

$$M_{B^-} = (E_{D^0} + E_{\pi^-}) -$$
$$(\vec{P}_{D^0} + \vec{P}_{\pi^-})$$

the beam-constrained mass defined as

$$M_{B^-} = (E_{beam}) -$$
$$(\vec{P}_{D^0} + \vec{P}_{\pi^-})$$

is plotted. In e^+e^- collisions, the centre-of-mass and lab frame are identical. For two-body production of $B\bar{B}$ from the $\Upsilon(4S)$, the energy of the B is just equal to the engy of the beam, as measur- by the magnetic field of the ring dipoles, which

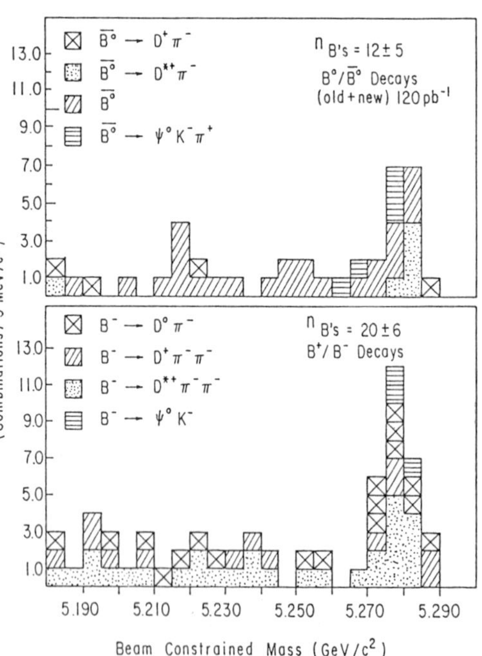

Fig. 12 Beam-constrained mass distributions for charged and neutral B mesons in several exclusive final states with charm.

is very accurately determined. A factor of nearly 10 improvement in mass resolution is achieved by using the
beam energy instead of the measured energy of the combination. Further by restricting the energy of the combinations to be close to the energy of the beam, a major reduction in background is achieved. Fig. 12 shows the
summed beam-constrained mass distribution for the neutral
combinations(top) corresponding to the final states $D^0\pi^-$,
$D^+\pi^-\pi^-$, $D^{*+}\pi^-\pi^-$ and ψK^- (top) and the charged combinations (bottom) corresponding to the final states $D^+\pi^-$,
$D^0\pi^+\pi^-$, $D^{*+}\pi^-$ and $\psi K^-\pi^+$. A clear enhancement corresponding to the neutral $\bar{B}^0$ and charged B^- in the mass region
above 5.270 GeV/c is evident. The charge conjugate
states are also included. The background is expected to
be flat and using a linear extrapolation of the background below 5.270 GeV/c , CLEO has calculated the individual branching ratios. The preliminary measurements
of the branching ratios of the charged B into the final
exclusive states $D^0\pi^-$, $D^+\pi^-\pi^-$, $D^{*+}\pi^-\pi^-$ and ψK^- are
$(0.2 \pm 0.1)\%$, $(0.5 \pm 0.4)\%$, $(0.2 \pm 0.1)\%$ and (< .13)% at
90% C.L. respectively. The preliminary measurements of
the branching ratios of the neutral B   into the final
exclusive states $D^+\pi^-$, $D^0\pi^+\pi^-$, $D^{*+}\pi^-$ and $\psi K^-\pi^+$ are
$(0.1 \pm 0.1)\%$, $(0.9 \pm 0.7)\%$, $(0.2 \pm 0.1)\%$ and (<0.25)%
at 90% C.L. respectively.

B. Masses

    Using only the first three final states in Table 1,
preliminary measurement of the B masses is given below

Mass (B^-) = $5276.2 \pm 1.2 \pm 3.0$ MeV/c
Mass (B^0) = $5279.5 \pm 1.2 \pm 3.0$ MeV/c
Mass Difference = $3.3 \pm 1.7 \pm 3.0$ MeV/c
Average Mass = $5277.5 \pm 0.9 \pm 3.0$ MeV/c

C. Decays Not-involving Charm

    B mesons will decay into
final states not involving
charm through the buW^- coupling
as shown in Fig. 13. Observation of states not involving
charm will be also be interpretted as evidence for the
buW^- coupling. CLEO has looked

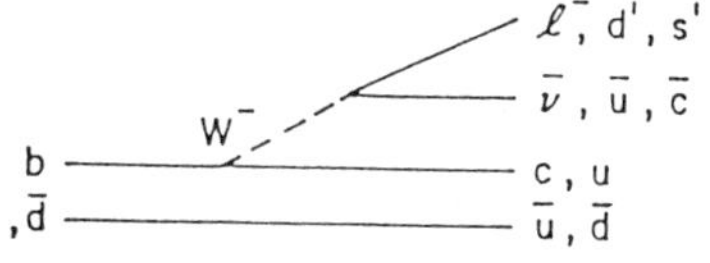

FIG. 13 Spectator Diag-
ram for buW^- decays.

for such states in 23.4 pb of new data ('85) by plotting
the beam-constrained masses with a cut on the energy difference for various final states not involving charm   . No
signal for B mesons was observed and the 90% C. L. upper
limits on the branching ratios $\bar{B}^0 \to \pi^+\pi^-$, $B^- \to \rho^0\pi^-$, $B^- \to \rho^0 A_1^-$,

and $B^- \rightarrow \rho^o A_2^-$ have been measured to be 0.02%, 0.02%, 0.08% and 0.04% respectively.

5. MEASUREMENT OF $B^o - \bar{B}^o$ MIXING

$B^o - \bar{B}^o$ mixing which proceeds through the second-order weak interaction box diagram show in Fig. 14, is described

FIG. 14   Box diagram for $B^o - \bar{B}^o$ Mixing

in terms of the mixing parameter r_B as

$$r_B = \{ \Delta M_B / \Gamma_B(tot) \} / [2 + \{ \Delta M_B / \Gamma_B(tot) \}^2]$$

where $\Delta M = M(B^o_L) - M(B^o_S)$ and $\Gamma_B(tot)$ is the total decay width of the B^o. In the Standard Model

$$\Delta M_B / \Gamma_B(tot) \propto B_B f_B m_t^2 |U_{tb} U_{td}^*|$$

which depends on B_B the bag-parameter, $f_B = 12 |\psi(0)|^2 /M(B)$, m_t the top quark mass besides the K-M matrix elements U_{tb} and U_{td}, all of which are poorly measured. Theoretical estimates of r_B vary from zero to a few percent. Soni[11] estimates r_B to vary from 8% to 30% for $m_t < 40$ GeV/c^2 and where experimental values of $\tau(D^+)/\tau(D^o)$, Γ_B (semileptonic), $\Gamma(b \rightarrow u l \bar{\nu})/ (b \rightarrow c l \bar{\nu})$ and ε were used to constrain the K-M angles θ_2, θ_3, and the CP phase δ.
Experimentally, $B^o - \bar{B}^o$ mixing is measured in terms of the mixing parameter y_B defined as

$$y = \{ N(B^o B^o) + N(\bar{B}^o \bar{B}^o) \}/N(B^o \bar{B}^o)$$

which is related to the theoretically defined mixing parameter r_B by the relation

$$y_B = 2 r_B /(1 + r_B^2)$$

Note $y_B = 0$ for $r_B = 0$ and $y_B = 1.0$ for $r_B = 1.0$
In $\tau(4S)$ decays to $B^o \bar{B}^o$, mixing will lead to observation of same sign dileptons according to the reactions:

$$\tau(4S) \rightarrow B^o B^o \rightarrow X l^- \bar{\nu} + X' l^- \bar{\nu} \rightarrow l^- l^- X''$$

$$\tau(4S) \rightarrow \bar{B}^o \bar{B}^o \rightarrow X l^+ \nu + X' l^+ \nu \rightarrow l^+ l^+ X''$$

and $y_B = \{ N(l^+ l^+) + N(l^- l^-) \}*$
$$[\{ B^+(SL)/B^o(SL) \}*\{ f(B^+)/f(B^o) \} + 1]$$

where $B^+(SL)/B^o(SL)$ is the ratio of the charged and neutral B semileptonic branching ratios and $f(B^+)/f(B^o)$ is the ratio of the $\Upsilon(4S)$ decay into charged and neutral $B\bar{B}$ pairs.

From a sample of 78 pb^{-1} of data at the $\Upsilon(4S)$ ('85), CLEO obtains

$$N(l^+l^+) + N(l^-l^-) = 4.3 \pm 5.7$$

and $N(l^+l^-) = 141.9 \pm 12.8$

after subtracting background contributions from parallel decay processes ($B{-}{>}Xl\bar{\nu}$, and $B{-}{>}$ $XD \to Kl\bar{\nu}$) and from hadrons misidentified as leptons. Assuming $B^+(SL)/B^o(SL) = 1.0$ and using the partially measured value of $f(B^o)=0.4$ and $f(B^+)=.6$, CLEO measures

$$y_B < .28 \quad (90\% \text{ C. L.})$$

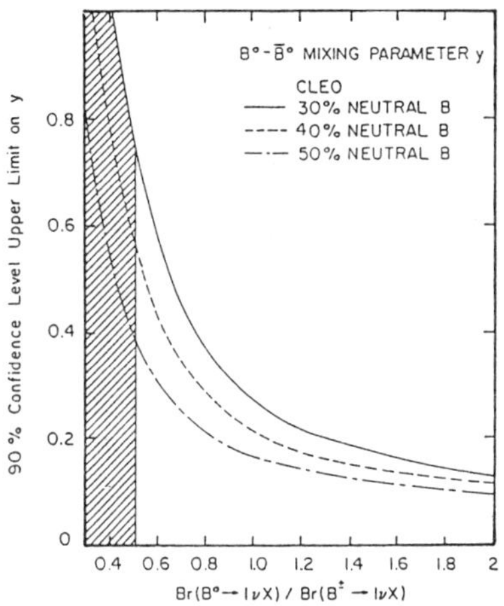

In general, y_B depends on $B^+(SL)/B^o(SL)$ and $f(B^o)$. Thus in Fig. 15 , the 90% C. L. upper limit on y_B is plotted as a function of $B^+(SL)/B^o(SL)$ for $f(B^o)=0.3$, 0.4 and 0.5 . A very preliminary measurement of $B^+(SL)/B^o(SL)$ as lying between .5 and 1.9 rules out the shaded area in the figure.

FIG. 15 Measurement of the Mixing Parameter y_B

CONCLUSION

The emerging picture from B decays seems to confirm the expectations from the Standard Model. Exotic models with no top quark have essentially been ruled out . The weak decays of the b-quark is primarily mediated through the charged weak current and no evidence for flavour-changing neutral current decays is observed since ($B \to e^+e^-X$)/ ($B \to Xl\bar{\nu}$) is measured to be less than 3% (90% C.L.). No evidence for buW^- coupling has been seen in either exclusive final states not involving charm or in studies of the lepton spectrum from B meson decays. From the study of the lepton spectrum, it appears that branching ratio $B(buW^-)/B(bc W^-)$ is less than a few percent at the 90% confidence level. Evidence for the bcW^- coupling is confirmed with the observation of copious production of charmed mesons D^o and D^{*+} and , for the first time, F mesons. It also seems that B meson decays into scalar D mesons are supressed relative to vector D^* mesons. A signal of B decays into ψ mesons is seen, but the measured branching ratio of about 1% may

not be interpretted as evidence of color-suppression in B
decays since colorless models can also give such values.
Reconstructed charged and neutral B mesons have been seen
using   states involving D^o, D^+, D^{*+} and ψ and one or more
kaons or pions. No statistically significant signal has
been seen in any specific exclusive state and the branching ratio into specific final states are turning out
to be of the order of few tenths of a percent instead of
few percent as reported earlier. Finally, we are beginning to approach meaningful measurements for the B^o-$\bar{B}^o$
mixing parmater y_B to confront the theoretical prejudices.
It may be mentioned that almost ten years after the observation of the first $T(1S)$ state, the world sample of
reconstructed B mesons is less than one hundred events.

REFERENCES

1.   J. Ellis , et al., Nucl. Phy. B131, 285 (1977)
2.   J. Green, et al., Phy. Rev. Lett., 51, 347 (1983)
     S. E. Csorna, et al., Phy. Rev. Lett., 54, 1894 (1985)
3.   D. Andrews et al., Nuc. Inst. & Meth., 211, 47 (1983)
4.   J. P. Leveille Proc. CLEO Collab. Workshop on B Meson
     Decay, 37-67 (1981)
5.   M. Suzuki, Phy. Rev. D31, 1158 (1985)
6.   A. Chen et al., Phy. Rev. Lett., 51, 634 (1983)
     H. Albrecht et al., Phys. Lett., 153B, 343 (1985)
7.   H. J. Kuhn & R. Ruckl, Phy. Lett., 135B, 477 (1984)
8.   P. Haas, et al., Phy. Rev. Lett., 55, 1284 (1985)
9.   ARGUS Collaboration
10. R. Giles et al., Phy. Rev. D, 30, 2279 (1984)
11. A. Soni, Phy. Rev. Lett., 53, 1407 (1984)

MEASUREMENT OF THE τ AND D^o LIFETIMES

Brian K. Heltsley

Cornell University, Ithaca, NY 14853 USA

on behalf of the CLEO Collaboration
(SUNY-Albany, Carnegie-Mellon, Cornell, Florida, Harvard,
Ithaca College, Ohio State, Rochester, Rutgers, Syracuse, Vanderbilt)

Using the CLEO detector operating at CESR, the lifetimes of the τ lepton and the D^o meson have been measured to be $\tau_\tau = (2.92\pm0.23\pm0.30) \times 10^{-13}$ sec and $\tau_{D^o} = (4.55\pm0.93\pm0.70) \times 10^{-13}$ sec.

The lifetimes of fundamental fermions are expected to scale with the fifth power of the ratio of their masses to that of the muon. Using this rule the decay distances of high energy τ leptons and charmed particles are hundreds of microns in the laboratory, allowing such lifetimes to become accessible to measurement in the current generation of experiments. In the case of the τ lepton the standard model assumptions can be directly tested, whereas the presence of additional quarks in charmed particles introduces complications. Final state quark interactions and W-annihilation effects cause charmed meson and baryon lifetimes to deviate from the simple mass scaling. In particular, present data[1] indicate that the D^+ lifetime is a factor of two larger than that of the D^o. Measurements of the lifetimes of the τ lepton and D^o meson are presented here, using data collected with the CLEO detecor operating at the Cornell Electron-Positron Storage Ring (CESR).

The measurements described here rely primarily on the charged particle tracking capability of the CLEO detector.[2] The vertex detector (VD) covers the region from 8 to 16 cm from the beam axis with 800 drift cells arranged in 10 cylindrical layers. The drift

476

chamber (DR) covers from 17 to 95 cm from the beam line with 5304 drift cells in 17 cylinders. Wires in the VD are 0.70 m in length and are parallel to the beam; in the DR wires are 1.93 m long, and the layers alternate axial with stereo ($\pm3^{\circ}$ tilt) wires. A gas mixture of 50% Ar-50% CH_6 flows through both chambers at atmospheric pressure (for the second half of the data sample used the VD was at 21 psia). The point setting error in the r-ϕ plane for each sense wire in μ-pair events is 90 μm in the VD and 180 μm in the DR; this resolution is degraded by 10% in hadronic events. Particles entering the VD must penetrate a beryllium beam pipe and the carbon filament inner VD wall which together constitute 1% r.l. at normal incidence. Reconstructed charged particle trajectories require that at least 23 of the 27 layers contain a hit and allow for a kink in the track due to multiple scattering in the 1% r.l. of material between the two chambers. A superconducting solenoid surrounding the chambers provides a 1 T axial magnetic field for momentum determination; a momentum resolution of $\delta p/p$=0.7%p (p in GeV/c) is obtained.

The τ lifetime determination used 3-prong decays taken from a sample enriched in "1-versus-3" $e^+e^-\rightarrow\tau^+\tau^-$ events, those in which the decay products of the other τ contained just one charged particle. Selection criteria based on the tracking information required that (i) exactly four tracks with a net charge of zero were found, with one track more than 90° away from other three; (ii) no pair of oppositely charged tracks had invariant mass less than 0.15 GeV/c^2, assuming both particles were electrons; (iii) the sum of the track energies exceeded 60% of the beam energy; (iv) the invariant mass of the 3-prong side was between 0.65 and 1.5 GeV/c^2; (v) the acollinearity of the 1-prong momentum with the combined 3-prong momentum was less than 55°; and (vi) the average angle between pairs of tracks on the 3-prong side was between 15° and 55°. Cut (ii) rejects 2-prong events such as μ-pairs which in addition have a photon converting in the material preceding the vertex detector. Cut (iii) rejects two-photon events, in which most of the center-of-mass energy is carried away by the final state electron and positron at small angles. Cuts (iv)-(vi) help to reduce the background from hadronic events, which constitute the dominant background. Finally, to further eliminate e-pairs (μ-pairs), events that contained two or more identified[3] electrons (muons) were rejected from the sample.

In 111 pb^{-1} of data taken near $\sqrt{s}$=10.58 GeV the τ-pair selection criteria yield a final sample of 5867 events. Application of the cuts to hadronic events simulated by Monte Carlo[4] yields a background of (18±3)%, where the error receives roughly equal contributions from uncertainties in τ branching fractions, the absolute two-jet cross section, and accuracy of the Monte Carlo simulation. All other backgrounds are negligible. The sample contains about 25% of all 1-versus-3 τ-pair events produced in this luminosity sample.

Determination of the decay distance requires knowledge of the center of the beam spot, which in CESR is known to have an rms width of about 600 μm in x and 50 μm in y. The beam location was found for each run to within ±40 μm from the intersection points of all tracks in hadronic events from that run. Then 10-20 nearby runs were averaged to obtain the beam spot center to within ±10 μm (this averaging was not performed across run boundaries where there were abrupt shifts in beam position).

For the 3-prong side in each event the most probable intersection point (x,y) and its error matrix (σ_{xx}, σ_{yy}, σ_{xy}) were computed from the track-fit parameters, which incorporate multiple scattering, beam size, and spatial resolution of the vertex and drift chambers. The most probable decay length, L_{xy}, projected in the plane transverse to the beam direction, and the proper decay length, $c\tau$, are given by

$$L_{xy} = \frac{x\sigma_{yy}t_x + y\sigma_{xx}t_y - \sigma_{xy}(xt_y + yt_x)}{\sigma_{yy}t_x^2 + \sigma_{xx}t_y^2 - 2\sigma_{xy}t_xt_y}, \qquad c\tau = \frac{L_{xy}}{\beta\gamma\sin\theta} \qquad (1)$$

where t_x, t_y are the direction cosines and θ the polar angle of the 3-prong system, $\gamma = E_{beam}/m_\tau$ represents the Lorentz boost to the τ center-of-mass, and $\beta = v_\tau/c$. The average error on $c\tau$ is 280 μm.

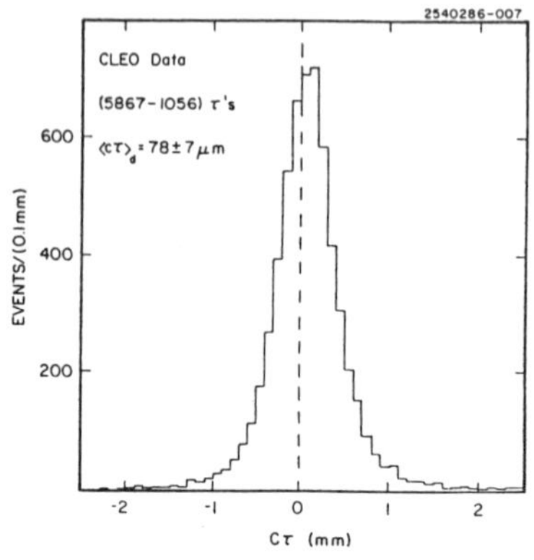

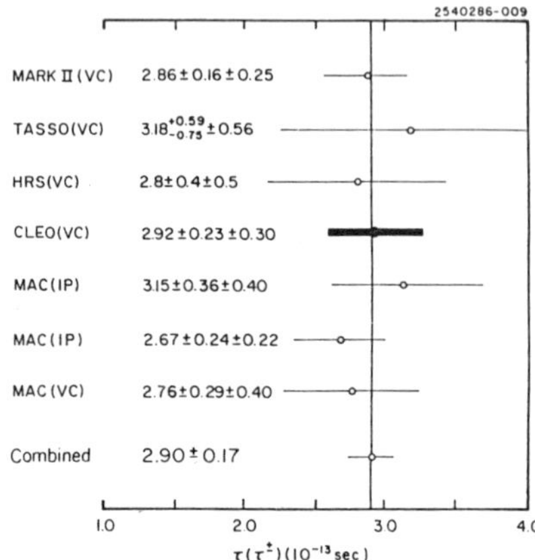

<u>Fig. 1</u>: Distribution of proper decay lengths for the τ sample.

<u>Fig 2</u>: Summary of τ lifetime measurements.

The distribution of the values of $c\tau$ is shown in Fig. 1. The peak of the distribution is clearly shifted above zero; the simple mean of the decay lengths is $\langle c\tau \rangle_d = (78\pm7)\mu$m. The presence of hadronic background tends to wash out the observed effect because tracks in such events emerge from closer to the vertex. The lifetime is calculated accounting for this background and other systematic effects with:

$$\langle c\tau \rangle_d = (1-f_{bgd})\langle c\tau \rangle_\tau + f_{bgd}\langle c\tau \rangle_{bgd} + \alpha_{cuts} + \alpha_{fit}, \qquad \tau_\tau = \langle c\tau \rangle_\tau/c, \quad (2)$$

where $f_{bgd} = (18\pm3)\%$ is the fraction of hadronic background, the mean decay length of such background events is $\langle c\tau \rangle_{bgd} = (34\pm13)\mu$m as determined from the Monte Carlo, and the a's represent systematic offsets attributable to the cuts and fitting procedure. The cuts were varied over reasonable ranges with less than 5 μm variation in $\langle c\tau \rangle_d$; hence $a_{cuts}=(0\pm5)\mu$m. On Monte Carlo τ-pair events generated with $\langle c\tau \rangle_\tau=100$ μm this analysis yielded the correct result, with no observed dependence on whether a decay included a neutral pion in addition to the three charged pions. When the beam position was shifted by several hundred microns, $\langle c\tau \rangle_d$ changed by less than 5 μm. These tests resulted in the assignment $a_{fit}=(0\pm5)\mu$m. With these values the CLEO measurement of the lifetime of the τ lepton is

$$\tau_\tau = [2.92 \pm 0.23 \text{ (statistical)} \pm 0.30 \text{ (systematic)}] \times 10^{-13} \text{ sec.}$$

This result is plotted with other measurements[1] in Fig. 2. The CLEO lifetime closely agrees with the world average of $(2.90\pm0.17)\times10^{-13}$ sec and the value[5] predicted by e-μ-τ universality in the standard model of $(2.8\pm0.2)\times10^{-13}$ sec.

For the D^O lifetime measurement additional requirements were imposed to reduce background: track momenta had to lie in the range from 0.3 to 4.0 GeV/c, and the average residual in the DR had to be less than 240 μm. Candidate D^O's were found in hadronic events by their presence in the decay chain $D^{*+} \rightarrow D^O \pi^+$, $D^O \rightarrow K^- \pi^+$ (and charge conjugates). All 3-track combinations with charges $(\mp,\pm,\pm)$ were assigned (K,π,π) masses. Then $M(K\pi\pi)-M(K\pi)$ was required to agree

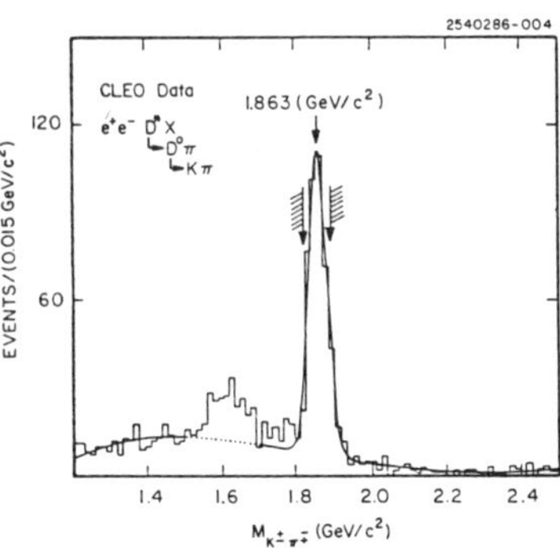

Fig. 3: Kπ mass combinations before the final mass cut.

with the known D^*-D^O mass difference, lying between 143.4 and 147.4 MeV/c^2. To further reduce combinatoric background, the D^O candidate was required to have momentum greater than 2.5 GeV/c and the cosine of the angle between the K in the D^o center-of-mass and the D^o boost direction was required to be less than 0.9. The resulting distribution of the Kπ invariant mass is shown in Fig. 3. Those Kπ combinations within 25 MeV/c^2 of the D^o mass were used in the lifetime measurement.

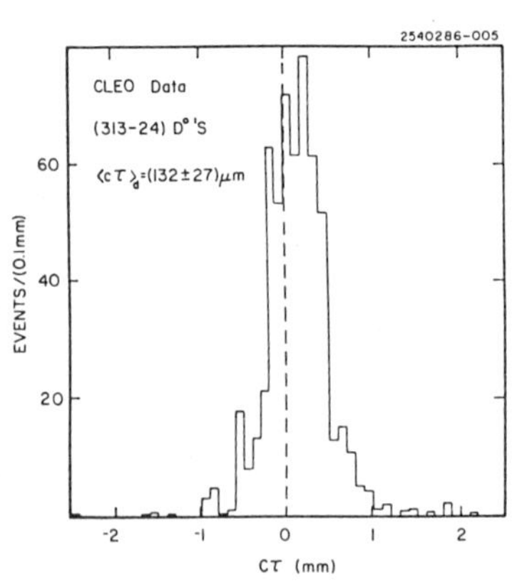

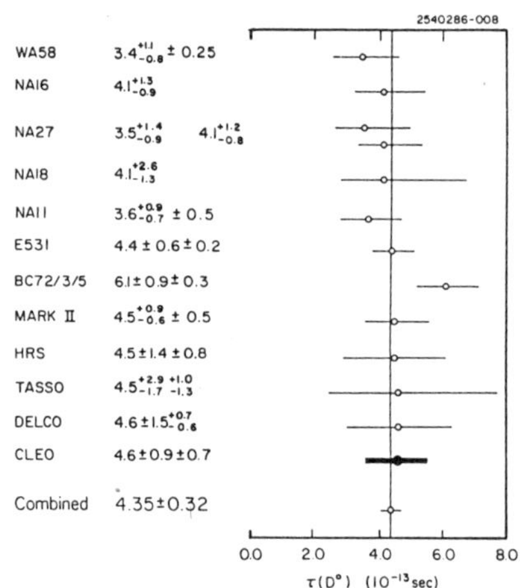

<u>Fig. 4</u>: Distribution of proper decay lengths for the D^o sample.

<u>Fig. 5</u>: Summary of D^o lifetime measurements.

The most probable decay point, proper decay length, and lifetime were calculated in a manner similar to the τ case using equations (1) and (2). The distribution of proper decay distance $c\tau$ is shown in Fig. 4; the average error on $c\tau$ was 315 μm. Again, there is a clear offset from zero; the error-weighted mean is $\langle c\tau \rangle_d = (132 \pm 27)\mu m$. The background was estimated as $f_{bgd} = (8 \pm 2)\%$ from Fig. 3, and events on the high-side D^o sideband were used to estimate $\langle c\tau \rangle_{bgd} = (77 \pm 77)\mu m$. By varying the cut values $\langle c\tau \rangle_d$ was found to change by as much as $\pm 15\mu m$; hence $a_{cuts} = (0 \pm 15)\mu m$. As with the τ lepton measurement, $a_{fit} = (0 \pm 5)\mu m$. An additional ± 15 μm error was added because of differences between the simple and error-weighted means which may reflect systematic difficulties with the error matrix still under study. The result is

$$\tau_{D^o} = [4.55 \pm 0.93(\text{statistical}) \pm 0.70(\text{systematic})] \times 10^{-13} \text{ sec},$$

which agrees well with other measurements[1] as shown in Fig. 5.

The installation of a precision vertex detector has given the CLEO detector the capability to measure particle lifetimes of the order of 10^{-13} sec. The τ lifetime measurement is competitive with other experiments, and, as one of the rites of passage for e^+e^- detectors measuring lifetimes, gives some confidence to proceed on to charmed particles. The CLEO D^o lifetime agrees closely with other measurements as well. Particle identification by dE/dx in the drift chamber has already resulted in the isolation of relatively clean samples of $D^{\pm}$, $\Lambda_c^{\pm}$, and $F^{\pm}$. Therefore in the near future CLEO will have lifetime measurements on an entire family of charmed particles.

References

1. See E.H. Thorndike, "Weak Decays of Heavy Fermions", Rapporteur's talk, International Symposium on Lepton and Photon Interactions, Kyoto, Japan, Aug. 19-24, 1985 (Univ. of Rochester preprint UR-935), and references therein.

2. D. Andrews, et al., Nucl. Instrum. and Meth. <u>211</u>, 47 (1983).

3. The electron identification uses information from the drift chamber track, energy loss in the dE/dx chambers, the time-of-flight counters, and shower chambers. Muon detection relies on penetration through at least 1 m of iron to the outer muon tracking chambers. Both are explained in: A. Chen, et al., Phys. Rev. Lett. <u>52</u>, 1084, (1984).

4. To generate hadronic events the Monte Carlo used a variant of the Lund group program: B. Andersson and B. Gustafson, Z. Phys. <u>C1</u>, 105 (1979). Then the geometry, materials, resolutions, and other known properties of the CLEO detector were simulated to produce events that were analyzed the same way as the real data.

5. C.A. Blocker, et al., Phys. Lett. <u>109B</u>, 119 (1982).

SEARCH FOR SHORT LIVED PARTICLES IN RADIATIVE $\Upsilon(1S)$ DECAYS

J. Kandaswamy

Newman Laboratory of Nuclear Studies

Cornell University, Ithaca, N.Y. 14853

We report here on a search made for a neutral particle 'a' in the mass range $2m_e < m_a < 2m_\mu$, which decays predominantly into an e^+e^- pair in radiative $\Upsilon(1S)$ decays. This was carried out using data from e^+e^- annihilations with the CLEO detector [1] at CESR (Cornell Electron Storage Ring). Such a search has been motivated by the observation of narrow structures in both e^+ and e^- spectra in heavy ion collisions. [2] These structures have been interpreted as evidence for the production of a neutral object with a mass of about 1.8 MeV/c^2 which decays into an e^+e^- pair.

The decay chain $\Upsilon(1S) \to \gamma + a$, $a \to e^+e^-$ would appear as a mono-energetic photon γ recoiling against an e^+e^- pair. The opening angle between e^+ and e^- in the CLEO detector would be very small for 'a' in the mass range of interest for this search. Therefore it would have the same topology as the non-resonant QED process $e^+e^- \to \gamma\gamma$, where one of the two photons converts to an e^+e^- pair in the beam pipe. We avoid this background problem by looking at $\Upsilon(1S)$ events [3] produced in the decay chain $\Upsilon(2S) \to \Upsilon(1S)\pi^+\pi^-$. We use the $\pi^+\pi^-$ pair to tag such events.

The data sample used in this analysis corresponded to 22.2 pb^{-1} of luminosity taken at $\Upsilon(2S)$ resonance. This gave a sample of (14600 ± 500) tagged [3] $\Upsilon(1S)$ decays. In CLEO, charged particles were tracked through a uniform magnetic field using a three-layer multiwire proportional chamber surrounded by a 17-layer cylindrical drift chamber, inside a superconducting coil. Photons were identified using the shower counters outside the coil. The shower counters cover a solid angle of 47% of 4π and have an energy resolution of $0.17\sqrt{E}$ in

GeV. A 0.1 radiation length thick lead converted was placed between the proportional chamber and the drift chamber; the magnet was operated at a lower field 0.35 T (instead of the normal 1.0 T) because of a concurrent search for low energy photon transitions in $T(2S)$ decays using converted e^+e^- pairs.

To identify the decay chain $T(2S) \to \pi^+\pi^- T(1S)$, $T(1S) \to \gamma a$, $a \to e^+e^-$, we require an event to have three or four charged tracks to allow for the possibility that the a decay into e^+e^- pair may produce only one instead of two distinguishable tracks. Any two of these tracks were required to be consistent with a $\pi^+\pi^-$ system recoiling against an object of mass lying within 35 MeV/c^2 of $T(1S)$ mass. The events were further required to have a shower with energy greater than 50% of the beam energy, consistent with the decay $T(1S) \to \gamma a$. After removal of background[4] from beam gas, cosmic rays and QED processes, there were no candidate-events corresponding to the decay chain of interest. We therefore present our results as upper limits at the 90% confidence level for the product branching ratio $B(T(1S) \to \gamma a) \times B(a \to e^+e^-)$.

The detection efficiency ε depends on both the mass m_a and proper lifetime τ_0 of the particle a. For a spin 0 particle of mass 1.8 $MeV/c^2 = \tau_0 = 10^{-13}$ sec, $\varepsilon = 0.15$. For a fixed mass m_a, as the life-time of a increases, the particle decays further from the production vertex. This results in both a decrease in charged track reconstruc-tion efficiency and a rise in the upper limit as a function of τ_0. Similarly for a given lifetime τ_0, as the mass m_a increases, the decay length in the detector decreases because of a smaller Lorentz-boost; consequently the detection efficiency increases. In figure (1) our results are shown as an upper limit at 90% confidence level for $B(T(1S) \to \gamma + a) \times B(a \to e^+e^-)$. CLEO has previously[5] searched for $T(1S) \to \gamma a$, where a decays beyond the active volume of the detector. The combination of these two searches places an upper limit of 5×10^{-4} for lifetimes $< 5 \times 10^{-13}$ sec and 3×10^{-4} for lifetimes $> 3 \times 10^{-11}$ sec. An overall upper limit of 2×10^{-3} is applicable to all life-times.

484

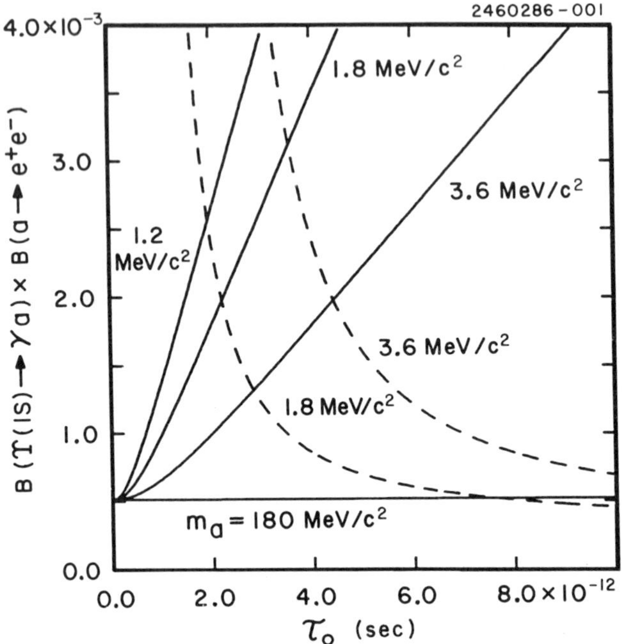

Figure 1. The upper limit on $B(\Upsilon(1S) \to \gamma a) \times B(a \to e^+ e^-)$ at 90% confidence level as a function of the proper lifetime, τ_0, for this experiment (solid line) and on $B(\Upsilon(1S) \to \gamma a)$ from our previous measurement (Ref. 5) (broken line). At long lifetimes the upper limit from Ref. 5 is 3×10^{-4}.

A light neutral pseudoscalar particle called the axion has been previously proposed[6] to explain the absence of P and CP violations in strong interaction gauge theories[7]. Several experiments[8] have unsuccessfully searched for such an object at various mass values. The standard axion theory[6] has two higgs doublets and a free parameter x, the ratio of the vacuum expectation values. The value of x determines the mass as well as the coupling of the axion to various quarks. The standard theory predicts that the axion would decay predominantly into $e^+ e^-$ pairs if its mass m_a were in the range $2m_\mu > m_a > 2m_e$. It also predicts the radiative decays of both Υ and

J/Ψ to an axion and γ. For $T(1S)$, the prediction[9] is

$$B(T(1S) \to \gamma a) = (2.7 \pm 0.7) \times 10^{-4} x^{-2}$$

For the case of three quark doublets and a mass of 1.8 MeV/c^2, the standard model gives $x = 0.04$ and a proper lifetime of 4×10^{-12} sec. This in turn gives a decay length of 3.4 m in $T(1S)$ decays and a branching ratio of (0.16 ± 0.04), in contradiction with previous CLEO searches since most of the decays would be outside the active volume of the detector. The present search can be used to set limits for short-lived axions which have not been completely ruled out[10] by other experiments. In the case of $m_a = 3.6$ MeV/c^2, $x = 0.02$, which gives a decay length = 0.18 m. The branching ratio for this case would be 0.67 ± 0.17. This value is about three orders of magnitude larger than our measured upper limits.

In summary, CLEO has searched for evidence of the production of light, short lived, spin 0 particles in radiative $T(1S)$ decays and has placed stringent upper limits on the product branching ratio $B(T(1S) \to \gamma a) \times B(a \to e^+ e^-)$ for masses $2m_e < m_a < 2m_\mu$. I would like to thank D. Besson, E. Blucher, T. Bowcock, A. Silverman and other colleagues of CLEO collaboration for valuable discussions. Thanks are also due to the efforts of CESR staff which made this work possible. This work was supported by the National Science Foundation.

References

1) D. Andrews et al., Nucl. Instrum. Methods, 211, 47 (1983).

2) M. Clemente et al., Phys. Lett. 137B, 41 (1984); T. Cowan et al., Phys. Rev. Lett. 54, 176. (1985); Phys. Rev. Lett. 56, 444 (1986).

3) D. Besson et al., Phys. Rev. D 30, 1433 (1984).

4) T. Bowcock et al., CLEO Preprint CLEO-86-2 (to be published).

5) M.S. Alam et al., Phys. Rev. D 27, 1665 (1983).

6) S. Weinberg, Phys. Rev. Lett. 40, 223 (1978); F. Wilczek, Phys. Rev. Lett. 40, 279 (1978).

7) R.D. Peccei and H.R. Quinn, Phys. Rev. Lett. 38, 1440 (1977); Phys. Rev. D 16, 1791 (1977).

8) T.W. Connelly et al., Phys. Rev. D $\underline{18}$, 1607 (1978); J.F. Cavignac
et al., Phys. Lett. $\underline{121B}$, 193 (1983).

9) We have not included the QCD corrections; See M.I. Vysotsky,
Phys. Lett. $\underline{97B}$, 159(1980); J. Ellis et al., Phys. Lett. $\underline{158B}$,
417 (1985).

10) G. Mageras, Max-Plank-Inst. fur Physik und Astrophysik, Munchen,
MPI-PAE/EXP. EL. 157 (to be published).

ARGUS PHYSICS FROM D MESONS

Malcolm Goddard

Physics Department , Carleton University
Ottawa , Ontario
CANADA   K1S 5B6

INTRODUCTION

In this talk I present two recent results from the ARGUS collaboration . Both of them rely on the excellent ability of the ARGUS detector to identify D^0 and D^* mesons produced from the continuum . However they shed light on two quite different aspects of particle physics . I shall first present ARGUS's observation of W-exchange in charmed meson decay , and secondly talk about our discovery of a new high-mass charmed meson .

DIRECT EVIDENCE FOR W-EXCHANGE IN CHARMED MESON DECAY

The Problem

In the spectator model of charmed meson decay[1] the charmed quark undergoes the weak decay with the light quark acting as a 'spectator', (see figure 1) . In this simple model the partial decay widths are therefore independent of the light quark flavor (ignoring phase space effects) . In particular , it predicts equal lifetimes for the D^+ ($c\bar{d}$) and the D^0 ($c\bar{u}$) mesons .

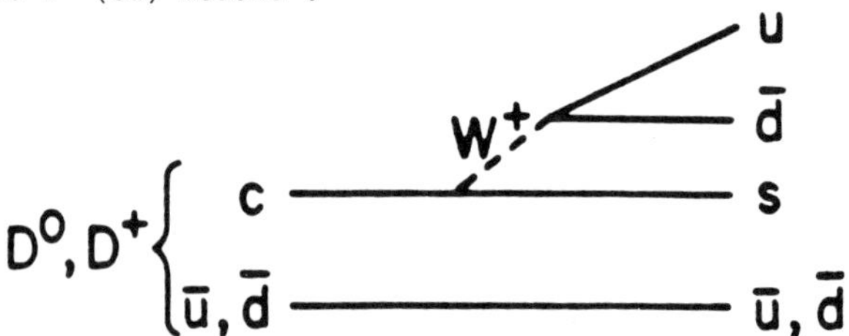

Figure 1 : Spectator diagram for D decay .

However , since 1979 direct measurements of the lifetimes show that the D^+ lives substantially longer than the D^0 . The current world average is[2] : 2.2 $\pm$ 0.3 , which is clearly at odds with the simple spectator prediction .

This inequality in the lifetimes means that either the D^0 decays are enhanced or the D^+ decays are suppressed , and in 1980 there were many mechanisms proposed[3,4] to accomplish this . However , recent measurements[5] of the D^+ and D^0 semi-leptonic branching ratios have shown agreement with the spectator model prediction for D^+ , while the measured D^0 branching ratio is considerably smaller than predicted . In addition by comparing these measured semi-leptonic branching ratios with the corresponding lifetimes , we conclude that the semi-leptonic D^+ and D^0 widths are equal . This means that the D^+ seems to be well described by the spectator model (including gluon corrections) and also the D^0 semi-leptonic decays are as expected . Therefore to account for the observed

D^+ and D^0 lifetime difference , the D^0 non-leptonic decays must be enhanced relative to the D^+ .

The W-exchange process then becomes the favored mechanism[4] to produce this enhancement (see figure 2) , because the coupling between the W and the light quarks makes this process available to the D^0 but not to the D^+ . If an $s\bar{s}$ quark pair are subsequently pulled out of the vacuum , we obtain the specific decay mode $D^0 \to K^0 \phi$. The experimental observation of this decay channel would then be unambiguous proof for the existence of W-exchange in D^0 decay[6].

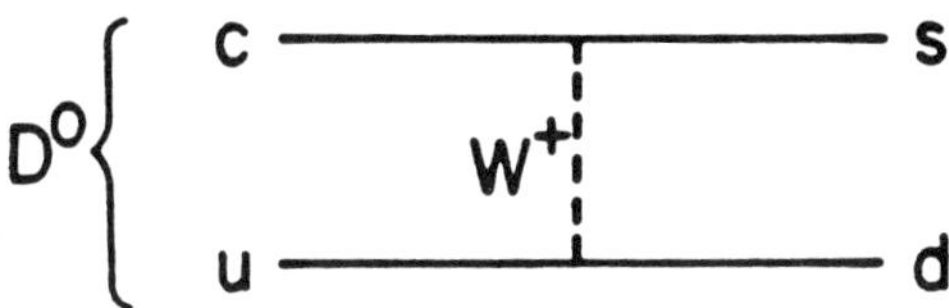

Figure 2 : W-exchange diagram for D^0 decay .

The Solution

In the summer of 1985 , the realization of the importance of this particular D^0 decay in understanding the D^+/D^0 lifetime difference and the fact that previous searches had produced only an upper limit[7] , led the ARGUS collaboration to look into its data for this decay . The data sample consisted of 82.2 pb^{-1} taken on the Upsilon 1S , 2S and 4S . Particle identification used both dE/dX and TOF information to form a combined particle identification probability[8].

The method was straightforward , it involved looking for a D^0 peak in the $K^0_S \phi$ invariant mass distribution . The $K^0_S K^+ K^-$ invariant mass distribution , for those combinations having $X_p > 0.3$ and with the $K^+ K^-$ mass in the phi band $(1.01 < \text{mass}(K^+ K^-) < 1.03 \text{ GeV}/c^2)$ is shown in figure 3 . It shows a beautiful peak at the D^0 mass above a small background .

Since the phi is produced from the decay of a spin 0 object , together with a spin 0 K^0_S , it must be in a state of zero helicity . Thus its decay-angular distribution , in its helicity frame , should be $\cos^2\theta$. Figure 4 shows the decay angle of the K^+ in the phi rest frame , relative to the K^0_S direction . It is consistent with expectations .

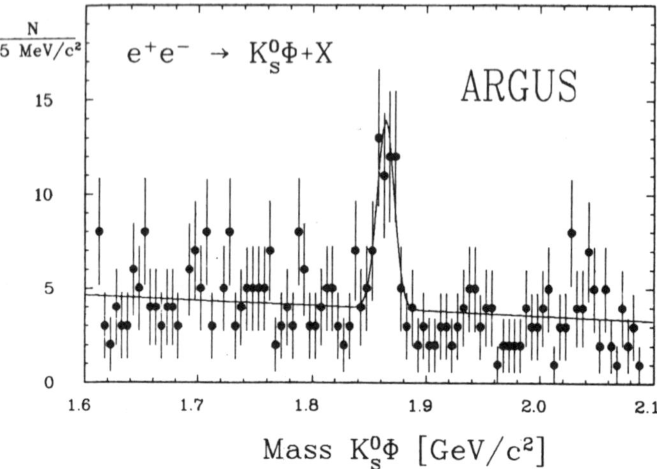

Figure 3 : K^0_S Φ mass distribution , with cuts as described in text .

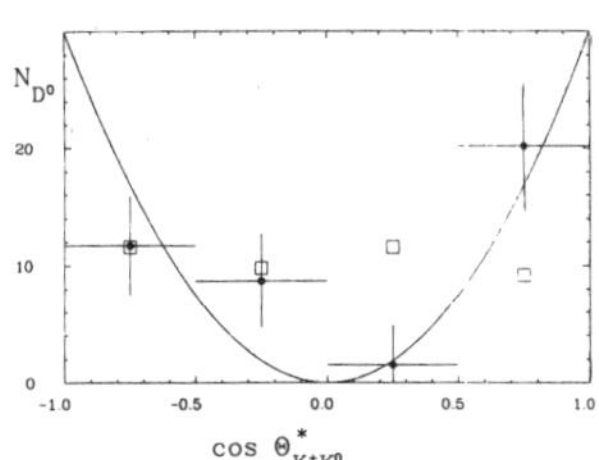

Figure 4 : Decay angle of the K^+ in the phi helicity frame . Solid curve is a fit to $\cos^2\theta$.

The $D^0 \rightarrow K^0\phi$ branching ratio is obtained from the number of events in the K^0_S ϕ peak (corrected for unseen ϕ decays) compared to the number in the $K^0_S\pi^+\pi^-$ peak :

$$\text{Br}\ (D^0 \rightarrow K^0\phi) = \varepsilon \cdot \frac{N\ (D^0 \rightarrow K^0_S\phi) \cdot \text{Br}\ (D^0 \rightarrow K^0\pi^+\pi^-)}{N\ (D^0 \rightarrow K^0_S\pi^+\pi^-)}$$

where ε is the relative efficiency for $D^0 \rightarrow K^0_S\pi^+\pi^-$ compared to $D^0 \rightarrow K^0_S\phi$ and the Br $(D^0 \rightarrow K^0\pi^+\pi^-)$ is well-known[7]. The result is :

$$Br\ (D^0 \rightarrow K^0 \phi\ ) = (1.4\ \pm 0.4 \pm 0.2\ )\ \%$$

to be compared to MARK III's previous upper limit of 1.7 % [7].

The W-exchange contribution to D^0 decay can , in principle , produce many different final states of which the $K^0_S \phi$ is the easiest to observe experimentally . The obvious question to ask at this point is whether this observed branching ratio can explain the D^+/D^0 lifetime difference . Bigi has calculated[9] that to completely explain the observed lifetime difference requires a $K^0 \phi$ branching ratio of between 0.5 to 2 % . The exact value depending on how the gluon corrections are applied . Therefore , the observed $D^0 \rightarrow K^0 \phi$ branching ratio is large enough to completely explain the D^+/D^0 lifetime difference . In fact in the future , when the error has been reduced , it may be possible to use this experimentally measured branching ratio to help determine the correct way to include gluon effects in these weak decays .

OBSERVATION OF A NEW CHARMED MESON

In 1976 the first charmed meson states were observed at SPEAR[10]. These were S-wave bound states of a charmed and light quark (u or d) either in a singlet (D meson) or triplet spin state (D^* meson) . In addition to these observed low-lying states , potential models predict the existence of higher-mass charmed mesons , either as radial excit- ations of the ground states , or as states with orbital angular moment- um between the quarks[11]. Figure 5 shows the predictions from one such · model .

Figure 5 : Spectroscopy of charmed mesons predicted in ref 11 .

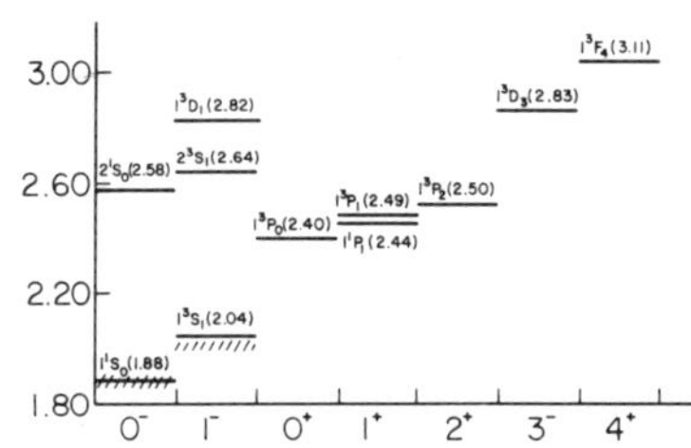

To search for these predicted high-mass charmed mesons the ARGUS strategy was to look for a possible D^{**} signal decaying via $D^{*+}\pi^-$ to all-charged final states :

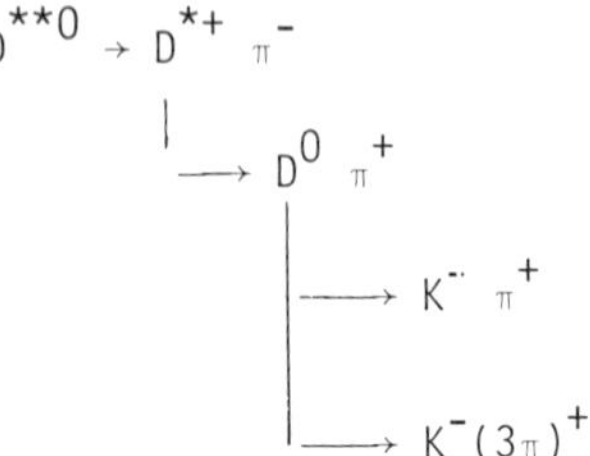

plus the charge conjugate reactions . These decays are experimentally the easiest and quickest to analyse since only charged particles are used and the background is greatly reduced by selecting the D^* .

The energy region of the Upsilon resonances , where ARGUS takes data , has some unique advantages in studying continuum charmed particle production . At this energy the hadronic cross section is large , compared to higher energies , while at the same time the energy is low enough so that the charmed mesons are produced relatively slowly in the lab , and their decay products are therefore well-measured .

The data sample used in this analysis consisted of 82.2 pb^{-1} from the Upsilon 1S , 2S and 4S regions with particle identification using both dE/dX and TOF information[8].

To 'tag' the D^* , D^0 candidates in the $K^-\pi^+$ and $K^-(3\pi)^+$ decay modes were first selected . Then each candidate was combined with a π^+ and the difference , DM , between the two invariant masses :

$DM = Mass(D^0\pi^+) - Mass(D^0)$, was calculated . Since both the D^0 and D^* widths are dominated by the experimental resolution , the error on this mass difference depends strongly on the error in measuring the low-momentum pion . Since this is well-measured , the error in DM is very small (see figure 6) and this allows a 'clean' sample of D^*'s to be selected .

In addition to the $K^-\pi^+$ and $K^-(3\pi)^+$ decay modes of the D^0 , we obtain (for free) contributions from the decays :

$$D^0 \to K^- \rho^+ \qquad \text{and} \qquad \to K^{*-} \pi^+$$
$$\quad \ \hookrightarrow \pi^+ \pi^0 \qquad\qquad\qquad\qquad \hookrightarrow K^- \pi^0$$

where the π^0 is not observed . In both of these cases the vector
particle must be produced in a helicity zero state and therefore will
decay with a $\cos^2\theta$ angular distribution . This means that a large
fraction of the time the π^0 decays against the rho or K^* line of flight .
When this happens it ends up nearly at rest in the D^0 rest frame and
contributes not much more than its rest mass to the $K^- \pi^+ \pi^0$ invariant
mass . Therefore it appears as a bump , at about 1.6 GeV/c^2 (the so-
called satellite peak) , in the $K^- \pi^+$ invariant mass distribution[12] .

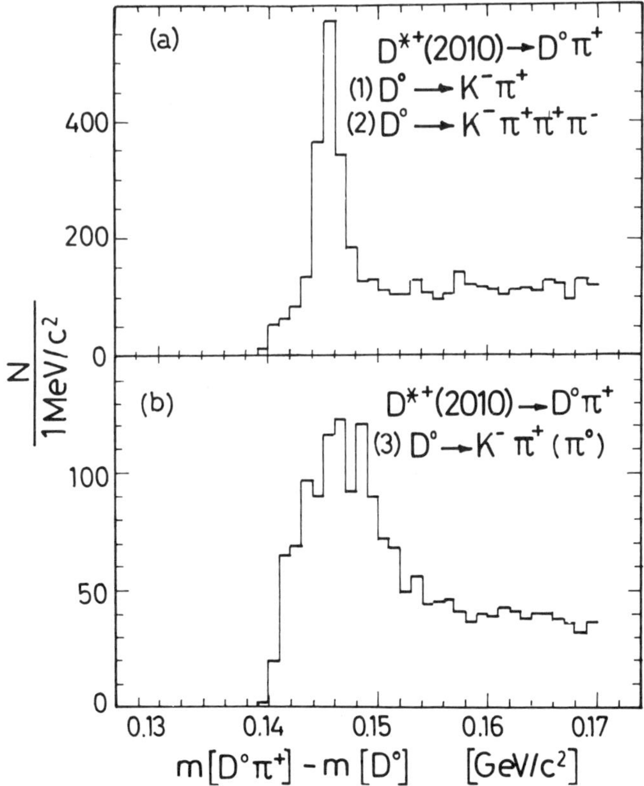

<u>Figure 6</u> : DM distribution including the cut $x_p(D^0 \pi^+) > 0.45$.

Figure 6 (a) shows the DM distribution for the $K^-\pi^+$ and $K^-(3\pi)^+$ decays of the D^0 and figure 6 (b) is for the D^0 satellite peak . The 'tagged' D^{*+} sample was obtained by the requirement : 144 < DM < 147 MeV/c^2 and the looser constraint for the satellite peak of : DM < 152 MeV/c^2 .

This mass-difference technique was then repeated by combining a π^- with each D^{*+} and computing : $DM^* = $ Mass $(D^{*+}\pi^-) - $ Mass (D^{*+}) . The DM^* resolution , in the region of 2.4 GeV/c^2 , is about 15 MeV/c^2 (25 MeV/c^2 for the satellite D^0) , and this is about an order of magnitude better than the resolution in the $D^{*+}\pi^-$ mass .

Since the charmed quark has a 'hard' momentum spectrum , the background was reduced by requiring $x_p(D^{*+}\pi^-) > 0.6$. This corresponds approximately to $x_p(D^{*+}) > 0.45$, and the effect of this on the DM distribution is shown in figure 6 . In addition , various background processes produce pions , at small angles to the jet axis and with momenta of hundreds of MeV/c . Thus these pions have about the same velocity as the D^* (for large x_p) and therefore produce a low $D^*\pi$ invariant mass . This simulates a $D^*\pi$ decay with a backward emitted pion . To eliminate this background , we accepted only events with $\cos\theta < 0$, where θ is the angle between the D^* and the $D^*\pi$ line of flight in the $D^*\pi$ rest system . Figure 7 shows the resulting DM^* distribution after these cuts . There is an almost five standard deviation peak . Fitting to a Breit-Wigner plus a polynomial times threshold factor gave the following parameters for this new resonance , which we call the $D^*(2420)$:

$$M = 2420 \pm 6 \text{ Mev/c}^2$$
$$\Gamma = 70 \pm 21 \text{ Mev/c}^2$$

with $135 ^{+34}_{-29}$ events in the peak .

In the old published SPEAR data[13], there is a bump in the D^0 recoil mass distribution at a mass of 2440 MeV/c^2 and with a width of 100 MeV/c^2 . It's quite possible that this was the first observation of the $D^*(2420)$.

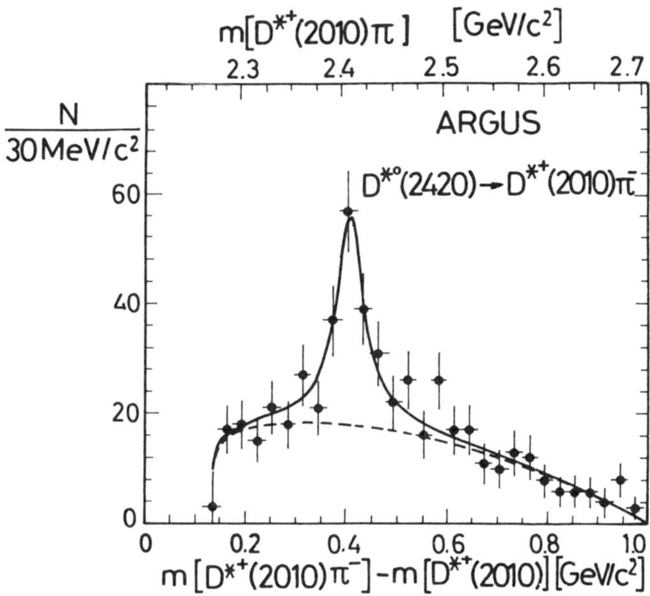

Figure 7 : DM* distribution with the cuts described in the text .

Figure 8 shows the fragmentation function for the D*(2420) .
It is 'hard' , as expected for charmed meson fragmentation and gives
acceptable fits to the predictions of two theoretically motivated
models[14].

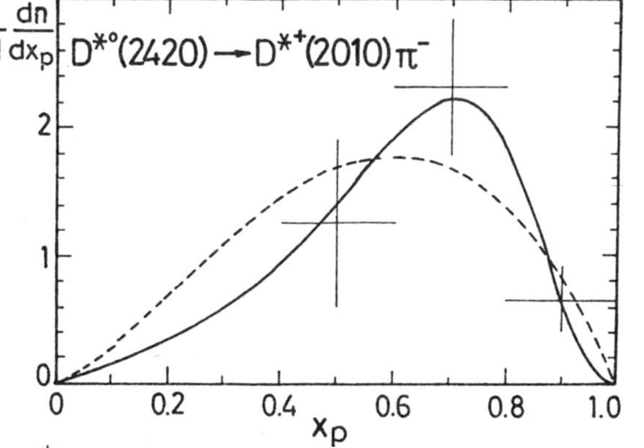

Figure 8 : D*(2420) fragmentation function . The curves are the
results of fits to the Peterson (solid) and Kartvelishvilli .

To determine the production cross section for the $D^{*0}(2420)$, we first calculated that ($24 \, {}^{+\ 8}_{-\ 6} \pm 8$) % of the D^{*+} production originates from the $D^{*+}\pi^-$ decay of the $D^{*0}(2420)$. Then , correcting for the unseen $D^{*0}\pi^0$ decay and using the previously measured D^{*+} production cross section[15] , we obtained :

$$\sigma \, (\, D^{*0}(2420) \,) \cdot Br \, (\, D^{*0}(2420) \to D^*\pi \,) = 340 \, {}^{+\ 190}_{-\ 180} \ pb \ .$$

where the statistical and the systematic errors have been combined .

According to the potential models (see figure 5) , our observed $D^{*0}(2420)$ is probably a P-wave state . The $D^*\pi$ decay mode then restricts the possibilities to the ${}^1P_1$ ($J^P=1^+$) , or ${}^3P_{1,2}$ ($J^P=1^+,2^+$) states and since charge congugation is not a good quantum number in this system there can be mixing between the two 1^+ states . If the $D^*(2420)$ is produced in a completely unpolarized state , its decay to $D^*\pi$ is isctropic and only the subsequent D^* decay contains information about its spin . This is currently under study .

CONCLUSIONS

These two results from ARGUS are significant contributions to our understanding of the fundamental forces of nature . The observation of the W-exchange decay of the D^0 meson seems to offer a convincing explanation for the discrepancy in the D^+/D^0 lifetime . The $D^*(2420)$ is a long-awaited addition to the D , D^* family and provides important input for QCD-inspired potential models .

REFERENCES

1) M. K. Gaillard , B. W. Lee and J. L. Rosner , Rev. Mod. Phys. 47 , 277 (1975)

J. Ellis , M. K. Gaillard and D. V. Nanopoulos , Nucl. Phys. B100 , 313 (1975)

A. Pais and S. B. Treiman , Phys. Rev. D15 , 2529 (1977) .

2) S. Stone , Invited Talk at the Xth International Conference on Weak Interactions , Savonlinna , Finland , 1985 .

3) For example :

B. Guberina et al. , Phys. Letts. 89B , 111 (1979)

M. Bander et al. , Phys. Rev. Letts. 44 , 7 (1980) , 962(E)

H. Fritzsch and P. Minkowski , Phys. Letts. 90B , 455 (1980) .

4) I. I. Bigi , Z. Physik C5 , 313 (1980) .

5) D. Hitlin , CALT-68-1230 .

K. Schubert , Invited Talk at the XIth International Conference on Neutrino Physics and Astrophysics , Nordkirchen , June 1984 .

6) I. I. Bigi and Fukugita , Phys. Lett. 91B , 121 (1980) .

7) R. H. Schindler , Proceedings of the XXIInd International Conference on High Energy Physics , Leipzig (1984) , Vol 1 171 .

8) H. Albrecht et al. (ARGUS collaboration) , Phys. Lett. 134B , 137 (1984) .

9) I. I. Bigi , Aachen preprint PITHA 85/14 , August 1985 .

10) G. Goldhaber et al. , Phys. Rev. Lett. 37 , 255 (1976)

I. Peruzzi et al. , Phys. Rev. Lett. 37 , 569 (1976)

J. E. Wiss et al. , Phys. Rev. Lett. 37 , 1531 (1976) .

11) S. Godfrey and N. Isgur , Phys. Rev. D32 , 189 (1985) .

12) G. Goldhaber , Proceedings of the XVIIth Rencontre de Moriond, La Plagne March 13-19 1983 .

13) G. Goldhaber et al. , Phys. Lett. 69B , 503 (1977) .

14) C. Peterson et al. , Phys. Rev. D27 , 105 (1983)

W. Kartvelishvilli et al. , Phys. Lett. 76B , 615 (1978) .

15) H. Albrecht et al. (ARGUS collaboration) , Phys. Lett. 150B , 235 (1984) .

B MESON PRODUCTION AND DECAY IN e^+e^- ANNIHILATION AT THE $\Upsilon(4S)$ RESONANCE.

The ARGUS Collaboration.

presented by

R. S. Orr

University of Toronto / IPP Canada
Toronto, Ontario M5S 1A7 CANADA

ABSTRACT

The status of B-meson studies with the ARGUS detector at DORIS is given. The published ARGUS results on the inclusive decay of $B \to \psi X$ are updated, using the most recent data. The inclusive momentum spectrum and branching ratio for Λ production from B decays is presented. Finally, the current status of fully reconstructed B-mesons, in ARGUS, is discussed.

1. INTRODUCTION.

During 1985 ARGUS collected 43pb^{-1} on the $\Upsilon(4S)$, and 17pb^{-1} on the continuum just below the resonance energy. This, together with the small amount of earlier $\Upsilon(4S)$ running, gives a total luminosity of 59pb^{-1} on the $\Upsilon(4S)$ and 23pb^{-1} in the continuum.

The ARGUS detector is a conventional, but precise, solenoidal detector. It is fully described in Ref.(1). A review of the present experimental status of, and theoretical interest in, the B-meson system may be found in Ref.(2). The B mesons are of interest in that they contain the 5th species of quark, the b-quark, and thus open the possibility of tests of the Kobayashi - Maskawa six quark model. In particular, it is important to measure the relative strengths of the transition of the b-quark to the c-quark, and the u-quark, viz. $|V_{bu}/V_{bc}|$. It is also important to search for mixing in the $B^0\overline{B^0}$ system, and possible CP-violation effects, even at the present level of statistics.

In addition to these questions in the area of electroweak interactions, it is of great interest to study the interplay of the strong and weak interactions in the decay of mesons heavier than the charmed mesons. The simplest picture of heavy flavour decay is the spectator model. However, it has become clear that the simplest spectator picture is deficient. Colour suppression does not seem to be active in charm decays[3], and W-exchange diagrams seem to be appreciable[4]. It is important to investigate these questions in the b-quark systems.

These latter questions can be approached most cleanly by measuring exclusive B-decay branching ratios. In the second part of this talk the status of fully reconstructed B-decays in ARGUS will be described.

2. MEASUREMENT OF INCLUSIVE B $\to \psi$X

A study of the inclusive decay of the B meson to the J/ψ addresses the question, mentioned in the introduction, of how the presence of QCD gluons modifies the naive spectator picture of heavy flavour decays. The process is "colour suppressed" in that colour matching has to be achieved between the c-quark from the b-quark decay and the $\bar{c}$-quark from the W, see Fig.1. If gluons altered the colour flow, the "suppression" could be removed. A measurement of the branching ratio also gives information on the final states produced, for example whether the decay is predominantly two body or not[5]. ARGUS has already published an analysis based on the early data sample[6]. Here I will give an updated result based on the full $\Upsilon(4S)$ sample available at present. The results are preliminary.

The inclusive J/ψ signal is observed through the dilepton decay, using both muons and electrons. The analysis is described in detail in Ref.6). Briefly, muons were identified by requiring at least one hit in the outer muon chambers of the detector. Candidate muons were required to deposit less than 2.5 times minimum ionizing in the shower counters, and the momentum measured in the drift chamber was required to be greater than 850MeV/c, in order to ensure an efficiency of order 100%. Electrons were identified using the information from the specific ionization measurement in the drift chamber, the time of flight counters, and the shower shape in the shower counters. Details of the electron identification procedure can be found in Ref.7). The efficiency for detecting two muons from the J/ψ, with momenta less than 2.0 GeV/c is about 50%, while the efficiency for detecting the e^+e^- decay is of order 30%. The resulting dilepton mass distribution is shown in Fig.2. There is a clear signal for the J/ψ. Fitting a gaussian signal and a smooth polynomial background results in (65 ± 12) events in the signal corresponding to a preliminary, improved branching ratio of:

$$\mathrm{Br}(\mathrm{B} \to \mathrm{J}/\psi\mathrm{X}) = (1.10 \pm 0.24)\%$$

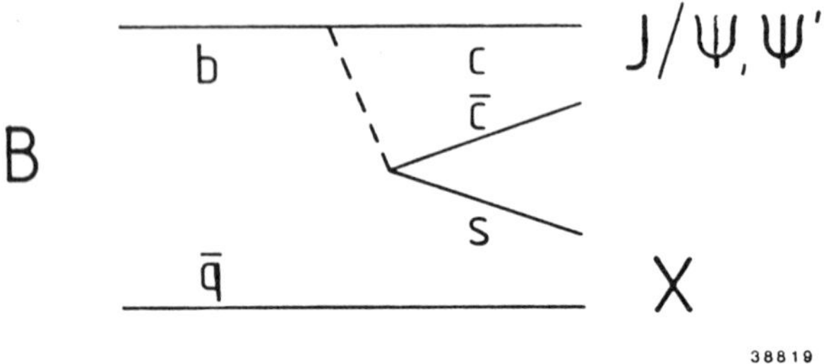

Fig.(1) Quark diagram for the weak decay $B \to J/\psi\, X$.

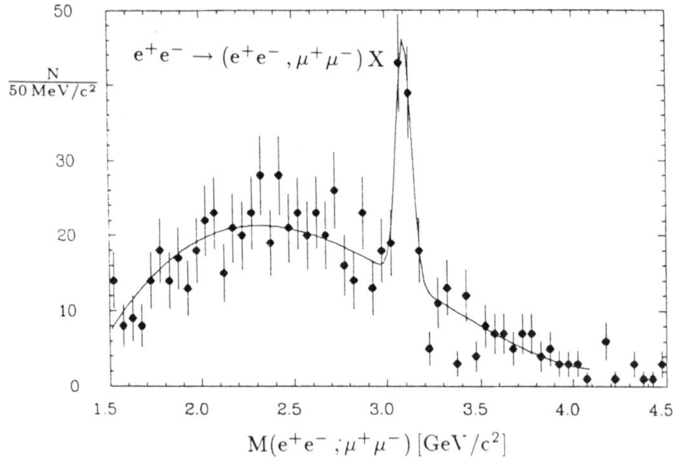

Fig.(2) Invariant mass distribution for dileptons in $e^+e^- \to (e^+e^-$ and $\mu^+\mu^-)\, X$ with $E_{CMS} \approx M(\Upsilon(4S))$.

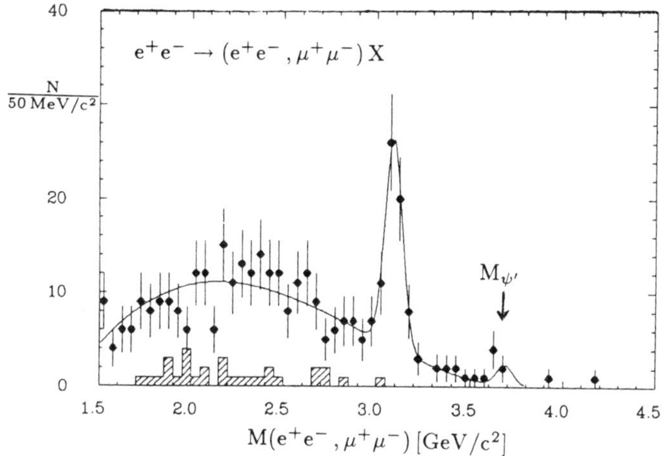

Fig.(3) The invariant mass distribution, as in Fig.(2), but with the topological cut $H_2 \leq 0.3$ as described in the text. The shaded histogram corresponds to continuum data from just below the $\Upsilon(4S)$.

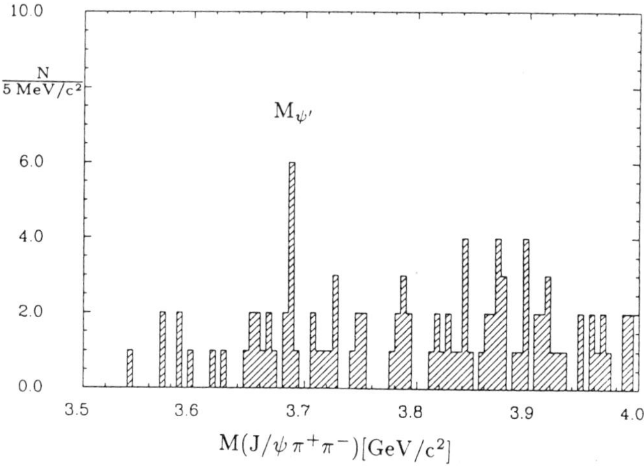

Fig.(4) The invariant mass distribution of J/ψ combined with a $\pi^+\pi^-$ pair.

A cleaner sample of inclusive J/ψ from B decays can be obtained by applying a topological cut, in order to reduce the two - jet continuum background. Specifically, by requiring the Fox − Wolfram moment $H_2 \leq 0.3$ we obtain the dilepton mass spectrum shown in Fig.3. The shaded histogram in this figure shows the corresponding distribution from the continuum data taken at energies just below the $\Upsilon(4S)$. In addition to the J/ψ, this distribution shows evidence for about 5 events coming directly from the ψ'. Combining J/ψs with $\pi^+\pi^-$ results in the mass distribution shown in Fig.4. There is indeed a signal at the ψ' mass. This corresponds to a branching ratio of $B \rightarrow \psi'X$ of about 0.4%, i.e. around 20% of the J/ψs in inclusive B decays originate from the ψ'.

If the light quark in B decays to the J/ψ behaved purely as a spectator, one might expect 2 - body decays such as $B \rightarrow J/\psi K$ and $B \rightarrow J/\psi K^*$ to dominate. This would lead to a hard momentum spectrum for the J/ψ recoiling against the light meson. The experimental momentum spectrum is shown in Fig.5, with the region expected to be populated by 2 - body decays indicated. The momentum spectrum is softer than one would expect on this simple picture.

Finally, having seen a clear signal for inclusive J/ψ production, ARGUS has searched for exclusive decay modes of the B meson involving the J/ψ or ψ'. More details on the procedure for reconstructing exclusive B decay channels is give in the next section. However, the resulting mass distribution is shown in Fig.6. There is an enhancement at the expected mass of the B meson.

3. INCLUSIVE B DECAYS TO Λ's

Baryon production from B decays proceeds, for example, via the diagram in Fig.7. Since only one $u\bar{u}$ − quark pair needs to be pulled from the vaccuum, one expects a relatively high rate[8] of baryon production from the B.

The reduced momentum spectrum for Λ production, after continuum subtraction, is shown in Fig.8. ARGUS measures a preliminary branching ratio of:

$$Br(B \rightarrow \Lambda X) = (3.1 \pm 0.7 \pm 0.5)\%$$

4. FULLY RECONSTRUCTED EXCLUSIVE B DECAYS.

Although tens of thousands of B mesons have been produced at DORIS and CESR, up until this year only of order 20 have been fully reconstructed[2]. ARGUS now has a sample of 71, combined with the CLEO sample this represents a fourfold increase in the total world sample of fully reconstructed B mesons.

While the decays of the $\Upsilon(4S)$ are a copious source of B mesons, the kinematics of their production makes it extremely difficult to fully reconstruct any appreciable number of them. Each B meson is produced with a rather small mo-

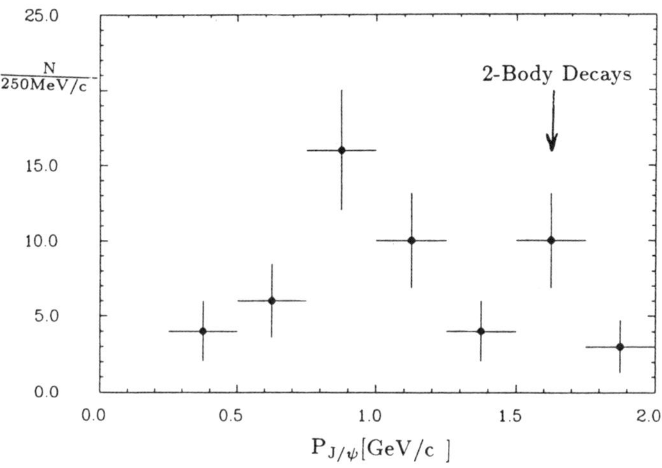

Fig.(5) The momentum spectrum of the J/ψ. The region populated by 2-body decays of the B meson is indicated.

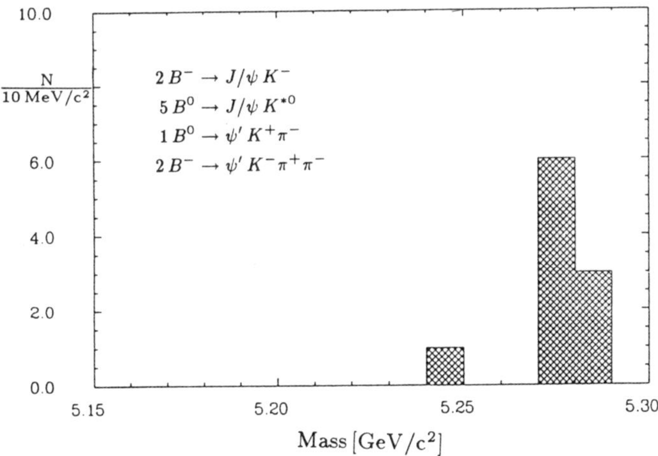

Fig.(6) Exclusive decays of the B meson containing the J/ψ or ψ'.

504

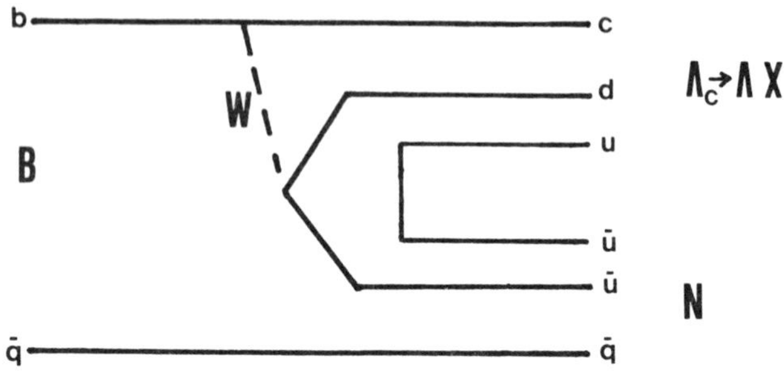

Fig.(7) An example quark diagram for baryon production in B meson decay.

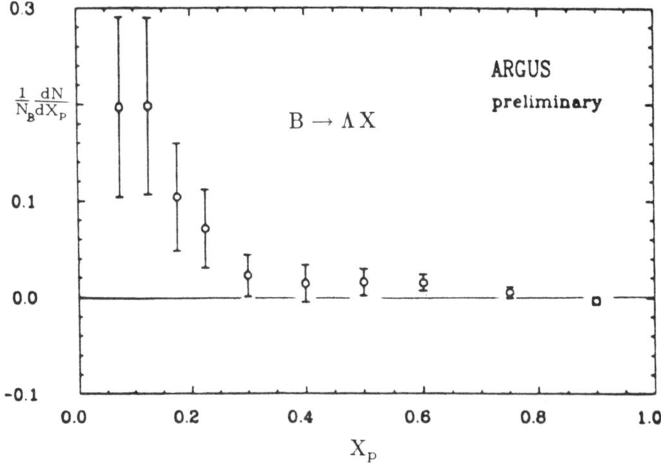

Fig.(8) Inclusive Λ reduced momentum spectrum. The non-resonant continuum has been subtracted.

mentum in the laboratory frame, viz. approx. 400 MeV/c. There is thus no jet -
like structure in the events which would help to attribute final state particles to
a given parent B. In addition, the final state multiplicity is high, and there are
many available decay channels. These facts lead to an enormous combinatorial
background, which is the main reason for the small number of B decays which
have been fully reconstructed to date.

A typical decay chain is:

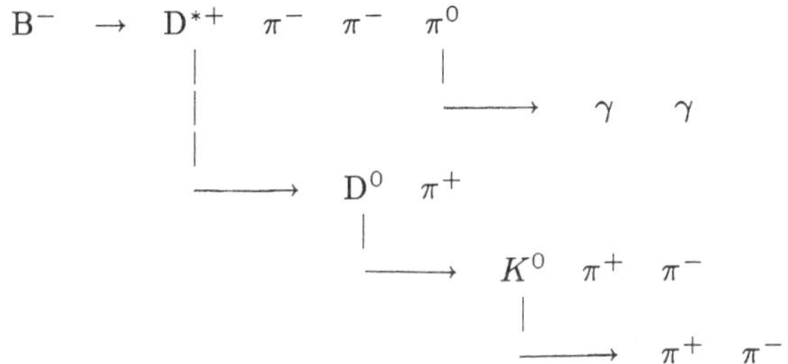

There are nine final state particles involved in this decay chain, and each
would have to be correctly identified and measured in order to fully reconstruct
the parent B meson. Since there will usually be an equally complex decay chain
from the other B in the event, and since there is no easy way of attributing
the final state particles to a particular parent B, the problem of combinatorial
background is readily apparent.

Clearly, the excellent particle identification properties of the ARGUS detector are vital in tackling this problem, as they will help to reduce the number of
mass hypotheses to be tried for each measured track, thus reducing the combinatorial background problem.

All the B decay channels discussed in this section involve the cascade decay:
$B \rightarrow D^{*+} \rightarrow D^0$. This cascade can be identified by means of the well known
technique[9] of exploiting the fine resolution inherent in measuring the D^{*+}, D^0
mass difference. In general, reconstructing the intermediate states involved in a
decay chain is also a powerful means of reducing combinatorial background. In
the example decay chain shown above, all the intermediate state particles, (viz.,
D^{*+}, D^0, K^0, π^0) have their masses constrained by kinematically fitting the
daughter momenta.

To approach the search for various decay channels by means of many different
cuts tuned for each channel and set of final state particles would lead to a very low,
and difficult to calculate, efficiency. In the ARGUS analysis only two principal

cuts are applied to the data in a consistent fashion, whatever the channel under consideration. These cuts are:

(1) A cut on the overall χ^2 derived from the particle identification, and the intermediate kinematic fits. The χ^2 probability is cut at 1%.

(2) Since $M(\Upsilon(4S)) = 2.0 \cdot E_{beam}$, the energy of each B meson candidate from $\Upsilon(4S)$ decay must be equal to the beam energy. Specifically, we first select candidates which satisfy: $|E_B - E_{beam}| \leq 3.0\,\sigma_{Erecon}$ and $H_2 \leq 0.45$. A fit is then performed constraining the total energy, of the B candidate, to be equal to that of the beam.

About 35% of events have 2 or more B candidates sharing some of the same tracks. For example, in the process:

$$\overline{B}^0 \rightarrow D^{*+} \quad \pi^- \quad \pi^- \quad \pi_1^+$$
$$\big|$$
$$\longrightarrow D^0 \quad \pi^+$$
$$\big|$$
$$\longrightarrow K^- \quad \pi^+ \quad \pi^- \quad \pi_2^+$$

π_1^+ and π_2^+ could be exchanged, or one or other could be substituted by another low momentum π. Such ambiguities are resolved on the basis of choosing which ever combination yields the best χ^2, when the beam energy fit is included.

In what follows the following channels have been considered, where the charge conjugate reactions are always implied.

$$\overline{B^0} \rightarrow D^{*+}\pi^-$$
$$\rightarrow D^{*+}\pi^-\pi^0 \qquad \text{and} \qquad B^- \rightarrow D^{*+}\pi^-\pi^-$$
$$\rightarrow D^{*+}\pi^-\pi^-\pi^+ \qquad\qquad\qquad \rightarrow D^{*+}\pi^-\pi^-\pi^0$$

$$\text{with} \qquad D^{*+} \rightarrow D^0\pi^+$$
$$\big|$$
$$\rightarrow K^-\pi^+$$
$$\rightarrow K^0\pi^+\pi^-$$
$$\rightarrow K^-\pi^+\pi^+\pi^-$$
$$\rightarrow K^-\pi^+\pi^0$$

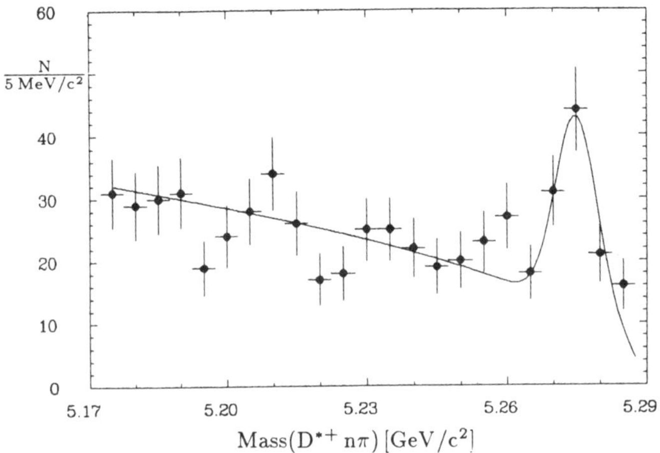

Fig.(9) Combined signal for neutral and charged B mesons. The background is as described in the text.

The mass spectrum resulting from this procedure is shown in Fig.9. There is an enhancement at the expected B meson mass of (71 ± 11) events.

The same procedure applied to events from the non-resonant continuum just below the $\Upsilon(4S)$ is shown in Fig.10(a). There are no events in the region of the B, indicating that the background in Fig.9 arises mainly from combinatorics. To estimate the background on the $\Upsilon(4S)$ we have used two methods, which give consistent results.

a) An "event mixing" background has been generated by combining a D^{*+} from one event with π's from another event. This is shown in Fig.10(b)

b) A wrong sign background has been generated by making combinations which could not arise from B decay, such as $D^{*+}\pi^-\pi^+$ rather than $D^{*+}\pi^-\pi^-$. This is shown in Fig.10(c).

The fitted curve in Fig.10(b) and Fig.10(c) is of the form:

$$dN_{bg}/dM = k.M.\sqrt{(1 - M^2/E^2_{beam})}$$

This satisfactorily describes the uncorrelated background, and has been used in Fig.9. Furthermore, the absence of any structure in the region of the B meson in these background distributions, gives confidence that the E_{beam} fitting procedure does not generate spurious peaks close to the kinematic limit. The enhancement in Fig.9 is due to a mixture of charged and neutral B mesons. In Fig.11(a)

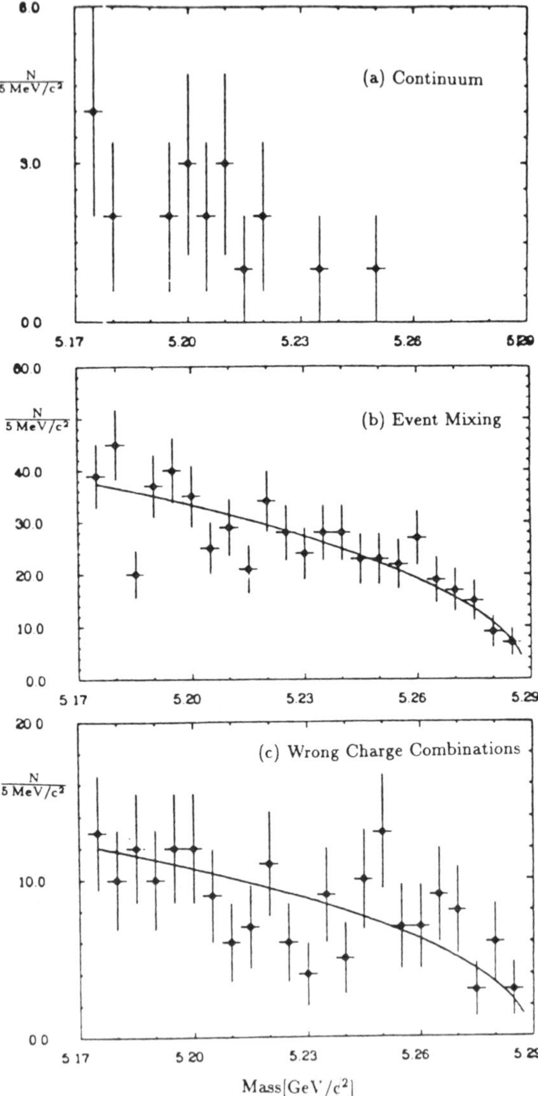

Fig.(10) Investigation of the background under the fully reconstructed B signal. The reconstruction technique as applied to:

(a) Data collected on the non-resonant continuum just below the $\Upsilon(4S)$.

(b) Event mixing background, as described in the text.

(c) Wrong sign combinations.

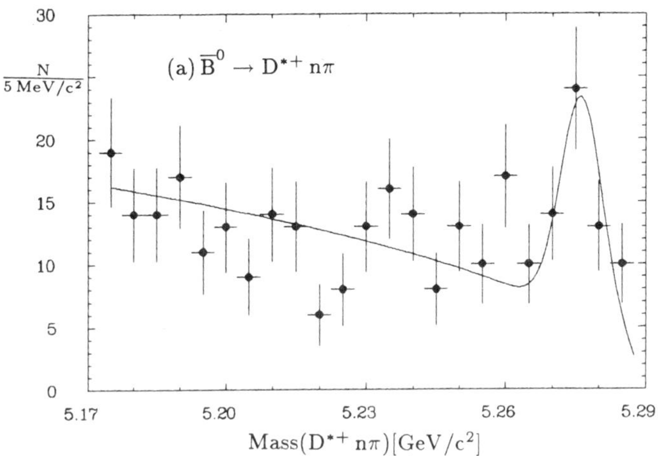

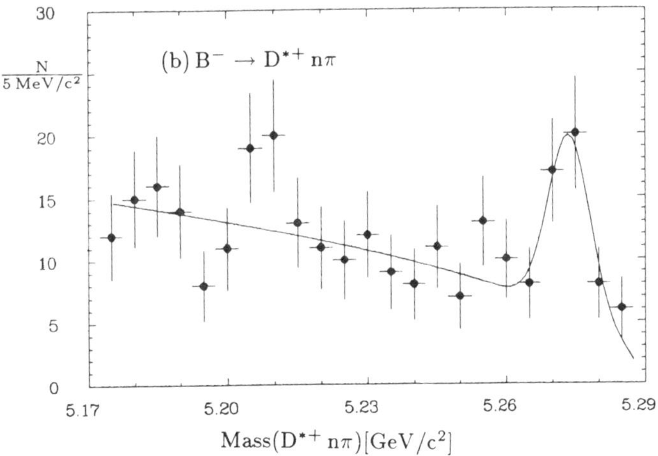

Fig.(11) The mass distribution for fully reconstructed B mesons, (a) $\overline{B}^0$ and (b) B^-.

510

and Fig.11(b) the distributions for neutral and charged B mesons are shown separately. There are (40 ± 8) $\overline{B}^0$ at a mass of $(5276.2 \pm 1.0 \pm 3.0)$ MeV/c^2, and (32 ± 7) B$^-$ at a mass of $(5273.8 \pm 1.3 \pm 3.0)$ MeV/c^2

At the time of writing the final acceptance calculations are not complete, and so no branching ratios are given here. However, it is already apparent that the branching ratios are lower than those found in the literature. For example, the branching ratio for $\overline{B}^0 \to D^{*+}\pi^-$ is of order $(0.2 - 0.5)\%$ compared with the CLEO[10] value of $(1.7 \pm 0.5 \pm 0.5)\%$.

REFERENCES.

[1] H.Albrecht et al. (ARGUS Collaboration), Phys. Lett. **134B** (1984), 137.

[2] E.H.Thorndike. Ann. Rev. Nucl. Part. Sci. **35** (1985), 195.

[3] J.H.Kühn and R.Rückl, Phys. Lett. **135B** (1984) 477.

[4] H.Albrecht et al. (ARGUS collaboration), Phys.Lett. **158B** (1985), 552.

C.Bebek et al. (CLEO) Cornell Preprint CLNS-86/715, Feb. 1986.

D.H.Coward (Mk.III) SLAC-PUB-3818, Oct. 1985 (Invited Talk at APS, Eugene, Oregon, August 1985).

[5] H.Fritzsch, Phys. Lett. **86B** (1979) 164

J.H.Kühn, S.Nussinov and R.Rückl, Z.Phys **C5** (1980) 117.

M.B.Wise, Phys. Lett. **89B** (1980) 229.

T.A. DeGrand and D.Toussaint, Phys. Lett. **89B** (1980) 256.

I.I.Bigi and A.Sanda, Nucl. Phys **B193** (1981), 85.

P.H.Cox, S.Hovater, S.T.Jones, and L.Clavelli, Phys. Rev **D32** (1985), 1157.

[6] H.Albrecht et al. (ARGUS Collaboration), Phys. Lett **162B** (1985),395.

See Also. P.Haas et al. (CLEO) Phys. Rev. Lett. **55** (1985), 1248.

[7] D.B.MacFarlane (ARGUS Collaboration) Proceedings of XXth Rencontre de Moriond, Les Arcs, Savoie (1985) Vol.II, 207-218

S.Weseler, "Untersuchung der semileptonische Zerfälle von B(5270)-Meson mit dem ARGUS-Detektor", Ph.D. Thesis, Universität Heidelberg (1986).

[8] I.I.Bigi Phys. Lett. **106B** (1981), 510.

[9] S.Nussinov. Phys. Rev. Lett. **35** (1975), 1672.

G.J.Feldman et al., Phys. Rev. Lett. **38** (1977), 1313.

[10] R.Giles et al., (CLEO) Phys. Rev. **D30** (1984), 2279.

A.Chen et al., (CLEO) Phys. Rev. **D31** (1985),2388.

Review of B Meson Lifetime Measurements at PEP*

RICHARD DUBOIS

University of Victoria and TRIUMF
Victoria, BC   V8W 2Y2

Abstract

Results from measurements of the average B-hadron lifetime performed at the PEP e^+e^- storage ring are reviewed. The lifetime is approximately 1 ps and leads to constraints on the mixing angles between the u, c, and b quarks.

---

* Work supported in part by the National Research Council and the National Science and Engineering Research Council

512

1. INTRODUCTION

With the advent in the past few years of e^+e^- storage rings with energy sufficiently higher than $B - \bar{B}$ meson threshold, it has become possible to measure lifetimes of B hadrons. Experiments have been unable to exclusively reconstruct B hadrons, so the lifetime measurements are an average over the produced hadrons. The primary interest in the lifetime measurements has been from the information they give on the transitions of the b quark to u or c quarks via the $W^\pm$.

An outline of the discussion is:

1. weak mixing angles
2. obtaining enriched B hadron samples
3. the impact parameter method
4. checks on the method
5. summary of results
6. conclusions

The purpose of this review is to give the status of the measurements at PEP and provide a guide to the problems involved in them. Examples are taken from the DELCO[1] experiment.

2. WEAK MIXING ANGLES

In the Standard Model description of quarks, the b quark is in a weak isospin doublet with the t quark. Were this the entire story, the b quark would be stable. The quarks are indeed not mass eigenstates of the weak interaction; the observed quarks are a mixture of the weak interaction eigenstates. This allows transitions between the generations of quarks. The lifetime of b-quark containing hadrons gives information on the transition of the third generation to the first two.

Kobayashi and Maskawa[2] first discussed the possibility of a six quark world and introduced a 3×3 matrix to parametrize the mixing of the three generations. The observed quarks can be obtained from the weak eigenstates by

$$\begin{pmatrix} d' \\ s' \\ b' \end{pmatrix} = V \begin{pmatrix} d \\ s \\ b \end{pmatrix}$$

where

$$V = \begin{pmatrix} V_{ud} & V_{us} & V_{ub} \\ V_{cd} & V_{cs} & V_{cb} \\ V_{td} & V_{ts} & V_{tb} \end{pmatrix}$$

The elements V_{ub} and V_{cb} are the only ones involved in the transitions of the b quark. Maiani[3] has suggested an alternative description of the mixing matrix,

$$V = \begin{pmatrix} c_\beta c_\theta & c_\beta s_\theta & s_\beta \\ -s_\gamma c_\theta s_\theta e^{i\delta} - s_\theta c_\gamma & -s_\gamma s_\beta s_\theta e^{i\delta} + c_\theta c_\gamma & s_\gamma c_\beta e^{i\delta} \\ s_\theta s_\gamma e^{-i\delta} - c_\theta c_\gamma s_\beta & -c_\theta s_\gamma e^{-i\delta} - s_\theta c_\gamma s_\beta & c_\beta c_\gamma \end{pmatrix}$$

where the s and c refer to *sin* and *cos*. The mixing angles are small, so we can approximate c_β by unity. Hence the transitions essentially give information on s_β and s_γ.

Figure 1 shows spectator semileptonic decays of the b quark to u and c quarks. The coupling at the $b\bar{u}$ vertex can be described by

$$\Gamma(b \to ue\nu) \propto \text{(Phase Space)} \, V_{ub}^2$$

$$\propto s_\gamma^2$$

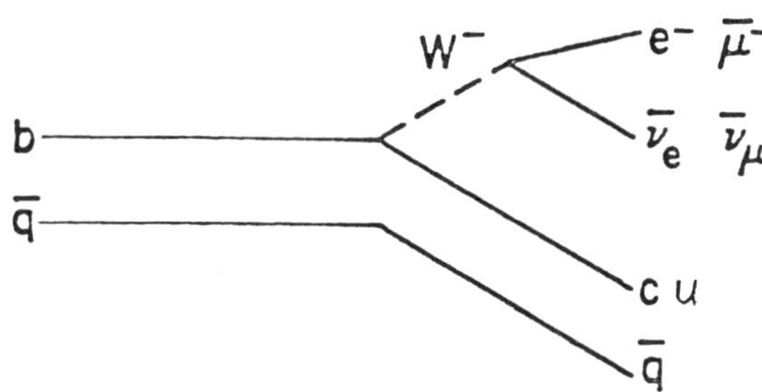

Figure 1. Semileptonic decay diagrams for b decays. Spectator diagram for b decay into either c or u quarks and either an electron or muon.

Similarly,

$$\Gamma(b \to ce\nu) \propto \text{(Phase Space)} \, V_{cb}^2$$

$$\propto s_\beta^2$$

With all decays, the total decay rate is given by

$$\Gamma_{tot} = \Gamma_0(a \, s_\gamma^2 + b \, s_\beta^2)$$

where a and b contain phase space factors and QCD corrections and

$$\Gamma_0 = \frac{G^2 m_b^5}{192\pi^3}$$

is the analogue of the μ decay rate. As an example,

$$\Gamma_0 = \frac{m_b^5}{m_\mu^5}\Gamma_\mu$$

and so, $\tau_0 = 6.4 \times 10^{-15}\ s\ = 6.4 \times 10^{-3}\ ps$. This means that if the b quark mixed fully with the first two generations, its lifetime would be very short.

3. OBTAINING ENRICHED B-HADRON SAMPLES

All experiments to date have used high-p_t semi-leptonic decays as a tag for heavy quarks. Table 1 lists the techniques used in identifying the leptons.

Table 1. Lepton ID Techniques

Lepton ID in PEP/PETRA Detectors	
Detector	Description
DELCO	Čerenkov, Pb-scint
Mark II	μ, Pb-Liq-Ar
MAC	Pb-gas
JADE	Pb-glass
TASSO	Čerenkov, μ, Pb-Liq-Ar

For c-quark containing hadrons, with $m_c \approx 2\ GeV/c^2$, $p_t^{max} \approx 1\ GeV/c$, while for $m_b \approx 5\ GeV/c^2$, $p_t^{max} \approx 2.5\ GeV/c$. Therefore, p_t relative to the parent direction can be used as a discriminator between B and C hadrons. Figure 2 illustrates the difference in p_t distributions for b and c semileptonic decays; good separation is obtained for $p_t > 1\ GeV/c$.

The DELCO experiment was optimized for clean electron detection[4]. Table 2 shows the expected relative parentage of direct electrons with $p_t > 1\ GeV/c$, indicating that this B-enriched region is fairly pure and free of background not from semileptonic decays. The Mark II and MAC experiments obtained fractions of $b : c$:background of 0.64: 0.16: 0.30 and 0.50: 0.20: 0.30 in their respective B regions.

4. THE IMPACT PARAMETER METHOD

As B hadrons are not reconstructed explicitly, one does not have access to the production vertex to measure the lifetime by flight path. In a method originated by

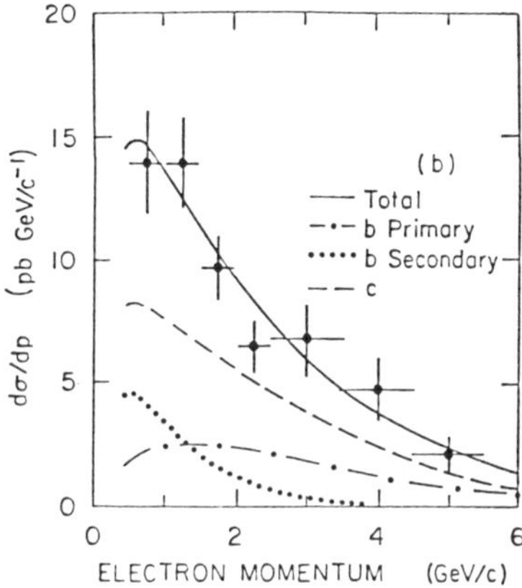

Figure 2. Momentum distribution in semileptonic decays of b and c quarks. The p_t distributions show the difference in mass of the parent. Separation can be done in a statistical sense based on p_t.

Table 2. Parentage of Electrons with p_t 1 GeV/c

Electron Parentage	
Process	Fraction
$b \rightarrow e$	0.70
$b \rightarrow c \rightarrow e$	0.17
$c \rightarrow e$	0.09
background	0.04

the Mark II[5] and MAC[6] groups, the impact parameter of the decay lepton relative to the beam is used as a measure of the parent lifetime. Figure 3 illustrates the method.

In this technique, it is necessary to have knowledge of the production and decay kinematics of the parent mesons as well as the relative populations in (p, p_t) of the daughter leptons. With an assumed value of the meson lifetimes (note that one must assume the relative abundances of B_u, B_s, B_d and B_c mesons), an intrinsic impact parameter distribution can be obtained.

One must then use the knowledge of spatial resolutions, multiple scattering and beam size to convolute the exact impact parameter distribution to a smeared one which can be compared with the measured values. A maximum likelihood fit is

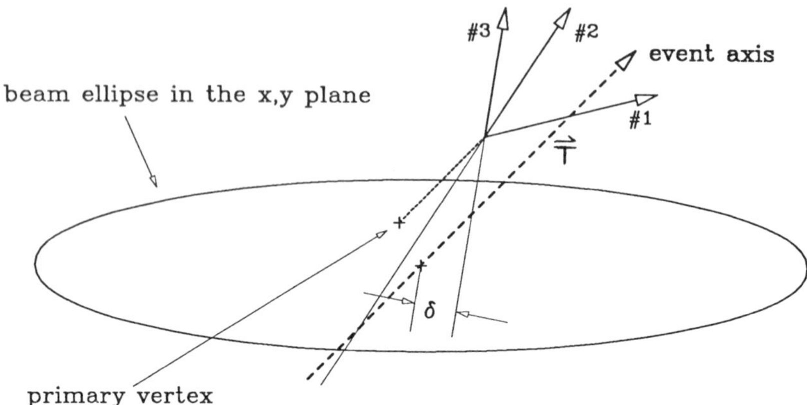

Figure 3. Impact parameter method The impact parameter of the decay lepton relative to the nominal beam position is measured. This length carries a sign determined by $\vec{\delta} \times \vec{T}$. $\vec{T}$ is given by the sphericity axis of the event and is estimated to match the parent direction within a few degrees.

used to vary the lifetime to match the observed distribution.

The lifetime is obtained by maximizing the product of probabilities over all events,

$$P(\delta) = f_b P_b(\delta) + f_c P_c(\delta) + f_{bc} P_{bc}(\delta) + f_{bkg} P_{bkg}(\delta) \qquad 4.1$$

where the f's are the populations of leptons from the decay chains $b \to e$, $c \to e$, $b \to c \to e$ and from backgrounds. The P's are the convolution of a resolution function (usually taken to be Gaussian) with the intrinsic impact parameter distribution as obtained by Monte Carlo simulation. Both the f's and P's depend on (p, p_t) and are evaluated on an event by event basis, depending on δ.

4.1 SOURCES OF ERROR: THE RESOLUTION FUNCTION

Four sources of error must be considered as affecting the measurement of the impact parameter. They are tracking resolution, beam size, multiple scattering and track confusion in 'busy' events. It is possible to use different topologies to isolate these effects from each other.

4.1.1 Tracking resolution

The tracking resolution can be isolated using high momentum tracks from the Bhabha process, $e^+e^- \to e^+e^-$. These tracks suffer little from multiple scattering and give a very uncluttered topology. In addition, the tracks come from a common production point, so that a measurement of the separation distance between the two tracks gives the tracking resolution at the beam origin. The DELCO result for this resolution was approximately 230 μm.

4.1.2 Beam size

In conventional e^+e^- storage rings the beam sizes are macroscopic compared to the impact parameter sizes considered. This is especially true for the horizontal size which is dominated by fluctuations in emission of synchrotron radiation, resulting in beam sizes on the order of $500 \times 50 \ \mu m^2$ in x and y respectively.

Again, Bhabha events can be used to isolate this effect: vertical-going Bhabhas trace out the horizontal beam extent, while horizontal tracks give the vertical size. The tracking resolution is removed in quadrature from the measured sizes to yield the true beam size. The azimuthal dependence of the beam size is accounted for by using a two dimensional Gaussian to describe the beam. DELCO found that the error from the beam size was about 315 μm, averaged over ϕ.

4.1.3 Multiple scattering

Multiple scattering can be dealt with by parametrizing the error in terms of

$$\sigma_{MS} = A/p \cos^{3/2} \theta$$

to account for incident angle dependence on material thickness and extrapolation error to the interaction point. In the DELCO case, material was kept to a minimum (to reduce γ conversions in the beam pipe and drift chamber wall), and amounted to about 3% of a radiation length, concentrated around 9 cm from the beam origin. Accounting for polar angle dependence of the thickness, for an (average) 1 GeV/c track, multiple scattering contributed about 260 μm to the error.

4.1.4 Track confusion

Once all these previously discussed effects are accounted for, the error on isolated tracks is understood. However, in complicated $e^+e^- \rightarrow$ *hadrons* events, the tracking resolution is degraded due to mistakes in assigning the correct drift chamber hits to the tracks. This can be investigated by comparing the width of the impact parameter distribution for hadron tracks that pass the lepton criteria with the expected width from the known error sources. DELCO found that these tracking mistakes introduced about an additional 20% to the resolution as well as non-Gaussian tails to the resolution function.

5. CHECKS ON THE METHOD

The final step in the chain is to verify that the measurement does not introduce any artificial offsets to the impact parameter distribution and that the method can reproduce a known lifetime. It has become a standard practice to use the τ lifetime

518

as that standard, since it is well known from flight path determinations and is the same order of magnitude as the B lifetime.

5.1 CONTROL SAMPLES: ZERO LIFETIME CHECK

Three event topologies are generally used to check that the method reproduces a zero (or small) lifetime where one is expected. These are the so-called 2γ or $e^+e^- \rightarrow e^+e^-e^+e^-$ events; hadron tracks satisfying B region criteria (but *not* containing a decay lepton) and hadron tracks satisfying the complementary criteria in p_t. The DELCO results on these tests are shown in Table 3 along with the expected values. These show that the method is not generating spurious non-zero values.

Table 3. Checks of Scale and Zero in Lifetime Measurement

Method Checks		
Topology	Measured δ (μm)	Expected δ (μm)
2γ	-10 ± 14	0
B region hadrons	46 ± 5	38 ± 11
C region hadrons	42 ± 2	43 ± 4
τ lifetime	0.263 ± 0.046 ps	0.282 ± 0.018 ps

5.2 τ LIFETIME

Using a sample of 1-3 topolgy τ events, which can be obtained very cleanly at these energies, the impact parameter distribution for each track in the event is shown in Figure 4, with a comparison with a standard τ Monte Carlo. The distribution leads to a lifetime of $\tau_\tau = (2.63 \pm 0.46) \times 10^{-13}$ s, in good agreement with previously measured values and with predictions of the Standard Model assuming the τ to be a sequential lepton.

6. RESULTS

Figure 5 (a-c) show the impact parameter distributions for leptons in the B region for the DELCO[1], Mark II[7] and MAC[8] experiments, with results of the fits for the first two also shown. All distributions show a clear positive shift, and have means varying from 80-220 μm which is a result of the differing definitions of the B region. In the DELCO case, where the (p, p_t) cutoffs are lowest, one can get to wider angles with respect to the parent direction and then one is essentially measuring $\beta\gamma c\tau$ for the parent. The lifetime values which result are listed in Table 4 and are on the order of 1 ps. For comparison, the results from JADE[9] and TASSO[10] are also listed.

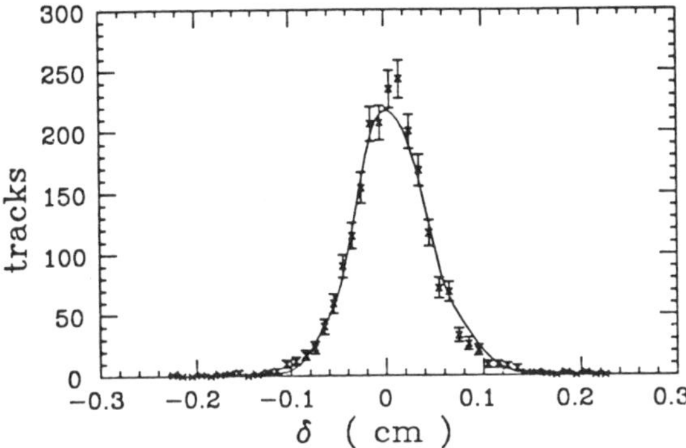

Figure 4. Impact parameters for τ tracks. The impact parameter relative to the nominal beam position is shown. The solid line is from a standard τ Monte Carlo.

7.   SOURCES OF SYSTEMATIC ERROR

There are several sources of systematic error which must be considered in these measurements. They break down into two categories: 'theoretical' uncertainties in obtaining the intrinsic impact parameters and uncertainties in the determination of the resolution function. As examples of these systematic uncertainties, the sources of error in the DELCO measurement will be given below.

7.1   THEORETICAL UNCERTAINTIES

7.1.1   Relative population fractions: error in f_j

The f_j used in equation 4.1 were derived from fits to the semileptonic (p, p_t) distributions and are generally known to about 15-20%. This level of variation results in a 0.3% spread in the lifetime. The effect is small as the region used in the fit is almost pure in its B meson content.

7.1.2   Heavy quark fragmentation functions

The impact parameter depends strongly on the momentum spectrum of the parent heavy meson. The spectra are obtained from measurements of D^* production for c quarks and from semileptonic decays for the b quarks. A one-σ variation in the mean parent momentum gives a 10% variation in the lifetime.

7.1.3   Intrinsic impact parameter distribution

The intrinsic impact parameter distribution depends to some extent on the cuts applied. An approximation to the cuts applied to the data is made when

520

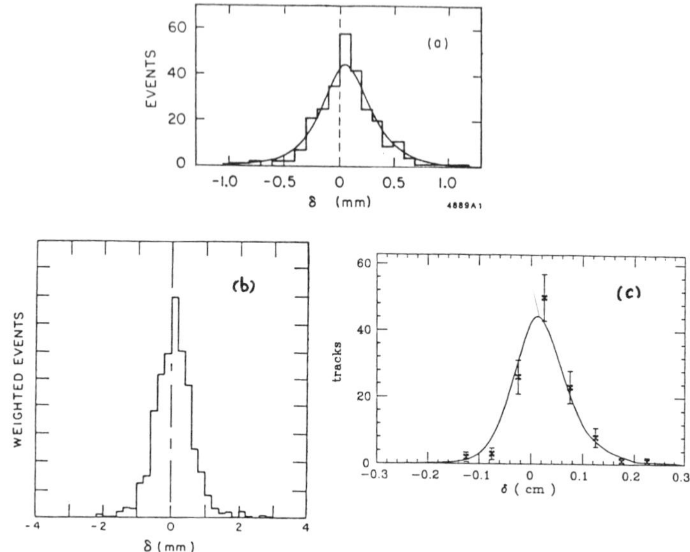

Figure 5. Impact parameter distributions from the Mark II, MAC and DELCO experiments. All the experiments see a positive shift to the impact parameter distribution. The solid curves on the Mark II and DELCO plots are from the maximum likelihood fits to the data.

generating the intrinsic distribution. The level of approximation can be checked by fully reconstructing a sizable enough number of generated events and checking how the distribution changes when using the approximate cuts against using the same cuts as applied to the data. No significant difference in the distributions were found and no systematic error attributed to this source.

Adding these theoretical uncertainties linearly, one obtains a 15% contribution to the lifetime error in the DELCO measurement.

Table 4. Results on B Lifetime from PEP/PETRA Experiments

Lifetime Results	
Experiment	Lifetime
Mark II	$0.85\pm0.17(stat)\pm0.21(syst)$
MAC	$0.81\pm0.28(stat)\pm0.17(syst)$
DELCO	$1.17^{+0.27}_{-0.22}(stat)^{+0.17}_{-0.16}(syst)$
JADE	$1.8^{+0.5}_{-0.3}(stat)\pm0.4(syst)$
TASSO	$1.83^{+0.38}_{-0.37}(stat)^{+0.37}_{-0.34}(syst)$

7.2 MEASUREMENT UNCERTAINITIES

The resolution function is convoluted with the intrinsic impact parameter distribution to obtain the expected distribution. It has a strong effect on the shape of the final spectrum. For DELCO, the only significant systematic uncertainty in the error function came from the tracking confusion error. This 20% correction was felt to be known to about half its value, introducing a 10% uncertainty to the error. This resulted in a 5% variation in the lifetime.

The total systematic uncertainty (added linearly) comes to 15%.

8. MIXING ANGLES REVISITED

Gaillard and Maiani[11] have calculated the phase space and QCD corrections to the decay rate and found

$$\Gamma_{tot} = \Gamma_0(2.75\ V_{bc}^2 + 7.69\ V_{bu}^2)$$

Since the b quark is relatively massive the QCD corrections are small; then the coefficients above are mainly due to phase space and reflect the difference between the masses of the final state u or c quark containing hadron.

Further information can be obtained from the limit on the ratio $\frac{|V_{bu}|}{|V_{bc}|}$ which results from measurements of the endpoint energy in semileptonic decays of B mesons. The limit[12] is

$$\frac{|V_{bu}|}{|V_{bc}|} < 0.14 \qquad\qquad 8.1$$

Coupled with the lifetime measurement, this limit determines V_{bc} and places a limit on V_{bu}, as shown in Fig. 6. Recently, there have been suggestions[13] that the neglect of Fermi motion in the heavy hadrons weakens the limit in Eq 8.1.

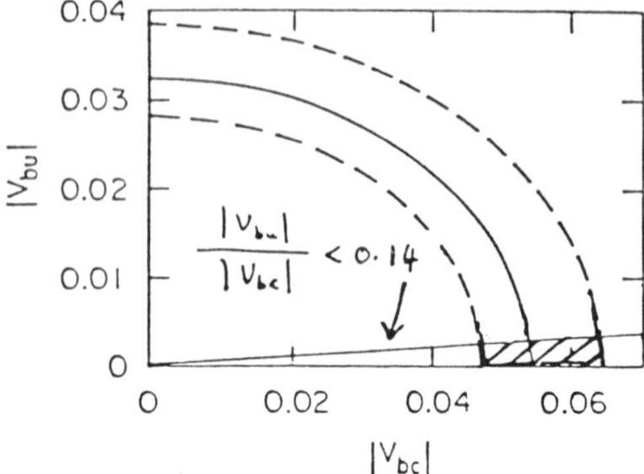

Figure 6.   Allowed region in V_{bu} vs V_{bc}. Eq 8.1 gives the ellipse defined by these two matrix elements. The limit on $b \to u$ vs $b \to c$ further restricts the allowed values.

However, unless the limit is severely undermined, the conclusion on V_{bc} remains much the same,

$$|V_{bc}| = 0.049 \pm 0.005$$
$$|V_{bu}| < 0.006$$

Note that Γ_0 contains a factor m_b^5; this introduces a large systematic error on these angles depending on how much one believes the b mass used.

9. FUTURE PROSPECTS

At present energies, the B lifetime measurements are limited by the systematic uncertainties. These are primarily due to the present knowledge of the fragmentation functions and intrinsic impact parameter distribution.

Alternatives to this may present themselves at the next generation of e^+e^- colliders. The Mark II Upgrade[14] group has studied the question of measuring the lifetime at SLC, and conclude that the systematic errors will be reduced to the 5% level because of the expected cleanliness of the B meson sample that will be obtained at the Z^0.

There will also exist the possibility of measuring the lifetime by a direct flight path technique. This eliminates the problems which arise from the sources mentioned above. Several important improvements are required to make this possible: there must be a copious source of B's, a good B tag and good particle identification to reduce false combinations and allow reconstruction of explicit final B states. At the Z^0, one expects to obtain about 3×10^5 b jets from a canonical one year run yielding 10^6 Z^0's. One might expect samples on the order of one to two hundred reconstructed B mesons, depending on assumptions on branching ratios and efficiencies.

The SLD and DELPHI detectors have the particle identification capabilities, while SLD will be able to tag B's well using the high precision CCD vertex detector made possible by the small beam pipe at the SLC.

10. CONCLUSION

The average B meson lifetime has been measured by the Mark II, MAC and DELCO detectors at PEP, yielding a result near 1 ps. This long lifetime implies that the b quark mixes into the lower generation quarks much less than the d and s quarks mix (cf the Cabibbo angle). Even if the CLEO and ARGUS limit on $\frac{b \to u}{b \to c}$ limit doubles, $V_{bc} \approx 0.05$.

Present measurements are becoming limited by systematic errors in the impact parameter method and we will probably have to wait until SLC comes on to see better measurements.

524

REFERENCES

1. DELCO collaboration: W. B. Atwood, P. H. Baillon, B. C. Barish, G. R. Bonneaud, H. DeStaebler, G. J. Donaldson, R. Dubois, M. M. Duro, E. E. Elsen, S. G. Gao, Y. Z. Huang, G. M. Irwin, R. P. Johnson H. Kichimi, J. Kirkby, D. E. Klem, D. E. Koop, J. Ludwig, G. B. Mills, A. Ogawa, T. Pal, D. Perret-Gallix, R. Pitthan, D. L. Pollard, C. Y. Prescott, L. Z. Rivkin, L. S. Rochester W. Ruckstuhl, M. Sakuda, S. S. Sherman E. J. Siskind, R. Stroynowski, S. Q. Wang, S. G. Wojcicki, H. Yamamoto, W. G. Yan, C. C. Young

   See PhD thesis of D.E. Klem (Stanford, 1986) for details on the DELCO measurement.
2. Kobayashi, M. and Maskawa, T. Prog Theor Phys $\underline{49}$(1973)652
3. Maiani, L. Proc 8th Int Symp on Lepton and Photon Interactions at High Energies, Hamburg(1977)867
4. Koop, D.E. *et al*, Phys Rev Lett $\underline{52}$(1984)970
5. Lockyer, N. *et al*, Phys Rev Lett $\underline{51}$(1983),1316
6. Fernandez, E. *et al* Phys Rev Lett $\underline{51}$(1983)1022
7. Jaros, J. Int Conf on Physics in Collision IV, Santa Cruz, 1984
8. Ford, W.T. Aspen Winter Conf, COLO-HEP-87 (1985)
9. Steffen, P. Conf on High Energy Physics, Leipzig, 1984
10. Wu, S.L. APS Meeting, Washington, D.C., 1985
11. Gaillard, M. and Maini, L. Proc Summer Inst on Quarks and Leptons, Cargese (1979),433
12. Chen, A. *et al*, Phys Rev Lett $\underline{111B}$(1984)1084
13. Grinstein, B. *et al*, Phys Rev Lett $\underline{56}$(1986)298
14. Jonathan Dorfan, private communication

STATUS OF THE CDF EXPERIMENT

N. Lockyer

University of Pennsylvania*

Abstract

A brief review of the status of the Collider Detector Experiment
at Fermilab (CDF) is reported. The CDF experiment is an ~ 4π solid
angle multi-purpose detector designed to study proton-antiproton
collisions in the 2 TeV center of mass range. This last fall the first
few collisions were observed in an engineering run with a partially
assembled detector. The Fall 1986 run will be ~3 months in duration
with a fully assembled detector.

INTRODUCTION

CDF is a large, multi-purpose detector being assembled at the B∅
intersection region at the Fermilab Tevatron. In the fall of 1985, the
first collisions at 1.6 TeV center of mass energy were observed with a
partially assembled and instrumented detector. The first 'physics run'
will take place in the fall of 1986 with a complete detector. The beam
energy will be ~1.6 TeV and the expected luminosity will be up to
10^{29} $cm^{-2}sec^{-1}$.

GENERAL DESCRIPTION OF THE DETECTOR

An elevation view of the forward half of the detector is shown in
Fig. 1. There is a Central Detector and Forward-Backward Detectors.
The Central Detector consists of charged particle tracking chambers
surrounded by a superconducting solenoid magnet, electromagnetic and

528

MUON DETECTION

There are two muon detection systems, one behind the central
calorimetry and the other behind the Forward–Backward Hadron
calorimeters. The Central Muon system is four planes of large drift
tubes operated in limited streamer mode for good charge division used
to measure the Z coordinate. They cover an angular range from 50°
$< \theta < 90^\circ$. The forward muon detector consists of two magnetized
steel toroids and three sets of drift chambers. This system covers the
angular range between $2^\circ < \theta < 17^\circ$ and the expected momentum
resolution is $\Delta P/P \simeq 20\%$. A large gap exists in the angular range
between $17^\circ < \theta < 50^\circ$. Upgrades are presently being considered to
improve the muon coverage.

CALORIMETRY

The layout of the calorimetry detectors can be seen in Figure 1.
Electromagnetic (EM) and hadronic calorimeters provide energy
measurements over the full solid angle. The calorimeters cover 2π in
the azimuthal coordinate $\emptyset$ and ± 4 in pseudorapidity, $\eta = -\ln \tan(\theta/2)$.
They are sampling calorimeters and have a projective tower geometry.

In the Central region $30^\circ < \theta < 150^\circ$, the EM shower counters use
lead scintillator and the hadronic calorimeters use steel scintillator
– both are readout by phototubes. The Forward–Backward detectors use
proportional tubes rather than scintillator since radiation effects are
more severe in this region, $\theta < 30^\circ$. The EM energy resolution varies
from ~$15\%/\sqrt{E}$ at $\theta = 90^\circ$ to ~$30\%/\sqrt{E}$ at small angles. The hadronic
calorimeter resolution is ~$70\%/\sqrt{E}$ at $\theta = 90^\circ$ and ~$140\%/\sqrt{E}$ at small
angles. The tower size is given in the η–$\emptyset$ plane by roughly $\Delta\eta \times \Delta\emptyset$
~ 0.1×0.09 for the proportional tube pad system and 0.1×0.26 for
the scintillator calorimeters. Care has been taken to provide good
gain monitoring in all calorimetry by use, in most cases, of imbedded
radioactive sources. An absolute calibration of gain to ~1% is expected
in all calorimetry subsystems. Much data and extensive study of the
calorimeters have been taken over the last two years in test beams at
Fermilab in order to better understand the calorimeters response to
pions and electrons over a wide range of momenta and angles.

Components of the calorimetry are:

1. The Central Calorimeters consist of 4 arches each of which is sub-
divided into 12 wedges. A wedge contains both EM and hadron calorimeters. The sampling medium is scintillator. An arch is shown in Fig. 2
and its placement in the detector is shown in Fig. 3.

Figure 2 - One of four Central Detector Arches, containing 12 wedges.

2. The Endwall Hadron Calorimeters are mounted on the steel endwalls of
the magnet yoke and are part of the flux return path. An end view of
these calorimeters assembled in the Central Detector can be seen in
Fig. 3. The sampling medium is scintillator.

3. The Endplug Calorimeters, EM and Hadron, are of the proportional
tube type and can be best seen in Fig. 1.

Figure 3 — Placement of arch in Central Detector

4. The Forward—Backward EM and Hadron calorimeters contain proportional tubes and cover $\sim 2^{\circ} < \theta < 10^{\circ}$. A set of Forward Hadron Calorimeters can be seen in Fig. 4 as well as the muon toroids.

TRIGGER

The trigger is implemented as a three level system.

Level 1

The level 1 trigger decision was designed to introduce no deadtime between the 3.5μsec crossing rate. The interaction rate at nominal luminosity is ~50kHZ. The level 1 trigger should reduce this rate by a factor of ~100 without losing interesting physics events. The level 1 trigger consists of the following:

Figure 4 - Forward Hadron Calorimeter and Muon Toroid

1. Trigger counters. These hodoscope scintillation counters permit a small timing window around the beam collision time and enable the rejection of beam gas events by requiring in time hits on both ends of the intersection regions.

2. Muon Systems: Fast signals from both the central and forward systems will be available.

3. CTC: One or several tracks above a programmable threshold (typically 3-5 GeV/c) can be required.

4. Calorimetry: Scintillation and gas calorimetry tower sums are formed. EM and Hadronic sums are separated and 4 different thresholds

above which a channel's energy is summed can be defined per tower. A
total energy trigger can be formed from this data.

Level 2

The level 2 trigger takes at least ~4µsec and typically ~10µsec or
longer for more complex events. The level 2 trigger should reduce the 5
kHZ input rate to ~50Hz. The detector is dead during the level 2
decision time. The level 2 trigger can find clusters of energy, missing
transverse energy, muons, and associate combinations of these with each
other and high momentum charged tracks.

Level 3

The level 3 trigger system will be based on ~100 nodes of
microprocessors being developed at Fermilab around the Motorola 68020
chip. Level 3 will reduce the rate of tape writing to ~1 hertz.

DATA ACQUISITION SYSTEM

A cluster of VAX computers readout a FASTBUS based data acquisition
system. Front end electronics reside on the detector in RABBIT
(Readundant Analog Bus Based Information Transfer) crates. Here, charge
is converted to a voltage, stored on capacitors and eventually digitized
by a 16 bit ADC. Other front end electronics consist of amplification
and shaping circuits. These pulses are then sent via cable to FASTBUS
TDCs and FADCs. The control of high voltage and general detector
monitoring is handled in a CAMAC based system.

ENGINEERING RUN OF FALL 1985

This run tested some of the data acquisition system, in particular
the FASTBUS system. Much of the central calorimetry, muon system
several VTPCs and the level 1 trigger were operational. Forward
trigger counters and Silicon Strip Detectors were also present. All
these detector components worked well and information including
calibration data was written to tape.

Collisions were observed at a center of mass energy of 1.6 TeV. The luminosity was estimated at $\sim 10^{-23} - 10^{-24}$ $cm^{-2} sec^{-1}$. This was a great success for the Accelerator and Antiproton Source physicists. An example of one of 23 $\bar{p}p$ collisions recorded on tape is shown in Fig. 5. An expanded view of the VTPC tracking and a histogram of the energy deposited versus rapidity is shown. This is an example of a minimum bias event.

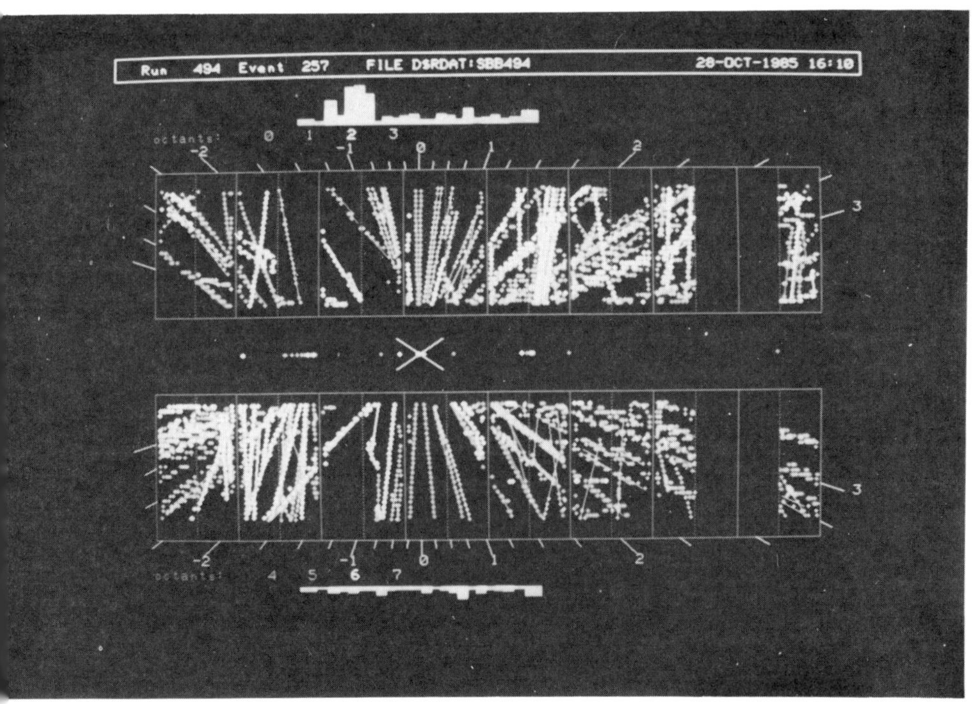

Figure 5 – Event display of the VTPC wire data and projection of calorimetry is shown. This event is from Fall 1985 Engineering run. The aspect ratio is distorted.

FUTURE AND CONCLUSIONS

The first physics run is scheduled for 3 months beginning December 1, 1986. A fully assembled and instrumented detector should be ready for data taking. The goal is to achieve a luminosity of 10^{29} $cm^{-2}sec^{-1}$ during this run. CDF hopes to record on tape a small sample of intermediate vector bosons, and a few hundred jets with $P_t > 200$ GeV. Discussions on upgrading the luminosity to 5 x 10^{31} $cm^{-2}sec^{-1}$ are now underway at Fermilab. At this luminosity, a few W pairs would be detected for study.

References: The Collider Detector at Fermilab, Hans B. Jensen, IEEE Nuclear Science Symposium, San Francisco, Oct. 23-25, 1985.

First Events and Prospects at the Fermilab Collider, Morris Binkley, Fermilab, Presented at Aspen Winter Particle Physics Conference, Aspen, Colorado, Jan. 12-18, 1986.

*CDF Collaboration

Argonne National Laboratory
R. Diebold, W. Li, L. Nodulman, J. Proudfoot, P. Schoessow, D. Underwood, R. Wagner, A. Wicklund

Brandeis University
J. Bensinger, C. Blocker, M. Contreras, L. DeMortier, P. Kesten, L. Kirsch, H. Piekarz, L. Spencer, S. Tarem

University of Chicago
D. Amidei, M. Campbell, H. Frisch, C. Grosso-Pilcher, J. Hauser, T. Liss, G. Redlinger, H. Sanders, M. Shochet, R. Snider, J. Ting, Y. Tsay.

Fermi National Accelerator Laboratory
M. Atac, E. Barsotti, P. Berge, M. Binkley, J. Bofill, A. Brenner, J. T. Carroll, T. Collins, J. Cooper, C. Day, F. Dittus, T. Droege, J. Elias, G. W. Foster, J. Freeman, I. Gaines, J. Grimson, D. Hanssen, J. Huth, H. Jensen, R. Kadel, H. Kautzky, R. Kephart, C. Nelson, C. Newman-Holmes, J. O'Meara, S. Palanque, A. Para, J. Patrick, R. Perchonok, D. Quarrie, S. Segler, D. Theriot, A. Tollestrup, K. Turner, C. van Ingen, R. Vidal, R. Wagner, V. White, R. Yamada, G. P. Yeh, J. Yoh

INFN - Frascati, Italy
S. Bertolucci, M. Cordelli, M. Curatolo, B. Esposito, P. Giromini,
S. Miozzi, S. Miscetti, A. Sansoni

Harvard University
G. Brandenburg, D. Brown, R. Carey, M. Eaton, A. Feldman,
E. Kearns, R. Schwitters, M. Shapiro, R. St. Denis

University of Illinois
G. Ascoli, S. Bhadra, R. Downing, S. Errede, L. Holloway, I. Karlinger,
H. Keutelian, U. Kruse, R. Sard, v. Simaitis, D. Smith, T. Westhusing

KEK, Japan
Y. Arai, Y. Fukui, S. Kikamo, M. Mishina

Lawrence Berkeley Laboratory
W. Carithers, W. Chinowsky, R. Ely, M. Franklin, C. Haber, R. Harris,
B. Hubbard, J. Siegrist

University of Pennsylvania
D. Connor, L. Gladney, S. Hahn, N. Lockyer, M. Miller, T. Rohaly,
R. Van Berg, J. Walsh, H. Williams

INFN - University of Pisa, Italy
G. Apollinari, F. Bedeschi, G. Bellettini, N. Bonavita, L. Bosisio,
F. Cervelli, R. Del Fabbro, M. Dell'Orso, E. Focardi, P. Gainnetti,
M. Giorgi, A. Menzione, R. Paoletti, G. Punzi, L. Ristori, A. Scribano,
P. Sestini, A. Stefanini, G. Tonelli, F. Zetti

Purdue University
V. Barnes, A. Byon, K. Chadwick, A. Di Virgilio, A. Garfinkel,
S. Kuhlmann, A. Laasanen, M. Schub, J. Simmons

Rockefeller University
S. Belforte, T. Chapin, G. Chiarelli, N. Giokaris, K. Goulianos,
R. Plunkett, S. White

Rutgers University
A. Beretvas, T. Devlin, U. Joshi, K. Kazlauskis, N. Pearson, T. Watts

Texas A and M University
J. Buchholz, S. Cihangir, D. DiBitonto, F. Marchetto, P. McIntyre,
T. Meyer, R. Webb

University ot Tsukuba, Japan
F. Abe, Y. Hayashide, M. Ito, T. Kamon, S. Kanda, Y. Kikuchi, S. Kim,
K. Kondo, M. Masuzawa, T. Mimashi, S. Miyashita, H. Miyata, S. Mori,
Y. Morita, T. Ozaki, M. Sekiguchi, M. Shibata, Y. Takaiwa, K. Takikawa,

A. Yamashita, K. Yasuoka

University of Wisconsin
J. Bellinger, D. Carlsmith, D. Cline, R. Handler, J. Jaske, G. Ott,
L. Pondrom, J. Rhoades, M. Sheaff, J. Skarha, T. Winch

Visitors
M. Sivertz - Haverford College, Haverford, PA
S. Kobayashi, A. Murakami - Saga University, Japan
Y. Muraki - ICRR, Tokyo University, Japan

CHARGED PARTICLE MULTIPLICITIES FROM CONSTITUENT COLLISIONS
IN VERY HIGH ENERGY HADRONIC INTERACTIONS

Serge Rudaz

School of Physics and Astronomy
University of Minnesota, Minneapolis MN 55455

and

Pierre Valin

Department of Physics
McGill University
Montreal, Quebec, Canada

I. Constituent Collisions and Soft Hadronic Interactions

At large momentum transfers, perturbative QCD calculations of elementary processes involving quarks and gluons leads to a very successful description of the jet phenomena observed at hadron colliders. On the other hand, there is no clear theoretical path relating constituent collisions to the the soft (low momentum transfer) processes which constitute the bulk of hadronic interactions; still, a few remarkable predictions of the additive valence quark model applied to soft processes (verified up to ISR energies) are worthy of note:

- the ratio of total cross-sections for mesons and baryons involving the same quark flavours

$$\sigma_T(\pi p)/\sigma_T(pp) = 2/3$$

- the inclusive yield of charged pions in the proton fragmentation region $(x \to 1)$

$$\frac{x}{\sigma} \frac{d\sigma(p \to \pi^+)}{dx} = F^{p \to \pi^+}(x) \propto u_p(x)$$

$$F^{p \to \pi^+}(x)/F^{p \to \pi^-}(x) = u_p(x)/d_p(x) \ .$$

The first result, based on a very simple incoherence assumption must surely follow from a future theory of soft QCD processes; the second is in accord with experiment (while a simple application of triple-Regge inclusive phenomenology fails).

Is it possible to understand how an impulse approximation to constituent collisions can describe the features of soft hadronic processes ? The first step in answering this question is to have an estimate of the average momentum transfer involved, and a simple empirical method for doing this has been described some time ago by Pokorski and Wolfram[1]. From the following reasonable assumptions

i) the average momentum transfer $|Q| \ll \sqrt{s}$, and is slowly varying with energy;

    ii) on average, all components of the four-vector Q^μ are compara-
ble (cf. Feynman's wee parton picture);

   iii) hadrons are produced fairly uniformly over most of the rapidity interval,

and using energy and momentum inclusive sum rules, these authors arrive
at the result

$$|Q| \simeq h(y=0) \langle p_T \rangle$$

where $h(y=0)$ is the height of the rapidity plateau (including neutral
particles) and $\langle p_T \rangle$ is the average transverse momentum. Over a wide

range of energies, this takes on a value of around one to a few GeV,
so that the inequality $|Q|R_p \gg 1$ is comfortably satisfied, and the

impulse approximation may perhaps be justified. Note that $|Q|$ changes
very slowly over the available range of energies, and so perturbative
QCD corrections to Bjorken scaling of structure functions may be deemed
irrelevant to the description of soft processes in terms of constituent
collisions.

The purpose of this article is to report on a picture of multiparticle production in pp and p̄p collisions based on elementary subprocesses involving quarks and gluons. Our purpose is frankly phenomenological, and we will state our (few) assumptions in due course: the main
result of the investigation is to ascribe the violation of KNO scaling
observed at the ISR and at the Sp̄pS Collider and the rise in the total
and inelastic cross-sections for pp and p̄p collisions to a single effect,
namely the onset of gluon-gluon collisions in addition to the basic
valence quark collisions which dominate low-energy (up to ISR) processes.

We first recall the statement of KNO scaling for multiplicity
distributions[2]: this was a very important step in correlating energy-
dependent observalbles in hadron scattering. Let $\langle n \rangle$ denote the
average charged particle multiplicity, σ_n the topological cross-section
for observing n charged particles in the final state (n is of course
even), and σ_{inel} the total inelastic cross-section ($\sigma_{inel} = \sum_n \sigma_n$).

Then, on the basis of exact Feynman scaling, and in the asymptotic limit
$\sqrt{s} \to \infty$, Koba, Nielsen and Olesen proved the following result: given the
above, form the function $\psi(z)$, $(z = n/\langle n \rangle)$

$$\psi(z) = \langle n \rangle P_n = \langle n \rangle \sigma_n / \sigma_{inel}$$

and this is a universal, energy-independent function. Remarkably, this
holds experimentally for energies up to the ISR, notwithstanding the
fact that Feynman scaling is violated, and that asymptopia certainly has
not been reached! Given this, it is important to find an alternate derivation of KNO scaling which does not depend on the original assumptions.
That this was possible was pointed out more than a decade ago by Eilam
and Gell[3], using a physical picture inspired by the success of the additive quark model in predicting cross-section ratios. In the following
section, motivated by this idea, we derive an analytic form for the
KNO scaling function which is in remarkable agreement with data up to

ISR energies. In the final section, we describe a two-component picture
in which the rise of the relative importance of gluon-gluon collisions
(with a softer x-distribution for gluons than for quarks) results in
the energy dependent violation of KNO scaling observed at colliders.
Barring no further contributions to hadronic inelastic cross-sections,
KNO scaling should be restored at the very high energies where the
gluon-gluon subprocesses overwhelm the quark-quark ones.

II. Additive Quark Model for Inelastic Processes up to ISR energies

We shall assume that below ISR energies (where the rise in total
cross-sections is observed) that $\sigma_{inel} = \sigma_q$ = constant is entirely due
to valence quark collisions, in keeping with the additive quark model.
Given two colliding quarks with respective momentum fractions x_1 and x_2
the basic ansatz is[3]

$$\frac{1}{\sigma_{inel}} \frac{d\sigma_{inel}}{dx_1 dx_2} = A \, f(x_1) f(x_2) \tag{1}$$

where A is a normalization constant to be determined later, and $f(x)$
is the valence quark structure function (we do not distinguish between
the structure functions of u and d quarks here: taking their difference into account will be seen to lead to an almost indistinguishable
result). The energy available for particle production is then

$$\hat{s} = x_1 x_2 s \tag{2}$$

while each event leads to n charged particles, written as

$$n = n' + n_o \tag{3}$$

with n_o representing the leading particles' contribution to the charged
multiplicity (due to the forward and backward going spectator diquarks).
A crude estimate is n_o = 1.3 (we do not attempt to fit this number, which
is in any event close to unity). We shall further assume that the
multiplicity distribution in the available energy at the constituent
level is <u>narrow</u>, and event-by-event, we shall take n' to be given by
a power law

$$n' = k \, \hat{s}^{\alpha} = k(x_1 x_2 s)^{\alpha} \tag{4}$$

so that the n-prong probability can be expressed as

$$P_n = \frac{\sigma_n}{\sigma_{inel}} = \frac{1}{\sigma_{inel}} \frac{d\sigma_{inel}}{dn}$$

$$= A \int dx_1 dx_2 \, f(x_1) f(x_2) \, \delta(n' - k(x_1 x_2 s)^{\alpha}) \tag{5}$$

The average charged particle multiplicity is readily computed:

540

$$\langle n \rangle = \frac{\int dn\ nP_n}{\int dn\ P_n} = \frac{\int dx_1 dx_2 k(x_1 x_2 s)^{\alpha} f(x_1) f(x_2)}{\int dx_1 dx_2 f(x_1) f(x_2)} + n_o \qquad (6)$$

so that in fact

$$\langle n \rangle = n_o + \langle n' \rangle \qquad (7)$$

where

$$\langle n' \rangle = ks^{\alpha} \langle x^{\alpha} \rangle^2 \qquad (8)$$

with the average computed as follows

$$\langle x^{\alpha} \rangle = \int_0^1 dx\ x^{\alpha}\ f(x) / \int_0^1 dx\ f(x) \qquad (9)$$

The parton momentum distribution function must clearly have a low-x behaviour such that these integrals are convergent for consistency of the approach. It is now simple to perform the integral in Eq. (5) to obtain the following KNO scaling result

$$\psi(z') = \langle n' \rangle P_n = \frac{A}{\alpha} \zeta^{1/\alpha - 1} \langle x^{\alpha} \rangle^2 \int_{\zeta^{1/\alpha}}^{1} \frac{dx}{x}\ f(x) f(\zeta^{1/\alpha}/x) \qquad (10)$$

where the scaling variable $z' = n'/\langle n' \rangle$, and ζ is simply z'/z'_{max},

$$\zeta = z'\ \langle x^{\alpha} \rangle^2 = z'/z'_{max} \qquad (11)$$

Thus, KNO scaling follows in this picture from the Bjorken scaling of the constituent structure functions (at an average momentum transfer Q^2 as given by the above empirical estimate). The charged multiplicity is given by a power of the centre-of-mass energy, and in fact, a good fit to this quantity below ISR energies is obtained by choosing $\alpha = 1/4$. The normalization factor A is readily computed from the well-known conditions

$$\int dz'\ \psi(z') = \int dz'\ z'\ \psi(z') = 2$$

to be

$$A = 2/(\int dx\ f(x))^2 \qquad (12)$$

Thus, given $\alpha = 1/4$ from experiment, and adopting the naive valence quark distribution function (in good agreement with deep inelastic lepton scattering data)

$$f_q(x) \propto \frac{1}{\sqrt{x}} (1-x)^3 \qquad (13)$$

we can obtain[4] an analytic expression for $\psi(z')$ in a straightforward way, using Equations (9), (10), (12) and (13):

$$\psi(z') = \frac{2450}{1089}\, \zeta \left[\frac{11}{3}(\zeta^{12}-1)+9\zeta^4(\zeta^4-1)-4\ell n\zeta(\zeta^{12}+1+9\zeta^4(\zeta^4+1)) \right] \quad (14)$$

with $\zeta = z'/K$, $K = z'_{max} = \langle x^{1/4}\rangle^{-2} = 1089/256 \simeq 4.25$. This rather baroque-looking analytic formula is quite unlike any previously given in the literature, and is compared with a compilation of data from both Fermilab and Serpukhov experiments in Figure 1. Note that given the leading particle subtraction, the model should probably be taken to describe full phase-space data, i.e. there should be no subtraction of "diffractive" events (the method for which varies from experiment to experiment anyway). The full curve is just obtained from Eq. (14), while the dashed curve is obtained by distinguishing up and down quark momentum distributions in the proton, as indicated by fits to deep inelastic lepton scattering data. The agreement with experiment is striking.

Another simply calculable quantity is the dispersion D_2,

$$D_2 = (\langle n^2\rangle - \langle n\rangle^2)^{1/2} = D'_2 = (\langle n'^2\rangle - \langle n'\rangle^2)^{1/2} \quad (15)$$

In fact, one readily obtains

$$\frac{D'_2}{\langle n'\rangle} = \left(\frac{\langle x^{2\alpha}\rangle^2 - \langle x^\alpha\rangle^4}{\langle x^\alpha\rangle^4} \right)^{1/2} = C \quad (16)$$

where the constant C is readily evaluated using $\langle x^{1/2}\rangle = 35/128$ and $\langle x^{1/4}\rangle = 16/33$ (with Eq. 13) so that

$$D_2 = C (\langle n\rangle - n_o) \quad (17)$$

with $C = 0.594$: this is in fact an ancient result known as the Wroblewski relation, known to follow once KNO scaling is assumed. Note that C is precisely calculable here, and Eq. 17 is compared with a compilation of data in Figure 2 (full line with $n_o = 1.3$, dashed with $n_o = 0.9$ which in fact gives a slightly better fit). Thus, this very simple model appears to be quite successful in describing the data up to ISR energies, where as is well-known, the pp cross-section is rising signalling the onset of a new regime. We now turn to an extension of this model to the energy range of the ISR and above.

III. The Breakdown of Valence Quark Dominance, Rising Cross-Sections and the Violation of KNO Scaling at ISR and Collider Energies

Many authors have ascribed the rise of the inelastic and total pp cross-sections to the onset of gluon-gluon interactions. Following a popular parametrization, we use the following formula to describe the energy dependence of the $p\bar{p}$ inelastic cross-section (at these

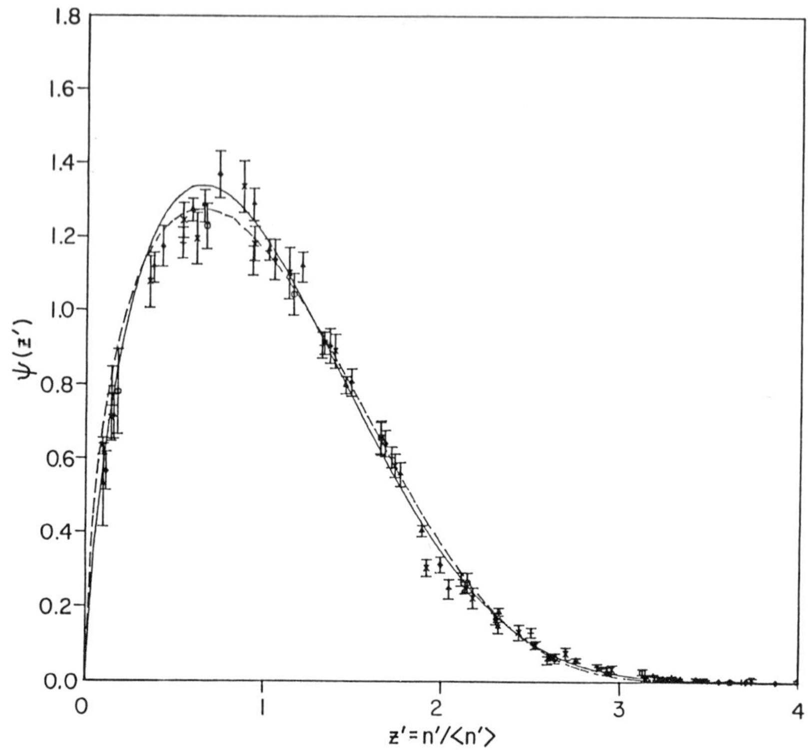

<u>FIGURE 1</u>

Evidence for KNO scaling in z' through the Serpukhov and
Fermilab energy range (lab momentum from 50 GeV/c to 405
GeV/c). The full curve is obtained from the analytic formula Eq. (14). The effect of distinguishing up and down
quark momentum distributions in the proton is shown in the
dashed curve (valence quark distribution functions as in
the article of Eichten <u>et al</u>., Rev. Mod. Phys. <u>56</u>, 579
(1984).

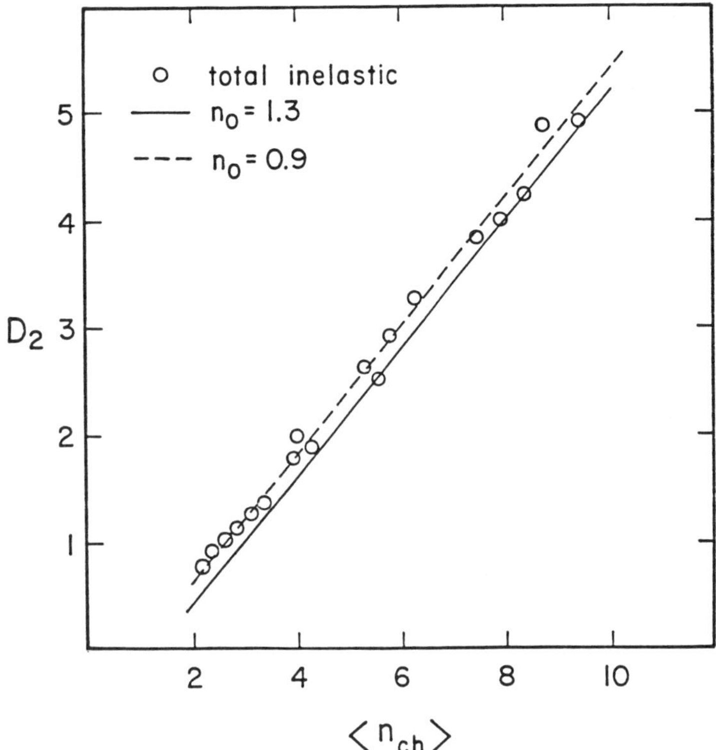

FIGURE 2

Data for the dispersion D_2 as a function of the average charged particle multiplicity, as compared to the Wroblewski relation (derived as Eq. (17) of the text, with $C = 0.594$) for two values of the leading-particle correction to the charged multiplicity, n_0. Data from a compilation by A. G. Ekspong, in Mesons, Isobars, Quarks and Nuclear Excitations, Progress in Particle and Nuclear Physics, edited by D. Wilkinson (Pergamon, London, 1984) Volume 11.

energies, both pp and $\overline{p}p$ cross-sections are similar, the annihilation part of the latter having died out as a power of the center-of-mass energy):

$$\sigma_{inel} = (32.6 + 0.32\ln^2 s/s_o) \text{ mb} \tag{18}$$

with $s_o = 244 \text{ GeV}^2$, to be interpreted as follows. Below s_o, we take $\sigma_{inel} = \sigma_q = $ constant, and above s_o, $\sigma_{inel} = \sigma_q + \sigma_g(s)$, with the subscripts q and g referring respectively to the contributions of the quark-quark and gluon-gluon collisions, identified with the first and second pieces of Eq. (18). This interpretation can perhaps be motivated following Gershteyn and Logunov as being due to a glucball production threshold. Indeed, taking for example $M_G \simeq 1.5 \text{ GeV}$ and an average momentum fraction $\overline{x} \simeq 1/10$, say, one estimates

$$s_{th} \simeq M_G^2/\overline{x}^2 = 0(200) \text{ GeV}^2 \tag{19}$$

which is certainly close to the phenomenological value of s_o. In fact, the UA(5) group reports a slow rise in the relative number of η mesons in the average event when going from ISR to SppS Collider energies: perhaps these are the dominant decay products of glueballs ?

From this two-component hypothesis for the inelastic cross-section, KNO scaling violation is easily seen to arise from the energy dependent superposition of two separately KNO scaling mechanisms with different probabilities. The question arises as to the choice of gluon momentum distribution: the naive parametrization diverges like $1/x$ as $x \to 0$ and is therefore unsatisfactory for our purposes. In fact, we require the low-Q^2 gluon distribution, the determination of which in QCD is a central problem and almost certainly a necessary step in our understanding of the Pomeron: there is no data below $x = 0.01$, and only sparse data for the interval $0.01 < x < 0.1$. In short, the low-x, low-Q^2 behaviour of the gluon distribution is unknown. To make further progress, we choose to parametrize the relevant gluon distribution as follows

$$f_g(x) = \begin{cases} f_{go}(x) = 3(1-x)^5/x, & x \geqslant x_c \\ \sqrt{x_c/x} \, f_{go}(x_c), & x \leqslant x_c \end{cases} \tag{20}$$

introducing a parameter x_c: for values of x above x_c, we have the naive gluon distribution, and below x_c, the behaviour mimics that of the valence quark distribution. We should note that this low-x behaviour at low-Q^2 has no repercussions on the high Q^2 behaviour abstracted from the Altarelli-Parisi equation and which is determined by SppS Collider two-jet cross-sections. It will be convenient to trade the parameter x_c, given the distributions in Eqs. (13) and (20) for an equivalent quantity η which we call the relative inelasticity for reasons soon to become clear:

$$\eta = \langle x^\alpha \rangle_g^2 / \langle x^\alpha \rangle_q^2 < 1 \tag{21}$$

the exponent α is still defined through Eq. (4) assumed to hold for both
quark-quark and gluon-gluon collisions, with the same constant k (in
fact, analysis of three-jet events in electron-positron annihilation
seem to indicate the same multiplicity for gluon as for quark jets).
The averages in Eq. (21) are calculated from Eq. (9), the subscript "q"
indicating that the valence quark distribution function (Eq. (13)) is
to be used, while "g" indicates the analogous thing for gluons (Eq. (20)).
The way the relative inelasticity η comes into the picture will be
sketched below, when a new formula for the charged particle multiplicity is derived in this picture: the inequality in Eq. (21) is expected
simply because the gluon distribution is certainly softer than that of
valence quarks.

The way to proceed now is to simply superimpose incoherently the
two distinct multiplicity distributions arising from quark and gluon
subprocesses. Details are given in Ref. 4: we limit ourselves here
to a broad outline. Working with the quantity n', one considers the
separate contributions of the quark-quark and gluon-gluon multiplicity
distributions, $F_{n'}(i)$, i = q or g:

$$F_{n'} = F_{n'}(q) + F_{n'}(g)$$

$$\sum_{n'} F_{n'}(q) = \sigma_q \ , \ \sum_{n'} F_{n'}(g) = \sigma_g \tag{22}$$

with each process characterized by an average multiplicity of its own,
$\overline{n}_i'$,

$$\overline{n}_i' = \frac{1}{\sigma_i} \sum_{n'} n' F_{n'}(i)$$

$$= k <x^\alpha>_i^2 \ s^\alpha \ , \quad i = q \text{ or } g \ . \tag{23}$$

Now it is easy to obtain the overall average charged particle multiplicity:

$$<n'> = \frac{\sum_{n'} n' F_{n'}}{\sum_{n'} F_{n'}} = \frac{1}{\sigma_{inel}} \sum_{i} \sum_{n'} n' F_{n'}(i) \tag{24}$$

so that one may write

$$<n'> = \overline{n}_q' P_q + \overline{n}_g' P_g$$

$$P_q = \sigma_q/\sigma_{inel}, \ P_g = \sigma_g(s)/\sigma_{inel} \tag{25}$$

Finally, the new formula for the charged particle multiplicity is

$$<n'> = B \ s^\alpha (P_q + \eta P_g) \tag{26}$$

with $B = k <x^\alpha>_q^2$ and η was defined before. Note that for $s < s_o$, where
only valence quark collisions are involved, $P_g = 0$ and one recovers
the earlier result, Eq. (8); conversely, in the limit of very high
energies, $<n'> \rightarrow \eta B s^\alpha$, a reduction as compared to the naive extrapolation
of the low energy formula by a factor η. It is in fact well-known that

an extrapolation of the $s^{1/4}$ fit to the low-energy data on average charged multiplicities leads to an overestimate for SppS Collider energies. The formula Eq. (26) allows us to understand this, as gluon-gluon collisions result on the average in a smaller energy available for particle production, and thus, to a smaller multiplicity.

Referring the interested reader to Ref. 4 for the details of the construction of the new KNO function $\psi(z')$, we just make a comment on the maximum value of the scaling variable $z' = n'/\langle n'\rangle$: at sub-ISR energies, this is simply obtained from Eq. (4) by setting $x_1 = x_2 = 1$ and the result is

$$K_q = \langle x^\alpha\rangle_q^{-2} = z'_{max} \tag{27}$$

(cf. Eq. (14) et seq., where the choice $\alpha = 1/4$ was made). When both quark and gluon subprocesses contribute to the inelastic cross-section, one obtains a new formula (using Eq. (26))

$$z'_{max} = \frac{ks^\alpha}{k\langle x^\alpha\rangle_q^2 s^\alpha (P_q + \eta P_g)}$$

i.e.

$$z'_{max} = \frac{K_q \sigma_{inel}}{\sigma_q + \eta \sigma_g(s)} \ . \tag{28}$$

We immediately see that our picture necessarily implies a <u>broadening</u> of the KNO function with the onset of gluon-gluon collisions, and this is indeed what is observed at Collider energies: for very high energies, we expect (barring new contributions to the inelastic cross-section, such as might signal compositeness of quarks and gluons, for example) a return to a KNO scaling regime characterized by an end-point of the KNO function given by $K_q/\eta > K_q$. Thus, the signal for gluon collisions is the broadening of the KNO function (relatively more high multiplicity events) with $z'_{max}(s)$ the increasing function of energy given in Eq. (28), and this in spite of the fact that on the average, gluon-gluon collisions yield a smaller number of charged particles than quark-quark collisions: a little thought shows there is of course no paradox here. The KNO scaling violations result from a slightly unconventional source!

The actual determination of parameters, once $\alpha = 1/4$ is fixed from low-energy multiplicity data, reduces to obtaining x_c in Eq. (20) (or equivalently, η) by fitting the UA(5) data on the violation of KNO scaling[5] at $\sqrt{s} = 540$ GeV, and B in Eq. (26) from the charged multiplicity. We find that $x_c = 0.03$ (and thus $\eta = 0.658$) and $B = 1.40$ provide satisfactory fits. The results are shown in Figures 3 to 6.

In Figure 3, we show the predicted KNO function at ISR energies[6]: the solid curve is the result of our calculations at $\sqrt{s} = 52.6$ GeV, and is decomposed into the energy-independent quark component (short-dashed line) and the considerably smaller energy-dependent (scale-violating) gluon component (long-dashed line). Figure 4 presents our fit to the UA(5) total inclusive data at $\sqrt{s} = 540$ GeV (same notation). Figure 5

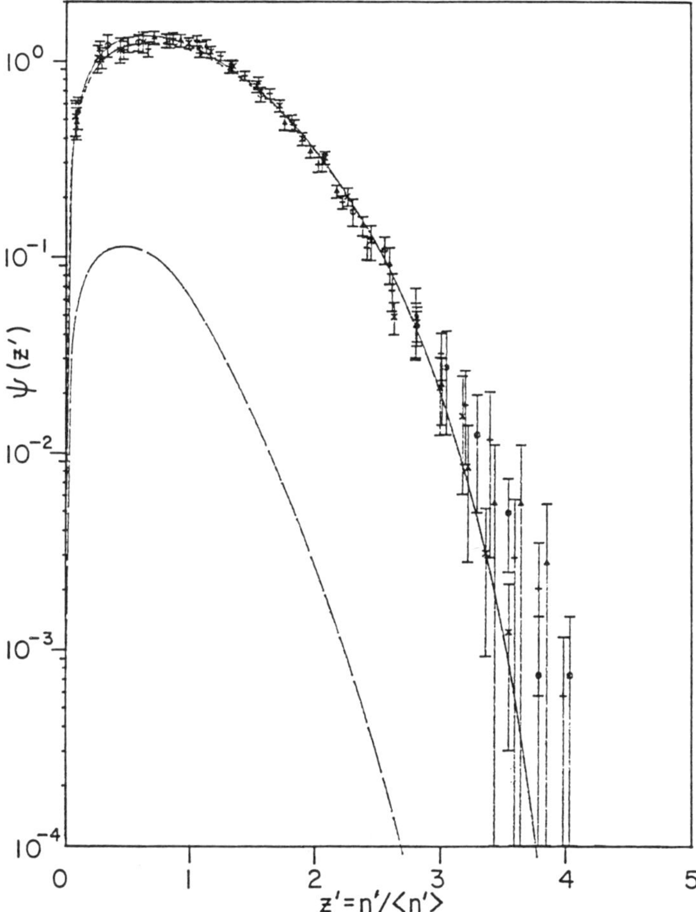

<u>FIGURE 3</u>

Evidence for KNO"scaling"through the energy range of the
ISR (data from Ref. 6). The solid curve is the prediction
of the model described in the text and Ref. 4 for $s^{1/2}= 52.6$
GeV, decomposed into an energy-independent quark component
(short-dashed line) and, at these energies, the much smaller
gluon component (long-dashed line).

548

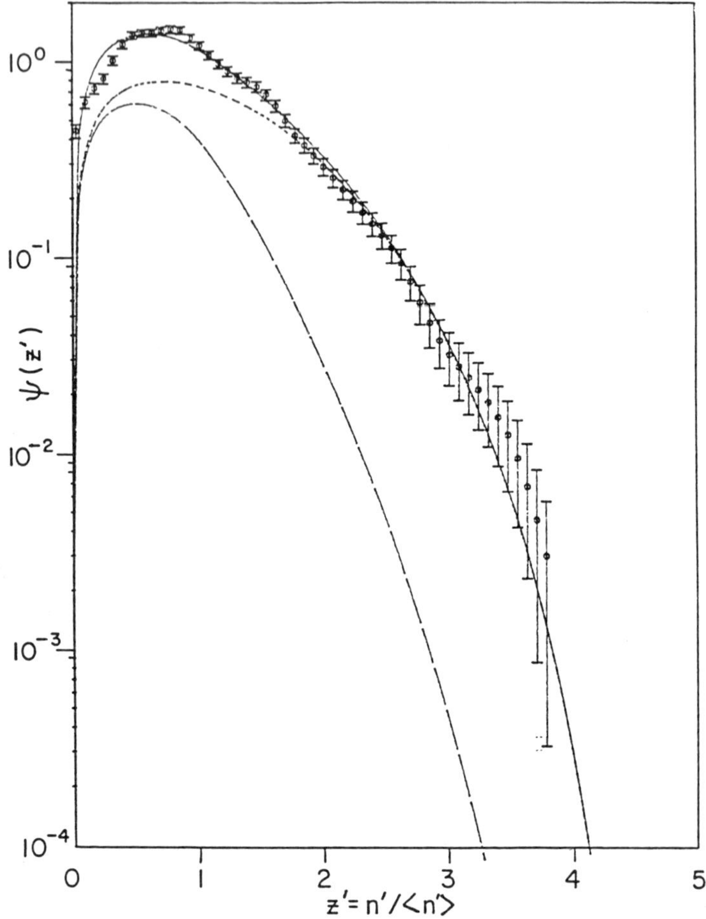

<u>FIGURE 4</u>

The KNO function at $S\bar{p}pS$ energies ($s^{1/2}$=540 GeV). Neither
the data (UA5 group) nor the theoretical curve are compatible with the "scaling" curve shown in Figure 3 for ISR
energies. Note the relatively larger gluon component and
the broadening effect indicated by Eq. (28). Data from
Ref. 5; see also G. J. Alner <u>et al</u>. (UA5 Collaboration),
Phys. Lett. <u>160</u>B, 199 (1985) for further references.

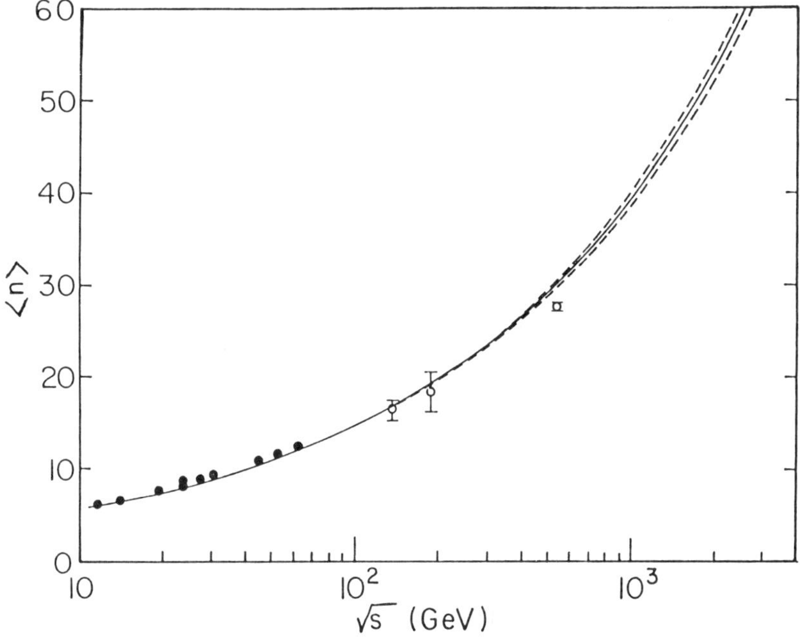

FIGURE 5

Fit to the average charged particle multiplicity, using Eq.
(26) and Eq. (7) with n_o = 1.3, as a function of energy
from $s^{1/2}$ = 10 GeV to the Tevatron Collider range. Parameter
values as indicated in the text after Eq. (28). Open circles
are cosmic-ray data (B. S. Chaudhary and P. K. Malhotra, Nucl.
Phys. B86, 360 (1975) and S. Takasa et al., Phys. Rev D25, 1765
(1982), while the highest energy point is from the UA5 collaboration.

is our fit to the average charged particle multiplicity, as given by Eq. (26). Note that we predict a considerably larger $\langle n \rangle$ at LHC-SSC energies than that inferred from the usual $a + b\ell ns + c\ell n^2 s$ fit to this quantity. Finally, Figure 6 is a plot of the multiplicity moments $C_j' = \langle n'^j \rangle / \langle n' \rangle^j = \langle z'^j \rangle$ which shows that the violation of KNO scaling is relatively mild. The data are a compilation of many experiments, and the dashed lines are our model's predictions.

In Ref. 4, we have also considered the question of the subtraction of "single-diffractive events" and its effects on the shape of the KNO function.

IV. Conclusions

It would seem that an interpretation of the bulk of the (soft) hadron collision cross-sections in terms of constituent collisions is possible, and yields a successful, physically motivated picture of multiparticle production. To test these ideas, it would clearly be desirable to have measurements of meson-baryon processes at very high energies: this would require planning for fixed-target set-ups at the very large hadron colliders currently under consideration.

In any event, we finally show, in Figure 7, our prediction for the evolution of the KNO function as a function of energy, with the broadest curve corresponding to the new scaling regime expected at truly asymptotic energies (gluon-gluon dominance of the constituent subprocesses). It is amusing to speculate that measurements of a global quantity such as the KNO function could possibly yield the first signals for hadron constituent compositeness, in the form of the onset of an unexpectedly large violation of KNO scaling at some energy threshold.

ACKNOWLEDGEMENTS

One of us (S.R.) wishes to thank the Organizers of the 1986 Lake Louise Winter Institute for the opportunity to participate in a pleasant and stimulating meeting. The work described here was supported in part by the U.S. Department of Energy under Contract No. DE-AC-02-83ER40105, by a Presidential Young Investigator Award administered by the National Science Foundation, as well by a supplemental grant by the Exxon Education Foundation (S.R.), and by the Ministere de l'Education du Quebec through its Chercheur-Boursier program, as well as by the National Sciences and Engineering Research Council (NSERC) of Canada (P.V.).

REFERENCES

1. S. Pokorski and S. Wolfram, Z. Phys. C15, 111 (1982)
2. Z. Koba, H. B. Nielsen and P. Olesen, Nucl. Phys. B40, 317 (1972)
3. G. Eilam and Y. Gell, Phys. Rev. D10, 3634 (1974)
4. S. Rudaz and P. Valin, Phys. Rev. D34, No. 5 (September 1986)
5. UA5 Collaboration, Phys. Rep., to be published; we wish to thank
   T. O. White and the UA5(Bonn-Brussels-Cambridge-CERN-Stockholm)
   Collaboration for communicating their results to us in tabular form.
6. A. Breakstone et al., Phys. Rev. D30, 528 (1984)

TWO PHOTON DECAY WIDTHS OF NON-STANDARD SPIN-0 BOSONS

R. Bates* and John Ng
TRIUMF and Physics Department, University of British Columbia
Vancouver, British Columbia, CANADA   V6T 2A3

P. Kalyniak
Physics Department, Carleton University
Ottawa, Ontario, CANADA   K1S 5B6

ABSTRACT

The two photon decay widths of non-standard Higgs
bosons are studied in a minimal broken supersymmetry
model, and the two Higgs doublet extension of the
standard model. A large enhancement of these widths
relative to the standard model is possible. Supersymmetry imposes a severe constraint on this possible
enhancement. The subsequent upper bounds on Higgs
boson production rates via the two photon fusion
mechanism are compared for ep and e^+e^- colliders.

I.   INTRODUCTION

Currently the leading candidates to extend the standard model are
theories which possess supersymmetry. We wish to learn more about the
important Higgs sector for these models, and will focus on their
neutral spin-0 Higgs bosons. The specific model chosen for study[1]) is
one of minimal broken supersymmetry. It is a specific example of a
model which requires two Higgs doublets. We would like to be able to
distinguish between the features of this supersymmetry model which
arise from either the new superparticle content, or because there are
additional Higgs fields present. Hence a general two Higgs doublet
model with no supersymmetry was also examined[2]) for comparison.

---

* presented by R. Bates

Details of these models can be found in the literature[3,4]) and it is their specific application to a calculation which is discussed below.

Additionally we would like to know what form of new physics to expect in the new particle accelerators being built, and if this physics is consistent with a supersymmetric description. We will thus examine the dominant modes of producing Higgs bosons, and the rates at which they are expected to occur in these machines. We concentrate on Higgs bosons in the intermediate mass range from 40 to 160 GeV/c^2. This was done because this is immediately interesting for HERA, SLC and LEP. However, it is quite straightforward to extend the analysis for larger Higgs masses. For the mass range studied, the primary decay mode of the Higgs boson is into a quark-antiquark pair. Such a signal would be difficult to filter out from the large jet backgrounds of hadronic colliders, and hence we shall only consider e^+e^- and ep machines.

The Higgs boson production modes that we will examine for these machines are the reactions

$$e^+e^- \rightarrow e^+e^-H^0 \tag{1}$$

$$e^-p \rightarrow e^-H^0X \tag{2}$$

where X denotes any hadronic states. Since the discussion is similar for both these reactions, we will focus on the subprocess for equation 2. The important diagrams for this reaction are the Z-boson exchange and two photon fusion mechanism (see reference 2). In the standard model the Z-boson exchange mechanism is the dominant process. However it produces Higgs bosons at essentially unobservable rates, which is also the case for supersymmetry models. On the other hand, the two photon fusion mechanism is more interesting. Although negligible in the standard model for energies up to $\sim$5 TeV, it is possible for the two photon fusion mechanism to become the dominant production mode in models with two or more Higgs doublets. The production cross sections can be expressed in terms of the Higgs 2γ-decay widths. We will find that a large enhancement of these widths can occur for two Higgs doublet models, but the addition of supersymmetry greatly restricts the size of this enhancement.

II. TWO HIGGS DOUBLET MODEL

This model[2,3] is a simple extension of the standard model, with
two doublets of Higgs fields rather than one. The particle content of
the two models differ only in the Higgs sector. The key result to note
is that the Higgs to fermion couplings differ from those in the
standard model by factors of $\tan\alpha$, which is the ratio of the vacuum
expectation values of the two Higgs fields. If $\tan\alpha$ is very different
from one, then there is the possibility of greatly enhancing these
fermion couplings, with important consequences for Higgs to two photon
decay process. The most important fermion loop contribution to the
Higgs 2γ-decay width will turn out to be the t-quark, and hence we
focus on enhancing its Higgs couplings with large $\tan\alpha$.

III. MINIMAL BROKEN SUPERSYMMETRY MODEL

Supersymmetry theories are specific examples of models with two
Higgs doublets, and they too will have the features just described.
Additionaly there is the new superparticle content to consider. We
find[1] however, that these superparticles do not significantly affect
the Higgs to 2γ-decay widths. Once again it is the additional Higgs
field content, leading to an enhancement of the Higgs-fermion
couplings, which has the largest effect on these widths.

The specific model studied was one of minimal broken supersymmetry.[4] The details of arriving at the mass eigenstates and
Feynman rules for this model are given in ref. 1. We must however note
a key result in the expression for the charged gaugino masses. This is
given by

$$
\tilde{M}_{\substack{1\\2}} = \frac{M_W}{\sqrt{2}} \left[\left(1 + \sin 2\alpha + \frac{\tilde{m}^2}{2M_W^2} \right)^{1/2} \pm \left(1 - \sin 2\alpha + \frac{\tilde{m}^2}{2M_W^2} \right)^{1/2} \right] \tag{3}
$$

where again $\tan\alpha$ is the ratio of the VEV's for the Higgs doublet, and $\tilde{m}$
is the bare mass parameter for the gaugino fields. This result allows
us to express $\tan\alpha$ in terms of the gaugino masses, which are themselves
bounded by experiment. Thus supersymmetry imposes a new constraint on
the maximum possible value of the enhancement factor $\tan\alpha$.

IV. TWO PHOTON DECAY WIDTHS

Many Feynman diagrams contribute to the scalar 2γ-decay width. There are the usual gauge boson and fermion loop contributions as in the standard model.[5]) In addition, two Higgs doublet models have a charged Higgs loop contribution. Finally the new superpartners of the supersymmetry model will give a charged gaugino and sfermion contribution. The corresponding diagrams for pseudoscalar 2γ-decay involve only fermion and gaugino loops. The matrix elements for all these diagrams can be computed and the importance of each contribution assessed.

In the mass range studied, the various scalar loop contributions were found to be negligible, leaving only the fermions and the gauge bosons. If the factor $\tan\alpha \sim 1$ there is no enhancement of the fermion loop contribution, and then as in the standard model the gauge boson loops would dominate. Again leading to a relatively unimportant 2γ-decay. Only if $\tan\alpha$ is large can the fermion loops dominate and cause the two photon fusion mechanism to produce Higgs bosons at observable rates. Below we will determine how large $\tan\alpha$ can be.

Bounds on the magnitude of $\tan\alpha$ can be obtained by low energy phenomenology, as discussed in reference 1. These give the bound $\tan\alpha \lesssim 40$. If one uses theoretical arguments concerning partial wave unitarity and perturbation theory one can impose the stricter bound of $\tan\alpha < 12$. For the supersymmetric model a more severe bound of $\tan\alpha \lesssim 5.7$ is the one imposed by gaugino masses as discussed above.

V. PRODUCTION IN ep AND e^+e^- COLLIDERS

Even though the cleanest signal would be for an e^+e^- machine, Higgs production rates at the SLC collider were found to be too low for observation. On the other hand, we did find promising results for the two Higgs doublet model, with production rates in the best case which could be readily observed at the HERA ep collider. Using $\tan\alpha \lesssim 40$, we found for one year running that one could have $\lesssim 65$ events for scalar Higgs and $\lesssim 176$ events for the pseudoscalar. However, these rates vary as $\tan^2\alpha$, and must be correspondingly reduced for the supersymmetry model. Again the rates become too low for observation for the minimum model. If a fourth family exists, the additional heavy fermions could increase the rate by roughly a factor of nine. This is up to about the TeV/c^2 range quite independent of the masses of the 4th family fermions. This would bring the production rate for supersymmetry models up to a barely observable 15-25 events per year at HERA.

VI. CONCLUSIONS

In models with additional Higgs doublets, there is the possibility for enhanced Higgs-fermion couplings if the ratio of the two VEV's is large. This in turn leads to enhanced rates which could readily be observed at HERA. Supersymmetry models impose an additional, and stronger constraint on this possible enhancement than do two Higgs doublet models in general. Consequently Higgs boson production in the minimal supersymmetric model would again be unobservable, although additional families could improve the rates. Hence for super-symmetry theories, the Higgs remains as elusive as in the standard model. Alternatively, if Higgs bosons are detected in the new colliders, then we can rule out the minimal broken supersymmetry model.

REFERENCES

[1] Bates, R., Ng, J.N. and Kalyniak, P., TRIUMF preprint TRI-PP-85-108, submitted to Phys. Rev. D.

[2] Bates, R. and Ng, J.N., Phys. Rev. D33, 657 (1986)

[3] Gunion, J. and Haber, H., Report No. SLAC-PUB-3404, 1984 (unpublished).

[4] Haber, H. and Kane, G., Phys. Rept. 117C, 76 (1985)

[5] Ellis, J., Gaillard, M.K. and Nanopoulos, D.V., Nucl. Phys. B106, 292 (1976)

GAUGINOS FROM $p\bar{p}$ COLLISIONS[*]

H. Baer[a], K. Hagiwara[b] and X. Tata[c]

[a]High Energy Physics Division, Argonne National Laboratory
Argonne, IL 60439

[b]Theory Group, DESY, Hamburg, West Germany

[c]Department of Physics, University of Wisconsin
Madison, WI 53706

ABSTRACT

We investigate signals for winos ($\widetilde{W}$) and zinos ($\widetilde{Z}$) when the decays $W \to \widetilde{W}\widetilde{\gamma}$, $Z \to \widetilde{W}\widetilde{W}$ and $W \to \widetilde{W}\widetilde{Z}$ are allowed at $p\bar{p}$ colliders. These processes lead to i) monojet and dijet plus missing transverse momentum ($\not{p}_T$) events, ii) various di- and tri-lepton events with little accompanying hadronic activity, and iii) events containing jets plus leptons plus $\not{p}_T$. Absence of such signals may allow new limits to be placed on $m_{\widetilde{W}}$ and $m_{\widetilde{Z}}$ of at least $m_{\widetilde{W}} + m_{\widetilde{Z}} \gtrsim m_W$, from CERN collider data.

1. INTRODUCTION

John Ellis has told you all about the reasons physicists love SUSY.[1] Supersymmetry solves the hierarchy problem, it is aesthetically appealing, it provides an avenue for unification with gravity, and, of course, it should manifest itself as the low energy limit of the superstring. So far we have no experimental indication of supersymmetry; we only know roughly where the supersymmetric

---

[*]Talk given by H. Baer at the Lake Louise Winter Institute in Particle Physics, Feb. 1986

particles don't live.[2] Recent work on production of strongly interacting SUSY particles at $p\bar{p}$ colliders seems to indicate that[3]

$$m_{\tilde{q}} > 65 - 80 \text{ GeV}$$

$$m_{\tilde{g}} > 60 - 80 \text{ GeV} ,$$

(1)

but this doesn't really bother theorists--the SUSY mass gap, and hence, the sfermion masses are expected to be $O(100-500 \text{ GeV})$ if SUSY is to provide a solution to the hierarchy problem. Most theorists professing devotion to SUSY favor models with light $\left(\sim O(\text{few GeV})\right)$ photinos ($\tilde{\gamma}$) with $\tilde{\gamma}$ as the lightest SUSY particle. In this case it is also expected on fairly general grounds that one of the supersymmetric counterparts of the gauge bosons--the wino ($\tilde{W}_1$) and zino ($\tilde{Z}_1$)--should have the following mass restrictions:[4]

$$m_{\tilde{W}_1} < m_W$$

$$m_{\tilde{Z}_1} < m_Z .$$

(2)

These sparticles may then be accessible to detection by the currently operating $p\bar{p}$ colliders! In the following calculations, we assume a small $\tilde{\gamma}$ mass, so the mass eigenstates $\tilde{W}_1$ and $\tilde{Z}_1$ contain large gaugino, but small higgsino, components. We also assume heavy $\left(\gtrsim O(100) \text{ GeV}\right)$ sfermion masses, and a ratio $r = v_1/v_2 = 1$ of Higgs field vacuum expectation values (this is the preferred choice of a wide class of SUSY models[5]). This talk is a summary of results presented in Ref. 6

2. PRODUCTION AND DECAY OF GAUGINOS

In $p\bar{p}$ colliders, gauginos can be produced via the decay of vector bosons,[7] e.g.

$$W \rightarrow \tilde{W}\tilde{\gamma} \; , \; \tilde{W}\tilde{Z}$$

$$Z \rightarrow \tilde{W}\tilde{\bar{W}} \; . \tag{3}$$

These channels dominate gaugino production if the above decays are kinematically allowed; t-channel production also occurs, but at very low rates. The gaugino-gaugino'-gauge boson coupling strengths are large. This leads to the substantial production cross-sections illustrated in Fig. 1.

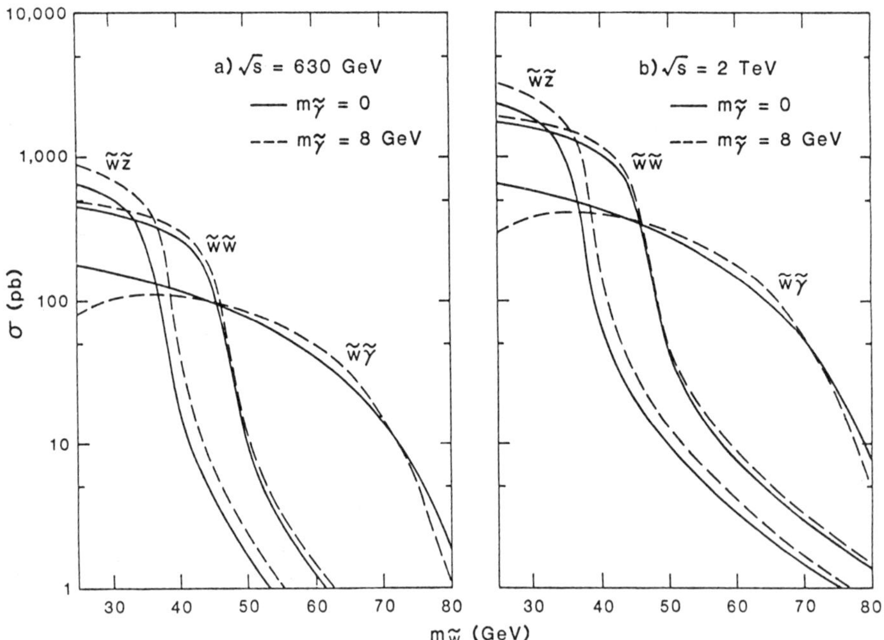

Fig. 1. Production cross-sections for $p\bar{p} \rightarrow \tilde{W}\tilde{\gamma}$, $\tilde{W}\tilde{\bar{W}}$, and $\tilde{W}\tilde{Z}$ versus $\tilde{W}$ mass at a) 630 GeV and b) 2 TeV for two choices of photino mass: $m_{\tilde{\gamma}} = 0$ and 8 GeV. The $\tilde{Z}$ mass is fixed by the other gaugino mass choices.

Assuming heavy sfermions, the gauginos have the following decay modes:

$$\widetilde{W} \rightarrow q\bar{Q}\widetilde{\gamma}\ (67\%) \qquad \text{or} \qquad \ell\bar{\nu}\widetilde{\gamma}\ (11\%)$$

$$\widetilde{Z} \rightarrow q\bar{q}\widetilde{\gamma}\ (46\text{--}61\%) \qquad \text{or} \qquad \ell\bar{\ell}\widetilde{\gamma}\ (13\text{--}18\%)\ . \tag{4}$$

The decay of $\widetilde{Z}$ into $\widetilde{W}$ is also allowed, but is small unless $m_{\widetilde{q}} \gtrsim$ 300 GeV.[6] Calculation of gaugino production and decay maintaining gaugino spin correlations requires the evaluation of many lengthy Feynman diagrams; we make the calculation manageable by evaluating each helicity amplitude as a complex number[8] and summing over final state helicities in our Monte Carlo programs.

A wide variety of spectacular signatures can be expected from gaugino production. All three modes, $\widetilde{W}\widetilde{\gamma}$, $\widetilde{W}\widetilde{W}$ and $\widetilde{W}\widetilde{Z}$, lead to mono- and multi-jet events plus $\not{p}_T$.[9] The $\widetilde{W}\widetilde{W}$ mode can give acollinear dilepton pairs (ee, $\mu\mu$ and eμ), and $\widetilde{W}\widetilde{Z}$ can lead to trileptons (eee, e$\mu\mu$, eeμ and $\mu\mu\mu$) and same or opposite sign dileptons if one of the above threesomes is lost or mis-identified. The size of these signals depends, in part, on the experimental cuts and selection criteria used. In our calculations, we don't try to match any particular year's cuts used by, say, UA1, but instead show that these signals can be observable if some set of typical cuts are used. We require

$$|p_{T_e}| \geqslant 10 \text{ GeV} ; \qquad\qquad |\eta_e| \leqslant 3.0$$

$$|p_{T_\mu}| \geqslant 3 \text{ GeV} ; \qquad\qquad |\eta_\mu| \leqslant 2.0 \tag{5}$$

$$|p_{T_{j_1}}| \geqslant 25 \text{ GeV} ; \qquad |p_{T_{j_2}}| \geqslant 12 \text{ GeV} ; \qquad |\eta_j| \leqslant 2.5 ,$$

and coalesce partons within $\Delta r = (\Delta\eta^2 + \Delta\phi^2)^{1/2} < 1$. More stringent electron cuts are used if final state quarks are present. We trigger on events with jets only when

$$\not{p}_T > \text{Max}(15, 4\sigma) , \quad \sigma = .7 \text{ GeV}^{1/2} \Big(\sum_{\text{partons}} E_T + E_T^S \Big)^{1/2} \tag{6}$$

where E_T^S is distributed according to[10]

$$\frac{dN}{dE_T} \sim \frac{4E_T}{\langle E_T \rangle^2} \exp\left(-\frac{2E_T}{\langle E_T \rangle}\right) \tag{7}$$

with $\langle E_T \rangle = 45$ GeV.

3. RESULTS

In Fig. 2, we plot the monojet and dijet plus $\not{p}_T$ cross-sections expected when both gauginos $\widetilde{W}$ and $\widetilde{Z}$ decay hadronically, as a function of $m_{\widetilde{W}}$ ($m_{\widetilde{Z}}$) on the lower (upper) scale.

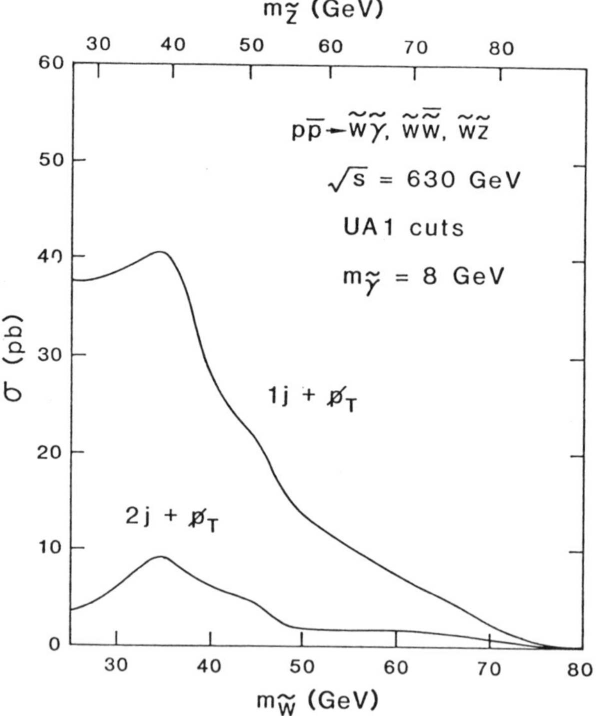

Fig. 2   Monojets and dijets + $\not{p}_T$ signals from gauginos.

We find monojet dominance for all values of wino mass, with a fairly constant monojet signal of $\sim$39 pb for $m_{\widetilde{W}} <$ 38 GeV. This is in contrast to the case of heavy (> 40 GeV) squark or gluino production where dijets dominate.[1,3] At this point $W \to \tilde{W}\tilde{Z}$ phase space is saturated and the monojet signal drops to ~20 pb at $m_{\tilde{W}}$ = 45 GeV, where the decay $Z \to \tilde{W}\tilde{W}$ cuts off. The signal for $m_{\tilde{W}} \gtrsim 46$ GeV comes almost entirely from $W \to \tilde{W}\tilde{\gamma}$; in this region it would be difficult to tell $W \to \tilde{W}\tilde{\gamma}$ from $W \to L\nu_L$, where L is a new fourth generation heavy lepton.

Fig. 3 illustrates signals expected when both gauginos decay leptonically.

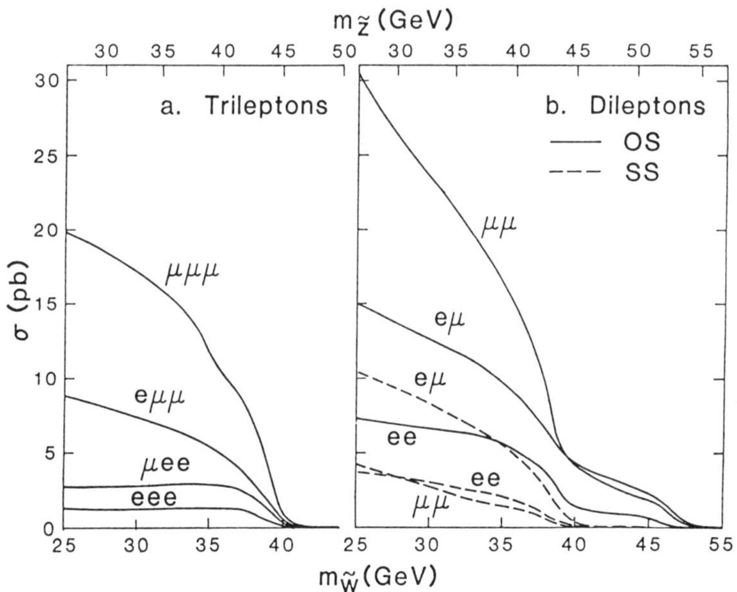

Fig. 3 Signals from $\tilde{W}\tilde{W}$ and $\tilde{W}\tilde{Z}$ production when both $\tilde{W}$ and $\tilde{Z}$ decay leptonically. $m_{\tilde{\gamma}}$ = 8 GeV.

In this case, we require no $\not{p}_T$ cut because the hard multi-leptons plus $\not{p}_T$ in association with only small amounts of hadronic debris (from W and Z production) are easily distinguishable from standard model backgrounds.[11] Isolated hard tri-leptons events (accompanied by $\not{p}_T$)

are a clear signal for $\widetilde{W}\widetilde{Z}$ production, and occur with substantial rates
if the decay $W \to \widetilde{W}\widetilde{Z}$ is allowed. These would be accompanied by same
and opposite sign dileptons plus $\not{p}_T$, where the azimuthal opening angle
$\phi_{\ell\ell}$ would be distributed between $10° < \phi_{\ell\ell} < 170°$. Non-observation of
these multi-lepton events could be used to easily rule out the decay
$W \to \widetilde{W}\widetilde{Z}$ in the theoretical framework in which we work. For
38 GeV $< m_{\widetilde{W}} <$ 46 GeV, one could still expect a few acollinear eμ and

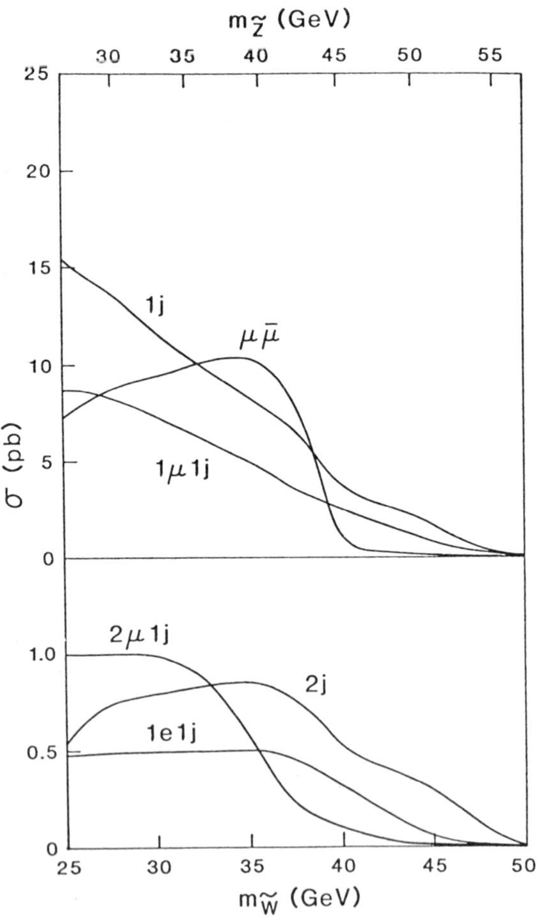

Fig. 4. Largest n jet + m lepton + $\not{p}_T$ signals from p$\bar{\text{p}}$ $\to$ $\widetilde{W}\widetilde{Z}$, $\widetilde{W}\widetilde{W}$
when one gaugino decays leptonically and one
hadronically, and $\sqrt{s} = 630$ GeV. Other signals are
also present but at much smaller levels.

$\mu\mu$ pairs from $Z \rightarrow \widetilde{W}\widetilde{\overline{W}}$, not yet enough to set rigorous (once detection efficiencies are taken into account) limits from the present accumulated $p\bar{p}$ collider data.

Finally, a set of n jet + m lepton + $\not{p}_T$ events would be expected in the data sample from the case where one gaugino decays leptonically, and one hadronically (see Fig. 4). Such events would have background problems from heavy flavor production ($t\bar{t}$, $t\bar{b}$,...).

4. CONCLUSION

We have performed detailed calculations of gaugino signals expected from N = 1 supergravity when sfermions are heavy $\left(\sim O(100 \text{ GeV})\right)$ and $\tilde{\gamma}$ is light ($m_{\tilde{\gamma}} < 10$ GeV), when $r = v_1/v_2 = 1$. Substantial monojets, trileptons, and acollinear dileptons are expected if $W \rightarrow \widetilde{W}\widetilde{Z}$ is allowed, and it is likely current collider experiments are probing at least $m_{\widetilde{W}} = 38$ GeV and $m_{\widetilde{Z}} = 43$ GeV. If $m_{\widetilde{W}} >$ 46 GeV, then the best channel for searching for gauginos is in $W \rightarrow \widetilde{W}\widetilde{\gamma}$ decays, where fat monojet + $\not{p}_T$ events offer the cleanest signal.

REFERENCES

1. See Ellis, J., Proceedings of this Institute.
2. See e.g. Komamiya, S., Proceedings of the International Symposium on Lepton and Photon Interactions at High Energies, Kyoto (1985).
3. Reya, E. and Roy, D. P., Phys. Lett. <u>166B</u>, 223 (1986); Barnett, R. M. Haber, H. E. and Kane, G., Nucl. Phys. <u>B267</u>, 625 (1986) and references therein.
4. Arnowitt, R., Chamseddine, A. and Nath, P., Phys. Rev. Lett. <u>49</u>, 970 (1982); Weinberg, S., Phys. Rev. Lett. <u>50</u>, 387 (1983); Chamseddine, Arnowitt, R., and Nath, P., Phys. Rev. Lett. <u>50</u>, 232 (1983).
5. See, e.g. Nilles, H. P., Phys. Rep. <u>110</u>, 1 (1984) and references therein.
6. Baer, H. Hagiwara, K., and Tata, X., ANL-HEP-PR-86-07 plus work in progress.

7. Nath, P., Arnowitt, R., and Chamseddine, A., Phys. Lett. <u>129B</u>, 445 (1983); Dicus, D., Nandi, S. and Tata, X., Phys. Lett. <u>129B</u>, 451 (1983); Barger, V., <u>et al</u>., Phys. Lett. <u>131B</u>, 372 (1983); Fayet, P., Phys. Lett. <u>133B</u>, 363 (1983); Dicus, D., <u>et al</u>., Phys. Rev. <u>D29</u>, 67 (1984); Altarelli G., <u>et al</u>., Nucl. Phys. <u>B245</u>, 215 (1984); Gottlieb S., and Weiler, T., Phys. Rev. <u>D32</u>, 1119 (1985).

8. Hagiwara, K. and Zeppenfeld, D., Nucl. Phys. B (in press).

9. Chamseddine, A., Nath, P., and Arnowitt, R., Northeastern preprint NUB 2681 (1985).

10. Halzen, F., <u>et al</u>., Z. Phys. <u>C14</u>, 351 (1982).

11. Baer, H., Ellis, J., Nanopoulos, D. V. and Tata, X., Phys. Lett. <u>153B</u>, 265 (1985).

SOME E_6 PHENOMENOLOGY: W PAIR PRODUCTION IN $e^+ e^-$ COLLISIONS

Pat Kalyniak and M.K. Sundaresan
Ottawa-Carleton Institute for Physics
Ottawa, Canada
K1S 5B6

ABSTRACT

We consider W pair production in $e^+ e^-$ collisions in the context of a rank 5 model of Wilson loop broken E_6. This process is sensitive to the existence of extra neutral gauge bosons, particularly to the mixing of an extra boson with the ordinary Z°.

The exceptional group E_6 has been motivated by superstring theories as a possible grand unified gauge group.[1] Dine et al. have given the subgroups of E_6, extensions of the standard model, which are candidates for the low energy gauge symmetry.[2] Common to all of the candidates and of probable importance to low energy phenomenology are two features: the existence of extra gauge bosons and, also, of extra neutral colour singlet fermions contained within the fundamental 27 supermultiplet. We note that both of these features can modify the standard model results for the process of W pair production in $e^+ e^-$ collisions. This particular process is a sensitive test of the gauge structure of the standard model via the delicate cancellations of the s-channel photon and Z° exchanges and the t-channel neutrino exchange shown in Fig. 1. As indicated in the Figure, extra "Z's" and neutral leptons, "N's", can contribute to the process. We investigate this possibility.

Of the many possible extended gauge groups, we focus on one rank 5 model. This is

$$G = SU_3^c \times SU_2^L \times U_1^L \times U_1^{\prime} \tag{1}$$

where the electromagnetic charge is given in terms of the U_1 hypercharges as

$$Q_{em} = I_3^L + \frac{Y^L}{2} + Y'$$ (2)

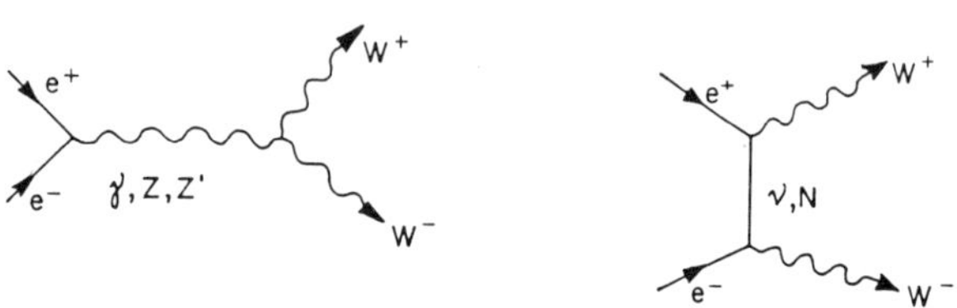

FIG. 1.   The s-channel neutral gauge boson and t-channel neutral lepton contributions to W pair production.

This assumes that E_6 has been broken via the Wilson loop breaking mechanism wherein the gauge fields transform nontrivially about some noncontractible paths in the six-dimensional multiply-connected manifold.[3] This is equivalent to an effective adjoint $\underline{78}$ of Higgs, Φ, acquiring vacuum expectation values (vevs). If only components of Φ corresponding to the Cartan subalgebra pick up vevs, E_6 breaks to a group of rank 6. We have previously[4] discussed an example of this Abelian Wilson loop breaking with the pattern

$$E_6 \longrightarrow SU_3^C \times SU_2^L \times U_1^L \times SU_2^N \times U_1^N$$ (3)

where the electromagnetic charge is given by

$$Q_{em} = I_3^L + I_3^N + \frac{Y^L + Y^N}{2} \tag{4}$$

Further breaking to a lower rank group occurs if a member of Φ not in the Cartan subalgebra acquires a vev. There are two SU_2^N doublets with $Y^N = \pm 1$, respectively, in the $(1,1,8)$ of the $\underline{78}$ which can break $SU_2^N \times U_1^N$ to U_1' where

$$Y' = I_3^N + \frac{Y^N}{2} \tag{5}$$

Here the notation $(1,1,8)$ corresponds to the $SU_3^C \times SU_3^L \times SU_3^N$ maximal subgroup of E_6 and the standard model is, consequently, left unbroken. There are two possibilities for the further breaking of G to $SU_3^C \times U_1^{em}$. The particle content of the superstring motivated E_6 models consists of $(N + \delta)\underline{27} + \delta\underline{\overline{27}}$ representations, where N and δ are model dependent. We list in Table 1 the fermion content of the $\underline{27}$ of E_6. Higgs representations for the breaking of G can be supplied from within the N generation supermultiplets. For example, two of the SU_2^L doublets from a generation $\underline{27}$ have the hypercharges as in a supersymmetric standard model. We denote these as H_g^1 and H_g^2 which correspond, respectively, to the fermion representations $(\nu_e\ E^-)$ and $(E^+\ \overline{N}_E)$ in Table 1. Higgs content could also be supplied from the $\delta(\underline{27} + \underline{\overline{27}})$ representations. Denoting these by ϕ, there is a possible mass term of the form $\overline{\phi}<\Phi>_0\phi$. Since the Wilson loop breaking occurs at the Planck scale, such ϕ masses are expected to be order the Planck mass. However, the mass of some components of ϕ can actually vanish for particular choices of the vevs of the effective $\underline{78}$ Higgs. This is the very attractive scheme of naturally massless Higgs content which can be used to break G at a low scale.[5]

Here we choose our Higgs content to include H_g^1 and H_g^2 from a generation $\underline{27}$ with vevs v_1 and v_2, respectively, and, in order to obtain improved phenomenology in terms of fermion masses, also a singlet Higgs H° with vev v_3 from ϕ. H° corresponds to N in Table 1.

We parametrize the three neutral gauge boson mass eigenstates as follows.

TABLE 1. The Fermion Content of a $\underline{27}$ Representation.

	I_3^L	Y^L	Y'	$Y_W = Y^L + 2Y'$
u	1/2	1/3	0	1/3
d	-1/2	1/3	0	1/3
$\bar{u}$	0	0	-2/3	-4/3
$\bar{d}$	0	0	1/3	2/3
h	0	-2/3	0	-2/3
$\bar{h}$	0	0	1/3	2/3
ν_e	1/2	-1/3	-1/3	-1
e^-	-1/2	-1/3	-1/3	-1
e^+	0	2/3	2/3	2
$\bar{n}$	0	2/3	-1/3	0
N	0	2/3	-1/3	0
ν_E	1/2	-1/3	-1/3	-1
E^-	-1/2	-1/3	-1/3	-1
E^+	1/2	-1/3	2/3	1
$\bar{N}_E$	-1/2	-1/3	2/3	1

$$A_\mu = sA_\mu^3 + cc'B_\mu + cs'C_\mu \qquad (6a)$$

$$Z_\mu = cc''A_\mu^3 - (sc'c'' + s's'')B_\mu + (c's'' - ss'c'')C_\mu \qquad (6b)$$

$$Z'_\mu = -s''cA_\mu^3 + (sc's'' - c''s')B_\mu + (ss's'' + c'c'')C_\mu \qquad (6c)$$

where

$$s = \sin\theta_W \qquad (7a)$$

$$s' = \sin\theta' \qquad (7b)$$

$$s'' = \sin\theta'' \qquad (7c)$$

In the above A_μ^3 is the neutral SU_2^L gauge eigenstate while B_μ and C_μ are the U_1^L and U_1' gauge bosons, respectively. A_μ is the massless photon. For $\theta'' = 0$, Z couples like the standard model Z° so the angle θ'' probes the mixing of the standard model Z° and the extra neutral massive gauge boson. In terms of the SU_2^L coupling, g, the U_1^L coupling g_1 and the U_1' coupling g_2, we have

$$\tan\theta_W = \frac{2g_1/g}{\sqrt{(1 + (2g_1/g_2)^2)}} \qquad (8a)$$

$$\tan\theta' = t' = 2g_1/g_2 \qquad (8b)$$

If the breaking of E_6 to G occurs at one scale then the U_1 couplings g_1 and g_2 are equal and eqns. (8a,b) reduce accordingly.

In terms of the mass eigenstates, the triple gauge boson coupling of relevance to W pair production are given in Fig. 2. In that Fig.,

$$N_{\mu\nu\lambda} = g_{\mu\nu}(k_1-k_2)_\lambda + g_{\mu\lambda}(k_3-k_1)_\nu + g_{\nu\lambda}(k_2-k_3)_\mu \qquad (9)$$

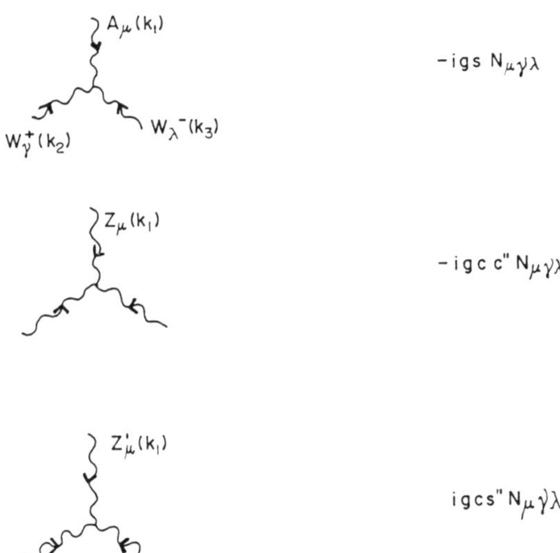

FIG. 2. Triple gauge boson Feynman rules.

The extra neutral boson Z' couples to W pairs providing there is some mixing with the standard model $Z°$.

We introduce the following abbreviated notation for the fermionic sector.

$$A = (u\ d) \qquad B = (\nu_e\ e^-) \qquad C = \bar{d} \qquad D = e^+ \qquad E = \bar{u}$$

$$F = N \qquad G = \bar{h} \qquad H = h \qquad J = \bar{n} \tag{10}$$

The E_6 allowed Yukawa terms are

$$<H^\circ>_o \{aCH\} \tag{11a}$$

$$<H_g^1>_o \{bBD + cAC\} \tag{11b}$$

$$<H_g^2>_o \{dAE + eBF\} \tag{11c}$$

The electron and u-quark get masses from the doublets H_g^1 and H_g^2, respectively. The d- and h-quarks mix. The neutral states ν_e and N mix while $\bar{n}$ remains massless. In the basis $\psi_j^\circ = (\nu_e, N)$, the neutrino mass terms are

$$-\frac{1}{2}(\psi^\circ)^T (M)\psi^\circ + h.c. \tag{12}$$

where

$$M = ev_2 \begin{pmatrix} 0 & 1 \\ 1 & 0 \end{pmatrix} \tag{13}$$

Defining two-component mass eigenstates χ such that

$$\hat{\chi}_i = N_{ij}\psi_j^\circ \tag{14}$$

this becomes

$$-\frac{1}{2}(\chi)^T (N^* M N^\dagger)\chi + h.c. \tag{15}$$

where $N^* M N^\dagger = N_D$ is diagonal. The diagonalizing matrix N, with phase chosen to yield positive masses is

$$N = \frac{1}{\sqrt{2}} \begin{pmatrix} 1 & 1 \\ -i & i \end{pmatrix} \tag{16}$$

so that the mass eigenstates are $\chi_1 = \frac{1}{\sqrt{2}}(\nu_e\ N)$ and $\chi_2 = \frac{i}{\sqrt{2}}(-\nu_e\ N)$. Constructing four-component Majorana spinors $X = (\chi\ \bar{\chi}) = (\chi\ i\sigma^2\chi^*)$ all the relevant couplings of leptons to gauge bosons can now be obtained. These are given in Fig. 3. In that Figure,

$$\alpha = -1 + 4s^2 + \gamma_5 \tag{17}$$

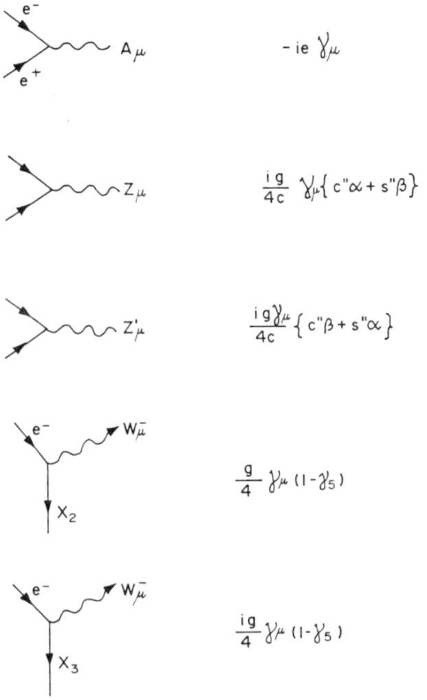

FIG. 3. Lepton-gauge boson Feynman rules.

and

$$\beta = s\left(\frac{t'^2-2}{t'}\right)\left(1 + \frac{1}{3}\gamma_5\right) \qquad (18)$$

The W pair production process occurs via the s-channel photon, Z, and Z' contributions and the X_1 and X_2 t-channel contributions. Neglecting the neutral lepton masses, the sum of the two t-channel contributions reduces to that of the single standard model Dirac neutrino. Following Alles, Boyer, and Buras[6] the cross section for W pair production at center of mass energy $\sqrt{s}$ can be expressed as the sum of the contributions of the individual diagrams and their interferences as follows.

$$\sigma(e^+e^- \longrightarrow W^+W^-) = \frac{\pi\alpha^2\beta}{8s\sin^4\theta_W} \sum_{ij} \bar{\sigma}_{ij} = \sum_{ij} \sigma_{ij} \qquad (19)$$

where i,j = γ,Z,Z',ν and $\beta = \sqrt{1-4M_W^2/s}$. The nonstandard contributions
of the Z' are suppressed by factors of sinθ". Durkin and Langacker[7]
have used low energy phenomenology to investigate the extra Z mass,
$M_{Z'}$, and its mixing with the standard model Z°, θ". Essentially, for
a model similar to ours they constrain $M_{Z'}$ to be greater than about
120 GeV and the mixing angle θ" to range between about -0.1 and +0.1.
We illustrate, in Fig. 4, the various standard model contributions
σ_{ij}, i,j = γ,Z,ν as a function of $\sqrt{s}$.

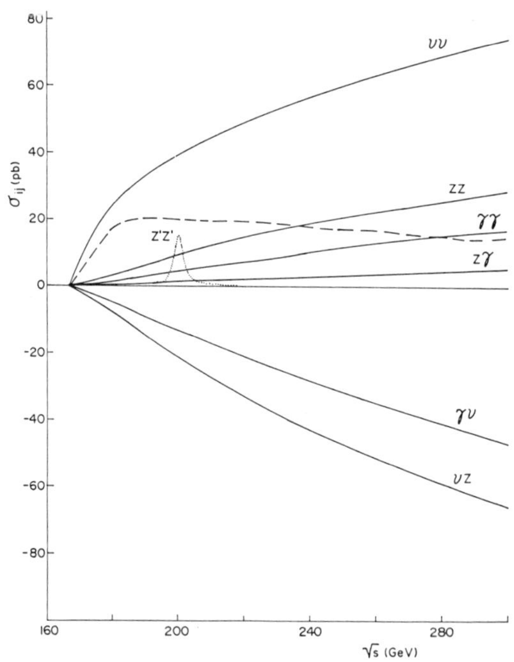

FIG. 4. Contributions to the W pair production cross section.

In Fig. 4, the dashed line is the sum of the standard model
contributions, that is, the total standard model cross section. The
dotted curve is the Z' contribution, $\sigma_{Z'Z'}$, for the case of a 200 GeV
mass Z' of width 3 GeV and a mixing θ" = -0.05 with the standard Z°.
The other nonstandard contributions are not shown since they are
small and cancel approximately. In Fig. 5, we illustrate the W pair

cross section as a function of $\sqrt{s}$ for three cases.

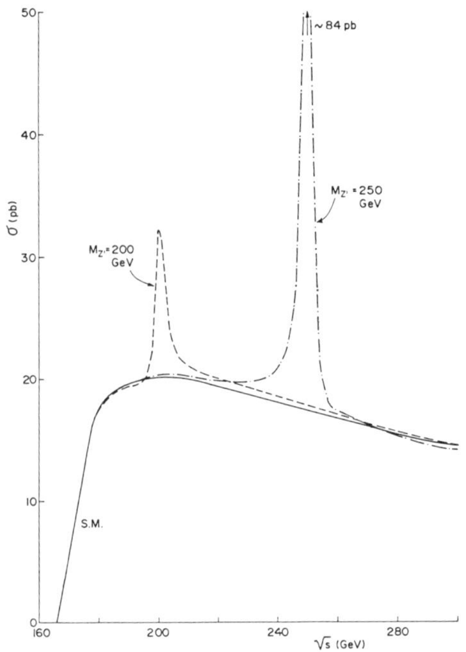

FIG. 5. The total cross section for W pair production.

In that Figure, the solid curve is the standard model result. The
dashed curve is the case of a 200 GeV mass Z' with $\theta'' = -0.05$. The
dash-dot curve is for a 250 GeV mass Z' with $\theta'' = +0.05$. A Z' width
of 3 GeV is used in each of the latter two cases. The height of the
Z' peak in the cross section is very sensitive to the angle θ''. For
instance, for the 200 GeV mass Z', the cross section peaks at 35 pb
for $\theta'' = -0.05$, at 78 pb for $\theta'' = -0.1$ and at 253 pb for $\theta'' = -0.2$.
The width dependence of the peak height can be extracted from μ pair
production, for instance, which is rather insensitive to θ''.

Finally, in Fig. 6, we illustrate the distribution in the W
scattering angle for the standard model and the case of a Z' with
mass of 200 GeV and mixing angle $\theta'' = -0.05$. The dominant extra Z'
s-channel contribution modifies the strong forward-backward peaking

effect of the neutrino t-channel contribution.

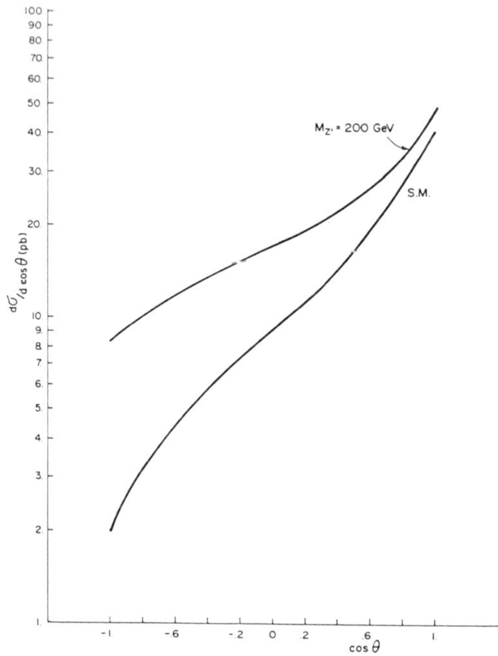

FIG. 6. The differential distribution in W scattering angle
for W pair production.

In conclusion, we find the process of W pair production to yield
interesting information on the existence of extra neutral gauge bosons.
In particular, this process is very sensitive to the mixing of such
possible extra gauge bosons with the standard model Z° and, thus,
would do well to be investigated if there were some preliminary hint
of the existence of extra Z's in, say, forward-backward asymmetries
in μ pair production. We have here considered W pair production via
extra Z bosons in e^+e^- collisions. We note that the process would
also be very interesting in this regard in hadronic collisions.

ACKNOWLEDGEMENTS

It is a pleasure to thank the organizers of the Lake Louise
Winter Institute and to acknowledge a useful conversation with Bruce
Campbell and John Ellis. We gratefully acknowledge the support of the
Natural Sciences and Engineering Council of Canada.

REFERENCES

1) B.Candelas et al., Nucl. Phys. B258, 46 (1985);
   E.Witten, Nucl. Phys. B258, 75 (1985).

2) M.Dine et al., Nucl. Phys. B259, 549 (1985).

3) A.Strominger and E.Witten, "Vacuum Configuration for
   Superstrings", NSF-TTP-84-170 preprint (1984).

4) P.Kalyniak and M.K.Sundaresan, Phys. Lett. 167B, 320 (1986).

5) J.D.Breit, B.A.Ovrut, and G.C.Segrè, Phys. Lett. 158B, 33
   (1985); A.Sen, Phys. Rev. Lett. 55, 33 (1985).

6) W.Alles, Ch.Boyer, and A.J.Buras, Nucl. Phys. B119, 125
   (1977).

7) L.S.Durkin and P.Langacker, University of Pennsylvania
   report UPR-0286-T (1985).

ASYMMETRIES IN e^+e^- COLLISIONS FROM E_6 GRAND UNIFIED THEORIES[*]

Stephen Godfrey and G. Bélanger

TRIUMF, 4004 Wesbrook Mall, Vancouver, B.C., Canada V6T 2A3

ABSTRACT

Motivated by the resurgence of interest in E_6 GUTs due to
their possible relevance as the low energy limit of superstring theories we studied the effects of the extra Z^0
bosons predicted by these theories on asymmetries in e^+e^-
collisions. We found that deviations from the standard
model predictions, consistent with low energy constraints on
the properties of the extra Z^0 bosons, can be quite substantial. In this paper we describe a method of unravelling the
underlying physics from experimental measurement, thereby
illustrating how asymmetries at e^+e^- colliders offer a
sensitive probe for new physics.

1. INTRODUCTION

Recently, there has been a resurgence of interest in E_6 grand
unified theories motivated by the possibility that they represent the
low energy limit of superstring theories.[1] Because in superstring
theories, E_6 is broken by a discrete symmetry to a rank 5 or 6 group,
one feature, common to all low energy limits of this theory, is the
prediction of at least one extra neutral gauge boson with the properties of the extra Z^0 bosons and their couplings to fermions dependent
on the details of the symmetry breaking.[2,3] In this context, several
groups have examined the constraints that low energy neutral current
data can put on the properties of the extra Z^0 bosons.[4,5] Beyond
this, one may ask, how might these extra Z^0's, which we will refer
to as Z''s, manifest themselves in the next generation of experiments
and how might we unravel their properties. In order to answer these

---

[*]Presented by S. Godfrey

questions we have studied the effects of the extra Z^0's of E_6 in e^+e^- collisions.[6,7] Specifically, we have examined the forward-backward and left-right asymmetries for representative allowed values of the Z' mass and Z^0-Z' mixing and compared them to the asymmetries expected from the standard model.

2. GROUP THEORETICAL CONSIDERATIONS

Since E_6 is a rank 6 group it has six diagonal generators in its Cartan subalgebra corresponding to the axes of the 6-dimensional root space.[3] At low energy E_6 must contain, at the least, the $SU(3)_C \times SU(2)_L \times U(1)$ symmetry of the standard model so two of the diagonal generators are accounted for by the diagonal generators of $SU(3)_C$ and two other directions in the group space correspond to the photon and the Z^0 gauge boson. This leaves two remaining directions of E_6 group space available for new physics and hence, extra Z^0 bosons. In principle, the Z' bosons can mix with the standard model Z^0 with the mixing angles constrained by experiment. We postpone a discussion of mixing effects for now and assume that the Z^0 couples as in the standard model to $I_{3L} - Q_{em} \sin^2\theta_w$. Since the charges of the extra Z^0 bosons, Q' and Q'', must be orthogonal to all generators of the standard model we can label the extra Z^0 charges using $U(1)_\chi$ and $U(1)_\psi$ which come from the subgroup chain:

$$E_6 \to SO(10) \times U(1)_\psi \to SU(5) \times U(1)_\chi \times U(1)_\psi \ . \tag{1}$$

where the generators of the standard model lie in the $SU(5)$ group. Therefore, the charges Q' and Q'' will be given by a linear combination of the generators $U(1)_\chi$ and $U(1)_\psi$ with respective charges Q_χ and Q_ψ giving:

$$Q' = Q_\chi \cos\theta_{E_6} + Q_\psi \sin\theta_{E_6} , \tag{2a}$$

$$Q'' = -Q_\chi \sin\theta_{E_6} + Q_\psi \cos\theta_{E_6} . \tag{2b}$$

Specific choices of θ_{E_6} correspond to different symmetry breaking patterns which can be related to different subgroup chains. As an example consider the case where E_6 is broken via a nonAbelian discrete symmetry to a rank 5 group

$$E_6 \to SU(3)_C \times SU(2)_L \times U(1)_L \times U(1) \ .$$

This is sometimes referred to as the minimal superstring model. Properly normalized, the charge operator of the Z' is given by

$$Q_\eta = -\sqrt{\frac{5}{8}}\, Q_\psi + \sqrt{\frac{3}{8}}\, Q_\chi$$

so in this model we find $\cos\theta_{E_6} = \sqrt{3/8}$ and $\sin\theta_{E_6} = -\sqrt{5/8}$.

Once the direction of the Z' generator in E_6 group space has been specified, the quantum numbers of all fermions in the fundamental representation of E_6 are uniquely determined. We list the couplings of the conventional fermions in Table 1. In addition to the 15 conventional fermions of each generation there are 12 new fermions including the right-handed neutrino, a new charge $-1/3$ quark, an iso-doublet of charged and neutral lepton, E^- and ν_E, and an extra neutrino. We do not consider these extra fermions further.

Table 1.

Quantum numbers and couplings of the conventional fermions in E_6

Fermion	Q_{cm}	$Q_\chi(f)$	$Q_\chi(f^c)$	$Q_\psi(f)$	$Q_\psi(f^c)$	C'_V	C'_A
e	-1	$\dfrac{3}{2\sqrt{10}}$	$\dfrac{-1}{2\sqrt{10}}$	$\dfrac{1}{\sqrt{24}}$	$\dfrac{1}{\sqrt{24}}$	$\dfrac{1}{\sqrt{10}}\cos\theta_{E_6}$	$\dfrac{1}{2\sqrt{10}}\cos\theta_{E_6} + \dfrac{1}{2\sqrt{6}}\sin\theta_{E_6}$
ν	0	$\dfrac{3}{2\sqrt{10}}$	$\dfrac{-5}{2\sqrt{10}}$	$\dfrac{1}{\sqrt{24}}$	$\dfrac{1}{\sqrt{24}}$	$\dfrac{2}{\sqrt{10}}\cos\theta_{E_6}$	$\dfrac{-1}{2\sqrt{10}}\cos\theta_{E_6} + \dfrac{1}{2\sqrt{6}}\sin\theta_{E_6}$
u	$\dfrac{+2}{3}$	$\dfrac{-1}{2\sqrt{10}}$	$\dfrac{-1}{2\sqrt{10}}$	$\dfrac{1}{\sqrt{24}}$	$\dfrac{1}{\sqrt{24}}$	0	$\dfrac{-1}{2\sqrt{10}}\cos\theta_{E_6} + \dfrac{1}{2\sqrt{6}}\sin\theta_{E_6}$
d	$\dfrac{-1}{3}$	$\dfrac{-1}{2\sqrt{10}}$	$\dfrac{3}{2\sqrt{10}}$	$\dfrac{1}{\sqrt{24}}$	$\dfrac{1}{\sqrt{24}}$	$\dfrac{-1}{\sqrt{10}}\cos\theta_{E_6}$	$\dfrac{1}{2\sqrt{10}}\cos\theta_{E_6} + \dfrac{1}{2\sqrt{6}}\sin\theta_{E_6}$

The neutral current Lagrangian is now written:

$$\mathcal{L}_{NC} = eA_\mu J^\mu_{em} + g_{Z^0} Z^0_\mu J^\mu_{Z^0} + g_{Z'} Z'_\mu J^\mu_{Z'}$$

where J^μ_{em} and $J^\mu_{Z^0}$ are as in the standard model and

$$J^\mu_{Z'} = \sum_f \bar{\psi}_f \gamma^\mu (C'_V - C'_A \gamma_5)\psi_f$$

with $C'_{V,A} = 1/2[Q'_f \mp Q'_{f^c}]$. We have only considered the minimal case of one extra neutral gauge boson, assuming that in the case of rank 6 groups where two extra neutral bosons are present, one of them is sufficiently heavier than the other so that its effects are suppressed for the energies being considered. There are therefore four phenomenological parameters in the analysis: θ_{E_6} – the Q_χ-Q_ψ mixing angle, ϕ – the Z^0-Z' mixing angle, M_Z' – the mass of the extra Z^0 boson, and $(g_{Z'}/g_{Z0})$ – the ratio of the Z' and Z^0 coupling constants. The ratio, $(g_{Z'}/g_{Z0})$, can be found in a specific symmetry breaking scheme using the renormalization group equations and is given by $(g_{Z'}/g_{Z0})^2 \leqslant \frac{5}{3} \sin^2\theta_w$.[5] It takes on its maximum value , which we use in what follows, when the extra $U(1)$ breaks off at the same scale as the breaking to $SU(3)_c \times SU(2)_L \times U(1)$. The angle ϕ is related to the standard model prediction for the mass of Z^0 (M_{SM}), the observed M_{Z0}, and M_Z'

$$\tan^2\phi = \frac{M_{SM}^2 - M_{Z0}^2}{M_{Z'}^2 - M_{SM}^2}$$

where $M_{SM}=M_W/\cos\theta_w$. With an accurate measurement of the physical Z^0 mass ϕ can be related to M_Z' and can therefore be removed as a free parameter. This leaves θ_{E_6} and M_Z' to be determined.

3. ASYMMETRIES FROM EXTRA Z^0's

Our goal is to study the effects of extra Z^0's on asymmetries in e^+e^- colliders and to unravel the underlying physics. With this in mind we will examine forward-backward asymmetries and left-right asymmetries in $e^+e^- \rightarrow \mu^+\mu^-$. We have also studied asymmetries which can arise in Bhabha scattering but do not include these results here, referring the reader to Ref. 6 for further details. With efficient flavour tagging of heavy quarks in jets one could also repeat the exercise for $e^+e^- \rightarrow q\bar{q}$.

For the asymmetries in $e^+e^- \rightarrow \mu^+\mu^-$ the processes which contribute to the cross sections are given in Fig. 1. Evaluating these amplitudes one obtains for the cross section of an unpolarized positron and a left-handed polarized electron:

$$\frac{d\sigma_L}{d\cos\theta} = \frac{\pi\alpha^2}{2s}\left\{ |b_{LL}|^2(1+\cos\theta)^2 + |b_{LR}|^2(1-\cos\theta)^2 \right\}$$

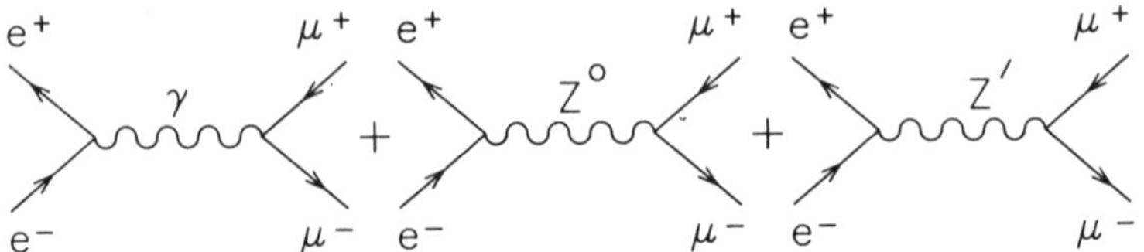

Fig. 1. Processes which contribute to the $e^+e^- \to \mu^+\mu^-$ cross section.

where

$$b_{ij} = 1 + \frac{C_i^e \, C_j^\mu}{4\cos^2\theta_w \sin^2\theta_w} \; \frac{s}{s - M_{Z0}^2 + i\Gamma_{Z0}M_{Z0}}$$

$$+ \frac{(g_{Z'}/g_{Z0})^2 \, C_i^{e'} C_j^{\mu'}}{4\cos^2\theta_w \sin^2\theta_w} \times \frac{s}{s - M_{Z'}^2 + i\Gamma_{Z'}M_{Z'}}$$

and i,j = L,R. In these expressions $C_{L,R} = C_V \pm C_A$. For the cross section with a right-handed electron one makes the replacement L $\leftrightarrow$ R. In our calculations we use the values $\sin^2\theta_w$ = 0.222, M_{Z0} = 92.5 GeV, and Γ_{Z0} = 2.5 GeV. The value of $\sin^2\theta_w$ was obtained by Barger et al.[4] from a fit to the low energy neutral current data and includes radiative corrections. Durkin and Langacker[5] found that $\sin^2\theta_w$ was not very sensitive to the model used to fit the low energy data (i.e. the value for θ_{E_6}) even when the Z^0-Z' mixing effects were included. The Z' width is given by:

$$\Gamma_{Z'} = \frac{g_{Z'}^2 M_{Z'}}{12\pi} \sum_f \left(c_V^{f'2} + c_A^{f'2} \right)$$

To obtain a value for $\Gamma_{Z'}$ we included only the fermions of the first 3 generations of the standard model. If the extra fermions of E_6 were discovered their contribution to the width could be included. In any case our results are rather insensitive to the exact value of the Z' width.

We have not included detector dependent radiative corrections such as bremsstrahlung off the initial or final legs and virtual photon graphs in our calculations. This is because the corrections from the soft bremsstrahlung and virtual photon diagrams depend on the detector resolution ΔE and those from the hard bremsstrahlung graphs depend crucially on the experimental acceptance cuts and can be large, up to 30% of the tree level contributions. These are best included by experimentalists via Monte Carlo simulations involving the details of their detectors.[8]

3.1 Left-right Asymmetry

The left-right asymmetry A_{LR} is defined by

$$A_{LR} \equiv \frac{\sigma_L - \sigma_R}{\sigma_L + \sigma_R}$$

In what follows we will display our results as the difference between the asymmetry with an extra Z^0 and that of the standard model, i.e.

$$\delta A_{LR} = A_{LR} - A_{LR}(SM) \; .$$

In Fig. 2 we plot δA_{LR} at $\sqrt{s} = 100$ GeV as a function of θ_{E_6}. We have set the Z^0-Z' mixing angle to zero. In Fig. 3 we show δA_{LR} as a function of $\sqrt{s}$ for the specific value of θ_η which arises in some superstring theories and for several values of $M_{Z'}$. Not unexpectedly the larger $\sqrt{s}$ and the lower the Z' mass the larger the deviation from

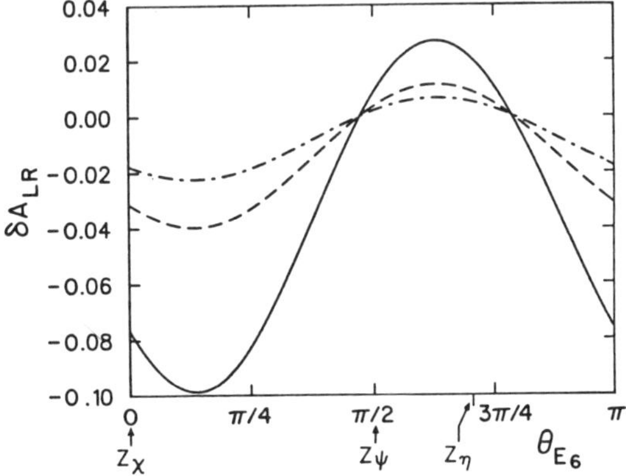

Fig. 2. δA_{LR} as a function of θ_{E_6}. The solid curve is for $M_{Z'}$=150 GeV, the dashed curve for $M_{Z'}$=200 GeV, and the dot-dashed curve for $M_{Z'}$=250 GeV. In all cases $\sqrt{s}$=100 GeV and ϕ=0. We point out the special angles corresponding to Z_χ, Z_ψ, and Z_η.

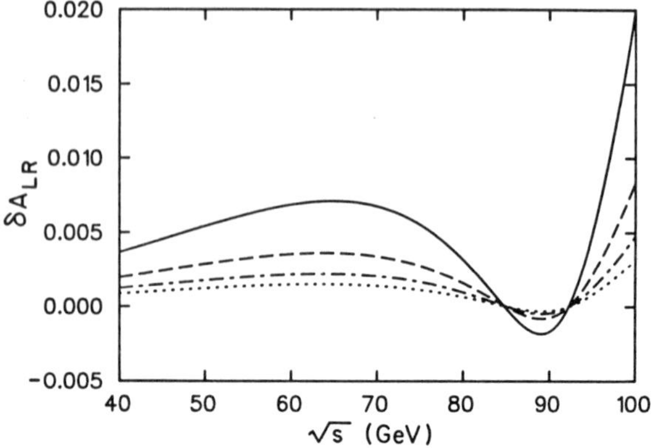

Fig. 3. δA_{LR} as a function of $\sqrt{s}$ for several values of M_Z' with $\theta_{E_6} = \theta_\eta$ and $\phi = 0$. The solid line is for $M_Z' = 150$ GeV, the dashed line for $M_Z' = 200$ GeV, the dash-dotted line for $M_Z' = 250$ GeV and the dotted line for $M_Z' = 300$ GeV.

the standard model. Also, without Z^0-Z' mixing the deviation at the Z^0 pole is negligible which simply reflects the dominance of the Z^0 pole. Much below the Z^0 pole, δA_{LR} is dominated by the γ-Z' interference term and is proportional to $C_V' C_A'$. As $\sqrt{s}$ increases the Z^0-Z' interference takes over and the dependence on C_V and C_A is more complicated. We will expand on this point below.

Finally, in Fig. 4 we consider the effect of Z^0-Z' mixing. We use the allowed bounds on Z^0-Z' mixing and $M_{Z'}$ obtained from the analysis of low energy data by Durkin and Langacker.[5] One sees that for the amount of mixing allowed by low energy phenomenology, the deviations from the standard model in the left-right asymmetry can be quite substantial. In fact, the Z^0-Z' mixing will be the dominant effect on asymmetries. The reason for this is that the Z^0 coupling is modified by the Z', i.e. $C_V \to C_V \cos\phi + C_V' \sin\phi$. The C_V' and C_A' terms are suppressed only by a mixing angle, not by a factor of $1/M_{Z'}^2$, as in the absence of mixing. The effects are therefore particularly significant around the Z^0 pole.

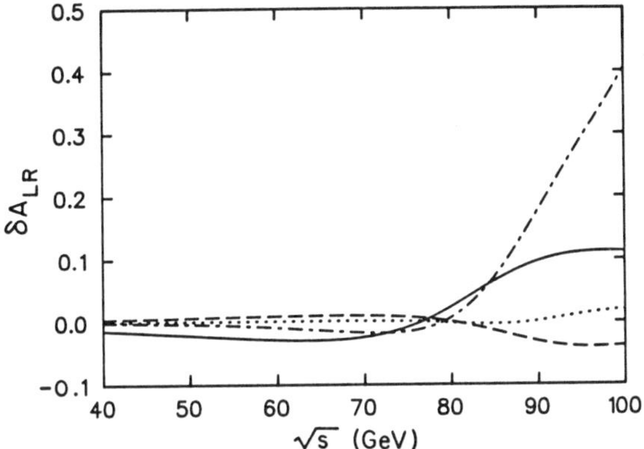

Fig. 4. δA_{LR} versus $\sqrt{s}$ for some representative parameter sets allowed by current data. The solid line is for Z_χ with $M_{Z'}$=260 GeV and ϕ=-0.08, the dot-dashed line is for Z_η with $M_{Z'}$=120 GeV and ϕ=-0.25, the dashed line for Z_η with $M_{Z'}$=200 GeV and ϕ=+0.04, and the dotted line is for Z_ψ with $M_{Z'}$=140 GeV and ϕ=+0.06.

3.2 Front-Back Asymmetry

The front-back asymmetry A_{FB} is defined by:

$$A_{FB} \equiv \frac{\int_0^1 d\cos\theta \, \frac{d\sigma}{d\cos\theta} - \int_{-1}^0 d\cos\theta \, \frac{d\sigma}{d\cos\theta}}{\int_0^1 d\cos\theta \, \frac{d\sigma}{d\cos\theta} + \int_{-1}^0 d\cos\theta \, \frac{d\sigma}{d\cos\theta}}$$

and $\delta A_{FB}=A_{FB}-A_{FB}(SM)$. In Fig. 5 we show δA_{FB} as a function of θ_{E_6} for $\sqrt{s}$ = 100 GeV with left-polarized, right-polarized, and unpolarized electrons, illustrating the effect of polarized beams on front-back asymmetries. In Fig. 6 we show δA_{FB} as a function of $\sqrt{s}$ for $\theta_{E_6} = \theta_\eta$ for several values of $M_{Z'}$. Examining the expression for A_{FB} we find that much below the Z^0 peak δA_{FB} is dominated by the γ-Z' interference term and is proportional to $C_A^2 C_V'^2$. Around the Z^0 peak the Z^0-Z' term dominates and is approximately proportional to $C_A^2 C_V'^2$. Again, in Fig. 7, we show some representative deviations of A_{FB} from the standard model for values of the parameters allowed by low energy phenomenology and once again they are found to be quite substantial.

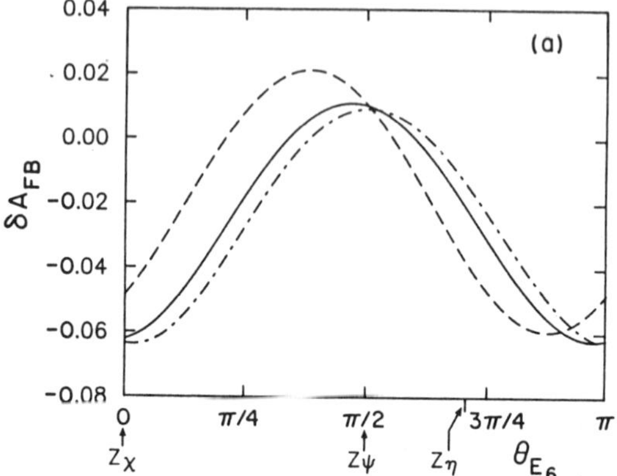

Fig. 5. δA_{FB} as a function of θ_{E_6} for $\sqrt{s}=100$ GeV, $M_{Z'}=150$ GeV and $\phi=0$. The solid line is for unpolarized electrons, the dashed line is for a right polarized beam and the dot dashed line is for a left polarized electron beam.

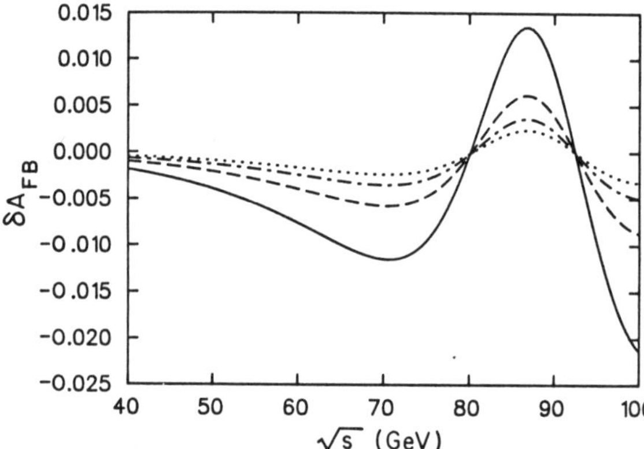

Fig. 6. δA_{FB} versus $\sqrt{s}$ for an unpolarized electron beam, $\theta_{E_6}=\theta_{\eta}$ and $\phi=0$. The labelling is the same as in Fig. 3.

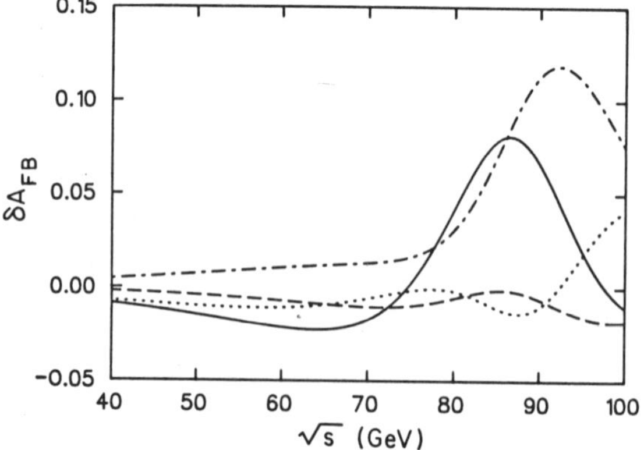

Fig. 7. δA_{FB} versus $\sqrt{s}$ with unpolarized electrons for some representative parameter sets allowed by current data. The labelling is the same as in Fig. 4.

4. DISCUSSION AND CONCLUSIONS

As can be seen from the various figures, the asymmetries due to
an extra Z^0 boson can deviate quite substantially from that of the
standard model. One can obtain relations between the asymmetries
and $M_{Z'}$, C_V' , and C_A' but in general it is non-trivial to extract
the parameters of the model from the expression for the various
asymmetries. An alternate method, while not as direct, is to make a
contour plot of the δA's as a function of θ_{E_6} and $M_{Z'}$. Since δA_{LR},
δA_{FB}, and δA_{LR}(Bhabha) have different functional dependencies on θ_{E_6},
$M_{Z'}$, and $\sqrt{s}$, by measuring the asymmetries one can constrain the
allowed region of the parameter space. If deviations are found the
measured values would restrict the properties of the Z' to a region
of the θ_{E_6}-$M_{Z'}$ plane. We illustrate this method in Fig. 8 where we
have plotted contours of constant δA_{LR}, δA_{FB}, and δA_{LR} (Bhabha) for
$\sqrt{s}$ = 100 GeV. Depending on flavour tagging efficiencies one could
also utilize asymmetries involving $c\bar{c}$ and $b\bar{b}$ pairs in the final
state. Because the quarks will in general have different couplings
to the Z' than the leptons, shifts in the asymmetries which differ
from those involving leptons would help unravel the underlying
theory.

To conclude, we have studied the asymmetries in e^+e^- colliders
that may arise from E_6 GUTs and have found that asymmetries provide
a sensitive probe of new physics. If we have accomplished nothing

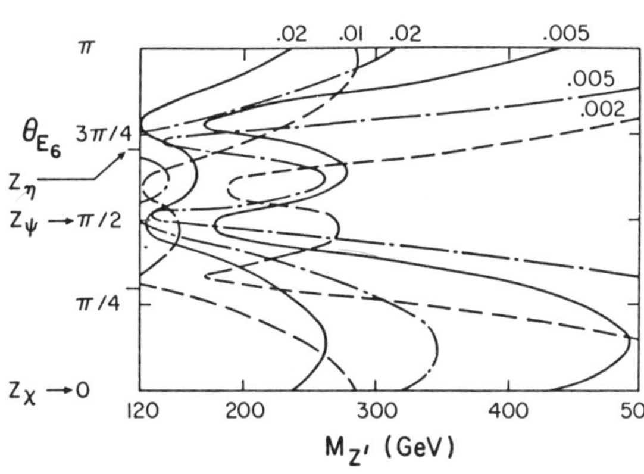

Fig. 8. Contour plot
of deviations from the
standard model for the
various asymmetries in
the θ_{E_6}-$M_{Z'}$ plane. The
solid lines are for
δA_{LR}, the dashed lines
for δA_{FB} and the dot-
dashed line for δA_{LR}
(Bhabha). All are the
$\sqrt{s}$=100 GeV and ϕ=0.

else we hope that we have convinced the reader that the deviations of asymmetries from the standard model allowed by current data are quite substantial and accurate measurements would severely restrict models proposed for new physics. Asymmetries measured at e^+e^- colliders could very well be our first indication of new physics beyond the standard model.

ACKNOWLEDGEMENTS

We would like to thank the organizers of the Institute for a stimulating conference and thank David London and John Ng for helpful discussions during the course of this work. This work was partially funded by the Natural Sciences and Engineering Research Council of Canada.

REFERENCES

1) We refer the reader to the review given by M. Dine in these proceedings for details and a complete set of references.

2) Various aspects of superstring phenomenology are discussed by J. Ellis in these proceedings.

3) J. Rosner, Comments Nucl. Part. Phys. 15, 195 (1986).

4) V. Barger, N.G. Deshpande, K. Whisnant, Phys. Rev. Lett. 56, 30 (1986); E. Cohen, J. Ellis, K. Enqvist, and D.V. Nanopoulos, Phys. Lett. 165B, 76 (1985).

5) L.S. Durkin and P. Langacker, Phys. Lett. 166B, 436 (1986).

6) For a more complete account of this work and a more detailed set of references see G. Bélanger and S. Godfrey, Phys. Rev. D34 (to appear); TRI-PP-86-18 (1986) submitted to Phys. Rev. D.

7) For a more complete survey of e^+e^- physics see J. Dorfan these proceedings. See also the "LEP Yellow book," CERN Report 86-02 (1986).

8) B.W. Lynn, M.E. Peskin, and R.G. Stuart, "LEP Yellow Book" CERN Report 86-02; B.W. Lynn and R.G. Stuart, Nucl. Phys. B253, 216 (1985).

THE LIGHT-CONE GAUGE AND THE AXIAL QED ANOMALY

P. Mathieu, Q. Ho-Kim and L. Marleau
Département de physique, Université Laval, Québec,
Canada G1K 7P4

ABSTRACT

Some properties of the light-cone gauge are briefly discussed. The calculation of the QED axial anomaly in that gauge is then sketched. It is shown that the anomaly equation in the light-cone gauge differs from the standard expression.

The light-cone gauge (l.c.g.) is certainly the gauge of the eighties. Indeed it plays a crucial role in recent achievements in supersymmetry and superstrings. First, as it is well known, superstrings can be quantized in a consistent way only in the l.c.g.[1]. This sole fact is sufficient to motivate a detailed study of all the aspects of the l.c.g. On the other hand, the l.c.g. is an essential ingredient in the proof of the finiteness of N=4 supersymmetry by Mandelstam[2] and Brink **et al.**[3]

The l.c.g. is a physical gauge in the sense that it deals only with the physical degrees of freedom[4]. But this simplicity has a price: in loop calculations there are spurious infinities, which call for a definite prescription to eliminate them. The origin of these infinities will be clarified shortly.

In QED, the l.c.g. is defined by the two conditions

$$n.A = 0 \quad , \tag{1}$$
$$n^2 = 0 \quad . \tag{2}$$

With the choice

590

$$n_\mu = \text{const.} \times (1,0,0,1) \quad , \tag{3}$$

the condition (1) can be written as

$$A^+ = 0 \quad , \tag{4}$$

where

$$A^\pm = \frac{1}{\sqrt{2}} (A^0 \pm A^3) \quad . \tag{5}$$

The gauge condition can be introduced either via a constraint in the Lagrangian or explicitly, by setting A^+ equal to zero and eliminating A^- by means of the field equation. In the former case, the Feynman rules are unmodified, except for the photon propagator which becomes

$$-\frac{i}{p^2} \left[g^{\mu\nu} - (\frac{n^\mu p^\nu + n^\nu p^\mu}{n \cdot p}) \right] \quad . \tag{6}$$

Clearly new infinities in momentum integrals will be induced by the factor $(p^+)^{-1}$. Hence this factor must be regularized. Two prescriptions have been used in the literature. The simplest one is the principal value prescription (PVP), which is defined by[4]

$$\text{PVP} : \frac{1}{p^+} \rightarrow \frac{1}{2} \left(\frac{1}{p^+ + i\varepsilon} + \frac{1}{p^+ - i\varepsilon} \right); \ \varepsilon > 0 \quad . \tag{7}$$

Although the results obtained with this prescription are consistent in the axial gauge ($n \cdot A = 0$, $n^2 \neq 0$)[7], recent calculations indicate that the PVP fails in the l.c.g.[5,8,9]. However, independently of its validity, the PVP has a clear drawback in the l.c.g. since it precludes a Wick rotation. The poles in the complex p^0-plane are determined by the equation

$$p^0 = -p^3 \pm i\varepsilon \quad . \tag{8}$$

Depending upon the sign of p^3, the poles lie either in the first and the fourth quadrant ($p_3 < 0$) or in the second and the third quadrant ($p_3 > 0$). Hence, there is necessarily a pole either in the first or in the third quadrant, which invalidates a continuous transformation from Minkowski space to Euclidean space. It is precisely this defect of the PVP that has motivated the search for a new prescription.

Mandelstam[3] and Leibbrandt[9] have independently proposed the following prescriptions:

$$\text{Mandelstam} : \frac{1}{p^+} \to \frac{1}{p^+ + i\epsilon p^-} \quad ; \qquad \epsilon > 0 \quad ,$$

$$\text{Leibbrandt} : \frac{1}{p^+} \to \frac{p^-}{p^+ p^- + i\epsilon} = \frac{1}{p^+ + i\epsilon/p^-} \quad ; \quad \epsilon > 0 \quad .$$

In both cases, the position of the poles is determined by the sign of p^- and so the Mandelstam-Leibbrandt prescription (MLP) can be written under the common form:

$$\text{MLP} : \frac{1}{p^+} \to \frac{1}{p^+ + i\epsilon \, \text{sgn} \, p^-} \qquad \epsilon > 0 \ . \qquad (9)$$

The poles in the complex p^0-plane are now determined by

$$p^0 = -p^3(1 - i\epsilon') \qquad \qquad . \qquad \qquad (10)$$

For $p^3 > 0$, there is a pole in the second quadrant while for $p^3 < 0$, it lies in the fourth quadrant. Wick rotation is then allowed and power counting theorems can be used.

There is an interesting one-loop calculation which could serve as a further test of the two prescriptions in the l.c.g., and this is the evaluation of the QED axial anomaly. This calculation has already been performed using the PVP, leading to inconsistent results[10]. This probably signals the failure of the prescription used. It is then useful to redo the calculation with the MLP. However a more fundamental reason motivates this exercise, which is to study the structure of anomalies in the l.c.g., with an eye on string theories.

Consider then the calculation of the QED axial anomaly in the l.c.g.. The first step is to write the QED Lagrangian for massless fermions

$$L = -\frac{1}{4} F_{\mu\nu} F^{\mu\nu} + i\bar{\psi} (\not{\partial} + ie\not{A}) \psi \qquad (11)$$

in terms of the physical degrees of freedom. One first decomposes the spinor field into two parts

$$\psi = \psi_+ + \psi_- \qquad (12)$$

where

$$\psi_{\pm} = \frac{1}{2} \gamma_{\pm}\gamma_{\mp}\psi \quad . \tag{13}$$

From the spinor field equations one finds that ψ_- can be expressed in a simple way in terms of ψ_+:

$$\psi_- = \frac{1}{2\partial^+} \left[\gamma_-\hat{\gamma} \cdot (\hat{\partial}+ie\hat{A}) \psi_+ \right] \tag{14}$$

where $\hat{a}.\hat{b} = a^i b^i$, $i = 1,2$. From this expression it can be seen that terms bilinear in ψ_- in the Lagrangian will induce a coupling of a photon to two ψ_+ spinors and also a two-photon-two-fermion coupling. On the other hand, from the vector field equations, A^- can be expressed in terms of the transverse photons and the spinor ψ_+:

$$A^- = \frac{1}{\partial^+} \left[\hat{\partial}.\hat{A} - \sqrt{2}e \frac{1}{\partial^+} \psi_+^*\psi_+ \right] \quad . \tag{15}$$

(the star indicates hermitian conjugate). Hence the term $(\partial^+ A^-)^2$ will generate a four-fermion coupling and contribute to the three-point vertex (one photon + two fermions).

The substitution of ψ_- and A^- into the Lagrangian (11) leads to a new Lagrangian from which one can extract the Feynman rules. These are:

$$-\frac{i}{q^2} g_{ij}$$

$$i\sqrt{2} \frac{p^+}{p^2}$$

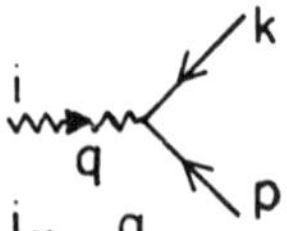

$$-\frac{ie}{\sqrt{2}} \left[\frac{2q^i}{q^+} + \frac{\gamma^i \not{p}}{p^+} + \frac{(\not{p}+\not{q})}{(p^++q^+)} \gamma^i \right]$$

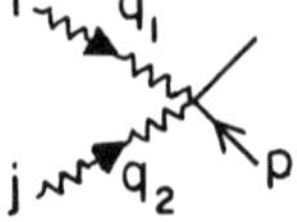

$$\frac{ie^2}{\sqrt{2}} \left[\frac{\gamma^i\gamma^j}{p^++q_2^+} + \frac{\gamma^j\gamma^i}{p^+ + q_1^+} \right]$$

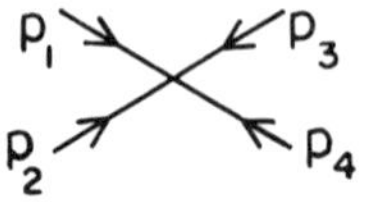

$$ie^2 \left[\frac{1}{(p_1^++p_2^+)(p_3^++p_4^+)} \right]_{\text{sym.}} \quad .$$

One then introduces the axial current

$$J_5^\mu = \bar{\psi}\gamma^\mu\gamma_5\psi \quad . \tag{16}$$

From the field equations, it follows that

$$\partial_\mu J_5^\mu = 0 \quad . \tag{17}$$

This is a direct consequence of the absence of a mass term in (11). However the conservation law (17) is violated by quantum corrections, namely

$$<0|\partial_\sigma J_5^\sigma \; A^\mu(p_1) \; A^\nu(p_2)|0> \neq 0 \quad . \tag{18}$$

This is the axial anomaly. The axial current can be decomposed as

$$J_5^+ = \sqrt{2}\;\psi_+^* \; \gamma_5 \; \psi_+ \tag{19a}$$

$$J_5^- = \sqrt{2}\;\psi_-^* \; \gamma_5 \; \psi_- \tag{19b}$$

$$J_5^k = \psi_+^* \; \gamma^0 \; \gamma^k \; \gamma_5 \; \psi_- + (-\leftrightarrow +) \quad . \tag{19c}$$

All three components contribute to the vertex but J_5^k also contains a three-point axial vertex while J_5^- has a further piece corresponding to a four-point (2 photons + 2 fermions) axial vertex.

The above Feynman rules, combined with the rules for the insertion of the axial current, show that in addition to the usual triangle diagram, three other diagrams must be considered:

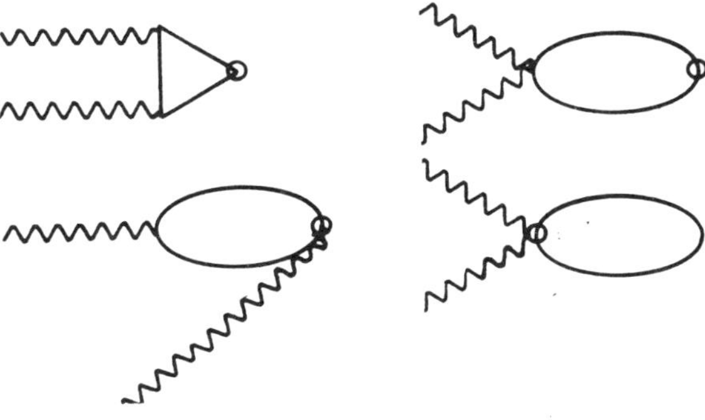

The calculation of the axial anomaly amounts to calculate

$$i(p_1+p_2)_\sigma \; R^{\sigma\mu\nu}(p_1,p_2) \quad , \tag{20}$$

where

$$R^{\sigma\mu\nu}(p_1,p_2) \equiv \langle 0|J_5^\sigma \, A^\mu(p_1) \, A^\nu(p_2)|0\rangle . \tag{21}$$

It is actually simpler to first differentiate (21) with respect to p_2 and then let $p_2 = -p_1 = -p$ [11]. In the l.c.g. this procedure is delicate but turns out to be correct. Hence one only has to evaluate $R^{\sigma\mu\nu}(p,-p)$, or more precisely $R^{\sigma ij}(p,-p)$. This still involves the calculation of many divergent integrals. The divergencies are regularized by the technique of dimensional regularization. Some care is required for the treatment of the γ_5. The safest approach is to avoid manipulating γ_5 and to evaluate the trace at the end of the calculation, when the infinities are cancelled. The results of this very long calculation are [12]

$$iR^{+ij}(p,-p) = \frac{e^2}{2\pi^2} \, \epsilon^{ij+-} p^+ \tag{22a}$$

$$iR^{-ij}(p,-p) = \frac{e^2}{2\pi^2} \, \epsilon^{ij+-} p^- \left(1 - \frac{p^2}{p^+p^-}\right) \tag{22b}$$

$$iR^{kij}(p,-p) = \frac{e^2}{2\pi^2} \, \epsilon^{ij+-} p^k \quad . \tag{22c}$$

Let us compare these results with those obtained in Ref. 10. The results for the first and the third equation (using the identity $\epsilon^{ik+-} p^j - \epsilon^{jk+-} p^i = \epsilon^{ij+-} p^k$) are the same (modulo signs). However for the second equation, Capper and Litvak have found extra terms inversely proportional to p^2, whose origin is difficult to figure out. Let us then compare our results with the standard expression for the axial anomaly in a covariant formalism [13] (which is distinguished by a subscript zero):

$$iR_0^{\sigma\mu\nu}(p,-p) = \frac{e^2}{2\pi^2} \, \epsilon^{\mu\nu\sigma\rho} \, p_\rho \tag{23}$$

which yields

$$iR_0^{+ij}(p,-p) = \frac{e^2}{2\pi^2} \, \varepsilon^{ij+-} p^+ \tag{24a}$$

$$iR_0^{-ij}(p,-p) = \frac{e^2}{2\pi^2} \, \varepsilon^{ij-+} p^- \tag{24b}$$

$$iR_0^{kij}(p,-p) = 0 \quad . \tag{24c}$$

Equations (22) and (24) are clearly different. However the source of this discrepancy can be easily pinpointed : the expression (23) does not satisfy the l.c.g. condition $A^+=0$ which requires that $R^{\sigma+\nu} = R^{\sigma\mu+} = 0$. (cf. the definition of $R^{\sigma\mu\nu}$ in eq. (21)). The form (23) is the most general pseudotensor constructed out of p^σ and $\varepsilon^{\mu\nu\lambda\rho}$ that satisfies Bose symmetry ($\mu\leftrightarrow\nu$, $p\leftrightarrow-p$) and Lorentz invariance. In the l.c.g. an additional vector is available, n , but further conditions must be satisfied:

$$n_\mu \, R^{\sigma\mu\nu} = n_\nu \, R^{\sigma\mu\nu} = 0 \quad . \tag{25}$$

These requirements completely fix the form of $R^{\sigma\mu\nu}$ to

$$iR^{\sigma\mu\nu} = A \, \varepsilon^{\mu\nu\lambda\rho} \, \frac{n_\lambda}{n.p} \, [p_\rho p^\sigma + a g_\rho{}^\sigma p^2 + b \frac{n^\sigma p_\rho}{n.p} \, p^2] \tag{26}$$

where A, a and b are constants. Indeed, a factor of n_λ must be contracted with one index of the ε-tensor in order to satisfy (25). But n_λ involves an arbitrary constant, which forces the presence of a factor of n in the denominator. The latter can only be contracted with a factor of p. The remainder must be even (Bose symmetry) and of degree two in the momentum (dimensional analysis), which fixes the form in the square bracket. (Two other terms could have been considered in the square bracket: $n^\sigma n_\rho p^4/(n.p)^2$ and $p^\sigma n_\rho p^2/(n.p)$. However in both cases two factors of n are contracted with the ε-tensor so they vanish trivially). It is easy to verify that the last two terms in (26) contribute in $R^{\sigma ij}$, only when $\sigma = -$, in which case they have the same form.

596

Now, how can we determine the two constants a and b? At first, notice that (23) satisfies $p_\mu R_0^{\sigma\mu\nu} = p_\nu R_0^{\sigma\mu\nu} = 0$. These equations are related to the preservation of gauge invariance. Imposing these two conditions on the form (26) yields a=0. On the other hand, one also observes that $p_\sigma R_0^{\sigma\mu\nu} = 0$. This condition applied to (26) gives b=-1. With these values and $(\mu,\nu) = (i,j)$, one gets

$$iR^{\sigma ij} = A \, \epsilon^{ij+-} [p^\sigma - n^\sigma p^2/n.p] \quad , \qquad (27)$$

and this corresponds exactly to the results (22). This analysis then provides a further confirmation of our results.

In summary, it has been shown that the covariant expression for the anomaly equation is not valid in the l.c.g.. The modified QED axial anomaly equation has been obtained by explicit calculations using the MLP. The structure of the resulting expression has also been inferred from general arguments.

This calculation raises the problem of the evaluation of a given prescription scheme in a specific gauge. Generally, prescriptions are tested by brute force through detailed calculations. However it is important to be able to determine the validity of a prescription at a more fundamental level. (For instance, why does the PVP works in the axial gauge but not in the l.c.g. and why does the MLP works in the l.c.g.?). This crucial issue must be resolved before one can do predictive calculations in physical gauges. (Partial results in that direction have been obtained in Ref. 14). Such an analysis could also have potential implications in string theories. For instance, the proof of finiteness of SO(32) superstring by Green and Schwarz[15] has clearly displayed the crucial role of a definite prescription for a delicate cancellation of infinities (see also Refs. 16,17). The uniqueness and the consistency of the regularization used there certainly requires a careful analysis.

This work was supported in part by grants from the Natural Sciences and Engineering Research Council of Canada.

References

1. Schwarz, J.H. , **"Superstrings"** (2 vol), World Scientific (Singapore 1985).

2. Mandelstam, S. , Nucl. Phys. $\underline{B213}$, 149 (1983).

3. Brink, L., Lindgren, O. and Nilsson, B.E.W., Nucl. Phys. $\underline{B212}$, 401 (1983).

4. For a review, see H.C. Lee, **"The light-cone gauge"**, talk given at the CAP Summer Institute on Quantum Field Theory 1985, London, Canada.

5. Capper, D.M., Dulwich, J.J. and Litvak, M.J. , Nucl. Phys. $\underline{B241}$, 463 (1984).

6. Cornwall, J.M., Phys. Rev. $\underline{D10}$, 500 (1974); Kummer, W., Acta Phys. Austriaca $\underline{41}$, 315 (1975).

7. Capper, D.M. and Leibbrandt, G., Phys. Rev. $\underline{D25}$, 1002 (1982); Lee, H.C. and Milgram, M.S. , J. Math. Phys. $\underline{26}$, 1793 (1985).

8. Lee, H.C. and Milgram, M.S., Z. Phys. $\underline{C28}$, 579 (1983).

9. Leibbrandt, G., Phys. Rev. $\underline{D29}$, 1699 (1984).

10. Capper, D.M. and Litvak, M.J., J. Phys. A: Math. Gen. $\underline{18}$, 955 (1983).

11. Jones, D.R.T. and Léveillé, J.P. , Nucl. Phys., $\underline{B241}$, 463 (1982).

12. Ho-Kim, Q., Marleau, L. and Mathieu, P. , Laval Univ. preprint.

13. See for instance K. Huang, **"Quarks, Leptons, and Gauge Fields"**, World Scientific (Singapore 1982).

14. Bassetto, A., Lazzizzera, I. and Soldati, R., Nucl. Phys. $\underline{B236}$, 319 (1984); Bassetto, A., Dalbosco, M., Lazzizzera, J. and Soldati, R., Phys. Rev. $\underline{D31}$, 2012 (1985).

15. Green, M.B. and Schwarz, J.H. , Phys. Lett. $\underline{151B}$, 21 (1985).

16. Frampton, P.H., Moxhay, P. and Ng, Y.J., Harvard Univ. preprint HUTP85/A059.

17. Moffat, J.W. , **"Superstring physics"**, talk at the CAP Summer Institute on Quantum Field Theory 1985, London, Canada.

PREDICTING THE FINE STRUCTURE CONSTANT
FROM HIGHER DIMENSIONAL GRAVITY*

G. Kunstatter

Physics Dept., University of Winnipeg
Winnipeg, Manitoba   R3B 2E9   CANADA

and

H.P. Leivo

Physics Dept., University of Toronto
Toronto, Ontario   M5S 1A7   CANADA

ABSTRACT

A non-technical review of Kaluza-Klein theory is presented,
illustrating why the classical theory fails to provide the desired
unification of the fundamental interactions. Recent calculations,
which use quantum corrections to predict the radius of the internal
dimensions and the effective fine structure constant, are described
and shown to yield gauge dependent, and hence physically meaningless,
predictions.

1.   INTRODUCTION

Kaluza-Klein theory[1] is an old and very appealing attempt to
unify gravity and electromagnetism within the geometrical framework of
pure gravity in five spacetime dimensions. More recently, this
approach has been generalized to include non-abelian gauge fields (for
a good review see ref. 2). Unfortunately, it is now known that the
Kaluza-Klein program in its original form, namely without any external
matter or gauge fields, cannot provide a phenomenologically viable
unification of the fundamental interactions, for a variety of reasons.
Recent developments in string theory[3], however, seem to indicate that
the problems associated with spontaneous compactification and higher
dimensional quantum gravity will remain an active area of research for
some time. (At least until the finiteness of superstring theories is
proved or disproved)[4]. Moreover, there has been a recent suggestion[5]
that superstring theories in ten dimensions might best be understood

---

*This work was supported in part by the Natural Sciences and
Engineering Research Council of Canada

in terms of a Kaluza-Klein mechanism in still higher dimensions. It
is thus worthwhile to investigate further the consequences of
postulating the existence of more than four dimensions, whether or not
the Kaluza-Klein program in its original form provides an accurate
description of the world in which we live.

In the following section, a brief review of Kaluza-Klein theory
will be presented for the non-expert, illustrating in simple terms why
pure Kaluza-Klein theory fails. In section 3, the relevance of
perturbative effects in quantum gravity on topologically non-trivial
spacetimes will be discussed, drawing heavily on the analogy with the
electromagnetic Casimir effect. In addition, a scheme for predicting
the radius of the internal dimensions and thus the lower-dimensional
gauge coupling constant will be described. Finally, a calculation of
the four-dimensional fine structure constant in five-dimensional
Kaluza-Klein theory will be shown to depend strongly on the gauge in
which the calculation is performed, rendering the predictions
meaningless. The implications of this result for perturbative
calculations in higher dimensional quantum gravity will be touched on
in the final section.

2.   KALUZA-KLEIN REVIEW

General relativity in four dimensions is based on the coordinate
invariant line element, or proper length:

$$- ds^2 = g_{\mu\nu}dx^\mu dx^\nu, \qquad (2.1)$$

where $g_{\mu\nu}(x)$ is the coordinate dependent metric tensor describing the
geometry of spacetime, which also plays the role of the "potential"
for the gravitational field strength. For example, the gravitational
field at the surface of the earth is:

$$F_G = - \frac{GM}{r^2} = \frac{\partial g_{00}}{\partial r} , \qquad (2.2)$$

just as in electromagnetism, $E = \partial\phi/\partial r$.

One important phenomenological restriction on higher dimensional
theories in general is that they must explain why only four spacetime

600

dimensions are visible experimentally. The Kaluza-Klein solution is
to consider general relativity in five dimensions but with the extra
dimension "curled up", so that the fifth coordinate x^5 is periodic,
with period $2\pi R$. Then, if the size of the compact dimension is
significantly less than the characteristic length scale in any
experiment, the fifth dimension will indeed be unobservable. The
following question now arises: Does the "invisible" dimension have
any observable consequences? As Kaluza and Klein discovered in the
1920's, it does in fact leave a striking signature. To see this, it
is necessary to decompose the five dimensional line element as
follows (A, B = 0, 1, 2, 3, 5; μ, ν = 0, 1, 2, 3):

$$- d\bar{s}^2 = g_{AB}dx^A dx^B = g_{\mu\nu}dx^\mu dx^\nu + \phi(\kappa A_\mu dx^\mu + dx^5)^2. \tag{2.3}$$

If we further assume that g_{AB} is independent of x^5, and that the
scalar field is constant, then the Einstein-Hilbert action in five
dimensions reduces to an effective four dimensional action of the
following form:

$$\begin{aligned}
\bar{I} &= \frac{1}{16\pi\bar{G}} \int d^5x \sqrt{-\bar{g}}\ \bar{R} \\
&= \frac{1}{16\pi G} \int d^4x \sqrt{-g}R(g) - \frac{1}{4} \int d^4x \sqrt{-g}F_{\mu\nu}F^{\mu\nu}
\end{aligned} \tag{2.4}$$

where $F_{\mu\nu} = A_{\mu,\nu} - A_{\nu,\mu}$ and we have made the following identifications:

$$G = \bar{G}/2\pi r \tag{2.5}$$

$$\kappa^2 = 16\pi\bar{G}/(2\pi r\phi) \tag{2.6}$$

where r is the invariant radius of the internal cylinder:

$$r = (2\pi)^{-1} \int dx^5 \sqrt{g_{55}} = R\sqrt{\phi}. \tag{2.7}$$

Thus, the "Kaluza-Klein miracle" occurs: the effective action is
precisely that of Einstein-Maxwell theory in four dimensions. In some
sense, then, we have a truly geometrical unification of gravity with
electromagnetism, in much the same way that Maxwell theory unifies E
and B within a single tensor $F_{\mu\nu}$. In the latter theory, once the
Galilean transformation group is enlarged to the physically correct
Lorentz group, it is possible to convert an electric field into a

magnetic field by performing a coordinate transformation (a boost).
Similarly, in Kaluza-Klein theory, the correct symmetry at high
energies is the group of coordinate transformations in five
dimensions, so that a gravitational field can be made to look like an
electromagnetic field by doing a boost in the fifth dimension. At low
energies, however, the effective action is given in Eq (2.4) so that
the only remnant of the symmetries involving the fifth direction is
precisely the gauge-invariance of the electromagnetic field.

In non-abelian Kaluza-Klein theory, the Yang-Mills fields are
self-coupled, so that it is possible to determine the gauge coupling
constant directly from the effective four-dimensional Einstein-Yang-
Mills action[6]. In the present case it is necessary to couple matter
to the five-dimensional geometry in order to determine the effective
coupling constant in four dimensions. One approach is to minimally
couple the matter in five-dimensions: that is, assume test particles
follow geodesics in the absence of external forces. If ϕ = constant,
the geodesic equation in five dimensions

$$\bar{m} \, \frac{D^2 x^A}{D\bar{s}^2} = 0 \tag{2.8}$$

can be written as:

$$p/\bar{m} = \bar{g}_{A5} \, \frac{dx^A}{d\bar{s}} = \phi \left(\kappa A_\mu \, \frac{dx^\mu}{d\bar{s}} + \frac{dx^5}{d\bar{s}} \right)$$

$$= \text{constant.} \tag{2.9}$$

$$\frac{D^2 x^\mu}{D\bar{s}^2} = \frac{p}{\bar{m}} \, \kappa F^\mu{}_\nu \, \frac{dx^\nu}{d\bar{s}} \tag{2.10}$$

Remarkably, the Lorentz force on the right hand side of Eq (2.10)
emerges purely from inertial forces in Eq. (2.8), with the momentum,
p, in the fifth direction playing the role of the electric charge.
Moreover, Klein showed in 1927[7] that since the momentum is quantized
($p = n\hbar/R$) due to the periodicity of the fifth coordinate, the
electric charge is automatically quantized as well. That is $q = ne$,
providing one makes the identification

$$e = \hbar\kappa/R \tag{2.11}$$

so that

$$\alpha = \frac{e^2}{\hbar c} = \frac{8\bar{L}p^3}{r^3} = \frac{16\pi Lp^2}{r^2} \qquad (2.12)$$

where we have used equations (2.5) and (2.6) and defined the four and five dimensional Planck lengths, respectively as $L_p = (\hbar G)^{\frac{1}{2}}$ and $\bar{L}_p = (\hbar\bar{G})^{1/3}$. Not only does this argument give rise to charge quantization from quantum mechanics but it also yields a value for the invariant radius of the fifth dimension of the order of the Planck length, which is definitely unobservable at present energies. Conversely, if one were able to predict the radius r dynamically, one would be able to compute from first principles the effective fine structure constant using Eq. (2.12). This is precisely what shall be attempted in the following section. But first some bad news:

The invariant mass $\bar{m}$ which appears in Eq. (2.8)-(2.10) is not the mass that would be measured by a neutral observer in four dimensions. Since we have identified charge with momentum in the fifth direction, a charged particle is necessarily in motion relative to a neutral observer. The corresponding Lorentz transformed mass as seen in the rest frame of the neutral observer is:[8]

$$m = \bar{m}\,\frac{ds}{d\bar{s}} = \left(\bar{m}^2 + \frac{p^2}{\phi}\right)^{\frac{1}{2}} = \left(\bar{m}^2 + \frac{e^2}{16\pi G}\right)^{\frac{1}{2}}$$

$$\geqslant \left(\frac{e^2}{16\pi G}\right)^{\frac{1}{2}} = \sqrt{\frac{\alpha}{16\pi}}\;M_{p1}\;, \qquad (2.13)$$

where M_{p1} is the Planck mass ($\sim 10^{-5}$ gm). In equation (2.13) the four dimensional proper length ds is defined by $-ds^2 = g_{\mu\nu}dx^\mu dx^\nu$.

Thus the theory cannot even classically describe low mass charged particles such as electrons, unless one admits the somewhat unphysical expedient of considering tachyonic particles in five dimensions (i.e. $\bar{m}^2 \leqslant 0$). The only other way to avoid this conclusion is to give up the assumption of minimal coupling in five dimensions, but then no predictions whatsoever are possible. The so-called chiral fermion

problem[9] is very much analoguous to the above, but there one is
looking for zero modes of the Dirac operator on compact internal
spaces with isometries. Again the solution is either to allow
tachyons in five dimensions or to give up minimal coupling. For
example one can couple the Dirac particle to topologically non-trivial
external gauge fields on the internal space[10].

There are further problems as well. When ϕ is not constant, m is
not constant, and if the scalar field is long range, this can lead to
macroscopically observable violations of the equivalence principle.[8]
In addition, when ϕ is long range, there is in principle an ambiguity
in the definition of the pseudotensor energy.[11] Moreover, spontaneously compactified solutions to the classical Einstein equations
yielding phenomenologically acceptable values of the effective cosmological constant in four dimensions do not exist in non-Abelian Kaluza
Klein theory unless there are some 10^{120} internal dimensions.[12]

The conclusion that can be drawn from the above is that classical
Kaluza-Klein theory is not viable. Of course, since the radius of the
internal dimensions is of the order of the Planck scale, it would not
be consistent to neglect quantum effects in any case. Unfortunately,
Einstein gravity does not yield a renormalizable quantum theory, so
that the calculation of quantum effects is inherently problematic.
There have nonetheless been attempts to solve the spontaneous
compactification problem by considering one-loop quantum gravity
calculations. These will be discussed in the following section.

3. SELF-CONSISTENT DIMENSIONAL REDUCTION

Quantum effects due to nontrivial boundary conditions have been
considered since 1949 when Casimir[13] calculated the vacuum energy
between two parallel, perfectly conducting plates. Since this is
analoguous to the quantum gravity calculation in Kaluza-Klein theory,
it is perhaps worthwhile to consider Casimir's calculation in a modern
context (see A. Chodos in ref. 2). The electromagnetic lagrangian:

$$L_{EM} = -\frac{1}{4} F_{\mu\nu} F^{\mu\nu} \tag{3.1}$$

604

yields the following field equations for the electromagnetic vector potential A_μ:

$$\Box A_\mu - \partial_\mu \partial^\nu A_\nu = 0. \tag{3.2}$$

Due to the gauge invariance of the lagrangian, however, not all four equations in (3.2) are independent. In order to quantize this system it is therefore necessary to fix a gauge. This can most easily be achieved by adding to the lagrangian a term which breaks the gauge invariance. For example, the modified lagrangian:

$$L = -\frac{1}{4} F_{\mu\nu} F^{\mu\nu} + \frac{1}{2\alpha} (\partial_\mu A^\mu)^2 \tag{3.3}$$

gives rise to four independent equations:

$$\Box A_\mu - (1 - 1/\alpha) \partial_\mu (\partial_\nu A^\nu) = 0. \tag{3.4}$$

Care must still be taken in quantization due to the contribution of the unphysical scalar and longitudinal modes to the propagator derivable from Eq. (3.4). These may be eliminated by restricting the space of physical states (Glupta-Bleuler quantization) or by using Faddeev-Popov ghosts (path integral quantization). We shall not discuss these technicalities here. We note only that Eq. (3.4) depends explicitly on the gauge parameter, α. In order for any calculational procedure to make sense it is necessary that all physical predictions be independent of the gauge fixing term added to the lagrangian. In the case of the electromagnetic Casimir energy calculation, standard theorems[14] ensure that this will indeed be true, so that we can consider the simplest case $\alpha = 1$. Quantization of each of the two physical modes of Eq. (3.4) then yields an infinite set of quantized harmonic oscillators with characteristic frequency:

$$\omega = \left(k_x^2 + k_y^2 + k_z^2 \right)^{\frac{1}{2}} \tag{3.5}$$

where the k's represent the components of the wave vectors. The boundary conditions imposed by the conducting plates, placed orthogonally to the z-axis, say, force the z-component of the wave vector to be quantized: $k_z = n\pi/R$, where R is the distance between the plates. The vacuum energy per unit area between the plates is:

$$\frac{E(R)}{AREA} = \sum_{n=-\infty}^{\infty} \tfrac{1}{2}\hbar\omega_n = \sum_{n=-\infty}^{\infty} \tfrac{1}{2}\hbar\,\frac{d^2k}{(2\pi)^2}\left(k_x^2 + k_y^2 + \frac{n^2\pi^2}{R^2}\right)^{\tfrac{1}{2}} \qquad (3.6)$$

Note that the summation over all integer n takes account of the two helicity states when $k_z = (n\pi/R) \neq 0$. Since expression (3.6) is divergent, it is necessary to introduce a regulator, such as a momentum cutoff, to render it finite. Noting that only energy differences are relevant in ordinary quantum field theory, one calculates:

$$V(R) = \frac{E(R) - E(R=\infty)}{AREA} = -\frac{\pi^2 hc}{720R^3} \qquad (3.7)$$

The energy difference per unit area is finite and independent of both the gauge fixing term and the regulator. The Casimir force obtained by taking the gradient of V(R) with respect to the plate separation R is the relevant physical quantity, which has been measured experimentally.[15]

Since periodic boundary conditions also appear in five-dimensional Kaluza-Klein theory, it is natural to ask whether an analoguous effect exists there. Indeed such a calculation[16] yields the following expression for the vacuum energy per unit four-volume, as a function of the invariant radius r of the internal cylinder:

$$V(\lambda) = +\frac{A}{r^4} \qquad (3.8)$$

where A is a negative, dimensionless number. Thus the quantum Casimir "force" tends to make the internal dimension collapse, rendering the Kaluza-Klein vacuum semi-classically unstable.

The basic idea of self-consistent dimensional reduction [17-21] is to arrive at a stable Kaluza-Klein vacuum by balancing the Casimir force with another. In particular, one adds a cosmological constant Λ to the five-dimensional Einstein action and calculates the one-loop effective potential due to fluctuations about a flat $M^4 \times S^1$ (Minkowski space x internal circle) background with invariant radius r. The result is of the form

$$V(R) = -\frac{2\pi r}{16\pi G}\left(2\Lambda + \frac{A(x)}{r^4}\right) \tag{3.9}$$

where the function A of the dimensionless variable $x = \sqrt{2\Lambda}\, r$ reduces to the constant A of Eq. (3.8) when $\Lambda = 0$. A stable radius is determined by demanding the absence of a net force,

$$\frac{\partial V}{\partial r} = 0 \tag{3.10}$$

For consistency one must simultaneously impose the condition

$$V(r) = 0 \; , \tag{3.11}$$

since V(r) plays the role of the effective cosmological constant of the four-dimensional theory, and Minkowski space is a solution only when this vanishes. These conditions yield the equation

$$x\,\frac{\partial A(x)}{\partial x} = 5A(x) \; , \tag{3.12}$$

a solution, x, of which then allows the radius and required cosmological constant to be expressed in terms of the Planck length as

$$\frac{r}{L_p} = \left(8\,\frac{A(x)}{x^2}\right)^{1/3} \tag{3.13}$$

and

$$2\Lambda \bar{L}_p{}^2 = x^2\left(8\,\frac{A(x)}{x^2}\right)^{-2/3} \tag{3.14}$$

As long as Eq. (3.12) has a suitable solution, Eq. (3.13) provides a dynamical prediction for the radius of the internal dimension. As shown in the previous section, it is then possible to predict the numerical value of the fine structure constant. To work consistently to one-loop order it is, however, necessary to consider more than just the effective potential.[20] Indeed, one must examine the effective action in order to take into account one-loop contributions not only to the vacuum energy or effective cosmological constant, but also to the gravitational constant and the normalization parameter κ of the electromagnetic field.

The general form of the one-loop effective potential is

$$\Gamma[g] = \int d^4x \ \sqrt{-g} \ \left(\frac{2\pi r}{16\pi\bar{G}} \ 2\Lambda - \frac{A(x)}{r^4} \right)$$

$$+ \int d^4x \ \sqrt{-g} \ \left(\frac{2\pi r}{16\pi\bar{G}} + \frac{B(x)}{r^2} \right) R(g)$$

$$- \frac{1}{4} \int d^4x \ \sqrt{-g}\phi\kappa^2 \left(\frac{2\pi r}{16\pi\bar{G}} - 4 \frac{C(x)}{r^2} \right) F_{\mu\nu}F^{\mu\nu} + \text{higher order} \quad (3.15)$$

Thus Eq. (2.5) is replaced by

$$G = \frac{\bar{G}}{2\pi r} \left(1 + \frac{8\bar{G} \ C(x)}{r^3} \right)^{-1} \quad (3.16)$$

while Eq. (2.6) becomes

$$\phi\kappa^2 = \left(\frac{2\pi r}{16\pi\bar{G}} - \frac{4C(x)}{r^2} \right)^{-1} \quad (3.17)$$

so that the expression for the fine-structure constant now reads

$$\frac{e^2}{hc} = \left(- \frac{A(x)}{x^2} + 4C(x) \right)^{-1} \quad (3.18)$$

It remains now only to calculate the functions $A(x)$, $B(x)$ and $C(x)$.

The Einstein action is gauge invariant under general coordinate transformations. It is therefore necessary to add a gauge fixing term just as in the electromagnetic case. Morever, the expressions for the quantum corrections to the classical action are formally divergent, so that a regulator is required. Recently, the calculation of the quantum effective potential $A(x)$ has been performed[21] in a one-parameter family of gauges obtained by adding the following harmonic gauge-fixing term to the Einstein action in five dimensions:

$$L_{gf} = \frac{1}{2\alpha} \ (\partial_A h^{AB} - \partial^A h)^2 \ , \quad (3.19)$$

where h_{AB} denotes the quantum fluctuations of the gravitational field in five dimensions and h its trace. This calculation had previously

been done only for the particular choice of gauge parameter $\alpha = 1$.
The result is that, unfortunately, not only A(x) but also r and Λ are
explicitly dependent on the gauge parameter α.

4. CONCLUSIONS

It has been shown[21] that the self-consistent dimensional
reduction procedure yields gauge-parameter dependent predictions for
the radius of the internal dimension in terms of the five- dimensional
Planck length. Although it is well known that the one-loop effective
potential is in general gauge-dependent off shell (i.e. at backgrounds
which do not solve the classical equations of motion), theorems
exist[14, 22] which prove that physical parameters derived from
consideration of the effective potential for Yang-Mills theories are
nonetheless gauge-independent. The gravitational analogue of these
theorems has not been proven and the calculations outlined above seem
to indicate that they do not hold. Of course, the full implications
for Kaluza- Klein theory at this stage are not clear, since only A(x)
has been calculated in the one-parameter family of gauges. It is, for
example, possible that C(x) has precisely the right gauge-dependence
to yield a gauge-invariant expression for the fine structure constant
in Eq. (3.18) even while the radius r remains gauge dependent. This
would be another Kaluza-Klein "miracle". This calculation is presently
underway.

References

[1] Th. Kaluza, Sitzungsbr. Preuss. Akad. Wiss., Phys. Math. K1
966, (1982); O. Klein, Z. Phys. <u>37</u>, 895 (1926).

[2] H.C. Lee (ed.) "An Introduction to Kaluza-Klein Theories"
(World Scientific, 1984).

[3] M.G. Green and J. Schwarz, Phys. Lett. <u>149B</u>, 117 (1984).

[4] For a review see J.W. Moffat in the "Proceedings of the CAP
Summer Institute on Quantum Field Theories", 1985 (to appear in Can.
J. Phys.)

[5] M.J. Duff, B.E.W. Nilsson and C.N. Pope, CERN preprint TH. 4217/85 (1985).

[6] S. Weinberg, Phys. Lett. B125, 265 (1983).

[7] O. Klein, Nature 118, 516 (1927).

[8] J. Gegenberg and G. Kunstatter, Phys. Letts. 106A, 410 (1984).

[9] E. Witten, in "Proceedings of the Shelter Island 1983 Conference on Quantum Field Theory", R. Jackiw, N.N. Khuri, S. Weinberg and E. Witten (eds.) MIT Press, (1985).

[10] R. Randjbar-Daemi, A. Salam and J. Strathdee, Nucl. Phys. B214, 491 (1983).

[11] L. Bombelli, R.K. Koul, G. Kunstatter, J. Lee, R. Sorkin (in preparation).

[12] J. Gegenberg and G. Kunstatter, Class. Quant. Gravity (in press).

[13] H.B.G. Casimir, Proc. Kon. Wed Akad. Wetenschap, Ser. B #51, 793 (1948); for a recent review, see G. Plunien, B. Müller and W. Greiner, Phys. Rep. 134 (1986).

[14] R. Fukuda and T. Kugo, Phys. Rev. D13, 3469 (1976).

[15] M.J. Sparnay, Physica 24, 751 (1958).

[16] T. Appelquist and A. Chodos, Phys. Rev. Lett. 50, 141 (1983); Phys. Rev. D28, 772 (1983).

[17] P. Candelas, and S. Weinberg, Nucl. Phys. B237, 297 (1984).

[18] A. Chodos and E. Meyers, Am. Phys. 156, 412 (1984).

[19] C.R. Ordonez and M.A. Rubin, U. of Texas report UTTG 18-84 (1984).

[20] D.J. Toms, Phys. Lett. 129B, 31 (1983); see also D.J. Toms in "Proceedings of the CAP Summer Institute on Quantum Field Theory", 1985 (To Appear in Can. J. Phys.)

[21] G. Kunstatter and H.P. Leivo, Phys. Lett. 166B, 321 (1986); U. of Winnipeg preprint (1986).

[22] N.K. Nielsen, Nucl. Phys. B101, 173 (1975).

ARE SUPERSTRING THEORIES FINITE?

John W. Moffat
Department of Physics
University of Toronto
Toronto, Ontario M5S 1A7
Canada

ABSTRACT

The problem of the finiteness of superstring theories is reviewed. Infinities are shown to occur in the multi-loop amplitude for superstrings that require to be cancelled order by order. Whether the supersymmetry of superstrings is capable of achieving this cancellation to all orders is discussed in detail.

1. DIVERGENCES IN GRAVITY THEORIES

One of the fundamental problems of modern physics is how to reconcile gravitation as a quantum theory with the standard models of particle physics, based on such renormalizable structures as $SU(3) \times SU(2) \times U(1)$. Einstein's theory of gravity is not a renormalizable theory when the gravitational field is coupled to matter[1]. For the vacuum field equation on the mass shell, the one-loop amplitude is renormalizable due to a symmetry imposed by the Gauss-Bonnet relation on the curvature tensor in four-dimensional Riemannian geometry. However, in the two-loop case there is no such symmetry and the amplitude diverges[2]. Thus Einstein gravity becomes divergent as soon as it can and no symmetry has been discovered in higher orders that can prevent the higher powers of the curvature tensor from entering as counterterms in the Lagrangian for gravity. It was thought for a while that supersymmetry could provide such a symmetry and theories like the $N = 1$ and $D = 10$ supergravity theory were constructed[3]. But after considerable effort it is now thought that these theories are in fact divergent at the three-loop and higher-loop orders.

We see that little success has been achieved in attempts to understand gravity using standard point-particle field theory. This has led to a renewed interest in string theories. A string is a one-dimensional extended object that sweeps out a two-dimensional surface, described by a time-like parameter $\tilde{\tau}$ and a space-like parameter σ. These string theories can be transformed to conformally flat space and by using a fixed gauge, like the conformal gauge or the light-cone gauge, the theories satisfy linear wave equations. In the light-cone gauge the string transverse coordinates X^i (i = 1,2,...d-2.) satisfy a linear wave equation in the variables σ and $\tilde{\tau}$, where d is the dimension of the embedding space-time. When the theories are quantized they are found to be ghost-free for bose and superstring theories only when d $= 26$ and d $= 10$, respectively. This could just mean that string theories are wrong, since they are not consistent in d $=$ 4 space-time. However, with the work of Green and Schwarz[4][5] it was found that anomaly cancellations occur for superstring theories only for the groups SO(32) and $E(8) \times E(8)$. Similar results were found to be true for the heterotic string model based on the group $E(8) \times E(8)$[6]. Thus, through this anomaly cancellation, a virtue of the superstring theory being only consistent in a d $=$ 10 space-time has been found, which has fostered new hopes that a consistent unified picture of the forces of nature has been discovered. At the same time, the theory could provide us with a consistent quantum theory of gravity.

The fundamental question still remains: Are these theories finite to all orders? Since it is thought that superstring theories contain gravity, albeit not in the form of standard Einstein gravity, due to the existence of many spin states including a dilaton field that does not easily permit itself to be converted into a massive mode, it would indeed be exciting if superstring theory were proved to be finite. As we shall see in the following this is a difficult question to answer. In the multi-loop amplitudes there occur infinities that require 'miracles' to cancel. Our past experiences with gravity have shown that such miracles are not obviously forthcoming. In the event that superstrings are not found to be finite to all orders, then a new way to solve the problem of gravitation must be sought.

2. SUPERSTRINGS

In the light-cone gauge, the Lagrangian for superstrings is given by (we use an imaginary time parameter $\tau = i\tilde{\tau}$):

$$L = \frac{1}{4\pi}\left\{\left(\frac{\partial X^i}{\partial \tau}\right)^2 + \left(\frac{\partial X^i}{\partial \sigma}\right)^2\right\}$$

$$-\frac{1}{2\pi}\left\{S_1\left(\frac{\partial}{\partial \tau} + i\frac{\partial}{\partial \sigma}\right)S_1 + S_2\left(\frac{\partial}{\partial \tau} - i\frac{\partial}{\partial \sigma}\right)S_2\right\}, \tag{1}$$

where the S_1 and S_2 are d-dimensional spinors. The strings fuse together or split at various interaction points , and a description of higher-loop amplitudes can be obtained given a set of external strings by summing over all the different topologies associated with light-cone diagrams. The different topologies are determined by giving the positions of interaction points, number of holes and handles, etc. The amplitude can be written as

$$\Sigma\, G^{N+g} \int DX \exp(i \int_\Sigma S), \tag{2}$$

where G is the string interaction strength, N is the number of external strings, g is the number of handles or loops being considered, and Σ is the Riemann surface.

We can evaluate the functional integral in (2) for a given loop order:

$$\int \mathcal{D}X \exp(-\int S_J)d\sigma d\tau =$$

$$(\det'\Delta)^{-(\frac{d-2}{2})}(\det'\not{p})^{\frac{d-2}{2}} \exp[-\frac{1}{4}\int (JN_\Sigma J + \Psi K\Psi)]\Pi(\text{vertex factors}), \quad (3)$$

where S_J denotes the gauge-fixed action, N_Σ is the Neumann function on the surface Σ, K is the Neumann function associated with the Dirac operator $\not{p}$, J is the external bose source, and Ψ is the external fermion source. We denote by Δ the laplacian on the Riemann surface Σ, and by $\det'\Delta$ the determinant of the laplacian without the zero mode.

The S-matrix is calculated using the functional integration over all world sheets traced out by strings. Unitarity is not obviously satisfied and has to be checked at every order. The mathematics is that of Riemann surfaces and automorphic functions. Conformal invariance and projective invariance play a fundamental role for the two-surfaces. Expressions for the amplitude for the superstring closed theory have been obtained by Mandelstam[7] and Taylor[8] and Restuccia and Taylor[8], and they have given a detailed analysis of the divergences in the multi-loop superstring amplitudes. The superstring amplitude takes the form

$$A_s = K(p_1,...,\varsigma_1,...) \int \prod_{r=1}^{N-1} |G'(z_r)|^2 d^2 z_r F(B) \prod \partial_{z_\alpha}\partial_{z_\beta} N(z)$$

$$\times \prod |G''(z_\alpha)|^{-1} f(a_i,b_i,K_i)d^2 a_i d^2 b_i d^2 k_i \exp(-\frac{1}{2}p_c p_s N(z_r,z_s)). \quad (4)$$

The factor $K(...)$ is a kinematic factor containing external momenta, polarization vectors, etc. of the external states. The second factor in (4) contains the Koba-Nielsen variables z_r and the associated Jacobian. The third factor $F(B)$ is the period matrix. The term $\partial_{z_\alpha}\partial_{z_\beta} N$ is associated with a set of Feynman lines on the world sheet. The fifth factor comes from the transformation of variables related to the interaction points in the light-cone variables to the Riemann surface variables in the upper-half plane of U. The next factor is the Jacobian of the transformation from the interacting point variables involved with loop or hole features into the variables of the conformally equivalent Riemann surfaces (moduli). The final factor is the exponential of the Neumann function raised to the power of the external invariants. The function $G(z)$ is determine by the geometry of the system. The interacting points are solutions of $G'(z) = 0$.

3. N-LOOP SINGULARITIES

The singuarities of the N-loop amplitude are associated with changes in the topology of the corresponding Riemann surface. In the single-loop case, the associated surface is a disk with a hole and a singularity is generated when the hole shrinks to zero. For N-loops, we expect $2^N - 1$ divergences from the shrinking of the individual circles and the touching of these circles. The singularities are analyzed by using projective transformation multipliers. These techniques were first introduced in the early seventies in the analysis of dual multi-loop amplitudes by Alessandrini[9], Lovelace[10], Kaku and Scherk[11] and others. The projective multipliers are defined by[11]

$$\frac{P(z) - x_1}{P(z) - x_2} = w\left(\frac{z - x_1}{z - x_2}\right), \tag{5}$$

where x_1, x_2 and w are known as invariant points or limit points and the multiplier of P, respectively. Groups whose limit points form a discrete set are called Schottky groups.

The Neumann function $N(z, z')$ must be invariant under the transformation $z \to wz$, so that it remains a single-valued function. Moreover N has singularities at $z = w^n z'$ and at $z = z'$. Any loop diagram, which is a Riemann surface with n handles, has a fundamental group generated by n A - cycles and n B-cycles. The cycle A_r intersects B_r once, no other cycles intersect. A-cycles are strings above the loops and B-cycles are cycles surrounding the loops. Break the string diagram along the A-cycles, and map it conformally onto the region of the complex plane exterior to n pairs of circles. Each pair of circles has a complex projective transformation $T_r (1 \le r \le n)$ that maps the circumferences onto one another. By multiplying $T_1...T_n$ and their inverses we get an infinite group. The infinite group of projective transformations represents the fundamental group of the string diagram. Any Riemann surface of genus n can be conformally transformed onto the complex plane with $2n$ holes. The transformation is unique up to a projective transformation and a modular transformation (Poincaré, Klein and Koebe).

An n-loop diagram (Riemann surface with n-handles) is characterized by $3n - 3$ complex invariants. The amplitude for n-loops is for bose closed-strings given by[7]

$$A_b = 4^{1-n}(-i\pi)(ig^2/4\pi)^{N+2n-2} \prod_{r,s} \int d^2 z_r d^2 z_{is} d^2 z_{2s} d^2 w_s (d^2 z_\alpha d^2 z_\beta d^2 z_\gamma)^{-1}$$

$$|(z_\alpha - z_\beta)(z_\beta - z_\gamma)(z_\alpha - z_\gamma)|^2 \prod_s |w_s(z_{1s} - z_{2s})|^{-4} |\mathrm{Im}\tau|^{-13}$$

$$\prod_{\{i \ne I\}} |1 - w_i|^{-48} \prod_{r > r'} |\phi(z_r, z_{r'})|^{-P_r \cdot P_{r'}}, \tag{6}$$

where $z_\alpha, z_\beta, z_\gamma$ are three arbitrary z's, z_1's or z_2's. A similar formula holds for open-strings. Singularities occur when the $w's \to 0$ e.g. if an intermediate string becomes infinitely short, or when two or more z's approach one another. The configurations where the z's approach one another are those where one or more loops shrink to a point.

4. SUPERSTRING AND BOSE MULTI-LOOP DIVERGENCES

We have learned that divergences can arise from two sources: 1) The coincidence of the Koba-Nielsen variables z; 2) Topology changes that cause factors in the integrand of Eq.(4) to become infinite. This can happen when (i) $\mathrm{Im}B \to 0$, (ii) $G''(z_r) \to 0$, (iii) Jacobian $\to 0$, and (iv) $|z_\alpha - z_\beta| \to 0$. The topology change that can cause these infinities is due to a change of genus of a Riemann surface by the disappearance of one or more holes or handles.

In the superstring multi-loop amplitude there are divergences coming from internal Feynman-like diagrams corresponding to possibility (iv), in addition to the usual smooth surfaces of bose strings. There can be eight internal fermion lines for an SO(8) model as well as two boson lines coming from vertex operators.

Divergences which arise from Feynman lines that issue from vertex operators are troublesome, for it is not clear how one should remove them in a string amplitude. The standard mass and coupling constant renormalization methods do not apply.

At one-loop you can, effectively, by projective transformations prevent the invariant points from meeting, leading to a finite one-loop result. Partial proofs of the finteness of the one-loop amplitude have been obtained by Green and Scwarz[12], Frampton et $al.$[13] and Clavelli[14] for the type I open-string theory with SO(32) internal symmetry. Clavelli and Harms[14] have shown that unitarity violating branch points occur at the location of each graviton pole for the parity violating, one-loop, six point function with external gauge bosons in type I theories. Such unitarity violating contributions would exclude type I superstring models as candidates for physical theories. A proof of finiteness for the one-loop amplitude in the heterotic string model has been obtained by Gross et $al.$[6].

All these proofs of finiteness in the one-loop case have potential ambiguities in the application of regularization techniques[5], a problem that still requires careful study. A proof of finiteness in the one-loop case will have used up all the projective invariance, and an analysis of the higher-order loops as functions on the modular surfaces and as the moduli vary, shows that one can have various loops disappearing at the boundaries corresponding to a change of topology. This can produce divergences coming from a loop disappearing or when two loops fuse. The Feynman lines between the ends of interaction points fuse together.

The folklore is that each divergence will correspond to splitting off a ring or torus, which has a $dilaton$-type divergence associated with the dilaton field ϕ. The vacuum expectation value of ϕ in the supersymmetric theory should be zero, and there may be a cancellation. The divergences could have zero coefficients multiplying them in the superstring case but this has to be proved at each order!

We are now in a position similar to that of supergravity theory. Is there enough supersymmetry to guarantee a "miraculous" cancellation for all N in the N-loop amplitude? This kind of cancellation to all orders will probably occur for only a very special kind of superstring theory e.g. it is not expected to happen for open-string theories of the type that have been investigated to date.

The hoped for uniqueness of superstring phenomenology does not appear to have been immediately fulfilled, because of the *ad hoc* nature of compactification methods needed to reduce the $d = 10$ theory to four-dimensional space-time. Perhaps a unique finite theory of superstrings will emerge. However, as we have seen, this remains to be demonstrated.

5. ACKNOWLEDGEMENTS

I thank Professor J. G. Taylor for stimulating discussions. This work was supported by the Natural Sciences and Engineering Research Council of Canada.

6. REFERENCES

1. G. t'Hooft and M. Veltman, Ann. Inst. Henri Poincaré <u>20</u>, 69 (1974).

2. M. H. Goroff and Sagnotti, A., preprint CALT-68-1289; LBL-19995 UCB-PTH-85/34, 1985.

3. P. van Nieuwenhuizen, Phys. Rep. <u>68</u>, 191 (1981); Bergshoeff, E., de Roo, M., de Wit, B., and van Nieuwenhuizen, P., Nucl. Phys. <u>B195</u>, 97 (1982); Chapline, G.F., and Manton, N.S., Phys. Lett. <u>120B</u>, 105 (1983).

4. M. B. Green and Schwarz, J.H. Phys. Lett. <u>149B</u>, 117 (1984).

5. For a review of superstring theory, see: J. W. Moffat, <u>Superstring Physics</u>. Review Lectures presented at the Quantum Field Theory Workshop, University of Western Ontario, London, Ontario, 1985. To be published in a special proceedings issue in Can. J. Phys.

6. D. J. Gross, Harvey, J.A., Martinec, E. and Rohm, R., Nucl. Phys. <u>B256</u>, 253 (1985).

7. S. Mandelstam, preprint UCB-PTH-85/47, 1985; preprint UCB-PTH-85/48, 1985.

8. J. G. Taylor, King's College preprint, London, U.K. 1986; Restuccia, A., and Taylor, J.G., King's College preprints, London, U.K. 1985.

9. V. Alessandrini, Nuovo Cim. <u>2A</u>, 321 (1971).

10. C. Lovelace, Phys. Lett. <u>32B</u>, 703 (1970).

11. M. Kaku and Scherk, J., Phys. Rev. <u>D3</u>, 430 (1971); <u>D3</u>, 2000 (1971).

12. M. B. Green and Schwarz, J.H., Phys. Lett. <u>151B</u>, 21 (1985).

13. P. Frampton, Moxhay, P., and Ng, Y.J., Phys. Rev. Lett. <u>55</u>, 2107 (1985).

14. L. Clavelli, Phys. Rev. D, to be published; Clavelli, L., and Harms, B., University of Alabama preprint, 1986.

NEW FRONTIERS IN PARTICLE PHYSICS:

A SUMMARY OF THE 1986 LAKE LOUISE WINTER INSTITUTE

Bruce A. Campbell

Department of Physics

University of Alberta

Edmonton, Alberta, T6G 2J1

Canada

Abstract

A summary is presented of the contributions to the 1986 Lake Louise Winter Institute. The themes of collider experiments, and superstring theories, serve as the unifing features defining the new frontiers.

Recent trends in the field of particle physics were the theme of the 1986 Lake Louise Winter Institute. On the experimental side, the push to higher energies at colliding beam facilities will be the leading order of business over the next decade. We can form circulating, colliding, beams, with any charged particle stable enough to have a useful beam lifetime. In practice this means either the electron (stable lightest particle with charge), or the proton (stable lightest particle with baryon number), and their antiparticles. So the possibilities for colliding beam experimentation are basically threefold: e^+e^-, $e^{\pm}p$, and $p\bar{p}$. Each of these combinations has been covered in courses of invited lectures.

J. Dorfan described [1] in detail the prospects for e^+e^- collisions in the Z^0 region and beyond. In the near future e^+e^- machines operating in the Z^0 mass range will use the large rate for Z^0 production and decay to yield large event samples. Production via the Z^0 gives us copious samples of any particle which experiences the electroweak interactions. Combined with the cleanliness of e^+e^- experiments, this should allow comprehensive and precise study of any new physics in this mass range. At energies above the Z^0 resonance (LEP II), e^+e^- colliders remain a clean, precise, way to explore new physics, but with the small event rate both good luminosity, and patience, will be required.

As described in the lectures [2] by A. Astbury, hadron colliders have exactly the reverse features. They have a large kinematic range, due to the relative ease of producing high energy beams of (anti)protons, but experimentation is difficult due to large backgrounds, and a priori unknown subprocess center of mass. The experience of the CERN and Fermilab Tevatron colliders should provide valuable guidance to the problems that will be faced at the LHC or SSC. Despite the experimental difficulties, the CERN collider has already produced an impressive harvest of results, as described by Dr. Astbury.

The third possibility, ep colliders, are expected to share features of both the e^+e^- and hadron colliders. The physics approached may be somewhat different, however, as with the exception of possible leptoquark production, new physics will be manifest in the effects of t-channel particle exchanges at large momentum transfer, rather than direct s-channel production. Having an electron (rather than a constituent of a second hadron) as one of the colliding particles allows these exchanges to be probed with greater precision. G. Wolf described [3] the machine, detectors, and expected physics, for the HERA ep collider. As HERA is the first ep collider ever undertaken, experience there should be most interesting from all points of view.

While experimentalists are planning these assaults on the physics of the electroweak scale ($O(10^2 \text{Gev})$), theorists in their search for unification have recently been focusing on the incorporation of gravity in unified theories at the Planck scale ($O(10^{19} \text{GeV})$). There has been an explosion of theoretical interest in superstring theories, which represent perhaps our first serious attempt at a comprehensive theory of the fundamental interactions. These theories, which are supersymmetric, (general) relativistic, ten dimensional, quantum theories, in which the fundamental quanta are one dimensional extended objects (strings), were described in the lectures [4] by M. Dine. J. Ellis explored [5] the possibilities for extracting, from a superstring theory unified at the Planck scale, predictions for new physics at the electroweak scale, which could be tested at the coming generation of colliders.

The general framework having been set by the lecture series above, the contributed sessions which followed contained results on: precision tests of the low energy standard model and its extensions, present collider experiments, theoretical predictions for new collider physics, and related theoretical, and astrophysical, considerations.

First let us review the presentations on precision low energy tests of the standard model. These can also act as probes for new physics beyond the standard model, since new particles at a large mass scale will contribute to low energy processes through their virtual exchange, the resulting contribution scaling inversely as a power of the heavy particle mass. In order for this contribution not to be overwhelmed by standard model contributions we must look at processes that are forbidden, or suppressed, in the standard model. Fortunately, the standard model has several exact or approximate conservation laws that may be violated by new physical processes. Baryon number, lepton number, the lepton flavours, are all exactly conserved in the standard model and so present an experimental window to possible new physics.

J. LoSecco discussed[6] precision tests in underground experiments, including proton decay and neutron oscillations (baryon number violation), neutrino oscillations (lepton flavour violation), and the existence of magnetic monopoles. The experimental limits on these processes, especially the lack of any clear evidence for proton decay, place severe restrictions on attempts to grand unify the gauge interactions.

D. Caldwell discussed[7] new particle searches by neutrinoless double beta decay, which can be used to set limits on Majorana neutrino masses, right handed weak currents, or pseudo-Goldstone bosons coupling to lepton number, or flavour. Constraints on massive neutrinos were also considered[8] by D. Britton, who used $\pi \rightarrow e\nu$ decays to limit the amount of weak charged current mixing to a heavy (60-130 MeV) neutrino.

D. Bryman reviewed[9] the searches for lepton flavour violation. These include muon decays and conversion, tau decays, and μe production in K decays. The non-observation of these processes gives us limits on the couplings and masses of extra Higgs bosons, leptoquarks, horizontal gauge bosons, and extended technicolour, or composite models.

G. Fogli[10] presented an analysis of radiative corrections for neutrino induced deep inelastic scattering. The consistency of the neutrino processes with the collider measurements of the W and Z masses when one loop corrections are included is an indication that quantum corrections are as expected in the electroweak gauge theory.

From the present generation of collider experiments, we had presentations from the "B-factory" experiments CLEO and ARGUS. From the CLEO collaboration M. Alam discussed[11] new results on decays of B mesons. Branching ratios for a variety of exclusive, and semi-inclusive, B meson decay modes have now been extracted. As these branching ratios become determined, one can begin to unravel the dynamics of B decays.

In particular the decay $\bar{B}^o \to F^+ K^-$ suggests the effect of non-spectator W exchange diagrams. A limit was presented on the amount of $\bar{B}^o$-B^o mixing from the ratio for same sign to opposite sign dileptons. B. Heltsey explained[12] in detail the CLEO measurement of τ and D^o lifetimes, in agreement with the world sample. J. Kandaswamy discussed[13] the CLEO search for axions in the mass range $2m_e < m_a < 2m_\mu$. For an axion of 1.6 MeV decaying to $e^+ e^-$, about which heavy ion experiments have caused some speculation, they quote the limit: $B(T \to \gamma a) \times B(a \to e^+ e^-) < 4. \times 10^{-4}$.

For the ARGUS experiment M. Goddard discussed[14] physics results on D mesons. They have discovered a new meson in this family, the D^{**o}, of mass 2420 ± 6 MeV, and width 70 ± 21 MeV, which they interpret as an orbital excitation of the $c\bar{u}$ system. ARGUS also finds a branching ratio for the exclusive decay $D^o \to \phi \bar{K}^o$ of $1.4 \pm 0.4 \pm 0.2\%$. This is evidence of a W exchange contribution to D^o decay, and relates to the lifetime difference between D^+ and D^o. R. Orr presented[15] the ARGUS results on B physics. They have now determined several semi-inclusive, and exclusive, decay modes of the B mesons, and are attempting to understand the implications for decay dynamics.

R. Dubois presented[16] a survey of recent PEP measurements of the B lifetime, and discussed possible improvements in precision one might obtain at $e^+ e^-$ machines at the Z^o.

On the hadron collider side, N. Lockyear discussed[17] the CDF detector at Fermilab's Tevatron collider, which had just achieved its first collisions in September 1985. One eagerly awaits results of their first physics run with a fully configured detector, beginning in the summer of this year.

On collider theory, large cross section soft scatterings, of the type shown by Lockyear, were the subject of S. Rudaz' presentation[18]. As well as being a traditional topic of hadron dynamics, the extrapolation of cross sections and multiplicities for soft collisions to supercollider energies raises important questions of pileup, multiple interactions per beam crossing, and the related problems of trigger design.

R. Bates discussed[19] two photon decay widths of non-standard Higgs bosons. Unfortunately, when used to estimate $\gamma\gamma$ Higgs production rates there is, in general, insufficient enhancement over the minimal Higgs rate to make $\gamma\gamma$ Higgs production useful.

H. Baer analyzed[20] the production of gauginos at hadron colliders. There are distinct signatures in events of the type: n leptons + m jets + missing p_T. Also there is sufficient rate at the Tevatron collider to explore an interesting range of $\tilde{W}$ and $\tilde{Z}$ masses. In particular, in the $\tilde{W}\tilde{\gamma}$ mode there is a useable rate for $\tilde{W}$ mass up to the W mass, and clean $1\not{p}_T$ or $jj\not{p}_T$ signatures.

S. Godfrey discussed[21] the effects of extra Z^os, such as those which might arise in E_6 unified theories, on forward-backward, and polarization, asymmetries in e^+e^- annihilation into fermion pairs. P. Kalyniak also considered[22] such Z^os, noting that if there is mixing with the usual Z^o, then there would be a $Z^{o'}$ pole contribution to W^+W^- production in e^+e^- collisions. These E_6 analyses were motivated by the appearance[5], in some superstring compactifications, of an E_6 "unification" group. However, in the superstring case the new Z^os are strongly constrained[5] by the nature of the symmetry breaking mechanism, and may not exhibit the variety of possibilities explored in these analyses.

Some theoretical issues arising from string theories were touched on in the contributions. P. Mathieu considered[23] the calculation of the axial anomaly in QED formulated in the light cone gauge. This should give us insight into how axial anomalies appear in the light cone gauge formulation of superstring theories.

G. Kunstatter discussed[24] different aspects of higher dimensional theories, especially focusing on the determination of the gauge coupling constant for gauge interactions arising from the Kaluza-Klein mechanism. He also noted that one might be able to explain the smallness of the cosmological constant in a theory with 10^{40} extra dimensions!

J. Moffat addressed[25] the question of the finiteness of the superstring loop expansion, and examined the technical issues involved.

622

Unfortunately, as of the time of this writing it is still unknown
whether any, or all, superstring theories are perturbatively finite.

The relationship between particle physics and astrophysics was
addressed[26] by M. Sundaresan, who discussed how resonant matter-enhan-
ced neutrino oscillations would affect the spectrum of solar neutrinos
seen by terrestrial underground experiments.

K. Olive discussed[27] the role of dark matter in cosmology; to
have viable astrophysics with it in the present epoch serves as a
constraint on the early universe when it arose. Dark matter represents
a key link between astrophysics, where it is an experimental fact, and
particle physics where it is a consequence in many theories. As discussed by Olive, low energy superstring inspired models give cosmic relic
densities of stable neutralinos of the right magnitude to explain the
dark matter observations, and to put the universe at the critical
density as required by inflation. So at least in this case there is a
tentative link from superstrings to experiment.

Let us hope that it is not the only such link, and that other
suggestions from superstring models will be confirmed by the new
generation of collider experiments. Let us also hope that we all have
the opportunity to again discuss these issues, in such beautiful
surroundings, at a future Lake Louise Winter Institute.

References

(1) J.M. Dorfan, these proceedings.

(2) A. Astbury, ibid.

(3) G. Wolf, ibid.

(4) M. Dine, ibid.

(5) J. Ellis, ibid.

(6) J.M. LoSecco, ibid.

(7) D.O. Caldwell, ibid.

(8) D.I. Britton, ibid.

(9) D. Bryman, ibid.

(10) G.L. Fogli, ibid.

(11) M.S. Alam, ibid.

(12) B.K. Heltsey, ibid.

(13) J. Kandaswamy, ibid.

(14) M. Goddard, ibid.

(15) R.S. Orr, ibid.

(16) R. Dubois, ibid.

(17) N. Lockyear, ibid.

(18) S. Rudaz, and P. Valin, ibid.

(19) R. Bates, J. Ng, and P. Kalyniak, ibid.

(20) H. Baer, K. Hagiwara, and X. Tata, ibid.

(21) S. Godfrey, and G. Bélanger, ibid.

(22) P. Kalyniak, and M.K. Sundaresan, ibid.

(23) P. Mathieu, Q. Ho-Kim, and L. Marleau, ibid.

(24) G. Kunstatter, and H.P. Leivo, ibid.

(25) J.W. Moffat, ibid.

(26) M.K. Sundaresan, ibid.

(27) K. Olive, ibid.